HANDBOOK OF MICROWAVE COMPONENT MEASUREMENTS

HANDBOOK OF MICROWAVE COMPONENT MEASUREMENTS

WITH ADVANCED VNA TECHNIQUES

Joel P. Dunsmore

Ph.D., Agilent Fellow, Agilent Technologies, USA

A John Wiley & Sons, Ltd., Publication

This edition first published 2012
© 2012 John Wiley & Sons, Ltd

Registered office
John Wiley & Sons Ltd, The Atrium, Southern Gate, Chichester, West Sussex, PO19 8SQ, United Kingdom

For details of our global editorial offices, for customer services and for information about how to apply for permission to reuse the copyright material in this book please see our website at www.wiley.com.

Library of Congress Cataloging-in-Publication Data

Dunsmore, Joel P.
  Handbook of microwave component measurements : with advanced VNA techniques / Joel P. Dunsmore.
    pages cm
  Includes bibliographical references and index.
  ISBN 978-1-119-97955-5
1. Microwave devices–Testing. I. Title.
  TK7876.D84 2012
  537.5′344–dc23

                                                                              2012011804

A catalogue record for this book is available from the British Library.

ISBN 9781119979555

Typeset in 10/12pt Times by Aptara Inc., New Delhi, India

To my dear wife Dana

Contents

Foreword

The electronics industry has undergone revolutionary changes in the past 20 years. System performance has significantly advanced, physical size of hardware has shrunk, quality and reliability have greatly improved and manufacturing costs have dramatically decreased. Underlying these advances has been the phenomenal growth in RF test and measurement capability. Modern-day RF test equipment has progressed to the point where it is not uncommon to measure signals below −100 dBm at milliseconds speed. Even more astounding is the ability to marry RF test capability with analysis software whereby test equipment can produce linear and non-linear models of the device under test to significantly improve the life of the design engineer, using this capability.

RF and microwave components have played an important role in this revolutionary change. Component size has shrunk, parasitics have been reduced, quality standards have greatly improved and costs have reduced ten-fold. At the same time, test fixtures and interconnects have improved to enable a higher level of precision during characterization and production measurement. In parallel with these advances, test equipment has improved to an extent where there has been a revolution in the capabilities to make precise and fast measurements of RF and microwave components. The success of a manufacturer of RF and microwave components is directly linked to the quality and capability of measuring component performance during the design, qualification and production phase of the product life cycle. From a practical point of view, the testing must be fast (1–2 seconds), the accuracy very precise (hundredths of a dB), with a high degree of repeatability. Each phase of the life cycle imposes its unique requirements for measurement accuracy and data collection.

During the design phase, full characterization of performance, including amplitude and phase, is a must in order to establish a reference for future production runs. So it becomes a necessary requirement to characterize and de-embed the test setup and test fixtures to isolate actual device performance. Fortunately, the modern vector network analyzer provides support in this regard. Consequently, the performance data obtained during the design phase becomes the gold standard for evaluating statistical variation obtained from future production lots; and these lots are accepted or rejected based on the statistical results, with sigma generally being the statistic most closely watched by the test or QA engineer to evaluate production lot acceptance.

When published specs are provided for a component, it is very important to understand that these specifications are simply markers which allow for a first impression or summary review of the component performance. However, only when performance data and performance graphs are provided for all parameters, including both amplitude and phase, would the "real" performance of the component be known. Furthermore, in characterizing components for use in customer evaluations, it is important to provide this data both within and outside the specified bandwidth. In non-linear components such as a frequency mixer, higher-order harmonics of the RF and LO signals are generated, and depending on the load impedance outside the specified frequency range, these higher-order harmonics can get reflected back into the mixer, causing an interaction between the desired signals and the unwanted harmonics. Fortunately, modern-day analyzers can make these harmonic measurements relatively quickly and easy.

Dr. Dunsmore has captured the essence of modern-day measurements. He provides a practical understanding of measurement capabilities and limitations. He provides a means for the test engineer to not only make measurements, but also to understand test concepts, anticipate measurement results and learn how to isolate and characterize the performance of the DUT, independent of potential errors inherent in the test environment. I am confident that this book will serve as a reference for understanding measurement methods, test block diagrams and measurement limitations so that correlation between the manufacturer and user would take place by using a common reference. This book has the potential to be an invaluable source to further the progress of the RF and microwave world.

Harvey Kaylie
President and Founder of Mini-Circuits

Preface

This book is a bit of mixture between basic and advanced, and between theoretical and practical. Unfortunately, the dividing lines are not particularly clear, and depend considerably upon the training and experience of the reader. While primarily a text about measurement techniques, there is considerable information about device attributes that will be useful to both a designer and a test engineer, as one purpose of device testing is to ascertain the attributes that do not follow the simplified models commonly associated with those devices. In practice, it is the unexpected responses that consume the majority of the time spent in test and troubleshooting designs, particularly related to active devices such as amplifiers and mixers.

The principal instrument for testing microwave components is the vector network analyzer (VNA), and recent advances have increased the test capabilities of this instrument to cover far more than simple gain and match measurements. As a designer of VNAs for more than 30 years, I have been involved in consulting on the widest range of microwave test needs from cell phone components to satellite multiplexers. The genesis and goal of this book is to provide to the reader a distillation of that experience to improve the quality and efficiency of the R&D and production test engineer. The focus is on modern test methods; the best practices have changed with changing instrument capability and occasionally the difference between legacy methods and new techniques is sufficiently great as to be particularly highlighted.

Chapter 1 is intended as an introduction to microwave theory and microwave components. The first half introduces characterization concepts common to RF and microwave work. Some important mathematical results are presented which are useful in understanding the results of subsequent chapters. The second half of Chapter 1 introduces some common microwave connectors, transmission lines and components, as well as providing some discussion of the basic microwave test instrumentation. This chapter is especially useful to engineers new to RF and microwave testing.

Chapter 2 provides a detailed look into the composition of common VNA designs along with their limitations. While this level of detail is not normally needed by the casual user, test engineers trying to understand measurement results at a very precise level will find it useful to understand how overall results are affected by VNA test configuration. While the modern VNA can make a wide range of measurements, including distortion, power and noise figure measurements, still the principal use is in measuring S-parameters. The second half of Chapter 2 illustrates many useful parameters derived from basic S-parameters.

Perhaps the most arcane aspect of using VNAs for test is the calibration and error-correction process. Chapter 3 is a comprehensive discussion of the error models for VNAs, calibration methods, uncertainty analysis and evaluation of calibration residuals. This chapter also introduces the idea of source and receiver power calibrations, about which, excluding this book, very little formal information is currently available. The chapter concludes with many practical aspects of VNAs that affect the quality of calibrated measurements.

Chapter 4 is likely the most mathematically rigorous, covering the very useful topic of time domain transforms used in VNAs. The topic of gating, its effects, and compensation methods is examined in particular. These first four chapters comprise the introductory material to microwave component measurements.

The remaining chapters are focused on describing particular cases for microwave component measurements. Chapter 5 is devoted to passive microwave components such as cables and connectors, transmission lines, filters, isolators and couplers. Best practices, and methods for dealing with common problems, are discussed for each component.

Chapter 6 is all about amplifier measurements, and provides the understanding needed for complete characterization. In particular, difficulties with measuring high gain and high power amplifiers are discussed, including pulsed RF measurements. Non-linear measurements such as harmonics and two tone intermodulation are introduced, and many of the concepts for distortion and noise measurements are equally valid whether using a spectrum analyzer or a modern VNA for the test receiver.

Chapter 7 extends the discussion of active device test to that of mixers. Because few engineers have experience with mixers, and they are often only superficially covered in engineering courses, the chapter starts with a detailed discussion of the modeling and characteristics of mixers and frequency converters. Measurement methods for mixers can be quite complicated, especially for the phase or delay response. Several key methods are discussed, with a new method of calibrating, using a phase reference, presented in detail for the first time. Besides the magnitude and phase frequency response, methods for measuring mixer characteristics versus RF and local oscillator power are presented, along with distortion and noise measurements. This chapter is required reading for any test engineer dealing with mixers or frequency converters.

Chapter 8 brings in the concept of differential and balanced devices, and provides complete details on the analysis and measurement methods for differential devices including non-linear responses, noise figure and distortion.

Chapter 9 provides a collection of very useful techniques and concepts for the test engineer, particularly with respect to test fixturing, including a complete discussion of creating in-fixture calibration kits.

Acknowledgments

Many of my colleagues assisted in the development and review of this book and I would like to acknowledge their help here. Henri Komrij, my R&D manager, has been a great supporter from the initial concept, as well as Greg Peters, VP and general manager of the Components Test Division. Many R&D engineers in our lab contributed to the review of the manuscript and their expertise in each field is sincerely appreciated: Keith Anderson, Dara Sarislani, Dave Blackham, Ken Wong, Shinya Goto, Bob Shoulders, Dave Ballo, Clive Barnett, Cheng Ning, Xin Chen, Mihai Marcu and Loren Betts. They did an excellent job and any remaining errors are entirely and regrettably my own.

Many of the new methods and techniques presented here rely on the difficult and precise implementation of measurement methods and algorithms and I'd like to thank our software design team, Johan Ericsson, Sue Wood, Jim Kerr, Phil Hoard, Jade Hughes, Brad Hokkanen, Niels Jensen, Raymond Taylor, Dennis McCarthy, Andy Cannon, Wil Stark, Yu-Chen Hu, Zhi-Wen Wong and Yang Yang, as well as their managers, Sean Hubert, Qi Gao and Dexter Yamaguchi for all their help over the years in implementing in our products many of the functions described here.

Finally I would like to remember here Dr. Roger Pollard, who as my Ph.D. adviser at University of Leeds and as a colleague during his sabbaticals at HP and Agilent Technologies, provided advice, mentoring and friendship; he will be greatly missed.

Joel P. Dunsmore
Sebastopol, CA

List of Acronyms

ACPL	adjacent channel power level
ACPR	adjacent channel power ratio
ADC	analog-to-digital-converter
AFR	automatic fixture removal
ALC	automatic level control
AM	amplitude modulated
APE	automatic port extension
arb	arbitrary-waveform generator
ATF	A-receiver transmission forward
ATS	automated test system
balun	BALanced-UNbalanced transformer
BTF	B-receiver transmission forward
BW	bandwidth
CMRR	common mode rejection ratio
CPW	coplanar waveguide
CSV	comma separated values
DANL	displayed average noise level
dBc	dB relative to the carrier
DDS	direct-digital synthesizer
DFT	discrete Fourier transform
DUT	device under test
DUTRNPI	DUT relative noise power incident
Ecal	electronic-calibration
EM	electromagnetic
ENR	excess noise ratio
ERC	enhanced response calibration
EVM	error vector magnitude
FBAR	film bulk acoustic resonator
FCA	frequency-converter application
FN	fractional-N
FOM	frequency offset mode
FPGA	field-programmable gate array
GCA	gain compression application
GPIO	general-purpose input/output

GUI	graphical user interface
IBIS	input output buffer information specification
IDFT	inverse discrete Fourier transform
IF	intermediate frequency
IFFT	inverse fast Fourier transform
IFT	inverse Fourier transform
IIP	input intercept point
IM	intermodulation
IMD	intermodulation distortion
IM3	third-order IM product
IP3	third-order intercept point
IMD	intermodulation distortion
IPwr	input power
KB	Kaiser-beta
LNA	low-noise amplifier
LO	local oscillator
LTCC	low-temperature cofired-ceramic
LVDS	low voltage differential signaling
MMIC	monolithic microwave integrated circuit
MUT	mixer under test
NF	noise figure
NFA	noise figure analyzer
NOP	normal operating point
NVNA	non-linear vector network analyzer
OPwr	output power
PAE	power added efficiency
PCB	printed circuit board
PIM	passive intermodulation
PMAR	power-meter-as-receiver
QSOLT	quick short open load thru
RBW	resolution bandwidth
RRF	reference-receiver forward
RMS	root-mean-square
RNPI	relative noise power incident
RTF	reference transmission forward
SA	spectrum analyzer
SAW	surface acoustic wave
SCF	source calibration factor
SE	single-ended
SMC	scalar mixer calibration
SMT	surface mount technology
SMU	source measurement unit
SNA	scalar network analyzer
SOLR	short open load reciprocal
SOLT	short open load thru
SRF	self-resonant frequency

SRL	structural return loss
SSB	single sideband
SSPA	solid-state power amplifier
STF	source transmission forward
SYSRNPI	system relative noise power incident
TD	time domain
TDR	time-domain reflectometer
TDT	time-domain transmission
TEM	transverse-electromagnetic
TOI	third-order intermodulation
TR	transmission/reflection
TRL	thru reflect line
TRM	thru reflect match
TVAC	thermal vacuum
TWT	traveling wave tube
UT	unknown thru
VCO	voltage-controlled oscillator
VMC	vector mixer/converter
VNA	vector network analyzer
VSA	vector signal analyzer
VSWR	voltage standing wave ratio
YIG	yttrium-iron-garnet
YTO	YIG-tuned-oscillator

1

Introduction to Microwave Measurements

"To measure is to know."[1] This is a text on the art and science of measurement of microwave components. While this work is based entirely on science, there is some art in the process, and the terms "skilled-in-the-art" and "state-of-the-art" take on particular significance when viewing the task of measuring microwave components. The goal of this work is to provide the latest, state-of-the-art methods and techniques for acquiring the optimum measurements of the myriad of microwave components. This goal naturally leads to the use of the vector network analyzer (VNA) as the principal test equipment, supported by the use of power meters, spectrum analyzers, signal sources and noise sources, impedance tuners and other accessories.

Note here the careful use of the word 'optimum'; this implies that there are tradeoffs between the cost and complexity of the measurement system, the time or duration of the measurement, the analytically computed uncertainty and traceability, and some previously unknown intangibles that all affect the overall measurement. For the best possible measurement, ignoring any consequence of time or cost, one can often go to national standards laboratories to find these best methods, but they would not suit a practical or commercial application. Thus here the attempt is to strike an optimum balance between minimal errors in the measurement and practical consequences of the measurement techniques. The true value of this book is in providing insight into the wide range of issues and troubles that one encounters in trying to carefully and correctly ascertain the characteristics of one's microwave component. The details here have been gathered from decades of experience in hundreds of direct interactions with actual measurements; some problems are obvious and common while others are subtle and rare. It is hoped that the reader will be able to use this handbook to avoid many hours of unproductive test time.

For the most part, the mathematical derivations in this text are intended to provide the reader with a straightforward connection between the derived values and the underlying characteristics. In some cases, the derivation will be provided in full if it is not accessible from existing literature; in other cases a reference to the derivation will be provided. There

---

[1] Lord Kelvin, "On Measurement".

Handbook of Microwave Component Measurements: With Advanced VNA Techniques, First Edition. Joel P. Dunsmore.
© 2012 John Wiley & Sons, Ltd. Published 2012 by John Wiley & Sons, Ltd.

are extensive tables and figures, with key sections providing many of the important formulas. The mathematical level of this handbook is geared to a college senior or working engineer with the intention of providing the most useful formulas in a very approachable way. So, sums will be preferred to integrals, finite differences to derivatives, and divs, grads and curls will be entirely eschewed.

The chapters are intended to self-standing for the most part. In many cases, there will be common material to many measurement types, such as the mathematical derivation of the parameters or the calibration and error-correction methods, and these will be gathered in the introductory chapters, though well referenced in the measurement chapters. In some cases, older methods of historical interest are given (there are many volumes on these older techniques), but by and large only the most modern techniques are presented. The focus here is on the practical microwave engineer facing modern, practical problems.

1.1   Modern Measurement Process

Throughout the discussion of measurements a six-step procedure will be followed that applies to most measurement problems. When approaching a measurement these steps are:

- **Pretest:** This important first step is often ignored, resulting in meaningless measurements and wasted time. During the pretest, measurements of the device under test (DUT) are performed to coarsely determine some of its attributes. During pretest, it is also determined if the DUT is plugged in, turned on and operating as expected. Many times the gain, match or power handling is discovered to be different than expected, and much time and effort can be saved by finding this out early.
- **Optimize:** Once the coarse attributes of the device have been determined, the measurement parameters and measurement system can be optimized to give the best results for that particular device. This might include adding an attenuator to the measurement receivers or adding booster amplifiers to the source, or just changing the number of points in a measurement to capture the true response of the DUT. Depending upon the device's particular characteristic response relative to the system errors, different choices for calibration methods or calibration standards might be required.
- **Calibrate:** Many users will skip to this step, only to find that something in the setup does not provide the necessary conditions and they must go back to step one, retest and optimize before recalibration. Calibration is the process of characterizing the measurement system so that systematic errors can be removed from the measurement result. This is not the same as obtaining a calibration sticker for an instrument, but really is the first step, the *acquisition* step of the error correction process that enables improved measurement results.
- **Measure:** Finally, some stimulus is applied to the DUT and its response to the stimulus is measured. During the measurement, many aspects of the stimulus must be considered, as well as the order of testing and other testing conditions. These include not only the specific test conditions, but also preconditions such as previous power states to account for non-linear responses of the DUT.
- **Analyze:** Once the raw data is taken, error correction factors (the *application* step of error correction) are applied to produce a corrected result. Further mathematical manipulations on the measurement result can be performed to create more useful figures-of-merit, and the

data from one set of conditions can be correlated with other conditions to provide useful insight into the DUT.

- **Save data:** The final step is saving the results in a useful form. Sometimes this can be as simple as capturing a screen dump, but often it means saving results in such a way that they can be used in follow-up simulations and analysis.

1.2   A Practical Measurement Focus

The techniques used for component measurements in the microware world change dramatically depending upon the attributes of the components; thus, the first step in describing the optimum measurement methods is understanding the expected behavior of the DUT. In describing the attributes and measurements of microwave components it is tempting to go back to first principles and derive all the underlying mathematics for each component and measurement described, but such an endeavor would require several volumes to complete. One could literally write a book on the all the attributes of almost any *single* component, so for this book the focus will be on only those final results useful for describing practical attributes of the components to be characterized, and then quote and reference many results without the underlying derivation.

There have been examples of books on microwave measurements that have focused on the metrology kind of measurements [1] made in national laboratories such as the National Institute for Standards and Technology (NIST, USA), or the National Physical Laboratory (NPL, UK), but the methods used there don't transfer well – or at all – to the commercial market. For the most part, the focus of this book will be on practical measurement examples of components found in commercial and aerospace/defense industries. The measurements focus will be commercial characterization rather than the kinds of metrology found in standards labs.

Also, while there has been a great deal written about components in general or ideal terms, as well as much academic analysis of these idealized components, in practice these components contain significant parasitic effects that cause their behavior to differ dramatically from that described in many textbooks. And, unfortunately, these effects are often not well understood, or are difficult to consider in an analytic sense, and so are only revealed during an actual measurement of a physical devices. In this chapter, the idealized analysis of many components is described, but the descriptions are extended to some of the real-world detriments that cause these components' behavior to vary from the expected analytical response.

1.3   Definition of Microwave Parameters

In this section, many of the relevant parameters used in microwave components are derived from the fundamental measurements of voltage and current on the ports. For simplicity, the derivations will focus on measurements made under the conditions of termination in real-valued impedances, with the goal of providing mathematical derivations that are straightforward to follow and readily applicable to practical cases.

In microwave measurements, the fundamental parameter of measurement is power. One of the key goals of microwave circuit design is to optimize the power transfer from one circuit to another, such as from an amplifier to an antenna. In the microwave world, power is almost always referred to as either an incident power or a reflected power, in the context of power traveling along a transmission structure. The concept of traveling waves is of fundamental importance to understanding microwave measurements, and to engineers who haven't had a course on transmission lines and traveling waves – and even to some who have – the concept of power flow and traveling waves can be confusing.

1.3.1   S-Parameter Primer

S-parameters have been developed in the context of microwave measurements, but have a clear relationship to voltages and currents that are the common reference for most electrical engineers. This section will develop the definition of traveling waves, and from that the definition of S-parameters, in a way that is both rigorous and hopefully intuitive. The development will be incremental, rather than just quoting results, in hopes of engendering an intuitive understanding.

This signal traveling along a transmission line is known as a traveling wave [2], and has a forward component and a reverse component. Figure 1.1 shows the schematic a two-wire transmission structure with a source and a load.

If the voltage from the source is sinusoidal, it is represented by the phasor notation

$$v_s(t) = \text{Re}\left(|V_s| e^{j(\omega t + \phi)}\right), \quad or \quad V_s = |V_s| e^{j(\omega t + \phi)} \tag{1.1}$$

The voltage and current at the load are

$$V_L = |V_L| e^{j\phi_L^V}, \quad I_L = |I_L| e^{j\phi_L^I} \tag{1.2}$$

The voltage along the line is defined as $V(z)$ and the current at each point is $I(z)$. The impedance of the transmission line is as described in Section 1.2.1, Eqs. (1.3), (1.4) and (1.5), provides for a relationship between the voltage and the current. At the reference point, the total voltage is $V(0)$, and is equal to V_1; the total current is $I(0)$. The power delivered to the load can be described as

$$P_L = P^F - P^R \tag{1.3}$$

Where P^F is called the forward power and P^R is called the reverse power. To put this in terms of voltage and current of Figure 1.1, the total voltage at the port can be defined as the sum of the forward voltage wave traveling into the port and the reverse voltage wave emerging from the port

$$V_1 = V_F + V_R \tag{1.4}$$

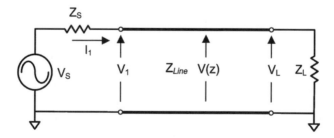

Figure 1.1   Voltage source and two-wire system.

The forward voltage wave represents a power traveling toward the load, or transferring from the source to the load, and the reflected voltage wave represents power traveling toward the source. To be formal, for a sinusoidal voltage source, the voltage as a function of time is

$$v_1(t) = V_1^P \cos(\omega t + \phi) = \text{Re}\left(V_1^P e^{j(\omega t + \phi)}\right) \tag{1.5}$$

From this it is clear that V_1^P is the peak voltage and the root-mean-square (RMS) voltage is

$$V_1 = \frac{V_1^P}{\sqrt{2}} \tag{1.6}$$

The $\sqrt{2}$ factor shows often in the following discussion of power in a wave, and it is sometimes a point of confusion, but if one remembers that RMS voltage is what is used to compute power in a sine wave, and is used to refer to the wave amplitude of a sine wave in the following equations, then it will make perfect sense.

Considering the source impedance Z_S and the line or port impedance Z_0, and simplifying a little by making $Z_S = Z_0$ and considering the case where Z_0 is pure-real, one can relate the forward and reverse voltage to an equivalent power wave. If one looks at the reference point of Figure 1.1, and one had the possibility to insert a current probe as well as had a voltage probe, one could monitor the voltage and current.

The source voltage must equal the sum of the voltage at port 1 and the voltage drop of the current flowing through the source impedance

$$V_S = V_1 + I_1 Z_0 \tag{1.7}$$

And defining the forward voltage as

$$V_F = \frac{1}{2}(V_1 + I_1 Z_0) \tag{1.8}$$

We see that the forward voltage represents the voltage at port 1 in the case where the termination is Z_0. From this and Eq. (1.4) one finds that the reverse voltage must be

$$V_R = \frac{1}{2}(V_1 - I_1 Z_0) \tag{1.9}$$

If the transmission line in Figure 1.1 is very long (such that the load effect is not noticeable), and the line impedance at the reference point is also the same as the source, which may be called the port reference impedance, then the instantaneous current going into the transmission line is

$$I_F = V_S \left(\frac{Z_0}{Z_0 + Z_S}\right) = \left.\frac{V_S}{2Z_0}\right|_{Z_0 = Z_S} \tag{1.10}$$

The voltage at that point is the same as the forward voltage, and can be found to be

$$V_F = V_S \left(\frac{Z_0}{Z_0 + Z_S}\right) = \left.\frac{V_S}{2}\right|_{Z_0 = Z_S} \tag{1.11}$$

And the power delivered to the line (or a Z_0 load) is

$$P_F = V_F I_F = \left(\frac{V_F^2}{Z_0}\right) = \frac{V_S}{4Z_0} \tag{1.12}$$

From these definitions, one can now refer to the incident and reflected power waves using the normalized incident and reflected voltage waves, a and b as [3]

$$a = \frac{V_F}{\sqrt{Z_0}}, \quad b = \frac{V_R}{\sqrt{Z_0}} \, provided \, Z_0 \, is \, real \tag{1.13}$$

Or, more formally as a power wave definition

$$a = \frac{1}{2}\left(\frac{V_1 + I_1 Z_0}{\sqrt{|\mathrm{Re}\,Z_0|}}\right), \quad b = \frac{1}{2}\left(\frac{V_1 - I_1 Z_0^*}{\sqrt{|\mathrm{Re}\,Z_0|}}\right) \tag{1.14}$$

where Eq. (1.14) includes the situation in which Z_0 is not pure real [4]. However, it would be an unusual case to have a complex reference impedance in any practical measurement.

For real values of Z_0, one can define the forward or incident power as $|a|^2$ and the reverse or scattered power as $|b|^2$, and see that the values a and b are related to the forward and reverse voltage waves, but with the units of the square root of power. In practice, the definition of Eq. (1.13) is typically used, because the definition of Z_0 is almost always either 50 or 75 ohms. In the case of waveguide measurements, the impedance is not well defined and it changes with frequency and waveguide type. It is recommended to simply use a normalized impedance of 1 for the waveguide impedance. This does not represent 1 ohm, but is used to represent the fact that measurements in waveguide are normalized to the impedance of an ideal waveguide. In (1.13) incident and reflected waves are defined, and in practice the incident waves are the independent variables and the reflected waves are the dependent variables. Consider Figure 1.2, a two-port network.

There are now sets of incident and reflected waves at each port i, where

$$a_i = \frac{V_{Fi}}{\sqrt{Z_{0i}}}, \quad b_i = \frac{V_{Ri}}{\sqrt{Z_{0i}}} \tag{1.15}$$

And the voltages and currents at each port can now be defined as

$$V_i = \sqrt{Z_{0i}}\,(a_i + b_i)$$

$$I_i = \frac{1}{\sqrt{Z_{0i}}}\,(a_i - b_i) \tag{1.16}$$

where Z_{0i} is the reference impedance for the ith port. An important point here that is often misunderstood is that the reference impedance does not have to be the same as the port impedance or the impedance of the network. It is a "nominal" impedance; that is, it is the

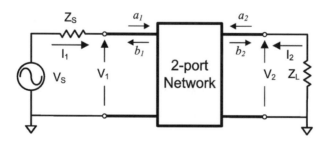

Figure 1.2   Two-port network connected to a source and load.

impedance that we "name" when we are determining the S-parameters, but it need not be associated with any impedance in the circuit. Thus, a 50 ohm test system can easily measure and display S-parameters for a 75 ohm device, referenced to 75 ohms.

The etymology of the term "reflected" derives from optics, and refers to light reflecting off a lens or other object with a index of refraction different from air, whereas it appears that the genesis for the scattering or S-matrix was derived in the study of particle physics, from the concept of wavelike particles scattering off crystals. In microwave work, scattering or S-parameters are defined to relate the independent incident waves to the dependent waves; for a two-port they become

$$b_1 = S_{11} a_1 + S_{12} a_2$$
$$b_2 = S_{21} a_1 + S_{22} a_2$$

(1.17)

which can be placed in matrix form as

$$\begin{bmatrix} b_1 \\ b_2 \end{bmatrix} = \begin{bmatrix} S_{11} & S_{12} \\ S_{21} & S_{22} \end{bmatrix} \cdot \begin{bmatrix} a_1 \\ a_2 \end{bmatrix}$$

(1.18)

Where a's represent the incident power at each port, that is the power flowing into the port, and b's represent the scattered power, that is the power reflected or emanating from each port. For more than two ports, the matrix can be generalized to

$$\begin{bmatrix} b_1 \\ \vdots \\ b_n \end{bmatrix} = \begin{bmatrix} S_{11} & \cdots & S_{1n} \\ \vdots & \ddots & \vdots \\ S_{1n} & \cdots & S_{nn} \end{bmatrix} \cdot \begin{bmatrix} a_1 \\ \vdots \\ a_n \end{bmatrix} \quad or \quad [b_n] = [S] \cdot [a_n]$$

(1.19)

From (1.17) it is clear that it takes four parameters to relate the incident waves to the reflected waves, but (1.17) provides only two equations. As a consequence, solving for the S-parameters of a network requires that two sets of linearly independent conditions for a_1 and a_2 be applied, and the most common set is one where first a_2 is set to zero, and the resulting b waves are measured and then a_1 is set to zero, and a second set of b waves is measured. This yields

$$S_{11} = \left. \frac{b_1}{a_1} \right|_{a_2=0} \qquad S_{12} = \left. \frac{b_1}{a_2} \right|_{a_1=0}$$

$$S_{21} = \left. \frac{b_2}{a_1} \right|_{a_2=0} \qquad S_{22} = \left. \frac{b_2}{a_2} \right|_{a_1=0}$$

(1.20)

which is the most common expression of S-parameter values as a function of a and b waves, and often the only one given for their definition. However, there is nothing in the definition of S-parameters that requires one or the other incident signals to be zero, and it would be just as valid to define them in terms of two sets of incident signals, a_n and a'_n and reflected signals b_n and b'_n

$$S_{11} = \left(\frac{b_1 a'_2 - a_2 b'_1}{a_1 a'_2 - a_2 a'_1} \right) \qquad S_{12} = \left(\frac{b_1 a'_1 - a_1 b'_1}{a_2 a'_1 - a_1 a'_2} \right)$$

$$S_{21} = \left(\frac{b_2 a'_2 - a_2 b'_2}{a_1 a'_2 - a_2 a'_1} \right) \qquad S_{22} = \left(\frac{b_2 a'_1 - a_1 b'_2}{a_2 a'_1 - a_1 a'_2} \right)$$

(1.21)

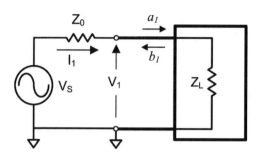

Figure 1.3   One-port network.

From Eq. (1.21) one sees that S-parameters are in general defined for a pair of stimulus drives. This will become quite important in more advanced measurements and in the actual realization of the measurement of S-parameters, because in practice it is not possible to make the incident signal go to zero due to mismatches in the measurement system.

These definitions naturally lead to the concept that S_{nn} are reflection coefficients and are directly related to the DUT port input impedance, and S_{mn} are transmission coefficients and are directly related to the DUT gain or loss from one port to another.

Now that the S-parameters are defined, they can be related to common terms used in the industry. Consider the circuit of Figure 1.3, where the load impedance Z_L may be arbitrary and the source impedance is the reference impedance.

From inspection one can see that

$$V_1 = V_S \left(\frac{Z_L}{Z_L + Z_0} \right), \quad I_1 = V_S \left(\frac{1}{Z_L + Z_0} \right) \tag{1.22}$$

which is substituted into (1.8) and (1.9), and from (1.15) one can directly compute a_1 and b_1 as

$$a_1 = \frac{V_S}{2\sqrt{Z_0}}, \quad b_1 = \frac{V_S}{2\sqrt{Z_0}} \left(\frac{Z_L - Z_0}{Z_L + Z_0} \right) \tag{1.23}$$

From here, S_{11} can be derived from inspection as

$$S_{11} = \frac{b_1}{a_1} = \frac{Z_L - Z_0}{Z_L + Z_0} \tag{1.24}$$

And it is common to refer to S_{11} informally as the input impedance of the network, where

$$Z_{In} = \frac{V_1}{I_1} \tag{1.25}$$

This is clearly true for a one-port network, and can be extended to a two-port or n-port network if all the ports of the network are terminated in the reference impedance, but in general, one cannot say that S_{11} is the input impedance of a network without knowing the termination impedance of the network. This is a common mistake that is made with respect to determining the input impedance or S-parameters of a network. S_{11} is defined for any terminations by Eq. (1.21), but it is the same as the input impedance of the network only under the condition that it is terminated in the reference impedance, thus satisfying the conditions for Eq. (1.20).

Consider the network of Figure 1.2 where the load is not the reference impedance; as such it is noted that a_1 and b_1 exist, but now Γ_1 (also called Γ_{In} for a two-port network) is defined as

$$\Gamma_1 = \frac{b_1}{a_1} \tag{1.26}$$

with the network terminated in an arbitrary impedance. As such, Γ_1 represents the input impedance of a system comprised of the network and its terminating impedance. The important distinction is that S-parameters of a network are invariant to the input or output terminations, providing they are defined to a consistent reference impedance, whereas the input impedance of a network depends upon the termination impedance at each of the other ports. The value of Γ_1 for a two-port network can be directly computed from the S-parameters and the terminating impedance, Z_L, as

$$\Gamma_1 = \left(S_{11} + \frac{S_{21}S_{12}\Gamma_L}{1 - S_{22}\Gamma_L} \right) \tag{1.27}$$

Where Γ_L, computed as in (1.24), is

$$\Gamma_L = \frac{Z_L - Z_0}{Z_L + Z_0} \tag{1.28}$$

or in the case of a two-port network terminated by an arbitrary load then

$$\Gamma_L = \frac{a_2}{b_2} \tag{1.29}$$

Similarly, the output impedance of a network that is sourced from an arbitrary source impedance is

$$\Gamma_2 = \left(S_{22} + \frac{S_{21}S_{12}\Gamma_S}{1 - S_{11}\Gamma_S} \right) \tag{1.30}$$

Another common term for the input impedance is the voltage standing wave ratio (VSWR) – also simply called SWR – and represents the ratio of maximum voltage to minimum voltage that one would measure along a Z_0 transmission line terminated in the some arbitrary load impedance. It can be shown that this ratio can be defined in terms of the S-parameters of the network as

$$VSWR = \left(\frac{1 + |\Gamma_1|}{1 - |\Gamma_1|} \right) \tag{1.31}$$

If the network is terminated in its reference impedance then Γ_1 becomes S_{11}. Another common term used to represent the input impedance is the reflection coefficient, ρ_{In}, where

$$\rho_{In} = |\Gamma_{In}| \tag{1.32}$$

And it is common to write

$$VSWR = \left(\frac{1 + \rho}{1 - \rho} \right) \tag{1.33}$$

Other terms related to the input impedance are return loss and it is alternatively defined as

$$RL = 20 \cdot \log_{10}(\rho), \text{ or } RL = -20 \cdot \log_{10}(\rho) \tag{1.34}$$

with the second definition being most properly correct, as loss is defined to be positive in the case where a reflected signal is smaller than the incident signal. But, in many cases, the former definition is more commonly used; the microwave engineer must simply refer to the context of the use to determine the proper meaning of the sign. Thus, an antenna with 14 dB return loss would be understood to have a reflection coefficient of 0.2, and that the value displayed on a measurement instrument might read -14 dB.

For transmission measurements, the figure of merit is often gain, or insertion loss (sometimes called isolation when the loss is very high). Typically this is expressed in dB, and similarly to return loss, it is often referred to as a positive number. Thus

$$Gain = 20 \log_{10} (|S_{21}|) \tag{1.35}$$

And insertion loss, or isolation, is defined as

$$Insertion\ Loss = Isolation = -20 \log_{10} (|S_{21}|) \tag{1.36}$$

Again, the microwave engineer will need to use the context of the discussion to understand that a device with 40 dB isolation will show on an instrument display as -40 dB, due to the instrument using the evaluation of (1.35).

Notice that in the return loss, gain and insertion loss equations, the dB value is given by the formula $20 \log_{10} (|S_{nm}|)$ and this is often a source of confusion, because common engineering use of decibel or dB has the computation as $X_{dB} = 10 \log_{10} (X)$. This apparent inconsistency comes from the desire to have power gain when expressed in dB equal to voltage gain, also expressed in dB. In a device sourced from a Z_0 source and terminated in a Z_0 load, the power gain is defined as the power delivered to the load relative to the power delivered from the source, and the gain is

$$Power\ gain = 10 \log_{10} \left(\frac{P_{To\_Load}}{P_{From\_Source}} \right) \tag{1.37}$$

The power from the source is the incident power $|a_1|^2$ and the power delivered to the load is $|b_2|^2$. The S-parameter gain is S_{21} and in a matched source and load situation is simply

$$S_{21} = \frac{b_2}{a_1}, \quad |S_{21}|^2 = \left| \frac{b_2}{a_1} \right|^2 = \frac{|b_2|^2}{|a_1|^2} = Power\ Gain \tag{1.38}$$

So computing power gain as in (1.37) and converting to dB yields the familiar formula

$$Power\ Gain_{dB} = 10 \log_{10} (|S_{21}|^2) = 20 \log_{10} (|S_{21}|) \tag{1.39}$$

A few more comments on power are appropriate, as power has several common meanings that can be confused if not used carefully. For any given source, as shown in Figure 1.1, there exists a load for which the maximum power of the source may be delivered to that load. This maximum power occurs when the impedance of the load is equal to the conjugate of the impedance of the source, and the maximum power delivered is

$$P_{max} = \frac{|V_S|^2}{4 \cdot Re\ (Z_S)} \tag{1.40}$$

But it is instructive to note that the maximum power as defined in (1.40) is the same as $|a_1|^2$ provided the source impedance is real and equals the reference impedance; thus the incident power from a Z_0 source is always the maximum power that can be delivered to a load. The actual power delivered to the load can be defined in terms of a and b waves as well

$$P_{del} = |a|^2 - |b|^2 \qquad (1.41)$$

If one considers a passive two-port network and conservation of energy, power delivered to the load must be less than or equal to the power incident on the network minus the power reflected, or in terms of S-parameters

$$|S_{21}|^2 \leq 1 - |S_{11}|^2 \qquad (1.42)$$

which leads the well-known formula for a lossless network

$$|S_{21}|^2 + |S_{11}|^2 = 1 \qquad (1.43)$$

1.3.2 Phase Response of Networks

While most of the discussion thus far about S-parameters refers to powers, including incident, reflected and delivered to the load, the S-parameters are truly complex numbers and contain both a magnitude and phase component. For reflection measurements, the phase component is critically important and provides insight into the input elements of the network. These will be discussed in great detail as part of Chapter 2, especially when referencing the Smith chart.

For transmission measurements, the magnitude response is often the most cited value of a system, but in many communications systems, the phase response has taken on more importance. The phase response of a network is typically given by

$$\phi_{S21} = \arctan \left[\frac{\text{Im}(S_{21})}{\text{Re}(S_{21})} \right] \qquad (1.44)$$

where the region of the arctangent is usually chosen to be $\pm 180°$. However, it is sometimes preferable to display the phase in absolute terms, such that there are no phase discontinuities in the displayed value. This is sometimes called "unwrapped" phase, in which the particular cycle of the arctangent must be determined from the previous cycle, starting from the DC value. Thus the unwrapped phase is uniquely defined for an S_{21} response only when it includes all values down to DC.

The linearity of the phase response has consequences when looking at its effect on complex modulated signals. In particular, it is sometimes stated that linear networks cannot cause distortion, but this is only true of single-frequency sinusoidal inputs. Linear networks can cause distortion in the envelope of complex modulated signals, even if the frequency response (the magnitude of S_{21}) is flat. That is because the phase response of a network directly affects the relative time that various frequencies of a complex modulated signal take to pass through the network. Consider the signal in Figure 1.4.

For this network, the phase of S_{21} defines how much shift occurs for each frequency element in the modulated signal. Even though the amplitude response is the same in both Figures 1.4(a) and (b), the phase response is different, and the envelope of the resulting output is changed.

Normal

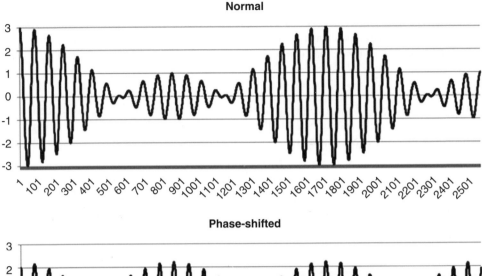

Phase-shifted

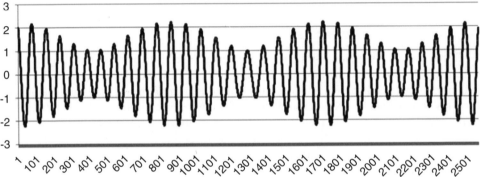

Figure 1.4 Modulated signal through a network showing distortion due to only phase shift.

In general, there is some delay from the input to the output of a network, and the important definition that is most commonly used is the group delay of the network, defined as

$$\tau_{GD} = -\frac{d\phi_{S_{21}}^{rad}}{d\omega} = \frac{-d\phi_{S_{21}}^{\circ}}{360 \cdot df} \tag{1.45}$$

While easily defined, the group delay response may be difficult to measure and/or interpret. This is due to the fact that measurement instruments record discrete values for phase, and the group delay is a derivative of the phase response. Using discrete differentiation can generate numerical difficulties; Chapter 5 shows some of the difficulties encountered in practice when measuring group delay, as well as some solutions to these difficulties.

For most complex signals, the ideal goal for phase response of a network is that of a linear phase response. Deviation from linear phase is a figure of merit for the phase flatness of a network, and this is closely related to another figure of merit, group delay flatness. Thus the ideal network has a flat group delay meaning a linear phase response. However, many complex communications systems employ equalization to remove some of the phase response effects.

Often, this equalization can account for first- or second-order deviations in the phase, thus another figure of merit is deviation from parabolic phase, which is effectively a measure of the quality of fit of the phase response to a second-order polynomial. These measurements are discussed further in Chapter 5.

1.4   Power Parameters

1.4.1   Incident and Reflected Power

Just as there are a variety of S-parameters, which are derived from the fundamental parameters of incident and reflected waves a and b, so too are there many power parameters that can be identified with the same waves. As implied above, the principal power parameters are incident and reflected – or forward and reverse – powers at each port, which for Z_0 real, are defined as

$$P_{Incident} = P_F = |a|^2, \quad P_{Reflected} = P_R = |b|^2 \tag{1.46}$$

The proper interpretation of these parameters is that incident and reflected power is the power that would be delivered to a non-reflecting (Z_0) load. If one were to put an ideal Z_0 directional coupler in line with the signal, it would sample or couple the incident signal (if the coupler were set to couple the forward power) or the reflected signal (if the coupler were set to couple the reverse power). In simulations, ideal directional couplers are often used in just such a manner.

1.4.2   Available Power

The maximum power that can delivered from a generator is called the available power or $P_{Available}$ and can be defined as the power delivered from a Z_S source

$$P_{Available} = P_{AS} = \frac{|a_S|^2}{\left(1 - |\Gamma_S|^2\right)} \tag{1.47}$$

where Γ_S is computed as in (1.24) as

$$\Gamma_S = \frac{Z_S - Z_0}{Z_S + Z_0} \tag{1.48}$$

This maximum power is delivered to the load when the load impedance is the conjugate of the source impedance, $Z_L = Z_S^*$.

1.4.3   Delivered Power

The power that is absorbed by an arbitrary load is called the delivered power, and is computed directly from the difference between the incident and reflected power

$$P_{del} = |a|^2 - |b|^2 \tag{1.49}$$

For most cases, this is the power parameter that is of greatest interest. In the case of a transmitter, it represents the power that is delivered to the antenna, for example, which in turn is the power radiated less the resistive loss of the antenna.

1.4.4   Power Available from a Network

A special case of available power is the power available from the output of a network, when the network is connected an arbitrary source. In this case, the available power is only a function of the network and the source impedance and is not a function of the load impedance. It represents the maximum power that could be delivered to a load under the condition that the load impedance was ideally matched, and can be found by noting that the available output power is similar to Eq. (1.47) but with the source reflection coefficient replaced by the output reflection coefficient of the network Γ_2 from (1.30) such that

$$P_{Out\_Available} = P_{OA} = \frac{|b_2|^2}{\left(1 - |\Gamma_2|^2\right)} \tag{1.50}$$

When a two-port network is connected to a generator with arbitrary impedance, the output scattered wave into matched load is

$$b_2 = \frac{a_S S_{21}}{1 - \Gamma_S S_{11}} \tag{1.51}$$

Here the incident wave is represented as a_S rather than a_1 as an indication that the source is not matched, and Γ_S is defined by Eq. (1.48). The output power incident to the load is

$$|b_2|^2 = \frac{|a_S|^2 |S_{21}|^2}{|1 - \Gamma_S S_{11}|^2} \tag{1.52}$$

Combining Eqs. (1.52) and (1.50), the available power at the output from a network that is driven from a generator with source impedance of Γ_S is

$$P_{OA} = \frac{|b_2|^2}{\left(1 - |\Gamma_2|^2\right)} = \frac{|a_S|^2 |S_{21}|^2}{|1 - \Gamma_S S_{11}|^2 \left(1 - |\Gamma_2|^2\right)} \tag{1.53}$$

with Γ_2 defined as in Eq. (1.30).

1.4.5   Available Gain

Available gain is the gain that an amplifier can provide to a conjugately matched load from a source or generator of a given impedance, and is computed with the formula

$$G_A = \frac{\left(1 - |\Gamma_S|^2\right) |S_{21}|^2}{|1 - \Gamma_S S_{11}|^2 \left(1 - |\Gamma_2|^2\right)} =$$

$$where \quad \Gamma_2 = \left(S_{22} + \frac{S_{21} S_{12} \Gamma_S}{1 - S_{11} \Gamma_S}\right) \tag{1.54}$$

Other derived values such as maximum available gain and maximum stable gain are discussed in detail in Chapter 6.

1.5 Noise Figure and Noise Parameters

For a receiver, the key figure of merit is its sensitivity, or ability to detect small signals. This is limited by the intrinsic noise of the device itself, and for amplifiers and mixers, this is represented as the noise figure. Noise figure is defined as signal-to-noise at the input divided by signal-to-noise at the output expressed in dB

$$NF \equiv N_{Figure} = 10 \log_{10} \left(\frac{Signal_{Input}/Noise_{Input}}{Signal_{Output}/Noise_{Output}} \right) = 10 \log_{10} \left(\frac{(S/N)_I}{(S/N)_O} \right) \tag{1.55}$$

and its related value, the noise factor, which is unit-less

$$N_F \equiv N_{Factor} = \left(\frac{Signal_{Input}/Noise_{Input}}{Signal_{Output}/Noise_{Output}} \right) = \frac{(S/N)_I}{(S/N)_O} \tag{1.56}$$

Here the signal and noise values are represented as a power, traditionally the available power, but incident power can be used as well with a little care. Rearranging (1.55) one can obtain

$$N_{Factor} = \frac{N_O}{Gain \cdot N_I} = \frac{N_{O\_Avail}}{G_{Avail} \cdot N_{I\_Avail}} \tag{1.57}$$

In most cases, the input noise is known very well, as it consists only of thermal noise associated with the temperature of the source resistance. This is the noise available from the source and can be found from

$$N_{Avail} = N_a = kTB \tag{1.58}$$

where k is Boltzmann's constant (1.38×10^{-23} joules/kelvin), B is the noise bandwidth and T is the temperature in kelvin. Note that the available noise power does not depend upon the impedance of the source. From the definition in (1.57), it is clear that if the temperature of the source impedance changes, then the noise figure of the amplifier using this definition would change as well. Therefore, by convention, a fixed value for the temperature is presumed, and this value, known as T_0, is 290 K.

This is the noise power that would be delivered to a conjugately matched load. Alternatively, the noise power can be represented as a noise wave, much like a signal, and one can define an incident noise (sometimes called the effective noise power) which is defined as the noise delivered to a non-reflecting non-radiating load, and is found as

$$N_{Incident} = N_E = N_A \left(1 - |\Gamma_S|^2 \right) \tag{1.59}$$

which is consistent with the definition of Eq. (1.47). Since the available noise at the output of a network doesn't depend upon the load impedance, and the available gain from a network similarly doesn't depend upon the load impedance, and the available noise at the input of the network can be computed as (1.58), the measurement of noise figure defined in this way is not dependent upon the match of the noise receiver. One way to understand this is to note that the available gain is the maximum gain that can be delivered to a load. If the load is not conjugately matched to Γ_2, both the available gain and the available noise power at the output would be reduced by equal amounts, leaving the noise figure unchanged and independent of the noise receiver load impedance. Thus for the case of noise measurements the available noise power and available gain have been the important terms of use historically.

Recently more advanced techniques have been developed and made practical based on incident noise power and gain. If the impedance is known, the incident noise power can be computed as in (1.59), and if the output incident noise power N_{OE} can be measured, then one can compute the output available noise as

$$N_{OA} = \frac{N_{OE}}{\left(1 - |\Gamma_2|^2\right)} \tag{1.60}$$

and substituting into (1.57) to find

$$N_F = \frac{1}{G_A} N_{OA} \frac{1}{N_{IA}} = \frac{|1 - \Gamma_S S_{11}|^2 \left(1 - |\Gamma_2|^2\right)}{\left(1 - |\Gamma_S|^2\right) |S_{21}|^2} \frac{N_{OE}}{\left(1 - |\Gamma_2|^2\right) (kTB)} \frac{1}{\left(1 - |\Gamma_S|^2\right) |S_{21}|^2 (kTB)} = \frac{|1 - \Gamma_S S_{11}|^2 N_{OE}}{\left(1 - |\Gamma_S|^2\right) |S_{21}|^2 (kTB)} \tag{1.61}$$

When the source is a matched source, this simplifies to

$$N_F = \frac{N_{OE}}{|S_{21}|^2 (kTB)} \tag{1.62}$$

Thus, for a simple system of an amplifier sourced with a Z_0 impedance and terminated with a Z_0 load, the noise factor can be computed simply from the noise power measured in the load and the S21 gain. However, Eq. (1.61) defines the noise figure of the amplifier in terms of the source impedance, and this is a key point. In general, although the 50 ohm noise figure is the most commonly quoted, it is only measured when the source impedance provided is exactly 50 ohms. In the case where the source impedance is not 50 ohms, the 50 ohm noise figure cannot be simply determined.

1.5.1 Noise Temperature

Because of the common factor of temperature in many noise figure computations, the noise power is sometimes redefined as available noise temperature

$$T_A = \frac{N_A}{kB} \tag{1.63}$$

From this definition, the noise factor becomes

$$N_F = \frac{T_A}{G_A 290} = \frac{T_{RNA}}{G_A} \tag{1.64}$$

where T_{RNA} is the relative available noise temperature, expressed in kelvin above 290 K.

1.5.2 Effective or Excess Input Noise Temperature

For very low noise figure devices, it is often convenient to express their noise factor or noise figure in terms of the excess power that would be at the input due to a higher temperature generator termination which would result in the same available noise temperature at the output. This can be computed as

$$T_e = 290(N_F - 1) \tag{1.65}$$

Thus an ideal noiseless network would have a zero input noise temperature, and a 3 dB noise figure amplifier would have a 290 kelvin excess input noise temperature, or 290 kelvin above the reference temperature.

1.5.3   Excess Noise Power and Operating Temperature

For an amplifier under test, the noise power at the output, relative to the kTB noise power, is called the excess noise power, P_{NE}, and is computed as

$$P_{NE} = N_F \, |S_{21}|^2 \, \frac{\left(1 - |\Gamma_S|^2\right)}{|1 - \Gamma_S S_{11}|^2 \left(1 - |\Gamma_2|^2\right)} \tag{1.66}$$

For a matched source and load, it is the excess noise, above kTB, that is measured in the terminating resistor and can be computed as

$$P_{NE} = \left(|S_{21}|^2 N_F\right) \tag{1.67}$$

which is sometimes called the incident relative noise or RNPI (as opposed to Available, or RNP). Errors in noise figure measurement are often the result of not accounting properly for the fact that the source or load impedances are not exactly Z_0. A related parameter is the operating temperature, which is analogous to the input noise temperature at the amplifier output, and is computed as

$$T_O = \frac{T_{OA}}{\left(1 - |\Gamma|^2\right)} \tag{1.68}$$

While the effect of load impedance may be overcome with the use of available gain, which is independent of load impedance, the effect of source impedance mismatch must be dealt with in a much more complicated way, as shown below.

1.5.4   Noise Power Density

The excess noise is measured relative to the kTB noise floor, and is expressed in dBc relative to the T_0 noise floor. However, the noise power could also be expressed in absolute terms such as dBm. But the measured noise power depends upon the bandwidth of the detector, and so the noise power density provides a reference value with a bandwidth equivalent to 1 Hz. Thus the noise power density is related to the excess noise by

$$P_{NoisePowerDensity} = \frac{N_{NE}}{B} = k(T_0 + T_e) \tag{1.69}$$

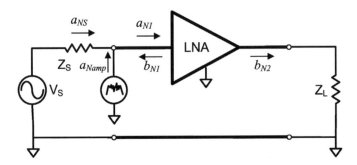

Figure 1.5   An amplifier with internal noise sources.

1.5.5   Noise Parameters

The formal definition of noise figure for an amplifier defines the noise figure only for the impedance or reflection coefficient of current source termination, but this noise figure is *not* the 50 ohm noise figure. Rather, it is the noise figure of the amplifier for the impedance of the source. In general, one cannot compute the 50 ohm noise figure from this value without additional information about the amplifier. If one considers the amplifier in Figure 1.5, with internal noise sources, the effect of the noise sources is to produce noise power waves that may be treated similarly to normalized power waves, *a* and *b*.

The source termination produces an incident noise wave a_{NS} and this adds to the internal noise created in the amplifier, which can be represented as an input noise source a_{Namp}. There are scattered noise waves represented by the noise emitted from the input of the amplifier, b_{N1}, and the noise incident on the load is b_{N2}. From this figure, one can make a direct comparison to the S-parameters, and see that reflected noise power might add to, or subtract from, the incident noise power and affect the total noise power. However, at the input of the amplifier, the noise generated inside the amplifier is in general not correlated with the noise coming from the source termination, so that they don't add together in a simple way. Due to this, the noise power at the output of the amplifier, and therefore the noise figure, depends upon the source impedance in a complex way. This complex interaction is defined by two real-valued parameters and one complex parameter, known collectively as the noise parameters. The noise figure at any source reflection coefficient may be computed as

$$N_F = N_{Fmin} + \frac{4R_n}{Z_0} \frac{\left|\Gamma_{opt} - \Gamma_S\right|^2}{\left|1 + \Gamma_{opt}\right|^2 \left(1 - |\Gamma_S|^2\right)} \tag{1.70}$$

where N_{Fmin} is the minimum noise figure, Γ_{Opt}, called "gamma-opt", is the reflection coefficient (magnitude and phase) that gives the minimum noise figure, and R_N, sometimes called the noise resistance, describes how the noise figure increases as the source impedances vary from the gamma-opt. The characterization required to determine these values is quite complex, and is covered in Chapter 6.

1.6   Distortion Parameters

Up to now, all the parameters described have been under the assumption that the DUT is linear. However, when a DUT, particularly an amplifier, is driven with a large signal,

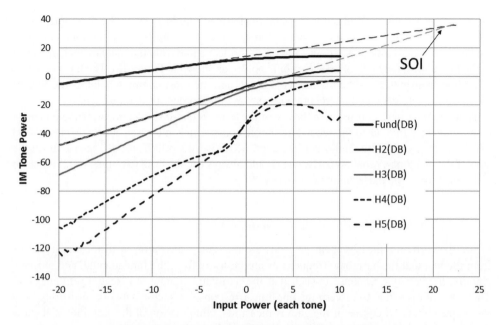

Figure 1.6   Output power of harmonics of an amplifier.

non-linear transfer characteristics become significant, leading to an entirely new set of parameters used to describe these non-linear characteristics.

1.6.1   Harmonics

One of the first noticeable effects of large signal drive is the generation of harmonics at multiples of the input frequency. Harmonics are described either by their output power, or more commonly by the power relative to the output power of the fundamental, and almost always in dBc (dB relative to the carrier), and their order. Second harmonic is short for second-order harmonic and refers to the harmonic found at two times the fundamental, even though it is in fact the first of the harmonic frequency above the fundamental; the third harmonic is found at three times the fundamental, and so on. Surprisingly, there are not well established symbols for harmonics; for this text we will use H2, H3 ... Hn to represent the dBc values of harmonics of order 2, 3 ... respectively. In Chapter 6, the measurements of harmonics are fully developed as part of the description of X-parameters, and utilize the notation $b_{2,m}$ to describe the output normalized wave power at port 2 for the mth harmonic. A similar notation is used for harmonics incident on the amplifier.

One important attribute of harmonics is that, for most devices, the level of the harmonics increases in dB value as the power of the input increases, and to a rate directly proportional to the harmonic order, as shown in Figure 1.6. In this figure, the x-axis is the drive power and the y-axis is the measured output power of the fundamental and the harmonics.

1.6.2   Second-Order Intercept

This pattern of increasing power as the input power is increased, but to the slope related to the order of the harmonic, cannot continue indefinitely or the harmonic power would exceed the fundamental power. While this is theoretically possible, in practice the harmonic power saturates just as the output power does, and never crosses the level of the output power. However, if one uses the lower power regions to project a line from the fundamental, and each of the harmonics, they will intersect at some power, as shown in Figure 1.6. The level at which these lines converge is called the intercept point, and the most common value is the SOI or second-order intercept, and intercept points beyond third order are seldom used.

There is sometimes confusion in the use of the term second-order intercept; while it is most commonly used to refer to the second harmonic content, in some cases, it has also been used to refer to the two-tone second-order intercept, which is a distortion product that occurs at the sum of the two tones. Most properly, one should always use the term two-tone SOI if one is to distinguish from the more common harmonic SOI.

1.6.3   Two-Tone Intermodulation Distortion

While the harmonic measurement provides a direct characterization of distortion, it suffers from the fact that the harmonic frequencies are far away from the fundamental, and in many circuits, the network response is such that the harmonic content is essentially filtered out. Thus, it is not possible to discern the non-linear response of such a network measuring only the output signal. Of course, if the gain is measured, compression of the amplifier will show that the value of S21 changes with input drive level. But it is convenient to have a measure or figure-of-merit of the distortion of an amplifier that relies only on the output signal. In such a case, two signals of different frequencies can be applied at the amplifier input, at a level sufficiently large to cause a detectible non-linear response of the amplifier. Figure 1.7 shows a measurement of a two-tone signal applied to the input of an amplifier (lower trace) and measured on the output of the amplifier (upper trace).

It is clear that several other tones are present at the output, which are the result of higher-order products mixing in the amplifier due to its non-linear response and creating other signals. The principal signals of interest are the higher and lower intermodulation (IM) products, PwrN_Hi and PwrN_Lo, where N is the order of IMD. Normally, IM products refer to the power of the IM product relative to the carrier, in dBc, and these terms are called IMN_Hi and IMN_Lo. For example, the power in the lower third-order tone is Pwr3_Lo; the level of the third-order tone relative to the carrier is called IM3_Hi. The frequency of the higher and lower tones are found at

$$f_{3Hi} = 2f_{Hi} - f_{Lo}, \ f_{3Lo} = 2f_{Lo} - f_{Hi} \tag{1.71}$$

And more generally

$$f_{mHi} = \left(\frac{m+1}{2}\right)f_{Hi} - \left(\frac{m-1}{2}\right)f_{Lo}, \ f_{mLo} = \left(\frac{m+1}{2}\right)f_{Lo} - \left(\frac{m-1}{2}\right)f_{Hi}\bigg|_{m\,odd}$$

$$f_{mHi} = (m-1)f_{Hi} + (m-1)f_{Lo}, \ f_{mLo} = (m-1)f_{Ho} - (m-1)f_{Lo}\big|_{m\,even} \tag{1.72}$$

In Figure 1.7, the amplifier is driven such that the fifth-order IM product is just visible above the noise floor in the upper trace.

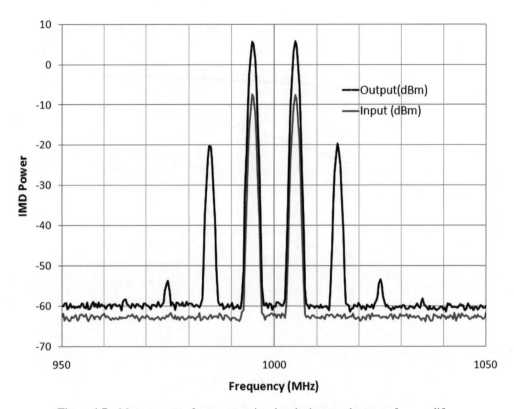

Figure 1.7   Measurement of a two-tone signal at the input and output of an amplifier.

IM products have the same attribute as harmonics with respect to drive power, and the power in the IM product (sometimes called the tone power, or PWRm for the mth-order IM power) increases in direct proportion to the input power and the order of the IM product. Thus, if the tone power is plotted along with the output power against an x-axis of input power, the plot will look like Figure 1.8, where the extension of the slope of the output power and IM tone powers at low drives will intersect. This point of intersection for the third-order IM product is known as the third-order intercept point, or IP3. Similarly, IP5 is the fifth-order intercept point, and so on.

It is also interesting to note that in general at high powers the IM tone powers may not increase but may decrease or have local minima. This is due to the effect of high-order IM products remixing, and creating significant signals that lie on the lower-order products and can increase or decrease their level, depending upon the phasing of the signals.

There is often some confusion about third-order IM products (IM3) and third-order intercept point (IP3) and both are sometimes referred to as "third-order intermod". For clarity, in this book the intercept point will always be referred to as IP.

Finally, for amplifiers used as a low noise amplifier (LNA) at the input of a receiver chain, it is often desired to refer the IP level to the input power which would produce an intercept point at the output. This is distinguished as the input intercept point (IIP), and in the case of ambiguity, the normal intercept point referencing to the output power should be most properly

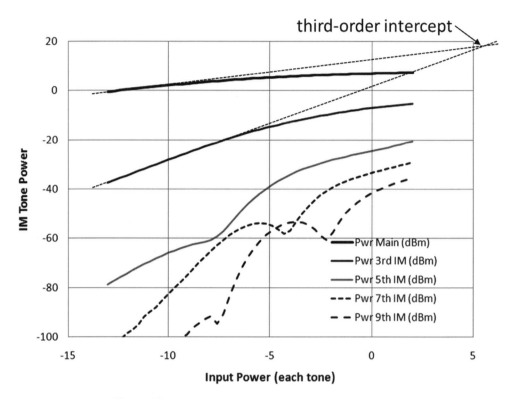

Figure 1.8 Output power and IM tone power vs input power.

referred to as the OIP. The most common intercept points are the third-order ones, OIP3 and IIP3. The input and output intercept points differ by the gain of the amplifier at drive level where the measurements are made.

Details of two-tone IM measurements are discussed at length in Chapter 6.

1.7 Characteristics of Microwave Components

Microwave components differ from other electrical devices in a few respects. The principal discerning attribute is the fact that the components' size cannot be ignored. In fact, the size of many components is a significant portion of a wavelength at the frequency of interest. This size causes the phase of the signals incident on the device to vary across the device, implying that microwave devices must be treated as distributed devices. A second, related attribute is that the reference ground for the device is not defined by a point, but is distributed as well. Indeed, in many cases the ground is not well defined. In some situations, grounds for a device are isolated by sufficient distance that signal propagation can occur from one device ground to another. Further, even if devices are defined as series only (with no ground contact), one must realize that there is always an earth ground available, so there can always be some impedance to this ground. In practice, the earth ground is actually the chassis or package of the device, or a power or other ground plane on a printed circuit board (PCB).

Finally, only in microwave components can one find the concept of wave propagation. In waveguide components, there is no "signal" and no "ground". Rather a wave of electromagnetic (EM) field is guided into and out of the device without regard to a specific ground plane. For these devices, even the transmission structures, a waveguide for example, are a large percentage of the signal wavelength. Common concepts such as impedance become ambiguous in the realm of waveguide measurements, and must be treated with special care.

1.8 Passive Microwave Components

1.8.1 Cables, Connectors and Transmission Lines

1.8.1.1 Cables

The simplest and most ubiquitous microwave components are transmission lines. These can be found in a variety of forms and applications, and provide the essential glue that connects the components of a microwave system. RF and microwave cables are often the first exposure an engineer has to microwave components and transmission systems, the most widespread example being a coaxial cable used for cable television (CATV, a.k.a. Community Antenna TeleVison).

The key characteristics of coaxial cables are their impedance and loss. The characteristics of coaxial cables are often defined in terms of their equivalent distributed parameters [5], as shown in Figure 1.9, described by the *telegraphers' equation:*

$$\frac{dv(z)}{dz} = -(r + j\omega l) \cdot i(z) \tag{1.73}$$

$$\frac{di(z)}{dz} = -(g + j\omega c) \cdot v(z) \tag{1.74}$$

where $v(z)$ and $i(z)$ are the voltage and current along the transmission line, and r, l, g and c are the resistance, inductance, conductance and capacitance per unit length.

For a lossless cable, the impedance can be computed as simply

$$Z = \sqrt{\frac{l}{c}} \tag{1.75}$$

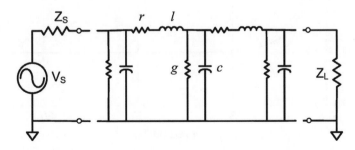

Figure 1.9 A transmission line modeled as distributed elements.

but it becomes more complicated when loss is introduced, becoming

$$Z_{lossy} = \sqrt{\frac{r + j\omega l}{g + j\omega c}} \tag{1.76}$$

In many applications, the conductance of the cable is negligible, particularly at low frequencies, so that the only loss element is the resistance per unit length, yielding

$$Z_{lossy} = \sqrt{\frac{r + j\omega l}{j\omega c}} \tag{1.77}$$

Inspection of Eq. (1.77) shows that the impedance of a cable must increase as the frequency goes down toward DC. Figure 1.10 demonstrates this with a calculation the impedance of a nominal 75 ohm cable, with a 0.0001 ohm/mm loss and capacitance of 0.07 pF/mm (typical for RG 6 CATV coax). In this case, the impedance deviates from the expected value at 300 kHz by over 10 ohms, and by 1 ohm at 1 MHz.

This low frequency response of impedance for any real transmission line is often unexpected by those unfamiliar with Eq. (1.77), and it is sometimes assumed that this is a result of measurement error. However, all real transmission lines must show such a low frequency characteristic, and verification methods must take into account this effect.

An "airline" coax consists of a cable with an air dielectric, sometimes supported by dielectric beads at either end or sometimes supported only by the center conductor of the adjacent connectors, as shown in Figure 1.11. This type of cable has virtually no conductance, so series resistive loss is the only loss element. The small white ring on the airline sometimes used to prevent sagging at the male end of the pin so that it may be more easily mated.

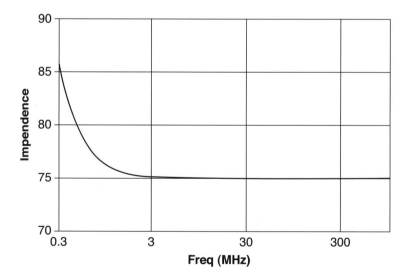

Figure 1.10   Impedance of a real transmission line at low frequency.

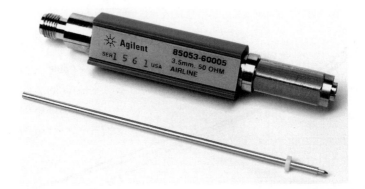

Figure 1.11 An airline coaxial transmission line. Reproduced by permission of Agilent Technologies.

In some special applications, such as using measurements of a transmission line loaded with some material to determine the properties of the material, none of the elements of the telegraphers' equation can be ignored.

At higher frequencies, the loss of a cable is increased due to skin effect which can be shown to increase as the square root of frequency [6]

$$r = \sqrt{\frac{\omega\mu}{2\sigma}} \tag{1.78}$$

Thus, the insertion loss of an airline coaxial cable depends only upon the resistance per unit length of the cable, and so the insertion loss (in dB) per unit length, as a function of frequency, can be directly computed as

$$Loss(f) = 8.68 \frac{r}{4\pi Z_0} \left(\frac{1}{R_a} + \frac{1}{R_b} \right) \tag{1.79}$$
$$= A \cdot f^{1/2}$$

where R_a and R_b are the inner and outer conductor radius and r contains the square root of frequency. Thus, all the attributes can be lumped into a simple single loss-term, A. Figure 1.12 shows the loss of a 10 cm airline as well as the idealized loss as described in (1.79), where very good agreement to theory is seen. However, the introduction of dielectric loading of the coaxial line will add some additional loss due to the loss tangent of the dielectric. This additional loss often presents itself as an equivalent conductance per unit length, and this loss is often more significant than the skin-effect loss. Because of the dielectric loss, the computed loss of (1.79) fails to fit many cables. The equation can be generalized to account for differing losses by modifying the exponent to obtain

$$Loss(f) = A \cdot f^b \tag{1.80}$$

Where the loss is expressed in dB, and A and b are the loss factor and loss exponent. From the measured loss at two frequencies it is possibly find the loss factor and loss exponent directly, although better results can be obtained by using a least-squares fit to many frequency points. Figure 1.12 shows the loss of a 15 cm section of 0.141 inch semi-rigid coaxial cable. The values for the loss at one-fourth and three-fourths of the frequency span are recorded. From

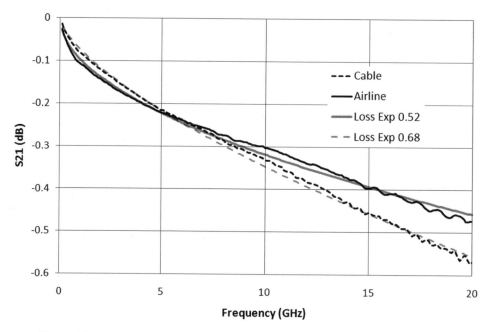

Figure 1.12   Loss of a 15 cm airline and a 15 cm semi-rigid Teflon-loaded coaxial line.

these two losses, the loss factor and exponent are computed as

$$L_1 = A \cdot (f_1)^b, L_2 = A \cdot (f_2)^b$$

Taking the log of both sides, this can be turned into a linear system as

$$\log(L_1) = \log(A) + b \cdot \log(f_1)$$
$$\log(L_2) = \log(A) + b \cdot \log(f_2)$$

(1.81)

And the this system of linear equations can be solved for the loss factor A and the loss exponent b

$$A = \exp\left(\frac{\log(f_1) \cdot \log(L_2) - \log(f_2) \cdot \log(L_1)}{\log(f_1) - \log(f_2)}\right)$$

(1.82)

$$b = \frac{\log(L_1) - \log(L_2)}{\log(f_1) - \log(f_2)}$$

(1.83)

The computed loss for all frequencies from (1.80) is also shown, with remarkably good agreement to the measured values over a wide range. Ripples in the measured response are likely, due to very small calibration errors, as discussed in Chapter 5.

The insertion phase of a cable can likewise be computed; in practice a linear approximation is typically sufficient, but the phase of a cable will vary with frequency beyond the linear slope due to loss as well.

The velocity of propagation for a lossless transmission line is

$$v = \frac{1}{\sqrt{l \cdot c}} \tag{1.84}$$

The impedance of a lossy cable *must* be complex from (1.77) and thus the phase response must deviate from a pure linear phase response, due to the phase velocity changing with frequency at lower frequencies. A special case for airlines, which have no dielectric loss is

$$v_{prop} \approx \sqrt{\frac{2\omega}{rc}}\Bigg|_{\omega \cdot l \ll r} \tag{1.85}$$

For cables in general, the dielectric loss will cause a deviation in the velocity of propagation similar to that seen for loss. So far the discussion has focused on ideal low-loss cables, but in practice cables have defects that cause the impedance of the cable to vary along the cable. If these defects are occasional, they cause little concern and are typically overlooked unless they are so large as to cause a noticeable discrete reflection (more of that in Chapter 5). However, during cable manufacturing it is typical that the processing equipment contains elements such as spooling machines or other circular equipment (e.g., pulleys, spindles). If these have any defects in the circularity, or even a discrete flaw like a dimple, it can cause minute but periodic changes in the impedance of the cable. A flaw that causes even a one-tenth ohm deviation of impedance periodically over a long cable can cause substantial system problems called "structural return loss" (SRL), Figure 1.13. These periodic defects add up all at one frequency and can cause very narrow (as low as 100 kHz BW), very high return loss peaks, and thereby cause insertion loss dropouts at these same frequencies. In practice the SRL test is the most difficult for low-loss, long-length cables such as those used in the CATV industry. Figure 1.14 shows a simulation of a structural return loss response caused by a 15 mm long, 0.1 ohm impedance variation, every 30 cm, and another −0.1 ohm variation every 2.7 m, each on the same 300 m coaxial cable with an insertion loss typical for mainline CATV cables. From the figure, two structural return loss effects are seen: a smaller effect every 50 MHz or so, due to the 2.7 m periodic variation and a much higher effect every 500 MHz or so due to the 30 cm impedance variation. The higher impedance variation occurs more often, and so the periodic error will have a greater cumulative effect resulting in a nearly full reflection as seen in the figure.

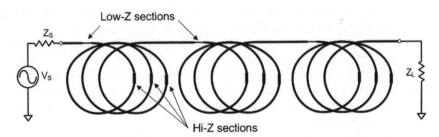

Figure 1.13   A model of a coax line with periodic impedance disturbances.

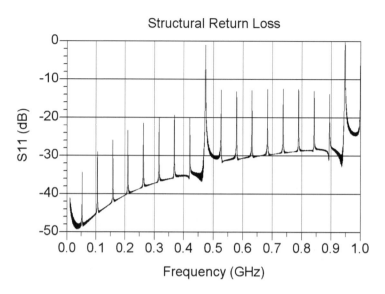

Figure 1.14   The return loss of a line with structural return loss.

1.8.2   Connectors

Connectors provide the means to transition from one transmission media to another. They are often not considered as part of the device or measurement system, but their effects can dominate the results of a measurement, particularly for low-loss devices. Connectors can be distinguished by the quality and application. One remarkable aspect of connectors is the great difficulty in measuring them with any kind of accuracy. This derives from the fact that most connectors provide a transition between different media, such as from a coaxial cable to a connector interface, or from a PCB to a connector interface. While the connector interface is often well defined, the "back-end" of the connector is poorly defined.

Connectors that are "in-series" provide transitions from male to female, and provide interconnections between components. These are easiest to characterize because the ports are well defined, and typically calibration kits are available and calibration methods are well understood. Connectors that are "between-series" are equally well defined, but until recently they have been difficult to characterize because there were not well defined standards for between-series adapters. Recent improvements in calibration algorithms have essentially eliminated any difficulty with characterizing these between-series adapters. Figure 1.15 shows some examples of in-series and between-series connectors.

For microwave work, there are some very commonly utilized connector types that are found on the majority of components and equipment. Table 1.1 shows a listing of these common connectors along with their normal operating frequency range. These are divided into three broad categories: precision sexless connectors, precision male-to-female connectors, and general purpose or utility connectors. These connectors are typically 50 ohms, but a few can be found as 75 ohm versions as well.

From Table 1.1 one can see that there are actually three frequencies associated with connectors: the generally understood operating frequency (often dictated by the calibration kit's

Table 1.1 Test connectors used for RF and microwave components

Name	Outer Conductor Diameter (mm)	Rated Frequency (GHz)	First Mode	Maximum Useable Frequency (GHz)
Type-N (50 ohm) Precision	7	18	18.6 GHz	26.5[a]
Type-N (50 ohm) Commercial	7	12	12.5 GHz	15
Type-N (75 ohm) Precision	7	18	18.6 GHz	18
Type-N (75 ohm) Commercial	7	12	12.5	15
7 mm	7	18	18.6 GHz	18
SMA	3.5	18	19 GHz	22
3.5 mm	3.5	26.5	28 GHz	33
2.92 mm ("K")	2.4	40	44 GHz	44
2.4 mm	2.4	50	52 GHz	55
1.85 mm ("V")	1.85	67	68.5 GHz	70
1 mm	1	110	120 GHz	120

[a]Some instrument manufacturers place this connector on 26.5 GHz instruments because it is very rugged; it has the same first modes as type-N and 7 mm.

maximum certified frequency), the frequency of the first mode and the maximum frequency determined by the waveguide propagating mode of the outer conductor. The operating frequency is always below the first mode, and usually by several percent. The first mode in many connectors is due to the support structure for the center pin. It is often of some plastic material, and thus has higher dielectric constant and a lower frequency to support a mode. In connectors and cables, modes are the term used to refer to non-transverse-electromagnetic (TEM)

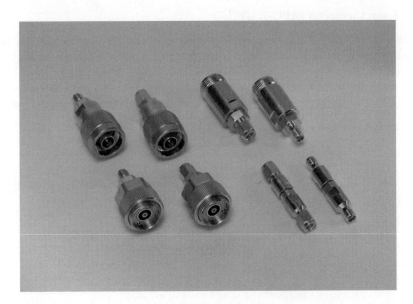

Figure 1.15 In-series and between-series connectors. Reproduced by permission of Agilent Technologies.

propagation that can occur in a circular waveguide mode defined by the inside dimension of the outer conductor. Adding dielectric in the bead that supports the center pin can theoretically lower the mode frequency, but if the bead is short, the mode will be evanescent (non-propagating) and may not affect the quality of the measurement. At a somewhat higher frequency, there will be a propagating mode in air for the diameter of the center conductor, but if the cable attached to the connector is sufficiently small, this mode may not propagate as well. It is the propagating modes that cause the significant dips in the response, and more importantly, these dips cannot be removed with calibration because they are not localized, and reflections in the mode of transmission far removed from the connector interface can interact with these connector modes, causing the frequency response of the mode effect to change when different devices are connected. If the response of the mode does not change when other devices are connected, it can be calibrated out.

Precision sexless connectors are now found only in metrology labs. Their chief benefit was a repeatable connector that has identical characteristics for each connector. As such, it was very easy to create a system calibration and any part with such connectors could be inserted between two cables in either direction. This was important because in the past it was difficult to deal with "non-insertable" devices from a calibration sense (a non-insertable device is one with the same sexed connector on each port, e.g., female-to-female). The 7 mm connector is often found on precision attenuators and airlines used as transfer standards. The 7 mm connector is also known as the GPC-7 for general precision connector, and very often the APC-7™ for Amphenol precision connector. Because these connectors are sexless, there is no need for adapters to provide interconnections between devices or between devices and cables.

1.8.2.1   7 mm Connector (APC-7, GPC-7)

The 7 mm connector has a couple of interesting attributes: the center pin has no slots but contains spring loaded center collets that protrude slightly from the mating surface, see Figure 1.16.

When mated, the collet from one connector floats against the other providing a good center contact. There is a very slight gap in the slotless outer sleeve of the center pin. As with almost

Figure 1.16   A 7 mm connector. Reproduced by permission of Agilent Technologies.

all RF connectors, the outer conductor forms the physical mating plane. On most connectors, there is a slip ring threaded sleeve surrounded by a coupling nut. To mate, the threaded sleeve is extended on one connector and retracted on the other. On the retracted connector, the coupling nut is extended to engage the other's sleeve and is tightened. Only one coupling nut should be tightened, although it is common but incorrect practice to also tighten the other coupling nut. In fact, tightening both coupling nuts can result in the center pins pulling apart, and a poorly matched contact. Occasionally, one sees parts that contain only a solid threaded outer conductor (serving the purpose of the threaded sleeve) and no coupling nut. These are more common on older test fixtures intended to mount directly the 7 mm connectors of network analyzer test sets.

1.8.2.2 Type-N 50 ohm Connector

The type-N-connector is very common in lower frequency and higher power RF and microwave work. It has the same outer diameter (7 mm) as the 7 mm connector, but is sexed. In fact, this connector has the unusual attribute of having the mating surface for the outer conductor (which is almost always the electrical reference plane) recessed for the female connector. The female pin protrudes (in an electrical sense) from the reference plane, and the male pin is recessed. Thus, the calibration standards associated with type-N-connectors have electrical models that are highly asymmetric for male and female standards.

The type-N-connector has precision forms, including ones with slotless connectors (metrology grade), ones with precision six-slotted collets and solid outer conductor sleeves (found on most commercial test equipment) and commercial forms with slotted outer conductor sleeves and four or even two slotted female collets. Slotless connectors have a solid hollow cylinder for the female connector with an internal four- or six-finger spring contact that takes up tolerances of the male center pin. As such the diameter of the female center pin does not depend at all on the radius of the male pin. Typical female contacts with collets expand or contract to accept the male pin, and thus their outer dimension (and thereby their impedance) varies with the diameter tolerance of the male pin.

The commercial forms are found on a variety of devices and interconnect cables. The male version of these commercial-grade parts present two common and distinct problems: there is often a rubber "weather-seal" O-ring in the base of the connector, and the outer nut of the male connector is knurled but has no flats to allow using a torque wrench. The first problem exacerbates the second, as the mating surface of the outer conductor of the male connector is often prevented from contacting the base of the female connector due to the fact that the outer (supposedly non-mating) surface of the female connector touches the rubber O-ring and prevents the male outer conductor from making full contact. If one can fully torque a type-N-connector, the rubber O-ring would compress, and the contact of the male outer conductor would occur, but as there are no flats for a torque wrench, it is very difficult to sufficiently torque the type-N-connector to get good repeatable connections. This one issue is the cause of hundreds of hours of retests when components don't pass their return-loss specs. The solution is quite simple: remove the rubber O-ring from the base of the male connector, always, before any measurement. A pair of tweezers and a pair of needle-nose pliers are indispensible for the process of removing this annoying O-ring. One will note that none of the precision versions of type-N-connectors contain such an O-ring. Figure 1.17 shows some examples of

Figure 1.17   Examples of type-N connectors, upper are commercial, lower are precision. Reproduced by permission of Agilent Technologies.

type-N-connectors; the upper two are commercial grade and the lower two are precision grade. Figure 1.18 shows the return loss and insertion loss of a male-to-male type-N adapter mated to a female-to-female type-N adapter for a precision pair and a commercial grade pair, where the loss is normalized to the length of the adapter. The commercial grade pair is operational only to about 12 GHz, due to moding in the connector. The precision type-N is mode free beyond 18 GHz.

1.8.2.3   Type-N 75 ohm Connector

Type-N-connectors also have a 75 ohm version, which has the same outer dimensions but a smaller center conductor. This is in some ways unfortunate as the smaller female collet of the 75 ohm version can be damaged when inserted with a 50 ohm male pin. There are a couple of versions of the 75 ohm female collet, one with short slots and six fingers, and one with long slots and four fingers. A precision slotless version is also available. The short slot version has the potential for better measurements, as the slots expand less so there is less uncertainty of the open capacitance. However, on many products with 75 ohm N-connectors, the long slot connector is used; the long slots were designed to accept a 50 ohm male pin, at least for a few insertions, without damage. Often the 75 ohm components have an extra machined ring or strip on the outer nut to help identify it. Versions of 75 ohm type-N-connectors are shown in Figure 1.19. An example of the insertion loss and return loss of a mated pair of a male-to-male

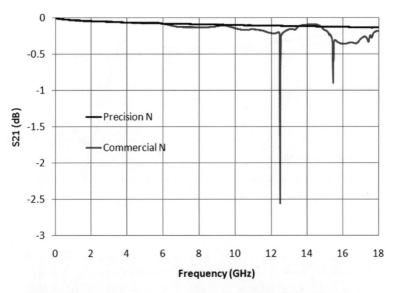

Figure 1.18   Performance of a precision and a standard type-N connector.

Figure 1.19   75 ohm type-N connectors, upper commercial, lower precision. Reproduced by permission of Agilent Technologies.

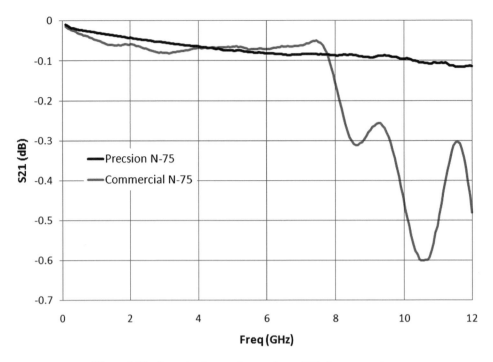

Figure 1.20   Insertion loss and return loss of 75 ohm connectors.

adapter with a female-to-female adapter is shown in Figure 1.20, where the loss is normalized for length of the adapter. The frequency limit of type-N 75 is often stated as 2 or 3 GHz, but that is only because the commercially available calibration kits were only rated to those frequencies. In practice these connectors could be used up to 7 or 8 GHz without difficulty. The response of the commercial grade connector is likely limited not due to moding (since the loss signature is quite low Q) but rather due to poor impedance control in the center pin support bead, causing impedance mismatch.

1.8.2.4   3.5 mm and SMA Connectors

The 3.5 mm connector is in essence half the scale of the N-connector, and provides higher frequency coverage. The center pin of the 3.5 mm connector is supported by a plastic bead, rather than solid dielectric, meaning that it has mode-free operation to a much higher frequency than SMA. Traditionally, 3.5 mm connectors are specified up to 26.5 GHz, but their first mode is nearly 30 GHz, and they are functional up to about 38 GHz. An interesting aspect of modes is that the first mode of a 3.5 mm connector is due to the bead (and its increased effective dielectric) but this mode is non-propagating so it is reasonable to use these connectors to even higher frequencies. The 3.5 mm female connector comes with several versions of center pin, the main varieties being a four-slot collet, and a slotless precision connection, found now on most calibration kits. Interestingly, even though the slotless connectors may have the center spring contact damaged due to oversized or misaligned male pins (under the microscope one

Figure 1.21   Upper: Left 3.5 mm (f) and (m), right SMA (f) and (m) connectors; Lower: 3.5 mm and SMA adapters. Reproduced by permission of Agilent Technologies.

or more fingers may be crushed back into the hollow of the female pin), the RF performance is almost unaffected due the robust solid outer conductor. In fact, one typically can only tell if a slotless connector is damaged by visual inspection, as the RF performance is substantially unchanged, as long as at least one finger is left to make contact.

The SMA connector is mechanically compatible with the 3.5 mm connector, but has a solid PTFE dielectric, and thus a lower operating frequency due to moding. SMA is traditionally considered to be an 18 GHz connector, but the first propagating mode is above 20 GHz, depending upon the type of cable that is connected to the SMA connector. The chief advantage of SMA connectors is very low cost, especially when mounted to semi-rigid coaxial cables. The dimensions are such that the center wire of the coax can be used as a connector pin for SMA, and only an outer conductor sleeve need be added to the coax outer conductor to form a male connector. But these cables are notoriously bad at maintaining the proper dimensions for the center pin, and often the center pins are poorly trimmed and improperly chamfered so that they cause mating problems with their female counterparts. This is particularly true when mating them to 3.5 mm female connectors, slotless ones in particular. Figure 1.21 shows examples of 3.5 mm and SMA connectors.

Figure 1.22 shows measurement plots of a mated pair of 3.5 mm male-to-male with a 3.5 mm female-to-female, as well as an SMA example of the same. The moding of the SMA connector is clearly seen above 25 GHz. The moding of the 3.5 mm connector is seen just above 27 GHz and again at 38 GHz.

1.8.2.5   2.92 mm Connector

The 2.92 mm connector is scaled down from the 3.5 mm connector and can be mechanically mated to both the 3.5 mm and the SMA connectors. The smaller diameter outer conductor

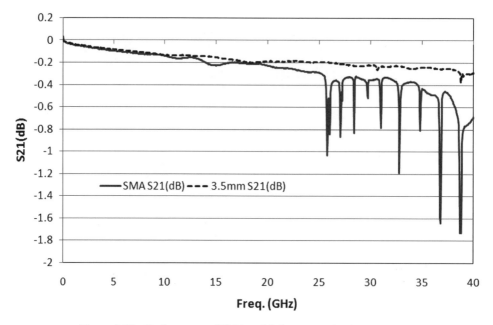

Figure 1.22   Performance of SMA and 3.5 mm mated-pair connectors.

means that its mode-free operation extends proportionally high, to 40 GHz, and is usable to perhaps 46 GHz. The female connector has a two-slot collet which provides sufficient compliance to mate with the center pin of the larger 3.5 mm and SMA connectors, but which makes it less suitable for precision measurements due to increased uncertainty of the contact point on and center pin radius, which now depends upon the radius of the pin that is inserted. A further point is that the metal wall of the female collet on the 2.92 mm connector is quite thin, and prone to damage if the mating pin is not well aligned or is oversize. It's not uncommon to find 2.92 mm female adapters missing one of the collet fingers. The 2.92 mm connector was popularized by the Anristu company (formally Wiltron) who introduced it as the K connector, and it is common to hear 2.92 mm connectors being referred to by that name.

Figure 1.23 shows some examples of 2.92 connectors. The key difference is in the diameter of the inside of the outer conductor. Figure 1.24 shows the insertion loss of a mated pair of 2.92 mm female-to-female adapter with a 2.92 mm male-to-male adapter, along with an example 3.5 mm mated adapter pair, with the moding of the 3.5 mm pair clearly seen.

1.8.2.6   2.4 mm Connector

The 2.4 mm connector is essentially a scaled version of the 3.5 mm connector, with an associated scaling in maximum frequency. It is used extensively on 50 GHz applications, though it can be used up to 60 GHz. This connector cannot be mated to any of the SMA, 3.5 mm or 2.92 mm, and in fact was designed to prevent damage if one tried to mate to these types. It comes with both slotted and slotless female center pins, much like the 3.5 mm connector.

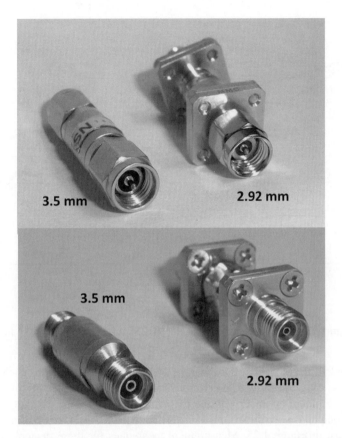

Figure 1.23 A 3.5 mm connector (right) compared with 2.92 mm, male and female.

1.8.2.7   1.85 mm Connectors

There are two variants of the 1.85 mm connectors, designed originally by Anritsu and Agilent. The Anritsu variety is called the V connector, and the Agilent variety is called the 1.85 mm connector. They are mechanically compatible and were originally designed for 67 GHz operation, usable to above 70 GHz. These connectors are mechanically compatible with the 2.4 mm connector.

1.8.2.8   1 mm Connector

The 1 mm connector is essentially a scaled version of the 1.85 mm connector, but cannot be mated to it. It is typically specified to 110 GHz performance, but is usable to above 120 GHz, with some versions being used up to 140 GHz.

1.8.2.9   PC Board Launches and Cable Connectors

For many design and measurement applications, the circuit of interest is embedded in a PCB. There are many types and styles of PCB launches, which typically have an SMA connector

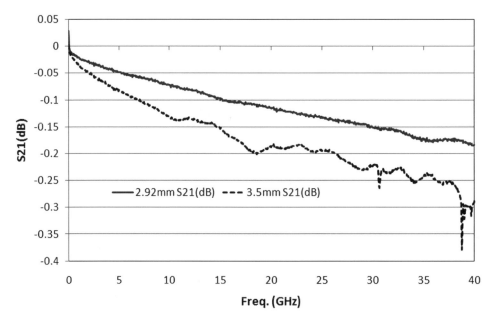

Figure 1.24   Performance of a mated pair, 2.92 mm compared with 3.5 mm.

on one end and PCB contacts at the other, as well as miniature versions such as the QMA connector. These can come in edge launch as well as right angle, and their performance depends greatly upon the mounting pattern on the PCB trace. These can be difficult to characterize because only one end is available in a standard connector. An example of a common PCB launch is shown in Figure 1.25. Measurement techniques for these devices, as well as the

Figure 1.25   PCB SMC launches. Reproduced by permission of Agilent Technologies.

method of removing their effects from the measurement of on-board PC components, are discussed in Chapter 9.

Connectors designed for coaxial cables provide similar challenges, as the cable to which they are attached affects the quality of the connection, and the common practice of attaching two connectors to each end of cable makes it difficult to separate the effects of one from the other. Time domain techniques can be applied to remove these unwanted effects as described in Chapter 5.

1.8.3   Non-Coaxial Transmission Lines

Transmission lines provide the interconnection between components, typically in a micro-circuit or a PCB. These are distinguished from a measurement perspective because they are typically much shorter, often not shielded, and the interface to them is not easy to make and sometimes not well defined. While there have been whole texts written on the subject, a short review of some common transmission line structures and their attributes is described below, with focus on attributes important for measurement. Transmission lines are characterized by the same three parameters: impedance, effective dielectric constant and loss.

1.8.3.1   Microstrip

Certainly the most widespread transmission line must be the microstrip line, Figure 1.26. This is found in planar structures such as PCBs and microcircuits. Consisting of a thin strip of metal on a dielectric substrate, over a ground plane, it is used for connection between components as well as for creating transmission line components such as couplers and filters [7].

The computation of the transmission parameters has been fully documented in many forms, but for measurement purposes these lines are typically 50 ohms (or the equivalent system impedance) even though as a design element they can take on any value. For most applications, the dielectric constant is 10 or less, and so the w/h ratio is greater than one for 50 ohms. The approximate impedance can be computed as [10]

$$Z_{\mu strip} = \begin{cases} \dfrac{60}{\sqrt{\varepsilon_{re}}} \ln \left(\dfrac{8h}{w} + \dfrac{w}{4h} \right) & for \ \dfrac{w}{d} \leq 1 \\[4mm] \dfrac{377}{\sqrt{\varepsilon_{re}} \left[\dfrac{w}{h} + 1.393 + 0.677 \ln \left(\dfrac{w}{h} + 1.444 \right) \right]} & for \ \dfrac{w}{d} \leq 1 \end{cases} \tag{1.86}$$

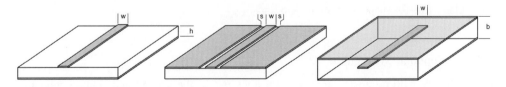

Figure 1.26   Planer transmission lines: (a) microstrip (b) coplanar waveguide, (c) strip line.

where ε_{re} is the effective relative dielectric constant, found from

$$\varepsilon_{re} = \left(\frac{\varepsilon_r + 1}{2}\right) + \left(\frac{\varepsilon_r - 1}{2}\right) \cdot \left(1 + 12\frac{h}{w}\right)^{-1/2} \tag{1.87}$$

The effective relative dielectric constant sets the velocity factor of the transmission line, but in microstrip some of the fields travel in the substrate and some in air. Due to this, the transmission is not purely TEM and some structures become more difficult to design, particularly coupled lines, the even and odd mode velocity factors of which are not the equal. Since the line is not pure TEM, at high frequency, dispersion effects will become apparent where the effective delay of the line is not constant with frequency.

The loss of microstrip lines is difficult to compute accurately because it depends upon many factors including the conductivity of the microstrip line and the ground plan, the dielectric loss of the substrate, radiated loss to the housing or shield, and losses related to both surface roughness and edge roughness. These roughness losses can be significant in PCB and low-temperature cofired-ceramic (LTCC) applications and are very dependent upon the particular processes used. While there are high-quality PCB materials (Duriod™ or GTEK™ are common trade names), the material known as FR4 is the most common, and the dielectric constant and loss of this PCB material can be uncertain. The finished substrate can be composed of layers of board material sandwiched together with glue, and the final thickness can depend upon processing steps, so it is best when evaluating microstrip transmission lines to produce sample structures that can help to determine the exact nature of the material.

One high-performance material used is single-crystal sapphire, and it has the unusual property of having a dielectric constant that has a directionality, with a higher constant of 10.4 in one of the three dimensions, and a lower constant of 9.8 in the other two. A second, common high-performance dielectric is ceramic, which is found in thin-film, thick-film and LTCC applications. It has a uniform dielectric constant typically between 9.6 and 9.8 depending upon the purity and grain structure of the ceramic.

1.8.3.2 Other Quasi-Microstrip Structures

For many applications, the size of 50 ohm microstrip line is not suitable for connections to very large devices. Common modifications include the *suspended substrate* microstrip line, where the ground plan has been removed some distance from the dielectric. This has the effect of lowering the effective dielectric constant and raising the impedance of the line. In this way a wider line can be used to connect to a wide component, and still maintain a matched impedance. A *shielded* microstrip line is entirely enclosed (the theoretical models of microstrip lines assume no top shield) and the top metal tends to lower the impedance of the line. This is particularly true for suspended microstrip lines.

1.8.3.3 Coplanar Waveguide

One difficulty with microstrip transmission lines is that the ground and signal conductors are on different physical planes; coplanar waveguide (CPW), as the name implies, provides

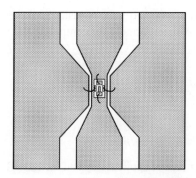 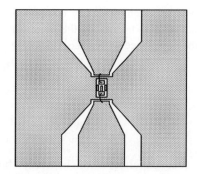

Figure 1.27   CPW mounted IC.

a coplanar structure of ground-signal-ground, as shown in Figure 1.26(b). An alternative is grounded coplanar, where the backside is a conductor as well, and in practice, all coplanar lines have associated package ground, but the ground may be ignored if there is a substantial air gap between the substrate and the package ground. The references provide some computations of coplanar waveguide impedance for various configurations [8, 9]. In microwave measurements, CPW is used extensively as a contacting means for on-wafer measurements, and is used to provide extremely low ground inductance for measuring microwave transistors and circuits as shown in Figure 1.27, with either topside grounds (left) or backside grounds (right). Note that since the impedance depends only on the scale of width to space, this allows contacts of large scale (such as probes) to be transitioned to very small scale such as IC devices.

CPW has some inherent problems due to the ground being on a surface plane or sheet. In many instances the CPW line is mounted in a metal package, and the ground plane is grounded at the package wall. If the distance from the package wall to the ground plane edge approaches a quarter-wavelength at the frequency of interest, or multiples thereof, then a transmission line mode can form such that the ground of the CPW appears, relative to the package ground, as an open. This concept of "hot grounds" for CPW has been observed in many situations and is sometimes avoided by periodically grounding one side to the other through a small via and cross connection on the backside of the CPW. Another method is to provide a lossy connection to the sidewall ground through absorptive material or thin film resist material to suppress energy in the unwanted mode. Another alternative is treating the CPW as a suspended substrate only under the gap between ground and conductor, and "stitch" the CPW ground to the backside ground through a series of conductive vias. The impedance of these structures is lowered by the added ground, so adjustment of the center line width is usually made to accommodate the additional ground paths.

1.8.3.4   Stripline

More common as a transmission line on an inner layer of a PCB, stripline consists of a thin strip or rectangle of metal sandwiched between two ground planes embedded in a uniform dielectric constant, as shown in Figure 1.26(c). The impedance of these lines is much lower than the equivalent width microstrip line, but they have the advantage of being fully TEM in nature, and so, often, design of components such as coupled lines is easier as the even and odd mode velocity factors are the same. An approximate formula for computing the value of a stripline impedance with a zero thickness strip is [10]

$$Z_{stripline} = \frac{30\pi}{\sqrt{\varepsilon_r}} \frac{b}{W_e + 0.441b} \tag{1.88}$$

$$W_e = \begin{cases} w & for\ w/b > 0.35 \\ (w - (0.35 - w/b)^2 \cdot b & for\ w/b < 0.35 \end{cases} \tag{1.89}$$

More complex formulas which include a broad range of applicability and include effects for finite strip thickness and asymmetric placement of the strip can be found in many references [11, 12].

1.9 Filters

Filters come in a variety of types including low-pass, band-pass, high-pass and band-stop. Multiport filters form diplexers or multiplexers, which are used to separate or combine signals of different frequencies from a common port to a port associated with the different frequencies of interest. Diplexers are sometimes called duplexers, but duplexing is a function of the operation of a communication system. That is, a system that can transmit and receive at the same time is said to be operating in a duplex mode. A diplexer is used to support the duplex operation by keeping the transmit signal from saturating the receiver.

The structure and variety of filters are almost endless but they all share these common attributes: low loss in the passband, low reflection in the passband, high reflection and high loss in the stopband. In very nearly every case, the goal of the design is to minimize unwanted loss, and this quality of a filter is often referred to as the Q of the filter. In microwave cases, filters are designed to operate into a matched impedance, and so there is always loss associated with power from the source being absorbed by the load. The Q of a filter in operation is fixed by loading of the ports, and can never be infinite. The quality of a filter is usually defined by its unloaded Q, which accounts for the (desired) power loss from the source to the load.

For many filters, the desired qualities are a tradeoff between creating a maximally flat passband and a maximally sharp cutoff. Thus, the measurement of the transmission response of the filter is critical in evaluating the quality of a filter design. For most filters used in communications, the transmission responses is desired to be equally flat (rather than maximally flat) across the passband, resulting in filters that have Chebyshev type response (equal ripple) in the passband [13]. The desire for sharp cutoffs has led to many filters employing an elliptic response which provides for finite zeros in the transmission response. Stopband performance of high-performance filters can also require careful consideration in measuring, with some requirements going beyond 130 dB of isolation over selected regions of the stopband. These extreme isolation requirements put tremendous burdens on the design of the filter, as well as design and use of the measurement systems.

In modern communications systems using complex modulation, the phase response of the filters is also critical, and a significant design parameter is controlling the phase of the filter to follow a linear response, with a key measurement parameter being deviation from linear phase. Closely aligned to that is maintaining a constant group delay through the passband. Equalization techniques are utilized that can remove higher-order phase responses, such that another measure of filter phase response is deviation from parabolic phase, where the phase is fitted to a second-order response, and the deviation of the phase from this second-order response is the measurement criteria. Some filters are used as part of a feed-forward or matched system network where their phase response as well as absolute phase and delay must be very carefully controlled.

The reflection response of filters is also a key measurement parameter. To the first order, any signal that is reflected is not transmitted, so that high reflections lead to high transmission loss. However, the loss due to reflection for most well-matched filters is much less than the dissipation loss. Still, low reflection at the test ports is required to avoid excess transmission ripple from concatenated components, and even moderate reflections from filters in a high power transmission path can cause damage to the preceding power amplifier. Thus, very low return loss is often a critical parameter of filters, and also a difficult parameter to measure well. This becomes especially true in the case of diplex and multiplex filters, where the loading of any port affects the return loss of the common port.

For high power applications, the filter itself can become a source of intermodulation distortion, and the attribute passive intermodulation (PIM) has become common in the measurement of these high-power filters. Poor mechanical contacts between components in a filter, poor plating on a filter or the use of magnetic materials in the plating or construction of the filter can lead to hysteresis effects that cause intermodulation distortion (IMD) to be created in an otherwise passive structure. The level of IMD typically found in these filters is less than −155 dBc, but this can be a difficult spec to meet without careful design and assembly.

Most of these high-performance communication filters are designed using coupled-resonator designs [14, 15]. Due to manufacturing tolerances, these filters cannot be manufactured to specification from the start, but require tuning of the resonators as well as the inter-resonator couplings. Techniques to optimize the response of these filters are highly sought and a key aspect of the filter measurement task, requiring fast precise response of the transmission and reflection response in real time.

Another type of filter commonly found in the IF paths of receivers is a surface acoustic wave (SAW) filter. The frequency of these SAW filters has been steadily increasing and they are sometimes found in the front end of a receiver. SAW filters can be made to very high orders, and can have very large delays (in the order of microseconds). Because of these long delays, special measurement techniques are required when attempting high-speed measurements. Another type of acoustic wave filter is the film bulk acoustic resonator (FBAR) filter; these are very small in size, and have been used as RF/TX duplexers in handset cell phones.

Ceramic coupled resonator filters are also used extensively in cell phone and radio applications. Because of manufacturing tolerances, the filters are often required to be tuned as part of the manufacturing process, and tuning consists of grinding or laser cutting electrodes until the proper filter shape is obtained. This presents some difficulty in coupled resonator filters as the tuning is often "one-way" and once the resonator frequency has been increased, it cannot be reduced again. This has led to the need for very high speed measurements to ensure that the latency between measurement and tuning is as small as possible.

Some examples of filters are shown in Figure 1.28.

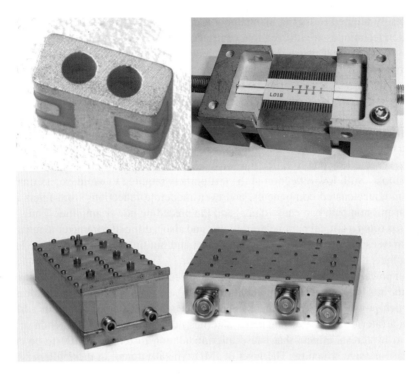

Figure 1.28   Examples of microwave filters: (upper left) cellular phone handset filter, (upper right) thin film filter, (bottom) cellular phone base station filters. Reproduced by permission of Agilent Technologies.

1.10   Directional Couplers

Directional couplers separate the forward and reverse waves in a transmission system (see Section 1.3). A directional coupler is classically defined as a four-port device, often with a good load on the fourth port, Figure 1.29, but in practice a load element is almost always permanently attached. The directional coupler has four key characteristics: insertion loss, coupling factor, isolation and directivity. In fact, directivity is related to the other three factors in a specific way

$$Directivity = \frac{Isolation}{Coupling \cdot Loss} \tag{1.90}$$

Most couplers have a nearly lossless structure, so that the directivity is nearly equal to isolation/coupling but for lossy structures, such as directional bridges, the definition above provides the proper description. In fact, consider the case of a directional coupler with 20 dB of coupling, 50 dB of isolation and 0.05 dB of insertion loss, setting the directivity at nearly 30 dB. If a 10 dB pad is added to the input, as shown in Figure 1.30, the isolation is increased by 10 dB, the loss is increased by 10 dB and the coupling stays the same. Thus, the simple but incorrect definition of directivity as isolation/coupling would yield an increase of the 10 dB.

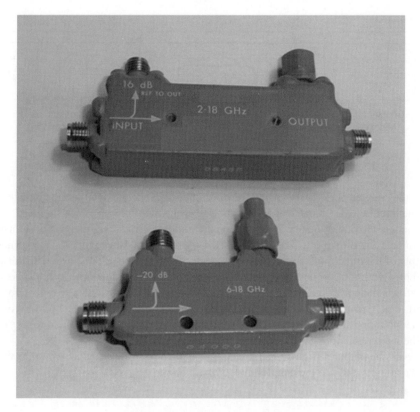

Figure 1.29   Directional couplers.

In fact, a better way of looking at directivity is the ability of the power at the coupled port to represent a change in reflection at the test port. Again considering Figure 1.30, if a signal of 0 dBm is injected into the input port, and a full reflection (an open or short) is applied to the test port, the coupled port will show a power of about −30 dB (10 dB loss, plus a full reflection, plus 20 dB coupling; here the isolation term is ignored for the moment). If a load is applied to the test port, the signal at the input sees a 10 dB loss and 50 dB isolation for a

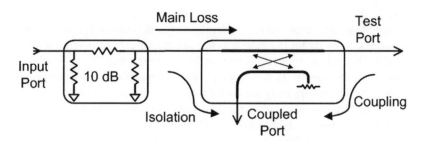

Figure 1.30   The effect of attenuation at the input of a coupler.

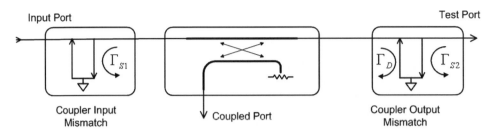

Figure 1.31   Coupler with mismatch after the test port flow graph.

value at the coupled port of -60 dBm. The difference between the open and the load is 30 dB, hence the directivity is 30 dB, and adding the pad at the input has no effect.

In practice, the output match of a directional coupler is critical and the test port mismatch can dominate the directivity. This signal flow is demonstrated in Figure 1.31. Mismatch at the output of the directional coupler affects directivity on a one-for-one basis. This mismatch is combined with the coupler input mismatch to create the overall source match. The source match affects the power measured at the coupled port when measuring large reflections at the test port. This "source mismatch" causes some reflected signal from the test port to re-reflect from the input port, reflect a second time off the test port termination and add to, or subtract from, the main reflection, causing error in the coupled port power.

However, the output mismatch is a direct error and causes reflection back into the coupler, thereby adding directly to the coupler directivity.

1.11   Circulators and Isolators

While most passive components are linear and bilateral (that is, the forward loss equals the reverse loss), a particular class of devices based on the ferromagnetic effect doesn't follow this rule. These devices comprise circulators and isolators. A circulator is a three-port device, with low loss in one direction between ports, say from port 1 to 2, port 2 to 3 and port 3 to 1. But it has high loss (called the isolation) in the reverse direction, from port 2 to 1, 3 to 2 or 3 to 1. An isolator is a special case of a circulator with a good load applied to port 3, such that it becomes a two-port device. Circulators pose a particular measurement difficulty as the isolation between ports depends upon the match applied to a third port. Thus, for good measurement quality, the isolation measurement requires very good effective system match on the ports.

Further, circulators are often tuned by magnetizing a permanent magnet attached to the circulator, and it's desired that the measurement system be able to determine the isolation of all three ports in a single connection step, and with good speed. Thus, multiport (greater than two-port) systems were developed to simplify the connections, and multiport calibration techniques were developed to satisfy the need for high-quality correction.

Even though they are passive devices, circulators and isolators are sometimes tested for their high-power response, such as compression and IMD. The ferromagnetic effect has hysteresis properties that can produce IMD and compression when driven with sufficiently high powers. Figure 1.32 shows an isolator (left) and a circulator (right), while Figure 1.33 shows the signal

Figure 1.32 An isolator and a circulator.

flow for a circulator. In Figure 1.32, the isolator on the left has an internal load mounted at the top, the circulator on the right has three ports, with an SMA connector in place of the isolator load.

1.12 Antennas

As the air-interface for all communications systems, antenna performance is the first (in a receiver) and the final (in a transmitter) characteristic that affects the overall system performance. An antenna can be very small and simple, such as a whip antenna found on a handset, or quite complicated such as those found in phased-array radar systems. Antennas have two key attributes: reflection and gain pattern.

Antenna reflection is essentially a measure of the power transfer efficiency from the transmitter to the over-the-air signal. Ideally, the antenna should be impedance matched to the transmitter's output impedance. In fact, it is typically the case that the antenna is matched to some reference impedance, typically 50 ohms, while the transmitter is likewise matched to the same reference impedance. This implies that while the two may be matched, in many cases they can be exactly mismatched if the phase of the antenna mismatch is not the conjugate of the phase of the transmitter's mismatch. The tighter the mismatch specification is for each, the

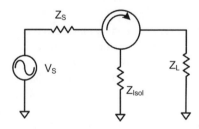

Figure 1.33 Schematic representation of a circulator.

less variation in transmitter power one sees when phasing causes the two mismatches to be on opposite sides of the reference impedance.

Further, simple antennas are matched to a rather narrow range of frequencies, and it is a significant aspect of antenna design to extend the impedance match across a broad range of frequencies. One common form is a bi-conical antenna, often found for use in testing the radiated emissions from electrical components. On the other end of the spectrum is the desire for a narrow band antenna to have a very low return loss over a small frequency range to minimize reflected power back to the high power transmitter.

Antenna gain, or antenna gain pattern, describes the efficiency of an antenna in radiating into the desired direction (or beam) relative to a theoretical omnidirectional antenna, often referred to as an isotropic radiator. This figure of merit is known as dBi or decibels relative to an isotropic antenna.

Antenna pattern measurements are the measurement of the antenna radiation pattern, typically plotted as a contour of constant dBi on a polar plot, where the polar angle is relative to the main beam or "bore-sight" of the antenna. Antenna pattern measurements can range from very simple gain measurements on an antenna on a turntable to near-field probing of complex multi-element phased array structure. While these complex measurements are beyond the scope of this book, many aspects of antenna return loss measurement, including techniques to improve these measurements, will be covered.

1.13   PCB Components

While a very wide ranging topic, the measurement of passive PCB components is focused on the measurement of surface mount technology (SMT) resistors, SMT capacitors and SMT inductors. These components comprise the majority of passive elements used in radio circuits, and also create some of the most undesirable side effects in circuits due to the nature of their parasitic elements. Below is a review of the models of these elements; during measurement the difficulty is in understanding the relative importance of aspects of these models and extracting the values of the model elements.

1.13.1   SMT Resistors

Resistors are perhaps the simplest of electronic elements to consider, and Ohm's law is often the first lesson of an electronic text

$$R = \frac{V}{I} \tag{1.91}$$

However, the model of an RF resistor becomes much more complex as frequencies rise and distributed effects and parasitic elements become dominant. In this discussion the focus will be on surface mount PCB components, because they are used almost exclusively today in modern circuits. Thin film or thick film hybrid resistors have similar effects, and although the parasitic and distributed effects tend to hold off until higher frequencies, much of this discussion applies to them as well.

A good model for a resistor consists of a resistive value in series with an inductance, both shunted by a capacitance. This is a reasonable model for an SMT resistor in isolation, but the

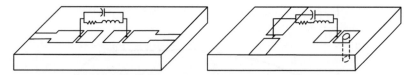

Figure 1.34 Models for a series resistor and a shunt resistor.

values and effects of the model are modified greatly by the mounting scheme of the component. For example, if it is mounted in series with a microstrip transmission line, and the impedance is such that the resistor is much narrower than the transmission line, then this model works well for predicting circuit behavior. On the other hand, if it is mounted on a very narrow line, then the contact pads will provide additional shunt resistance to ground, and the model must include some element to account for this effect. At lower frequencies, some shunt capacitance will do well, but at higher frequencies a length of low impedance transmission line might be a better choice.

A resistor used in shunt mode to ground can have an entirely different model, when it comes to parasitic effects, from that of a series resistance. While the RF value of the resistive element may stay almost the same as the series value (close to the DC value) the effective inductance can be substantially higher as the inductance of the ground via adds to that of the resistor in a microstrip configuration. A larger pad on the ground via, surprisingly, can add even more effective inductance as it resonates with the inductance of the via to increase the apparent inductance of the pair. Meanwhile the shunt capacitance of the resistor may be absorbed in the transmission line width. Figure 1.34 shows a model for a resistor mounted in the series and shunt configuration. Measurement examples to illustrate extracting these values will be shown in Chapter 9.

In many instances, one of the two parasitic elements will dominate the model for first-order high frequency effects. In fact, one can use some simple calculations to estimate a rough order of magnitude for these parasitic elements. Take for example an 0603 resistor, which has dimensions of approximately 0.8 mm width, 0.4 mm height (considering some excess plating, and some edge effect), and 1.6 mm length. If one considers the contact of the resistor wrapped around the body, one might reasonably divide the effective length by 3, to about 0.25 mm. Remembering that SMT resistors are often constructed on ceramic substrates, with a relative dielectric constant of about 10, then the capacitance can be computed as

$$C = \varepsilon_r \varepsilon_0 \frac{W \cdot H}{L} = 10 \cdot 8.85x10^{-15}(F/mm) \cdot \frac{.8 \cdot 0.4}{.25} = 0.11 \text{ pF} \qquad (1.92)$$

The actual value may be substantially greater or less depending upon the exact attributes of the electrodes, but this gives a starting estimate. For inductance one can look to the formula of a transmission line, and assuming that the resistor is mounted on a narrow line, such that its impedance is high, the inductance is from half the length, computed as

$$l = \mu_0 \cdot L = 4\pi \, 10^{-10}(H/mm) \cdot 0.8 \text{ mm} = 0.8 \text{ nH} \qquad (1.93)$$

Thus from the model one can compute the values of resistance for which the inductance or capacitive term dominates, at some frequency. For example, at 3 GHz, the inductance has a value of about 15 ohms reactance in series with the resistive element; the capacitance has a

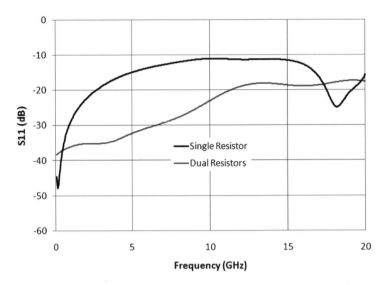

Figure 1.35 Input match of a single SMT resistor, and two in parallel.

value of about 1500 ohms reactive in shunt. At 50 ohms, the inductance value dominates, at 300 ohms; the capacitive value is the dominant parasitic effect. For low values of resistance and high frequency, the inductance becomes dominant and the series impedance is larger than expected, causing the loss through the resistance to be larger than expected due to this effect. At high values of resistance, the parasitic capacitance reduces the series impedance and the expected loss through the resistor is less than expected. The values change with the physical size of the component, thus crossover points differ in resistance and frequency, but with similar effects. This can be used to advantage as there exists a crossover point where the inductive and capacitive effects cancel somewhat, and the resistor behaves ideally to a higher frequency than for values below or above. Using this value of resistance in series or parallel arrangements can provide a range of resistances that avoid parasitic effects until higher frequencies. For the values above, a 50 ohm resistor terminated to ground will have about −18 dB return loss at 3 GHz; however, two 100 ohm resistors, terminated to ground will have about −36 dB return loss, thus providing a better RF resistance of 50 ohms than a single resistor. Thus, characterization of the parasitic effects, and proper compensation, can allow the use of SMT parts up to much higher than expected frequencies [16]. Figure 1.35 shows the effective impedance of a single 50 ohm SMT resistor and two 100 ohm SMT resistors in parallel, when used as a 50 ohm load. This effect also occurs for SMT inductors and capacitors.

1.13.2 SMT Capacitors

SMT capacitors have a different model from resistors. To a first order, their parasitic effects tend to be all in series, as shown in the model of Figure 1.36. The series inductance is primarily due to the package size, and is similar to that of a resistor. The series resistance is due to manufacturing characteristics of the capacitor, and thus cannot be easily estimated. However, its effect is typically very small in most wideband applications, where the capacitor is used as a series DC blocking capacitor or a shunt RF bypass capacitor. This is because the series

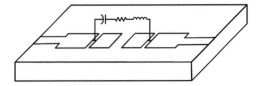

Figure 1.36   Model of an SMT capacitor.

inductance will dominate the series resistance in these cases, and it causes the impedance of the capacitor to rise with frequency (rather than go to zero). At very high frequencies, there may also exist a parasitic shunt capacitance across the entire package, that may cause the impedance to fall again.

The series resistance is of consequence when the capacitor is used in a tuned circuit, where the package inductance may be subsumed into the resonating inductor, and thus at resonance the series resistance adds to a degradation of the Q of the capacitor. With careful design, the capacitance value may be compensated for by including the effects of the series inductance; this effect is to make the capacitor look larger than its prescribed value. In fact, where the reactance of the parasitic inductance equals the reactance of the capacitance, the effective value of capacitance goes to infinity and the series impedance becomes just the parasitic resistance. So, for characterizing capacitors for use in tuned circuits, one must really assess their value near the frequency at which they will be having the most effect on a circuit. Consider a one-pole filter, where the cutoff of the filter starts to occur when the reactance of the capacitor reaches 50 ohms. In many cases, the inductance is quite significant and already altering the effective value of the capacitor. Thus, it is important to evaluate the effective capacitance near this point. A good rule of thumb is to evaluate a capacitor where the reactance is j50 ohms.

A further characteristic of capacitors that is significant is the internal assembly structure. Capacitors are typically formed by a set of interleaved parallel plates with alternate plates connected at each end to the terminals. The plates can be parallel to or to perpendicular to the PCB. In some cases, the capacitor body itself can form a dielectric resonator at high frequency, but below that the capacitor can act as a single, large conductive block on a PCB trace, typically resulting in a model that might best be considered a transmission line of somewhat lower impedance than the mounting line.

Capacitors used as bypass capacitors have an additional parasitic effect from the series inductance of the ground via, and from the pad above the ground via.

1.13.3   SMT Inductors

Inductors are perhaps the most complicated of the simple passive components. Because they are constructed of coils of very fine wire – sometimes multiple layers of coils – their parasitic elements are greatly affected by the details of their construction. Some inductors have the axis of the coil parallel to the PCB, and some are wound with the axis perpendicular. In both cases the model for the inductor is essentially the same as the resistor, as shown Figure 1.35, but with the value of the series inductance equal to the DC value of the inductor, and the series resistance equal to the DC resistance. Inductors, due to the nature of their construction, have very large relative parasitic capacitances. In cases where an inductor is used for a bias element (relying on its impedance to be high at high frequencies) one often finds that the parasitic capacitance will become the main effect over the band of interest. Thus, in many cases the value of inductance used is carefully selected, based on the overall effective inductance, and sometimes utilizes the shunt capacitance to provide a high impedance at a particular frequency of interest. It may be quite difficult to make a single inductor provide good RF performance over a wide band.

When inductors are used as elements in filters, it seems that the parasitic capacitance can often have significant effects, and for use in band-pass filters, the inductance must be evaluated for each use to find the effective value considering the parasitic capacitance.

A common figure of merit for inductors is the self-resonant frequency (SRF), above which they act more like a capacitor (impedance goes lower with increasing frequency) than an inductor. The value of the SRF can be estimated in one way by looking at the length of the wire used in making the inductor. The SRF will be less than the frequency for which the wire is one-quarter of a wavelength.

1.13.4   PCB Vias

The PCB via is perhaps the most common PCB component, yet often the most overlooked. The effect of a via depends greatly upon how it is structured in the circuit. A single via to ground in the center of a transmission line appears as almost a pure inductance. However, a via between RF traces can have aspects of inductance and some parasitic capacitance (due to pads around the via) that can cancel, in part or all, the inductive effect. When a via is used in a mounting pad for a shunt element, such as a resistor used as a load, or a bypass capacitor, the mounting pad and via form a resonant structure such that the size of the mounting pad can increase the effective impedance of the via. Further, several vias are often used in parallel to ground devices, sometimes to lower their effective inductance and sometimes to provide greater heat sinking of an active device. Putting vias in parallel does lower their effective inductance, but not in a simple way. Rather than halving the inductance, mutual inductance between vias means that the value of effective inductance doesn't reduce as expected. For example, putting two 100 ohm resistors at the end of a line to ground, placed in parallel, may show much larger inductive effect than the same two 100 ohm resistors place in a T-pattern, where the ground vias are separated and the mutual inductance is less.

1.14   Active Microwave Components

With a few exceptions, passive components follow some fundamental rules that greatly simplify their characterization: principally, they are linear, so their characterization doesn't depend on the power of the signal used to characterize them, but only on the frequency. Active components, on the other hand are sensitive to power, and their responses to both frequency stimulus and power stimulus are important. Often, passive components are operated well below any power level that causes a change in their response, but more and more active components are being driven into higher power operation to optimize their efficiency.

1.14.1   Linear and Non-Linear

In the measurement sense, one definition of a linear device is one in which the output power is a linear function of the input power. If the input power is doubled, the output power is doubled.

Almost all passive devices follow this rule, and many active devices as well. An alternative definition of linear is one for which only frequencies that are available at the input appear at the output. In practice, the first definition is more useful for system response. Some important characteristics of active components are discussed below.

1.14.2   Amplifiers: System, Low Noise, High Power

1.14.2.1   System Amplifiers

System amplifiers are simply gain blocks used to boost signal levels in a system, while providing reverse isolation. They can have higher noise figures than LNAs because they are used in a signal path where the signal is well above the noise floor. They often follow an LNA stage, frequently after some pre-filtering. They are also often used in the frequency converters as a local oscillator (LO) amplifier to isolate the RF signal from leaking out of the LO port, or as an isolation amplifier to prevent LO leakage out of the RF port. These tend to be very broadband amplifiers, with good input and output match, emulating an idealized gain block. The important figures of merit for such amplifiers are gain (S21), input and output match (S11, S22), and isolation (S12). Occasionally, directivity of an amplifier is defined as isolation (a positive number in dB) minus the gain (in dB), or S12/S21. It is a measure of the effects of a load apparent at the input of the amplifier, or of how the output impedance is affected by the source impedance [17], and it is important in cases where other system components have poor or unstable match. Since these amplifiers have wide bandwidths, it is important that they have good stability as they can have a variety of load impedances applied. Other figures of merit for system amplifiers can include gain flatness (deviation of the gain from nominal value), 1 dB compression point (the power at which the gain drops by 1 dB), harmonic distortion, and two-tone third-order intermodulation, sometimes expressed as third-order intercept point (see Section 1.3).

1.14.2.2   Low Noise Amplifiers

Low noise amplifiers are found in the front end of communications systems and are particularly designed to provide signal gain without a lot of added noise power. The key figure of merit is noise figure, along with gain. But, in a system sense, the noise-parameters (see Section 1.3) are quite important as they describe the way in which the noise figure changes with source impedance. LNAs are used in low power applications and their 1 dB compression point is not a key spec, but distortion can still be a limiting factor in their use so a common specification is the input referred the intercept point. A key tradeoff made in LNAs is between lowest noise figure and good input match. The source impedance where the LNA provides the lowest noise may not be the same as the system impedance, so a key design task for LNAs is to optimize this tradeoff.

1.14.2.3   Power Amplifiers

Many of the figures of merit for power amplifiers are the same as system and low noise amplifiers, but with an emphasis on power handling. In addition, the efficiency of the amplifier is one of the key specifications that one finds primarily with power amplifiers, implying that the DC drive voltage and current must also be characterized. Because power amplifiers are often used with pulsed RF stimulus, the pulse characteristics, such as pulse profile, including pulse amplitude and phase droop, are key parameters.

Power amplifiers are often driven into a non-linear region, and so the common linear S-parameters may not apply well to predict matching. Due to this, load-pull characterization is often performed on power amplifiers. Gain compression and output-referred intercept point are common for power amplifiers. Some amplifier designs such as traveling-wave-tube (TWT) amplifiers have a characteristic that causes the output power to reach a maximum and then decrease with increasing drive power, and the point of maximum power is called saturation. Gain at rated output power is another form of a compression measurement where rather than specifying a power for which the gain is reduced by 1 dB, it specifies a fixed output power at which the gain is measured.

Power amplifiers are often specified for their distortion characteristics including IMD and harmonic content. In the case of modulated drive signals, other related figures of merit are adjacent channel power ratio (ACPR) and adjacent channel power level (ACPL). A figure of merit that combines many others is error vector magnitude (EVM), which is influenced by a combination of compression, flatness and intermodulation distortion among other effects.

1.14.3   Mixers and Frequency Converters

Another major class of components is mixers and frequency converters. Mixers convert an RF signal to an IF frequency (also called down-converters) or IF frequencies to RF (up converters) through the use of a third signal, known as the local oscillator. Typically, the output or input that is lower in frequency is called the intermediate frequency (IF). The LO provided to the mixer will drive some non-linear aspect of the circuit, typically diodes or transistors, which are switched on and off at the LO rate. Using the second definition of linear devices from above, frequency converters and mixers would not be considered linear. In fact, in their normal operation, they are linear (under the first definition) with respect to the desired information signal, and ideally the frequency conversion does not change the linearity of the input/output transfer function.

The input is transferred to the output through this time-varying conduction and the output signal includes the sum and difference of the input and the LO. Mixers tend to be fundamental building blocks that are combined with filters and amplifiers and sometimes other mixers to form frequency converters. In practice, frequency converters are typically filtered at the input to prevent unwanted signals from mixing with the LO and creating a signal in the output band; and they are typically filtered at the output to eliminate one of the plus or minus products of the mixing process. Some converters, known as image reject mixers, have circuitry that suppresses the unwanted product, sometimes called the sideband, without filtering. These are typically created by two mixers driven with RF and LO signals, phased such that the output signals of the desired sideband are in phase and add together to produce a higher output, and the undesired sideband are out of phase, and so cancel and produce a smaller output. Monolithic microwave integrated circuits (MMIC) sometimes blur the line between converters and mixers as they may contain several amplifier and mixing stages, but they typically don't contain filtering.

Mixers have fundamental parameters that include conversion loss (or gain in the case of MMICs with amplifiers), isolation (of which there are 12 varieties, to be discussed in Chapter 7), compression level, noise figure (which for passive mixers is typically just the conversion loss), input and output match, and most of the other parameters found for amplifiers.

Mixers are sometimes referred to as passive mixers or active mixers. Passive mixers don't contain amplification circuits, and are typically constructed of diodes (with a ring or star configuration being the most common), and have baluns in some paths to provide improved isolation, and reduced higher-order products. Higher-order mixing products, also called spurious mixing products or spurs, refer to signals other than the simple sum and difference of the input and LO frequencies. They are often referred to by the order of harmonic related to creating them: for example, a 2:1 spur is created at the frequency of two times the LO plus and minus one times the input. The level of the spurious mixing products, which are sometimes called mixer intermod products (even though the two tones are not applied to the input), change with respect to the RF drive signal at the rate of the order of the RF portion of the spur. For example, a 2:1 spur will increase 1 dB in power for each dB in power that the RF signal increases. However, since the LO is creating the non-linearity in the mixer, as well as the spurious signal, it is difficult to predict how the spurious power will change with respect to LO power. In many cases, driving the LO power higher produces a higher spurious signal, as the relative magnitude of the RF signal to the LO power is reduced. In other cases the transfer impedance of the non-linear element becomes more consistent across the RF drive level. The spurious higher-order products of a mixer are sometimes defined as a spur table, which shows the dBc values of higher-order products relative to the desired output. Chapter 7 has more details on measuring these behaviors.

Mixers with baluns can suppress some of the higher-order products, with baluns on the RF port suppressing products that have even-order LO spurious, and baluns on the LO port suppressing products that have even-order RF spurious. These are called single-balance mixers, and it is very typical that the LO port is balanced. Double-balanced mixers have baluns on the RF or IF ports. Refer to Chapter 7 for more details on mixer configurations. Triple-balanced mixers are usually composed of a pair of double-balanced mixers, adding a balun to the IF port as well. Their main advantage is to divide the RF signal power between the two diode quads, lowering the RF relative to the LO, whereby the spurious signals created will be lower; then the outputs are combined to recover the power. This provides a mixer with the same conversion loss and lower spurious products at the same output power level. The disadvantage is that since the LO drive is also divided, higher LO drive power is needed to achieve the same linearity for each diode quad.

The creation of spurious mixer products is a key aspect of system design, with the goal of eliminating spurious signals from the IF output. Unfortunately, some frequency plans are such that the spurious products must fall into the band of interest. In such cases, system designers move to multiple conversion stages to create a first stage which produces an output free of spurious signals over the range of input signals of interest, and then has a second conversion stage which produces the frequency of interest at the output. This multiple conversion or "dual-LO" system is typically called a frequency converter, and often contains additional filtering and amplification, as shown in Figure 1.37.

Because of the multiple components used, the frequency response of converters often has gain ripple and phase ripple, which can distort the information signal. Key figures of merit for converters are gain flatness, group delay flatness and the related phase flatness which is also

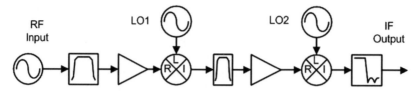

Figure 1.37   Dual LO frequency converter.

known as deviation from linear phase, and represents residual ripple after fitting the phase data to a straight line. Modern systems employ equalization techniques that can remove some of the flatness effects provided they follow simple curvatures; as such another specification found on converters is deviation from parabolic phase, which is the residual ripple in the phase data when it is fitted to a second-order curve.

Mixers often have quite poor input or output match, due to the switching nature of their operation, and so their effect on system flatness when assembled into a converter can be quite dramatic. Until recently, it was difficult to predict the effects of output load of a mixer on its input match, but mathematical tools to model mixers as system elements have been developed [18] that describe these relationships. Mixers that produce an output that is the sum of the input and output signals are relatively simple to describe, but mixers that produce an output that is the difference between input and LO have a more complicated behavior in the case where the input frequency is less than the LO frequency. These are sometimes called image mixers and their unusual characteristic is that as the input frequency goes up, the output frequency goes down; this also applies to phase: a negative phase shift of the input signal results in a positive phase shift of the output signal. How these special cases affect the system performance will be described in more precise mathematical terms in Chapter 7.

1.14.4   Frequency Multipliers and Limiters and Dividers

Mixers are not the only way to create new frequencies at the output, frequency multipliers are also used to generate high frequency signals, particularly when creating mm-wave sources. Frequency multipliers produce harmonics by changing a sine-wave input signal into non-linear wave. The basic doubler is a half-wave or full-wave rectifier, such as a diode bridge. A pair of back-to-back diodes turns a sine wave into a square wave, which is rich in odd harmonic content. This is essentially the same as a limiter.

The key figure of merit of a multiplier is the conversion loss from fundamental drive to the desired harmonic. Other important characteristics are fundamental feed through and higher-order harmonics.

Limiters have the key characteristic of maximum output power; that is, the power at which they limit. Also important is the onset of limiting, and the compression point. Ideal limiters are linear until the onset of limiting, and then they effectively clip the output voltage above that level.

Other multiplier types are step-recovery diodes and non-linear transmission lines which, when driven with a sine wave, effectively "snap" on to produce a very sharp edge. Depending upon the design, the on-time can be very short, which produces an output rich in harmonics. Some digital circuits can also be used to create very narrow pulses from a sine-wave input as a pulse generator. Such a pulse will also be rich in harmonics.

One aspect of a multiplier that is not easily discerned is the group delay through it. That is because for some change in input frequency one will see a multiplied change in output frequency. FM that passes through a doubler will have the same rate as the original FM, but twice the deviation. For this reason, doublers or multipliers are seldom used in the signal or communication path of RF or microwave systems, but they can be used in the base carrier paths and in the LOs of many systems.

1.14.4.1   Frequency Dividers

Frequency dividers provide for a lower value of frequency than the input frequency. Like multipliers they are highly non-linear and often produce square-wave outputs. Some key specifications of dividers are the minimum and maximum input power to ensure proper operation of the divider, output power and harmonics, as well as additive phase noise. Typically, for each divide-by-two stage, the phase noise is reduced by 6 dB. But noise or jitter in the divide circuitry can add additional noise to the signal; the added phase noise at the output relative to the phase noise at the input is called additive phase noise. This is also a concern with mixers, where the LO phase noise can be added to the output signal, and to a lesser extent amplifiers.

1.14.5   Oscillators

Oscillators in some ways represent the most non-linear of electrical circuits with frequencies created at the output with no input (other than noise). Oscillators have a wide variety of characteristics that are important to characterize which include output frequency, output power, harmonics, phase noise, frequency pushing (change in frequency with change in DC power), frequency pulling (change in frequency due to change in load impedance) and output match.

Voltage-controlled oscillators (VCOs) have the additional ability to control the frequency of the output due to a voltage change at the input. The voltage-to-frequency control factor is a key attribute of a VCO. A related microwave component is an yttrium-iron-garnet (YIG) oscillator, which uses a spherical YIG resonator as the frequency control element of the YIG-tuned-oscillator (YTO). The YIG resonator has the characteristic that the resonant frequency changes with magnetic field. YTOs have very wide tuning bandwidth (up to 10:1) and very low phase noise. Tuning is performed by changing the current in an electromagnet but can be very low bandwidth due to the large inductance of the magnet. YTOs often have a second, lower inductance coil (called the FM coil) which provides a small change to frequency but with high bandwidth.

As the focus of this book is stimulus/response measurements, the measurement of oscillators will not be covered.

1.15   Measurement Instrumentation

1.15.1   Power Meters

Perhaps the simplest and most common of microwave instruments is the power meter. It consists of a sensor, which absorbs or detects RF power and converts it to a DC signal, and measurement circuitry which accurately measures this DC signal and applies correction and calibration factors to it to produce a reading of the RF power level. Power meters come in a variety of forms and complexities, some of which are noted below.

1.15.1.1 Calorimeters

Often considered the most accurate and traceable of power measurement systems, the calorimeter consists of a thermally isolated RF load which absorbs the RF energy. This load is kept in a heat exchanger, and a thermopile is used to sense the change in temperature. Since the fundamental measurement is temperature, the traceability of these systems to fundamental SI units is very good. These systems can handle very large power but are slow to respond, heavy and are typically not used by practicing RF engineers, except in special cases.

1.15.1.2 RF Bolometers and Thermistors

An RF bolometer or thermistor is a system where the RF measuring element is a thermally sensitive resistor used as part of a DC bridge system. The DC bridge is electrically balanced, and when an RF signal is applied to the bolometer element, the element heats and its DC resistance changes. The DC bridge is nulled using an offset voltage, and the measure of the offset voltage can be related directly to the power absorbed by the bolometer. The key aspect of the bolometer is that it is equally sensitive to RF or DC power, so a precision DC source can be used to produce a known power at the bolometer, and the balancing circuit thus calibrated relative to the DC power absorbed. The heating effect of the RF power produces the same offsets as the DC power and thus is easily calibrated. Bolometers have a relatively small dynamic range (the range of input powers over which they operate) but have linearity (the ability to correctly measure differences in input power) derived from a bridge circuit using DC substitution. Typically, bolometers are only found in precision metrology laboratories and are not in common use among RF engineers.

1.15.1.3 RF Thermocouples

Until recently, RF thermocouples were the most common type of power sensor used. These thermocouples convert heat directly into a DC voltage, and due to their small size and thus small thermal mass, are much faster responding and have a larger dynamic range than either thermistors or calorimeters. As with other sensors, these require calibration with a precision source, but are typically DC blocked, so the source must be a low frequency AC source. These sensors are commonly used throughout the RF industry, but have the detriment of being somewhat slow responding (with response times in the several to tens of milliseconds range), but are extremely linear and relatively non-responsive to harmonics. That is, harmonic power will be detected as an RMS error of the power of desired signal. Since harmonics 20 dBc down or lower represent less than 1% of the power of the main signal, the error due to harmonics is quite low.

1.15.1.4 Diode Detectors

For modern power meter applications, the diode or multidiode power sensor is often the preferred choice. These sensors employ one or more diodes which rectify the RF signal and produce an equivalent DC signal. Occasionally, the DC signal is "chopped" or modified in such a way as to produce a square wave to the measurement portion of the power meter, typically a precision analog-to-digital-converter (ADC). Chopping the signal helps compensate for DC offsets in the ADC input.

Older diode detectors used only a single diode, and the top 20 dB of the detector range was often described as the "linear" range; below that range the diode would operate in "square-law" mode where the output voltage would be a function of the square of the input RF signal. In the low power range, the output voltage would be linearly related to the square of the input voltage of the RF signal, and thus be linearly proportional to the detected power. In such a region, they operated almost as well as the thermistor sensors, but with much faster speeds and much wider dynamic range. At the top of their measurement range, in the linear region, the output circuitry and measurement algorithms are adjusted to compensate for the change to the linear mode of operation. However, in the linear mode the power in the harmonics has a much greater effect, and a 20 dBc harmonic signal can have up to a 10% change in the measured power of the fundamental, even though it only contains 1% of the power. This is due to the peaking effect that the harmonic can have on the RF voltage. Out of the square-law region (also known as the linear region, which is in fact where the power meter is not as linear in the usual sense of the word) the power meter may not give accurate readings for complex modulated signals or signals with high harmonic content or high peak-to-average envelope power.

More modern diode sensors use a multitude (two or more) of embedded diode elements, some of which are padded with larger attenuation to allow them to operate at higher powers and still be in the square-law region. Complex algorithms in the power meter instrumentation detect when the power from one sensor exceeds the square-law region, and changes to take its readings for power from one of the attenuated diodes. This extends the useful range of the power sensor over more common older diode sensors.

1.15.2   Signal Sources

1.15.2.1   Analog Sources

While not a measurement instrument in their own right, signal sources or signal synthesizers, or simply sources, are used as accessory equipment in a variety of measurement tasks. They can provide CW signals in place of a mixer local oscillator, or provide an input signal to an amplifier or filter. These are typically called "analog sources" and their key attributes are frequency range, output power range (minimum and maximum), phase noise and spectral purity and frequency switching speed.

While the first two attributes are obvious, the phase noise and spectral purity are key attributes when making measurements close to the carrier such as IMD measurements, or when making other distortion measurements such as harmonics.

Switching speed becomes important in automated test systems (ATS) when using the source as a swept frequency stimulus. Commonly, stand-alone signal sources make a tradeoff between lower phase noise and slower switching speeds.

1.15.2.2   Vector Sources

Another class of signal sources is vector signal generators, which have an internal I/Q modulator that allows an almost infinite variety of signals to be created. Some of these vector sources (also called "digital sources" due to the fact that they can create signals using digital modulation techniques) have built in arbitrary-waveform generators (arbs) while others have broadband I/Q inputs to allow external arbs to drive their vector modulators directly.

With vector sources, the arbs can be used to create a wide variety of signals including extremely fast switching CW sources (within the bandwidth of the I/Q modulator), two-tone or multitone signals, pseudo-random noise waveforms and complex modulated signals following the formats used in digital communications and cellular phones.

Some key attributes of vector sources are the modulation bandwidth of the I/Q inputs, the modulation bandwidth or speed of the arbitrary waveform generator (if it is built-in), the memory of the arb (which affects the length of signals that can be created) and the I/Q fidelity or linearity of the modulator. This linearity limits the ability of the vector source to produce clean signals. For example, a two-tone signal can be created by doing a double sideband suppressed carrier modulation but if there is imbalance or non-linearity in the modulator, there will be carrier leakage between the two tones.

The output power amplifiers of vector sources are very important as their distortion will directly affect the modulated signal, causing third-order intermodulation (TOI) and spectral spreading of modulated signals.

1.15.3   Spectrum Analyzers

A spectrum analyzer (SA) is a specialized type of receiver, which displays the power of a signal on the y-axis versus the frequency of the signal on the x-axis. As such, it could be considered a frequency-sensitive power meter.

Key attributes of a spectrum analyzer are its displayed average noise level (DANL) and its maximum input power. The maximum input is set by the compression of the input mixer in the SA, and can be increased by adding input attenuation. However, adding attenuation degrades the DANL by an equal amount. A further limitation in measuring signals, for example TOI, is the self-generated distortion of the input mixer, which will generate TOI signals at the same frequencies as that of the TOI from the signal under test. The data sheet for an SA will typically specify the distortion in dBc relative to some input level at the mixer. This, coupled with the noise floor, will set the measurement range of the SA. Lower resolution bandwidth will lower the noise floor at the cost of speed of measurement. Similar effects are present for the measurement of harmonics.

Another key attribute of a spectrum analyzer is its frequency flatness and power linearity specifications. Flatness specifications of a spectrum analyzer are usually quite large, as much as ±2.5 dB for 26 GHz microwave version although typical performance is much better, and this flatness can be compensated for with an amplitude calibration. The large value for frequency response comes from the interactions of the preselector (which is usually a swept YIG filter) and the first converter in the SA. To a first order these are stable and can be corrected for, but there will still be a residual flatness error even after calibration related to the post-tuning drift of the preselector; that is, it does not always tune its peak value to the same frequency for the same settings. Another source of uncorrected error is the mismatch between the SA input and the output match of the signal source being measured. In some cases, this can be quite high, up to ±1 dB or more.

As the name implies, the key role of spectrum analyzers is in determining the quality of unknown spectrums. The use of spectrum analyzers in microwave component test applications is primarily for measuring the frequency response or distortion response of a system as a known stimulus is applied. These applications of spectrum analyzers are now being replaced by advanced vector network analyzers which have higher speed receivers, built-in sources and advanced calibration capability.

1.15.4   Vector Signal Analyzers

With the advent of digitally modulated signals for RF and microwave communications, spectrum analyzers have evolved into much more complex systems that include the ability to do wideband demodulation of these signals. These specialized spectrum analyzers are often called vector signal analyzers (VSA) and play an important role in component test.

For many active components, a key figure of merit is the distortion that they apply to the vector modulated signal in the form of amplitude or phase error relative to an ideal signal. The composite of all errors over a set of digital symbols is called the error vector, and the average magnitude of this error is the error-vector-magnitude (EVM). While EVM is a signal figure of merit, as it is compared to an ideal waveform, the EVM from an amplifier is a combination of the EVM of the input signal and the errors added by the amplifier. From this it is clear that EVM is not a microwave-component parameter; however, a related value, residual or added EVM is, which is described as the EVM at the output signal relative to the input signal. In practice, high-quality sources are used to produce the digitally modulated signals, so the input effect on EVM is small, but with higher data rates and wider bandwidths of modulation, these input effects are becoming more important. Thus, there is a need for a multichannel VSA which can compare input to output signals. A normal SA with a VSA capability does not provide such dual-channel capability, but some manufacturers supply a specialized dual-channel receiver for the VSA, while other implementations of a VSA use a wideband digitizer, or even a digital oscilloscope, to do direct digitization of the modulated signal. To date, up to four simultaneous channels have been reported in such a VSA.

1.15.5   Noise Figure Analyzers

An offshoot of a spectrum analyzer, the noise figure analyzer (NFA) is a specialized test instrument that is designed particularly for making noise figure measurement. NFAs started out as specialized spectrum analyzers, with improvements in the quality of the receiver and with electronically switched gains to allow the noise figure of the test equipment to be minimized relative to the signal being measured. Some of the things needed to accomplish this, such as adding high gain low noise amplifiers in front of the first converter, reduced the maximum input power of the instrument so that it was no longer suitable for general-purpose SA applications.

On the other hand, several spectrum analyzer manufacturers have added a noise figure personality to their SA offerings, so that there is quite a lot of overlap between the capabilities of the two systems. However, most SA implementations require the use of an added LNA, at least over some of the band. Newer SAs have an IF structure almost as flexible as an NFA to optimize the performance of the system.

All of these systems of NFA utilize the "hot/cold" or "Y-factor" method of measuring noise figure (more about this in Chapter 6) using as an input to the DUT a noise source that can be turned on and off. From careful measurement of the output noise the gain and noise figure of the DUT can be discerned.

More recently, vector network analyzers have been modified to operate as noise figure analyzers, utilizing an entirely different technique called the "cold-noise" method. In this method, the output noise power is measured, along with the gain, using the normal VNA measurement of gain, and the noise figure is computed from these values. No noise source is used in the measurement. This has the advantage of being faster (only one noise measurement is needed) but it does have the disadvantage of being sensitive to drift in the gain of the VNA noise receiver. The Y-factor method does not depend on the gain of the NFA receiver, but this advantage is often offset by the fact that the gain measurements of the NFA are very sensitive to match errors, as are the noise measurements, and these are not compensated for.

The ultimate in noise figure analysis is a noise parameter test system. This system properly accounts for all mismatch effects. Some systems use both a VNA and NFA, to measure the gain and the noise power respectively. All noise parameter systems include an input impedance tuner to characterize the change in noise power versus impedance value. Recently, tuners have been combined with VNA-based noise figure analyzers to produce very compact, high-speed noise parameter test systems. These newer systems provide the ultimate in speed and accuracy available today.

1.15.6   Network Analyzers

Network analyzers combine the attributes of a source, and a tracking spectrum analyzer, to produce a stimulus/response test system ideally suited to component test. These systems have been commercialized for more than 40 years, and provide some of the highest quality measurements available today. While there are many distinct manufacturers and architectures, network analyzers broadly fall into two categories: scalar network analyzers (SNA) and VNAs.

1.15.6.1   Scalar Network Analyzers

These instruments were some of the earliest implementations of stimulus/response testing, and often consisted only of a sweeping signal source (sometimes called a sweeper) and a diode detector, the output of which was passed through a "log-amplifier" which produced an output proportional the power (in dBm) at the input. This was sent to the y-axis of a display, with the sweep tune-voltage of the sweeper sent to the x-axis, thus producing frequency response trace. Later, the signal from the detector and the sweeper were digitized and displayed on more modern displays with marker readouts and numerical scaling.

Other SNA systems were developed by putting a tracking generator into a spectrum analyzer, so that the source signal followed the tuned filter of the SA. This produced a frequency response trace on the SA screen.

SNAs had the attribute of being very simple to use, with almost no setup or calibration required. The scalar detectors were designed to be quite flat in frequency response, and a system typically consisted of one at the input and one at the output of a DUT. However, for measurements of input and output match, or impedance, the SNA relied on a very high quality coupler or directional bridge. If there was any cabling, switching or other test system fixturing between the bridge and the DUT, the composite matches of all were measured. There was no additional calibration possible to remove the effects of mismatch. As test systems became more complex and integrated, scalar analyzers started to fall from favor, and there are virtually none sold today by commercial instrument manufacturers.

1.15.6.2   Vector Network Analyzers

For microwave component test, the quintessential instrument is the VNA. These products have been around in a modern form since the mid 1980s, and there are many units from that time still in use today. The modern VNA consists of several key components, all of which contribute to making it the most versatile, as well as the most complicated, of test instruments; these components are:

RF or microwave source: This provides the stimulus signal to the DUT. RF sources in a VNA have several important attributes including frequency range, power range (absolute maximum and minimum powers), automatic level control (ALC) range (the range over which power can be changed without changing the internal step attenuators), harmonic and spurious content and sweep speed. In the most modern analyzers, there may be more than one source, up to one source per port of the VNA. Older VNAs required that the source be connected to the reference channel in some way, as either the receiver was locked to the source (e.g., the HP 8510), or the source was locked to the receiver (e.g., the HP 8753). Modern VNAs, for the most part, have multiple synthesizers so that the source and receiver can be tuned completely independently.

RF test set: In older model VNAs the test set was a separate instrument with a port switch (for switching the source from port 1 to port 2), a reference channel splitter and directional couplers. The test set provided the signal switching and signal separation to find the incident and reflected waves at each port. Most modern VNAs have the test set integrated with the rest of the components in a single frame, but for some high power cases, it is still necessary to use external components for the test set.

Receivers: A key attribute of VNAs is the ability to measure the magnitude and phase of the incident and reflected waves at the same instant. This requires sets of phase synchronous receivers, which implies that all the receivers must have a common LO. In older RF VNAs the reference channel was common to ports 1 and 2, and the port switch occurred after the reference channel tap. Most modern analyzers have one receiver per port, which is required for some of the more sophisticated calibration algorithms. More about that in Chapter 3.

Digitizer: After the receiver converts the RF signals into IF baseband signals, they pass to a multichannel phase-synchronous digitizer which provides the detection method. Very old VNAs used analog amplitude and phase detectors, but since at least 1985, all VNAs have utilized a fully digital IF. In modern VNAs, the digital IF allows complete flexibility to change IF detection bandwidths, modify gains based on signal conditions and detect overload conditions. Deep memory on the IF allows complicated signal processing, and sophisticated triggering allows synchronization with pulsed RF and DC measurements.

CPU: The main processor of a VNA used to consist of custom-built micro-controllers, but most modern VNAs take advantage of Windows™-based processors, and provide very rich programming environments. These newer instruments essentially contain a PC inside, with custom programming, known as firmware, which is designed to maximize the capability of the instrument's intrinsic hardware.

Front Panel: The front panel provides the digital display as well as the normal user interface to the measurement functions. Only the spectrum analyzer comes close to the sophistication of the VNA, and in more modern systems, the VNA essentially contains all the functions of each of the instruments mentioned so far. Thus, its user interface is understandably more complex. Significant research and design effort goes into streamlining the interface, but as the complexity of test functions increases, with more difficult and divergent requirements, it is natural that the user interface of these modern systems can be quite complex.

Rear Panel: Often overlooked, much of the triggering, synchronization and programming interface is accomplished through rear-panel interface functions. These can include built-in voltage sources, voltmeters, general-purpose input/output (GPIO) busses, pulse generators and pulse gating, as well as LAN interfaces, USB interfaces and video display outputs.

The detailed operation of a VNA is described in Chapter 2.

Extensions to traditional VNAs allow them to create multiple signals for two-tone measurements, and to have very low noise figures for noise figure measurements. But the main attraction of VNAs is calibration. A key attribute is that since they measure the magnitude and phase of waves applied to their ports, they can use mathematical correction to remove the effects of their own impedance mismatch and frequency response in a manner that makes their measurements nearly ideal. The details of VNA calibration is covered in depth in Chapter 3.

Thus, even though there is a wide variety of test equipment available for microwave component measurement, by far the most widely used is the VNA, and while many of the topics of component measurements in this book are extensible to any of the previous instruments, the specific implementation and examples will be illustrated primarily using the VNA, as that has become the predominant component test analyzer in use today.

References

1. Collier, R.J. and Skinner, A.D. (2007) *Microwave Measurements*, Institution of Engineering and Technology, London. Print.
2. Marks, R.B. and Williams, D.F. (1992) A general waveguide circuit theory. *Journal of Research of the National Institute of Standards and Technology*, **97**(5), 535–562.
3. Agilent Application Note AN-95-1, http://contact.tm.agilent.com/Agilent/tmo/an-95-1/index.html, original form can be found at http://cp.literature.agilent.com/litweb/pdf/5952-0918.pdf.
4. Kurokawa, K. (1965) Power waves and the scattering matrix. *IEEE Transactions on Microwave Theory and Techniques*, **13**(2), 194–202.
5. Magnusson, P. (2001) *Transmission Lines and Wave Propagation*, CRC Press, Boca Raton, FL. Print.
6. Collin, R. (1966) *Foundations for Microwave Engineering*, McGraw-Hill, New York. Print.
7. Hong, J.-S. and Lancaster, M.J. (2001) *Microstrip Filters for RF/Microwave Applications*, Wiley, New York. Print.
8. Wen, C.P. (1969) Coplanar waveguide: A surface strip transmission line suitable for nonreciprocal gyromagnetic device applications. *IEEE Transactions on Microwave Theory and Techniques*, **17**(12), 1087–1090.
9. Simons, R.N. (2001) *Coplanar Waveguide Circuits Components and Systems*, John Wiley and Sons. Print.
10. Pozar, D.M. (1990) *Microwave Engineering*, Addison-Wesley, Reading, MA. Print.
11. IPC-2141A (2004) *Design Guide for High-speed Controlled Impedance Circuit Boards*, IPC, Northbrook, IL. Print.
12. Cohn, S.B. (1954) Characteristic impedance of the shielded-strip transmission line. *Microwave Theory and Techniques, Transactions of the IRE Professional Group on Microwave Theory and Techniques*, **2**(2), 52–57.

13. Zverev, A.I. (1967) *Handbook of Filter Synthesis*, Wiley, New York. Print.
14. Cameron, R.J., Kudsia, C.M., and Mansour, R.R. (2007) *Microwave Filters for Communication Systems: Fundamentals, Design, and Applications*, Wiley-Interscience, Hoboken, NJ. Print.
15. Hunter, I.C. (2001) *Theory and Design of Microwave Filters*, Institution of Electrical Engineers, London. Print.
16. Dunsmore, J. (1988) Utilize an ANA to model lumped circuit elements. Microwaves and RF, Nov 1988, p. 11
17. Mini-Circuits. Amplifier Terms Defined AN-60-038. Mini-Circuits. Web. 11 Feb. 2012. http://www.minicircuits.com/app/AN60-038.pdf.
18. Williams, D.F., Ndagijimana, F., Remley, K.A., *et al.* (2005) Scattering-parameter models and representations for microwave mixers. *IEEE Transactions on Microwave Theory and Techniques*, **53**(1), 314–321.

2

VNA Measurement Systems

2.1 Introduction

S-parameter measurements of devices provide the common reference for RF and microwave circuit and system analysis. While the basic methods of S-parameter measurements were developed decades ago, many advances have occurred in just the last five years that make obsolete the common understanding of the capabilities and limitations of these measurements. Vastly improved hardware and software capabilities provide for control of stimulus signals and analysis of response signals that allow measurement systems to extend the basic linear S-parameters to multiport, differential and non-linear characteristics. In the past, S-parameters were limited to two ports; now up to 32-port systems are readily available. In the past, measurements were limited to linear responses; now non-linear, distortion, noise and even load pull characterizations are possible. In the past, calibration techniques were restricted to a few limited sets of standards and algorithms, and limited to devices with the same input and output frequencies; now a wide range of calibration algorithms and applications can be applied to a variety of components, with very few restrictions.

A clear understanding of the underlying architecture of a VNA is necessary to understand the full capabilities and limitations of the modern VNA. The first part of this chapter deconstructs the VNA to discuss the individual block diagram elements, their attributes and deficiencies, and how they operate together to provide the capability and applications described in later chapters. In the history of VNAs, the HP 8753 and the HP 8510 were the industry-leading RF and microwave VNAs of the 1980s and 1990s, from which many of the principal understandings of capabilities and limitations were formed. For that reason, many of the characteristics of these analyzers are discussed below, to provide a context for the discussion of the modern VNA attributes. In almost all cases, many well-known limitations of these products no longer apply, and a key goal of the first section of this chapter is to illuminate to the reader these improvements.

By around the year 2000, an arms-race of sorts emerged in the world of VNAs with the nearly simultaneous introduction of the PNA and ENA family from Agilent, the Ballmann S100, the ZVR and ZVK from Rohde-Schwartz, and the Lightning™ and Scorpion™ from Anritsu, and the 3765 from Advantest. By 2010, Agilent and Rohde-Schwartz advanced to the modern generation of multifunction component test platforms, the PNA-X and the ZVA, while

Handbook of Microwave Component Measurements: With Advanced VNA Techniques, First Edition. Joel P. Dunsmore.
© 2012 John Wiley & Sons, Ltd. Published 2012 by John Wiley & Sons, Ltd.

Anristu's products remained mostly in the area of linear S-parameter test in the form of the Vectorstar^{TM}. As the author is a principal designer and architect of the Agilent products, the details of the VNA architecture, structure and capabilities described here are derived from this knowledge. But many of the factors that attend this discussion apply equally well to all measurement systems, regardless of manufacturer, including custom-built systems sometimes seen in university research labs or national standards laboratories. Because of these advances, many rules-of-thumb and common understandings based on the first generation of commercial VNAs are no longer relevant.

The second portion of this chapter describes the wide range of measurements and characteristics that can be derived from the basic measurements. In Section 2.3, the basic functionality for making measurements is described along with real-world issues and errors that affect these measurements. Particularly in VNA-based measurements, many of these errors can be characterized during a calibration process, and error correction can be applied to the results to remove, to a large extent, the effects of these errors. The calibration and error correction process will be described in detail in Chapter 3. Very detailed descriptions of measurements of particular devices are covered in subsequent chapters: linear devices (Chapter 5), amplifiers (Chapter 6), mixers (Chapter 7) and balanced devices (Chapter 8).

2.2 VNA Block Diagrams

The basic block diagram for a component test system is a stimulus source, which is applied to the input of the DUT, and a response receiver at the output of the DUT. For S-parameter measurements, the inputs consist of incident waves at all the ports, and the outputs consist of scattered waves at all the ports, so in general one would require a stimulus and two receivers at each port. In addition, there must be signal separation devices at each port to isolate the incident and scattered waves.

Early systems measured only transmission and/or reflection response, in only one direction, and thus consisted of at most a directional device (bridge or coupler) at the input and a receiver at the output. These systems were classified as transmission/reflection (TR) systems, and were most commonly found as scalar network analyzers, although lower-cost vector network analyzers sometimes had TR test sets as well. The advantage of a vector TR analyzer is that the errors in the directional device could be removed with calibration and error correction.

Figure 2.1 shows the block diagram of a TR system. For simplicity sake the reference receiver will be normally at port 1, measuring the $a1$ wave, and the test receivers were limited to the two ports as well; normally the test receiver at port 1 is the reflection receiver ($b1$) and the test receiver at port 2 is the transmission receiver ($b2$). The source is typically split using a two-resistor power splitter or a coupler to create a reference signal that is proportional to the incident signal on the DUT, followed by a directional coupler or directional bridge, whose coupled arm goes to the reflection test receiver, measuring $b1$. After the DUT, the transmission test receiver measures $b2$.

The two-resistor power splitter is critical choice here for getting good measurements, even though it may appear to be unmatched when looking back into the input port. This is because the measurements of gain ($b2/a1$), and return loss ($b1/a1$) are ratio measurements, and any reflections from the DUT will go back through the coupler, through the splitter and reflect off the source impedance. Even if the source is not well matched, the reflected signal from

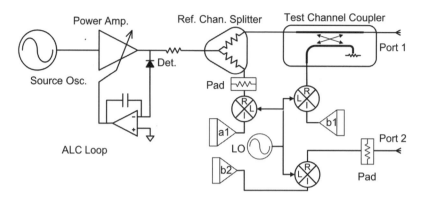

Figure 2.1   A TR network analyzer block diagram.

the internal source is also sampled by the reference channel. This has the effect of making the common node of the two-resistor splitter act as a virtual ground [1]. In such a case, the effective source match of this system, when making ratio measurements, is the value of the output resistance in the source arm of the splitter. Thus, for best match in a 50 ohm reference system, that impedance should be 50 ohms.

Reflected signals from the DUT will be split to go partially to reference channel and partially to the source to terminate in the source impedance, and may reflect off the source impedance Z_S if it is not well matched. This reflected signal will add equally to the signal that appears at the reference receiver, and to the $a1$ signal applied to the DUT. Through this action, the two-resistor power splitter acts to separate the reflected signal from the incident signal. In a similar manner, if a directional coupler is used to provide the reference signal (see Figure 2.3), re-reflected signals from the source would likewise add to both the reference receiver and the incident wave $a1$, and thus the receiver would still isolate the reflected wave from the incident wave. Any leakage in the isolation direction of the directional coupler is an error; in this way, the directivity of the reference coupler sets a limit on the source match of the VNA.

The combination of the reference-channel signal separation and the reflection coupler or bridge is sometimes called a reflectometer, and the ratio of powers at the reflectometer coupled arms directly provides the return loss of the DUT, ignoring source match, frequency response and directivity errors. When high-quality couplers are used, these errors can be quite small. When a splitter and a directional bridge are used, there is often an additional attenuator pad used in the reference path to compensate for the resistive loss of the bridge, such that the reference and reflected signals will have the same loss for a total reflection (open or short).

A full S-parameter system extends the block diagram of the TR system by adding a reflectometer at each port, and provides a source at each port. Older systems would switch the source between the ports using a test port switch. Two such block diagrams are shown in Figure 2.2. There are two distinct versions which use either one or two reference receivers.

The three-receiver version (upper diagram) was common in lower-cost or RF network analyzers in the past, but has largely been replaced with four-receiver versions. Having individual receivers for all the reference and test port channels provides for more and better calibration choices, as will be discussed in Chapter 3.

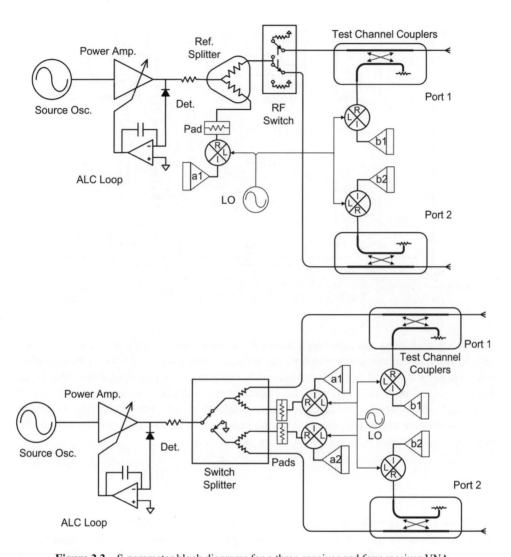

Figure 2.2 S-parameter block diagrams for a three-receiver and four-receiver VNA.

Older analyzers, such as the HP-8510, used separate external sources, and switched the source between the ports; others had internal sources, but the cost of the source was a major proportion of the instrument cost, and so a single switched source was used in these integrated analyzers. Often, the reference channel splitter was integrated into this switch as well. This provided a compact switch-splitter assembly and allowed a lower-cost alternative to individual splitters or directional couplers.

Modern network analyzers make use of a hybrid approach with two or more internal sources, such that more than one port at a time can have an output signal, as shown in Figure 2.3.

While there is no requirement for having more than one port active in traditional S-parameter measurements, advanced measurements such as two-tone IMD, active-load or

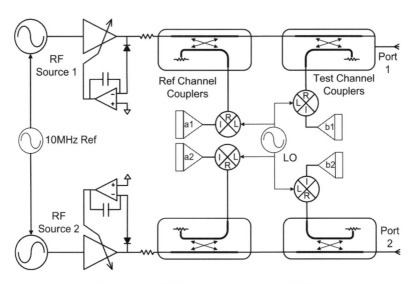

Figure 2.3   Multiple sources in a single VNA.

differential-device test can make good use of these extra sources. In these systems it is common to use a directional coupler in both the reference and test arms for lower loss from the source to the test port, allowing higher maximum test port power. One synthesizer (that is, the frequency generation unit) may be shared between two ports (of a four-port VNA) with an individual output amplifier and leveling circuits available at each of the ports. This system can provide outputs at any of the ports at the same time, or provide two different frequencies out of pairs of ports, which is useful for mixer test applications.

2.2.1   VNA Source

The VNA source provides the stimulus for the S-parameter measurement. In the original VNAs these were open-loop sweepers, but by about 1985 the use of frequency synthesizers had become the norm. Sweepers used open-loop swept-frequency oscillators to produce the stimulus signals; synthesizers replaced the open-loop control with fractional-N or multiloop signal generation where the output signal is digitally derived from a 10 MHz reference oscillator to a resolution of less than 1 Hz. Early sources were routed directly to the test sets of VNAs, which were also external, stand-alone instruments, often with the first converter assembly inside.

   More recently the source is provided internal to the VNA. In the first integrated VNAs, the signal quality of the internal source was less than that of external instrumentation sources, but had the advantage of being much faster at sweeping frequency. For applications such as filter tuning and test, very fast sweep times across wide frequency ranges were required. Common to VNA sources are the ability to vary power level over a prescribed range, usually called the automatic-loop-control (ALC) range which have values from 20 to over 40 dB. These ranges are often extended with integrated discrete step attenuators after the source.

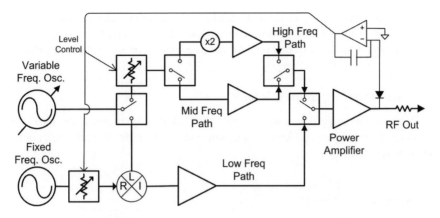

Figure 2.4 Example of a VNA source block diagram.

Most VNA sources also have a power flatness specification to provide constant power to the DUT. In many cases, this level is digitally corrected by a factory calibration to be quite accurate at the test port. In addition, a user-performed source power level correction is usually available to provide additional accuracy enhancement, as described later in Chapter 3.

A typical VNA source block diagram is shown in Figure 2.4.

Generally, there is a fundamental oscillator that provides a swept frequency response over one or more octaves. This output is often switched or split to use as an LO to a low-frequency heterodyne source stage; such a block diagram allows arbitrarily low frequencies to be created in the source mixer.

The swept frequency oscillator is typically phase locked to a lower-frequency fractional-N (FN) circuit or direct-digital synthesizers (DDS). In older VNAs, the phase locking was accomplished through the reference receiver, which reduced cost but required that the reference path signal remain present for all measurements. Modern analyzers separate the frequency-synthesis for sources from receivers for greater flexibility, and no longer require that any signal be present in the reference path. Due to this, the RF signal may be pulse modulated without losing phase lock on the synthesizer.

The output may also go to multiple stages of division or multiplication, followed by amplification and filtering. The final output signal is combined from each of the input signals, resulting in broad frequency coverage. Typically, this common output has some RF detector on it to provide for ALC loop operation, maintaining constant source power over the different frequency bands and compensating for amplifier flatness.

The level control circuits often use an amplitude modulator before the amplification chain to complete the ALC loops. In some modern VNAs, a pulse modulator is also added to provide for high-speed pulsed RF measurements. When a pulse modulator is used inside the ALC loop, the ALC function must be disabled because it will attempt to respond to the pulsed signals. In this open-loop mode, more sophisticated calibration or digital control must be used to control the source output power. Some older analyzers provided for analog inputs to the ALC circuit to allow AM signals or external detector inputs to be used, but most modern VNAs use only digital controls. In fact, recently, the use of a reference or test receiver as the ALC loop power control, rather than the internal diode-detector, has become more common. This form of receiver leveling provides very accurate results if the receivers are calibrated, much wider range of leveling than the diode detector can support and the ability to be programmatically controlled to correct for any external path loss or to provide a prescribed power profile.

2.2.2 Understanding Source Match

One of the most confusing issues with respect to VNA measurements is the idea of source match. In fact, there are three different and distinct source attributes that are often confused as the source match of a VNA.

2.2.2.1 Ratio Source Match

The ratio source match is that match which will affect the results of a ratio measurement, given a DUT that is not perfectly matched to the reference impedance. The value of the ratio source match, most commonly called the raw source match, or uncorrected source match, is derived from a combination of the quality of the reference channel signal-separation device, and any mismatch between this device and the input port of the DUT. This is always the value used to compute the uncertainty or accuracy of a gain or return loss measurement. But this match only applies to parameters that have a ratio of the some receiver to the reference receiver. The ratio source match can be determined during the calibration process, and is shown in Figure 2.5 for two cases of reference channel signal separation: the upper trace is using a two-resistor power splitter, and the lower trace is using a directional coupler. While the detailed response is different, the overall quality is quite similar between the two cases.

When a splitter is used, since the splitter uses equal 50 ohm resistors in most cases, the input match to the splitter (as it appears from the source) is nominally 50 ohms, and the loss through the splitter is about 6 dB.

2.2.2.2 Power Source Match

The power source match is that value which describes how the output power of the source varies with the applied load. If the power match is zero (perfectly matched) then the output forward wave, $a1$, would not be affected at all by the load. However, even in some ideally constructed S-parameter measurement architectures, the power source match is not zero. Consider the block diagram of Figure 2.1, simplified in Figure 2.6, where a two-resistor power splitter is used, and the source impedance is also 50 ohms.

Thus, from the test port one sees a series 50 ohm resistance (of the splitter), behind which is the 50 ohm source impedance in parallel with 100 ohms (50 ohms from the splitter, 50 ohms from the reference receiver, in series), to generate a power match of

$$Z_{PwrSrcMatch} = Z_{splitter\_main} + \cfrac{1}{\cfrac{1}{Z_S} + \cfrac{1}{Z_R + Z_{splitter\_R}}} = 50 + \cfrac{1}{\cfrac{1}{50} + \cfrac{1}{50 + 50}} = 83.3 \text{ ohms}$$

(2.1)

as the Thevenin equivalent impedance. From this it is clear that the for the two-resistor splitter case, even in an ideal case the power source match cannot be Z_0.

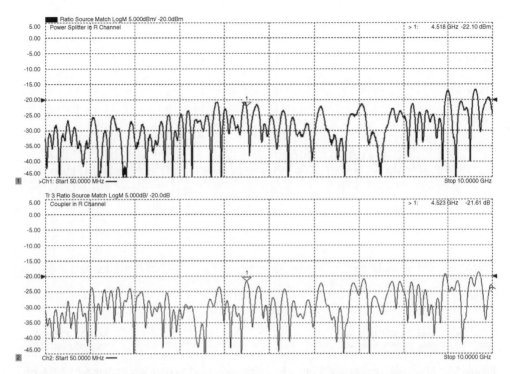

Figure 2.5 Ratio source match; upper trace is when using a power splitter, lower trace when using a directional coupler.

When a directional coupler is used in the reference channel, the nominal match may be much closer to Z_0. The result of this is that when the DUT is not matched, there will be a reflected signal that, while it will be detected by the reference channel and thus compensated for in gain measurements, will cause the $a1$ wave to vary from the value one sees when the port is terminated in 50 ohms and thus cause error or ripple in the drive power to the DUT, as illustrated in Figure 2.7. The figure shows, in the dark trace, the incident power

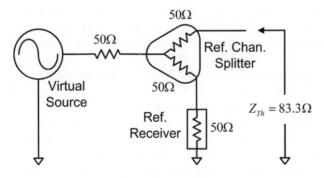

Figure 2.6 Simplified diagram of source power match.

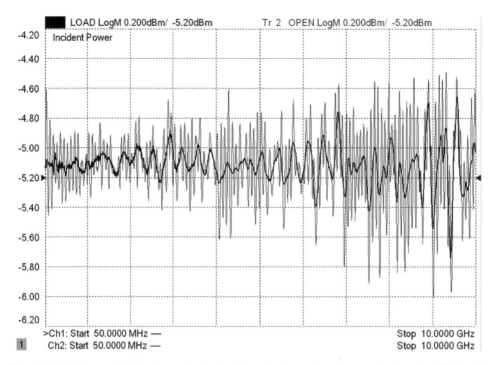

Figure 2.7   Measured incident power into a load termination and an open termination for a VNA with a coupler in the reference channel.

at *a1* for a load at the test port. It is not perfectly flat due to other mismatches in the system. The light trace shows the result of an open at the test port. This large ripple is due to the poor power source match and generates an error in the incident signal of nearly 1 dB. Since this is power measured at the *a1* receiver, it is related to the ripple in the output power but other loss or reflections between the reference splitter and the test port can affect the output power. Other reflections past the reference splitter will add to the power source match, but are not represented in the *a1* measurement. In the case of a linear device, which S-parameters presume, this is of no consequence, but in the case of non-linear devices, such as amplifiers in compression, this will directly affect the reported output power. In such a case, the drive power from the VNA can be higher, or lower, than the displayed power setting, so the amplifier will be further, or less, in compression and the power reported will be in error if based on the input drive level.

The power source match is also quite difficult to determine as it is only apparent when there is a mismatch applied at the test port. In essence, one must vary the impedance applied to the test port and measure the change in power coming out of the drive port to infer the power source match; it cannot be directly measured. Using a "long-line" technique, a "line-stretcher", a sliding mismatch, or an impedance tuner as a termination of the test port, and adding a coupler to sample the incident wave (also shown in Figure 2.8), one can determine the power source match from a series of measurements, where the line stretcher or tuner is changed.

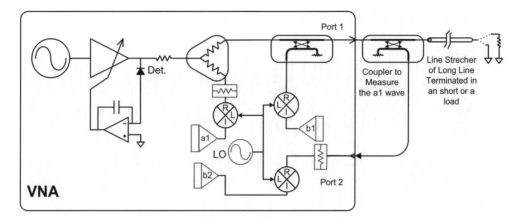

Figure 2.8   Block diagram for measuring power source match.

A line-stretcher is a transmission line structure, usually stripline, which allows the length of the transmission line to be varied. These are sometimes called "trombone-lines" because the center conductor is constructed as a trombone-like slider. An example of one is shown in Figure 2.9.

The traces shown in Figure 2.10 measure the *a1* wave from the test port output by adding an external coupler and routing the coupled arm to the *b2* receiver. The main arm of the coupler is connected to a power meter and the power set to obtain -10 dBm; then the *b2* receiver is calibrated to this output power. Next the power meter is removed and a long line terminated in full reflection (short or open) is put in its place. The ripple on the trace is an indication of the power mismatch. Though not shown, in this case the effective source match (or ratio source match) is very good, but the output power has ripple related to the power mismatch. When the line is terminated with a short or open, the peak-to-peak ripple is exactly the voltage standing-wave-ratio (VSWR) of the power source match. The upper trace is a measurement of a system with a power splitter used for the reference signal separator. The lower trace shows

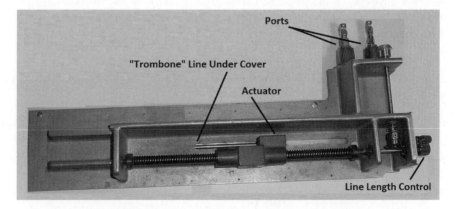

Figure 2.9   A line stretcher used for match measurements.

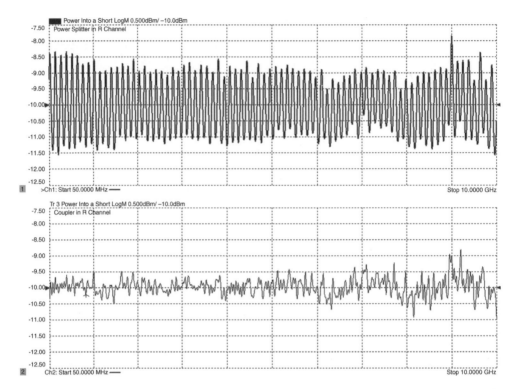

Figure 2.10 Measurement of long-line indicating power source match using an external coupler terminated in a short; upper is for a two-resistor power splitter, lower trace is for a coupler in the reference path.

the same measurement, but this time the power splitter is replaced with a directional coupler for the reference-channel signal separation. Clearly, there is an improvement in the power source match using the coupler.

At any frequency, the VSWR of the power source match can be determined from this response; the envelope of the response can be used directly, or the line stretcher may be adjusted to obtain a peak and valley at any frequency of interest, from which the power source match is computed as

$$SM_{PWR} = 20 \log \left(\frac{1 - 10^{\frac{VSWR}{20}}}{1 + 10^{\frac{VSWR}{20}}} \right) + 2 \cdot L_{CM} \tag{2.2}$$

Where $VSWR$ is the peak-to-peak ripple in dB found at the output of the monitoring coupler and L_{CM} is the loss in the main arm of the monitoring coupler. In the upper trace of the example above, the p-p ripple at low frequency is about 3.0 dB, and the mainline loss of the external coupler is about 1.6 dB, so the power source match is

$$SM_{PWR} = 20 \log \left(\frac{1 - 10^{\frac{3}{20}}}{1 + 10^{\frac{3}{20}}} \right) + 2 \cdot 1.6 = -15.3 + 3.2 = -12.1 \text{ dB} \tag{2.3}$$

This is almost exactly the power match expected from a 50 ohm splitter (83.3 ohms or −12.05 dB). The lower trace shows a power source match for a directional coupler of around −21.6 dB at low frequencies, and −18 dB at higher frequencies.

Extracting the effective match from a mismatch ripple is a technique that will be useful for many other analyses in component measurement. An alternative test method uses a mismatch pad connected to another of the VNA ports, and the ripple is measured at the second port. In such a case, the VSWR is computed into the value of the mismatch, and the return loss value of the mismatch pad is added to the measured VSWR of the source to find the true value. The ripple here is different from the reference channel ripple of the measured wave at the $a1$ receiver shown in Figure 2.7, as the effects of mismatch after the reference splitter are not apparent at the $a1$ receiver.

The power source match is often set by the ALC loop in the source, and as such the reflections from the DUT are sensed by the detector diode in the ALC loop such that the source power is adjusted to maintain a constant voltage at the detector.

2.2.2.3 Source Output Impedance

The effective output match of the VNA source is the same as the power source match over the region within the ALC loop bandwidth of the source signal. In that region, the ALC loop responses to reflection signals are as described above. Outside the ALC loop, the source presents a different output impedance. This is the reflection that would be measured in response to a signal not related to the source output signal. This source reflection is important in cases where the DUT presents other signals reflected into the source, such as from a mixer, or intermodulation products from an amplifier. This value represents the manner in which these other signals would be reflected back out of the source. It can be measured directly as any other reflection coefficient, using a separate VNA reflectometer; an example measurement is shown in Figure 2.11. Here, the source frequency is shown as a large spike at 1 GHz on the source output impedance measurement. This is again a case where the coupler in the R-channel used to sample the incident wave provides better match (lower trace) than a two-resistor splitter (upper trace).

2.2.3 VNA Test Set

2.2.3.1 Test Set Switch

In some VNAs the source is switched between ports using a test set switch, which can come before or after the reference channel splitter. The termination of this switch provides the load match of the port when the source is not active on that port. This load match is not the same as the source match (ratio or power) and so some advanced calibration techniques, which rely on the port match being consistent whether the port is a source or a load, must be modified, as discussed in the next chapter. If the switch comes before the reference channel splitter, there will be a reference channel receiver for each port (four-receiver VNA, Figure 2.2 lower). If the switch comes after the reference channel splitter (three-receiver VNA, Figure 2.2 upper) the reference channel is shared between ports. It only samples the source signal when the source is active. This three-receiver architecture does not support some calibration methods, such

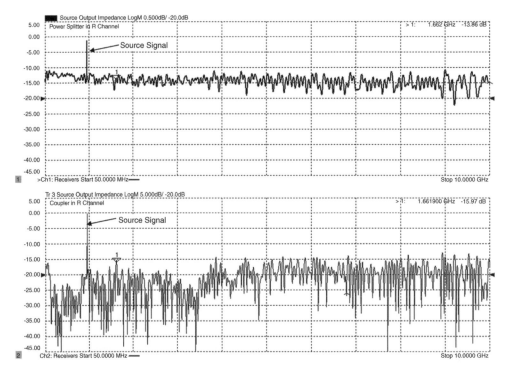

Figure 2.11 Measured source output impedance away from the source frequency; upper trace using a power splitter in the R-channel, lower trace using a coupler.

as thru-reflect-line (TRL) and so modifications and compromises to the calibration methods must be made.

The short explanation for the difficulty is that TRL calibration methods require measuring the load match of port 2 when port 1 is active. To do this measurement, the ratio of $a2/b2$ is acquired during the thru step. But there is no $a2$ receiver available in the three-receiver architecture. Modifications can be made that assume that the source match and load match of the port are identical, but this case is not common unless attenuation is added after the reference channel split. Attenuation added reduces the difference between source and load match at a port by twice the attenuation value. Pre-characterization of the difference in source and load match, called the delta-match, can be performed and removes the need for characterization at the time of calibration. This allows three-receiver architectures to support the same calibration as four-receiver architectures, and is found in some of the more modern low-cost analyzers.

2.2.3.2 Step Attenuator Effects

In some VNA designs, a step attenuator is added between the reference coupler and the test coupler, to allow a greater change in the source power setting beyond that which the source ALC circuit can produce. This step attenuator has the additional benefit of providing a good match to the test port. In cases where the power source match and ratio source match are not the same, the step attenuator reduces the difference between these values based on twice the attenuation value. Reducing the difference between power source match and ratio source match allows one to compute the error in source power from the ratio source match, which is determined as part of the normal calibration process. In general, the power source match is not characterized during any normal calibration.

Another issue to be concerned about with step attenuations is their effect on the quality of the measurement when the attenuation value is changed. In most newer VNAs, the nominal attenuator value is known, and the effective value of the reference receiver is compensated for when the attenuator value is changed. The source ALC power is also changed, so that changing the attenuator value causes only a slight change in the value of the power coming from the test port; the nominal attenuator value is usually within 0.25–0.5 dB of the actual attenuator value. Since the port power stays the same, the internal source power must be raised by the amount of increase in the step attenuator. Since the reference receiver comes before the step attenuator, it will see a larger signal value; as it is desired to have the reference receiver power display the same value as the port power, its reading is also decreased by the value of the step attenuator.

Placing the step attenuator at this point in the block diagram has a distinct advantage in that it allows a large signal in the reference channel even when a very small signal is needed at the test port, providing a low noise signal.

The loss of the step attenuator is well compensated for, but its effect on match is not. If the preset condition has the step attenuator set to 0 dB, the match of the port is terminated back in the test port switch. When even a single stage of test port attenuation is used, the predominant source and load match characteristic is set by the match of the attenuator, which is typically quite good. Thus, it is good practice to use some source step attenuation if the maximum test port power is not required for the measurement. And since the raw match is better for any attenuator setting other than the 0 dB step, the effect on a calibrated measurement if the attenuator is switched to a different, non-zero, value is smaller if the calibration is performed with some attenuation applied.

Some older VNAs did not allow changing the attenuation value after calibration and did not compensate for the nominal value of an attenuator change; error correction in these VNAs is often turned off for a step attenuator change.

In general, changing a step attenuator will change all the raw error terms on the port that has the step attenuator. Techniques are discussed in Chapter 3 which can compensate for much of this change.

2.2.3.3 Test Set Reflections

In addition to the source match effects produced by the source impedance, power source match and ratio source match, reflections from within the test set of the VNA will also exist, as well as from the test port cables and any fixtures that provide an interface from the VNA to the DUT. These sources of mismatch are common to all of the above source match effects and will add to them in a similar way. However, since they are common, their effects on port power and gain are also the same.

The reflection and mismatch between the reference channel split and the test-port coupler affect the incident signal, $a1$, but are not monitored by the reference channel receiver.

Reflections after the test-port coupler also affect the *a1* signal, but will be apparent in changes measured on the reflected signal, *b1*. However, their composite effect will add to the overall source match and their effects on measurements can be compensated for, provided they remain stable. In addition, mismatch and loss after the test-port coupler can be characterized in such a way that changes to these values, such as that due to drift in a test port cable, can also be compensated for in some cases. Mismatch correction in power measurements is discussed in detail in Chapter 3.

2.2.4   Directional Devices

One vital VNA component is the directional device used at the test port to separate the reflected wave from the incident wave. This is most often a directional coupler or directional bridge, although simpler structures have been proposed as well. These devices are characterized by their mainline loss (the attenuation of the *a1* signal), the coupled-arm loss (the attenuation of the *b1* signal) and their directivity (the ability to separate the *b1* signal from the *a1* signal). In addition, any mismatch in the directional device will contribute to the port match and source match. If there is mismatch before the directional device, it will have no effect on its directional characteristics (directivity or isolation). However, any mismatch after the directional device, such as in a test port cable or fixture, will contribute equally to mismatch and degradation of directivity, as described in Section 1.10.

2.2.4.1   RF Directional Bridges

Most RF VNAs make use of a directional bridge, which has the important characteristic of maintaining good coupling and isolation over very wide frequency ranges, even at very low frequencies. While the most common implementation of a bridge is a balanced Wheatstone bridge, this simple implementation can be modified to create a component that has characteristics very similar to a directional coupler, but with much wider frequency range and very low frequency of operation. A bridge is often used in metrology applications where balance in a DC resistive path provides a measure of some quantity such as the power absorbed by a load (see Section 1.15). To understand how these bridges can be configured as directional couplers, with low loss and high isolation, consider the diagram in Figure 2.12, which is a common representation of a Wheatstone bridge.

In this configuration, the signal from the source is applied across the top and bottom of the bridge, and if the ratio of R1/R2 is equal to R4/R3, the net voltage across Rdet (which in a common bridge represents the meter movement) will be zero, and no current will flow through the detector.

In a thermistor, all the resistors are 50 ohms, and one of them represents the RF input of the power sensor, typically R3; a DC signal is applied from the source across the bridge, and the imbalance is measured as the voltage difference across the Rdet resistor. In an RF bridge it is desired to isolate the bottom node of the bridge from ground, and so a transformer is added for this purpose, Figure 2.13. This 1:1 transformer performs the function of a BALanced-UNbalanced transformer, or balun, changing the unbalanced (or grounded) source into a balanced signal across the bridge. Doing this allows grounding a different leg of the bridge which, as will be seen, is key to making a bridge act as a directional coupler.

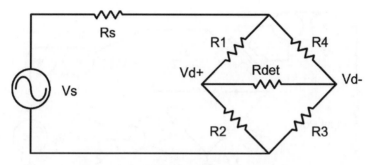

Figure 2.12 Schematic of a directional bridge.

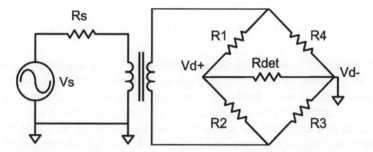

Figure 2.13 Adding a transformer between the source and the bridge.

From this modification, the RF implementation of the bridge can be better understood. Since the low side of the detector is now ground, the resistors represented by Rdet and R4 can be replaced with transmission line structures of similar impedances, representing the RF ports of the directional bridge, Figure 2.14. In this figure, the Rdet resistor is replaced with the coupled port of the bridge, and one can see that RF energy flowing from the source appears equally at both the center conductor and the ground of the isolated port.

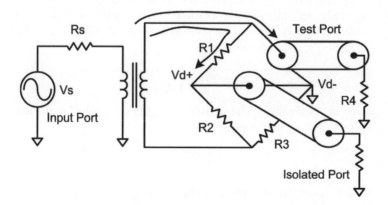

Figure 2.14 Replacing bridge elements with RF ports.

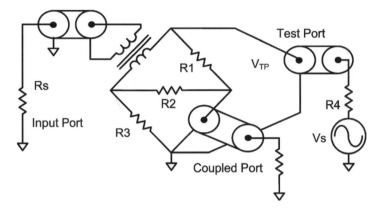

Figure 2.15   A bridge redrawn to show the coupling factor.

However, since the RF current appears at the test port, relative to ground, a portion of the RF signal will appear across R4; the relative value of the voltage on R4 to Vs/2 is the insertion loss of the directional bridge. If the bridge uses equal resistors, then each of R1, R2, R3 and R4, as well as Rs, are all 50 ohms. With these values, it is easy to see that Vs is applied equally to R1 and R2, as well as R3 and R4, so that the voltage across R4 is one-fourth of the source voltage. Therefore the loss of an equal resistor balanced bridge is one-half voltage applied at the bridge input, or −6 dB. In general, the insertion loss of a bridge, where $R_S = R_4 = Z_0$ is

$$L_{Bridge} = 20 \log \left(\frac{Z_0}{Z_0 + R_3} \right) \tag{2.4}$$

From this description we can see that in the case where the bridge is terminated in Z_0, there is no signal in the isolated port, demonstrating that this bridge isolates the incident signal. The first criterion of a directional device is satisfied. The second criterion is that the bridge does respond to the reflection signal from the test port. To understand how that occurs, it is useful to redraw the bridge, bringing the ground point of the test port down to the bottom of a redrawn circuit, Figure 2.15.

In this drawing, the source has been moved from the input to the output, but the bridge circuit is topologically identical to the previous figure. When driven from the test port (or when measuring a reflected signal), the isolated arm becomes the coupled arm, and the coupling factor of the coupled arm can be computed as

$$C_{Bridge} = 20 \log \left(\frac{Z_0}{Z_0 + R_1} \right) \tag{2.5}$$

For the case of an equal resistor bridge, the coupling factor is equal to the loss, −6 dB. If R1 is not equal to Z_0, R3 can be computed as

$$R_3 = \frac{(Z_0)^2}{R_1} \tag{2.6}$$

Note that the loss is directly proportional to the coupling as

$$L_{Bridge} = 20 \log \left(1 - C_{Bridge} \right) \tag{2.7}$$

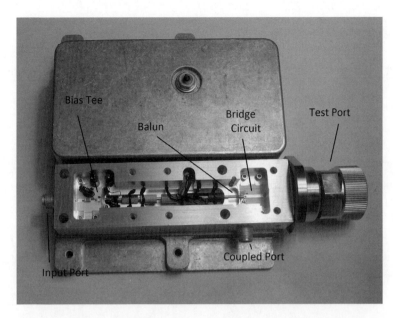

Figure 2.16   An example of a directional bridge from the HP 8753B.

For RF VNAs, it is common to use a directional bridge in the test set. Directional bridges of this type have been used since the 1970s and an example of such a bridge used in the HP 8753B is shown in Figure 2.16. This bridge has been modified to have an unequal coupling and loss, so the insertion loss is lower than normal (around -1.5 dB) and the coupling is higher than normal (around -16 dB) for a Wheatstone bridge.

The RF performance of such a microwave bridge is shown in Figure 2.17. The insertion loss increases with frequency due to the loss of the coax balun and increased coupling due to parasitic series inductance in R3. This same inductance causes a degradation of directivity in the bridge as frequency increases. Bridges are inherently lossy structures, where some of the power is absorbed by the resistive elements in the bridge. The power absorbed by the bridge is equal to the insertion loss of the bridge minus the power coupled to the coupled port.

Bridges of this type have been used successfully up to 27 GHz.

2.2.4.2   Directional Couplers

Directional couplers are more often used in higher microwave frequency ranges, due to the difficulty of maintaining good bridge performance at high frequencies. Directional coupler design is a very broad topic and much literature has been devoted to structures that can be used as couplers. However, for use in VNAs, there are some particular characteristics that are critical. In general, commercial directional couplers are designed to maintain a flat coupling factor over their bandwidth, and the bandwidth is limited by this coupling factor. Couplers used for VNA reflectometers require very wide bandwidths, so rather than a flat response, they are often designed with an equal-ripple or Chebyshev response. Ripple in the loss or coupling factor is not of much concern in a modern VNA, where calibration techniques can remove

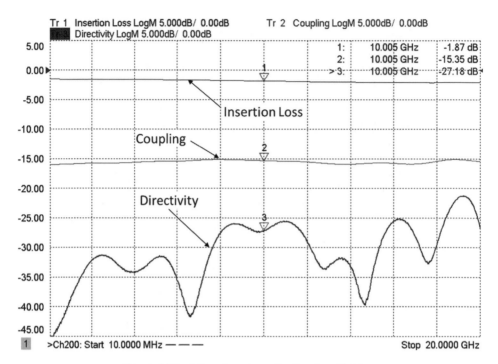

Figure 2.17 RF performance of a directional bridge.

almost any frequency response error. Isolation is an important criterion in VNA couplers. One attribute of directional couplers that distinguish them from bridges is that they are ideally lossless devices such that all the power applied is either coupled (to the coupled port or the internal load) or transmitted through the coupler. The relationship between insertion loss and coupling factor is

$$L_{Coupler} = 20 \log \left[1 - \left(C_{Coupler} \right)^2 \right] \tag{2.8}$$

Directional couplers typically come in one of three forms: waveguide couplers, microstrip couplers and stripline couplers.

Waveguide couplers are most common at mm-wave frequencies, but have the inherent limitation of narrowband operation due to the narrowband nature of waveguides. The structure of waveguide couplers is a four-port device with the main arm connected in such a way as to have irises (or holes) to a second waveguide. The second waveguide can have two ports or one port internally terminated. The nature of the coupler is symmetrical; in theory either port can be the coupled port, but in practice a load is often embedded in the coupled arm. Because of the fundamental function of a waveguide coupler, the forward-coupled wave comes out of the waveguide port nearest the test port. This often causes confusion in the symbols used.

A microstrip or stripline coupler uses a different EM configuration to perform coupling, and the coupled arm of these couplers is the one farthest from the test port. Microstrip couplers often suffer from the fact that there is some dispersion in microstrip lines, and since the even- and odd-mode waves in the coupled lines experience different effective dielectric constants,

Figure 2.18 A directional coupler used in VNAs.

they will have different velocities of propagation. This makes it more difficult to create microstrip couplers with good isolation. For this reason, many VNA couplers are in the form of stripline (or slabline, which is similar to stripline but with a rectangular center conductor thickness), suspended in air. These couplers are designed to have very stable coupling and isolation factors. For a VNA, it is not so important what the exact directivity is, as long as it is completely stable. Figure 2.18 shows an example of a directional coupler used in VNAs. The test port connector is one attribute that differentiates this from a commercially available directional coupler that might be used as a component in a different system. This connector is designed to be firmly mounted to the VNA front panel and to withstand numerous connections and reconnections. This coupler has an integrated load and so only exposes three ports.

2.2.4.3   1+Gamma

Another proposed reflectometer structure is a 1+Gamma structure, whose name comes from the block diagram architecture, Figure 2.19. As the name implies, the signal at the *b1* receiver is a combination of the incident (*a1*) and reflected (gamma) signal.

In this configuration, the signal in the test or *b1* receiver never goes to zero; rather it is minimum with a short, maximum with an open, and nominal 1 when there is a load attached. Also, the signal variation between an open and a short is about 14 dB less than that for a bridge or directional coupler. Put another way, the reflection gain of the 1+gamma bridge is lower than for a directional coupler or bridge. Consider the Smith chart in Figure 2.20; an open, short and load (all non-ideal with fringing capacitance and series inductance) are shown for each on a 1+gamma reflectometer.

The value of attenuation in the reference channel is adjusted to set the value of the open circuit reflection to 1. For a directional coupler, the load gives a zero reflection (ideally) and the short gives a −1 reflection. For the 1+gamma bridge, the open is also 1, but the short is +0.6, and the load is +0.75; thus the difference between the open and the short moves from 2 to only 0.4. These reflections are mapped to the full Smith chart through the error correction math, in such a way that the values from the reflections, and any instability, are multiplied by 5. Also, since the load condition has a large signal in the *b1* receiver, any instability in that

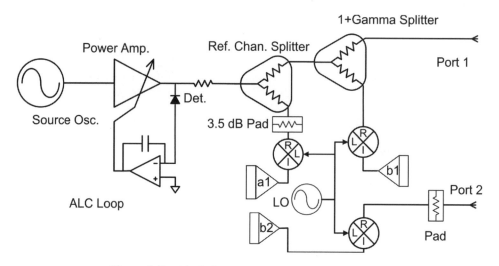

Figure 2.19   Block diagram of a 1+gamma reflectometer.

signal is apparent as a directivity error, which is also multiplied by 5. In theory, if directivity is defined as the average of the open/short response relative to the load response, then the directivity of a 1+gamma reflectometer is about 0 dB (remember that directivity for a coupler or bridge is always positive, often 20 dB or more).

Theoretically, any directivity error can be corrected for by a calibration, but in practice, certain unstable errors can cause uncorrectable errors when the directivity is very poor. Thus, 1+gamma structures have largely disappeared from use. Also, this same multiplying effect causes any slight drift in the test port cable to cause a considerable change in the measured reflection coefficient, after calibration.

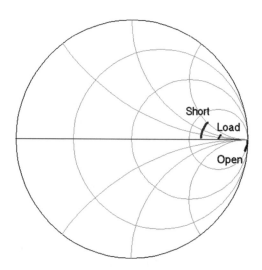

Figure 2.20   Smith chart showing reflections of a 1+gamma bridge with an open, short and load.

2.2.5   VNA Receivers

The final RF components in a VNA block diagram are the test and reference receivers. Dynamic range is a key specification of a VNA, and is sometimes referred to as the difference between the maximum signal level that the receiver can accept while still operating and the noise floor of the receiver. In most cases, the maximum damage level is significantly above the maximum operating level of the receiver, which is usually limited by the input compression level of the receiver. The maximum operating level is set by the structure of the components but for most modern VNAs, it is around −5 dBm at the receiver mixer input, or about +10 dBm at the test port, after considering the coupling factor of the test-port coupler. The noise floor of the receiver is set primarily by the type of mixing down converter used, of which the two principal types are sampling down-converters (or samplers) and mixers.

2.2.5.1   Samplers

The sampling down-converters are circuits that are driven by a low frequency pulse, which has very high harmonic content. The example circuit shown in Figure 2.21 is typical of older VNA sampling receivers such as found in the HP8753 or HP8510 VNAs. In the circuit, the diode pair acts like a switch, which is driven with a very short pulse from a pulse generator driven by a voltage controlled oscillator, operating at relatively low frequencies. The short conduction angle (the amount of a cycle which the diodes conduct) of the pulse means that the frequency content is very high – sometimes referred to as a harmonic-comb – and the sampler can convert frequencies much higher than the VCO drive.

It is not unusual to use harmonics of up to 200 times the VCO frequency. Because the conduction angle is so short, the effective input impedance is very high, essentially multiplied by the maximum harmonic number; this means the effective noise figure of the sampler is very high as well. Since the conduction angle does not depend upon the frequency being measured, the noise figure does not depend upon the frequency being measured. The conduction angle and conversion efficiency can be adjusted by adjusting the diode bias, so that they are not

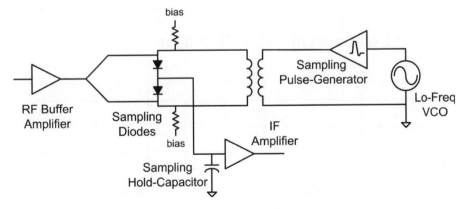

Figure 2.21   Schematic of a sampler.

quite turned on, and the pulse provided by the VCO gives the extra current to turn on the diodes fully.

Two advantages of samplers are that they do not require a high frequency local oscillator, as a mixer would, and they can simultaneously down-convert a signal and all its harmonics to the IF frequency. This capability is exploited in a sampling oscilloscope, and in some of the modern versions of sampling VNAs used for non-linear measurements. However, for the most part, samplers have been largely abandoned in VNAs due to a number of difficult problems that they present.

The foremost problem is the degraded noise floor in the sampler-based VNA. The effective noise floor is further reduced as the conversion efficiency of higher-order harmonics typically degrades near the top of the sampler frequency range. Almost all modern VNAs use some form of mathematical response correction on the sampler response, so that frequency response of the VNA receiver to a constant input power appears flat over its entire frequency range. This response correction, which removes the effect of roll-off in real conversion loss, has the consequence of increasing the apparent noise floor of the sampler at higher frequencies.

A secondary problem with samplers is that the rich harmonic content of the VCO dictates that the sampling receiver has many regions where it is sensitive to other signals, such as harmonics or spurious signals of the VNA source or DUT, and spurious signals present at the DUT output. This makes sampling receivers particularly poor at measuring mixers or frequency converters, where the sampling comb-tooth can cross mixer output signals at many different frequencies. Figure 2.22 shows an example of spurs from a source generating responses in a sampling receiver. In this case the source signal is generated by mixing a 3.8 GHz fixed RF signal with a 3.8–6.8 GHz swept YIG oscillator LO. The mixed product provides the desired 0–3 GHz source output, but spurious signals at $2 \cdot RF - LO$ and $3 \cdot RF - 2 \cdot LO$ do show up in the 0–3 GHz VNA measurement receiver band. While small, these spurious responses do degrade the S21 accuracy.

Another difficulty with sampling mixers becomes apparent when measuring filters in the stopband, and is caused by remixing of signals reflected off the DUT back into the input reflection receiver such as the $b1$ receiver. Because this effect has the appearance of a signal bouncing off the input reflection of the filter stopband, and then bouncing off the $b1$ mixer

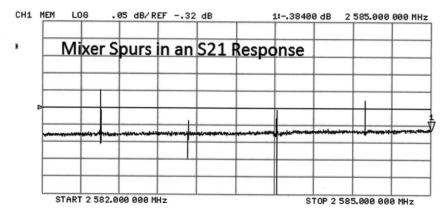

Figure 2.22   Spurs from a source crossing a harmonic of the VCO.

(at a different frequency), it is sometimes called "sampler bounce" (or mixer bounce in the case of mixers). Designs of these components must be very carefully considered to avoid these bounce signals, and the basic design of samplers makes them especially susceptible to this particular type of crosstalk.

For these reasons, and the fact that creating wideband mixers with full-band RF-frequency LOs has become much more cost effective, the use of samplers in VNAs has been phased out.

2.2.5.2 Mixers

Almost exclusively, modern VNAs use mixers as VNA receivers. The mixers are driven by a fundamental or low-harmonic-order LO. This provides a much larger conduction angle for frequency conversion, and as such provides a lower noise floor. The tradeoff of course is that the cost of the LO may be more due to its increased complexity.

Noise Floor

With mixers used as VNA receivers, the critical performance attributes are noise floor and input compression point. A fundamental LO provides best performance; using a third-order harmonic of the LO to obtain higher RF frequency response theoretically degrades the conversion loss by 9 dB, and fifth harmonic degrades it by 14 dB, based the idea that the LO drives the mixer into square-wave conversion. This degradation in conversion loss is represented as an increase in noise floor for most VNAs, due to the internal factory-based response correction.

Spurious Responses

Mixers have much lower unwanted spurious responses compared to sampler-based receivers. The primary spurious (or unintended) conversion occurs at the IF image of the desired RF signal. If the RF is above the LO (lo-side mixing), the image will be one IF below the LO (hi-side mixing). Because mixers used as VNA receivers have lower noise floor and lower spurious responses, a much wider class of measurements can be performed including noise figure, two-tone IMD, and even some modulated measurements.

2.2.5.3 Phase Noise

The LO distribution attributes are critical in the performance of mixer-based systems. The phase noise of the LO contributes directly to errors (primarily trace noise) in the phase of the measured signal. Even the trace-noise of the amplitude response can be degraded by phase noise as the signal moves across the IF bandwidth filter. Providing a common and coherent LO to all the mixers improves the trace noise of the measurement by allowing the effects of LO phase noise to be reduced when the measured parameter is a ratio, such as gain or return loss.

2.2.5.4 Isolation and Crosstalk

Isolation between VNA receivers is important when measuring high dynamic-range devices such as filters. In almost all measurements, the reference path has a large signal (as it measures the incident wave) and so is a constant source of leakage signals. Partly for this reason, it is very common to provide additional loss in front of the reference channel mixer (5 or 10 dB) to lower the incident signal level and provide greater reverse isolation for mixer bounce. This also helps to keep the reference channel mixer operating in its linear region for higher source power signals, avoiding compression in the reference mixer.

In fact, there are four primary crosstalk paths in VNAs:

1. RF signal from the internal source or reference receiver to the transmission test-port – this crosstalk signal is independent of the DUT characteristics, and its level remains constant regardless of the DUT properties, but will change with test frequency. Leakage of the reference mixer IF to the test mixer IF path has similar characteristics, but its value doesn't change with frequency.
2. RF signal from the reflection test-receiver ($b1$) that leaks to the transmission test-receiver – this signal depends upon the input reflection of the DUT; if the DUT is well matched, there will be no signal at the reflection receiver. Since this signal depends upon the DUT, correcting for it is more complicated.
3. RF signal that leaks from the test set switch, to port 2, and reflects off the DUT output match into the port 2 transmission test mixer – as this signal depends upon the DUT characteristics, correcting for it can also be more complicated. Modern VNAs that use separate sources instead of test set switches eliminate this source of crosstalk.
4. The final source of signal leakage is related to any test fixture or probing done to connect to the DUT. Leakage from port 1 to port 2 of the probes or fixtures is usually electric field radiation or magnetic field coupling between the ports. Since these fields are non-TEM, they do not remain constant with changes in the DUT characteristics, and their effect may not be well understood. Probe-to-probe isolation is key problem in measurements, but one that is not well accounted for. Careful fixture or probe design that includes shielding is perhaps the best solution to this final leakage effect.

In most modern VNAs, the design of the mixers and LO isolation networks are such that the level of the first three sources of crosstalk are at or below the noise floor of the receiver. As such they can be ignored except in special cases where extended dynamic range is desired, as discussed in Chapter 6. The fourth cause of crosstalk is inherent in the fixtures or probes, and it can sometimes be removed with calibration. But since the source is often due to radiation from one port to the other, this radiation pattern depends in a complex way on the actual loading of the port and the structure of the DUT. For example, in a probed situation, leaving the probes up as an "open" calibration standard can cause the probes to act as E-field antennas, and produce crosstalk between the probes. Grounding the probes, to produce a short, can cause magnetic field coupling between the probes, again producing crosstalk. Both of these crosstalk terms are non-TEM, meaning that they have E and H fields that propagate in the direction from port 1 to port 2. Normal calibration methodologies do not correct for non-TEM crosstalk as their values do not remain constant if the DUT configuration changes.

2.2.6   IF and Data Processing

The final hardware portion of the VNA block diagram is the IF processing chain. The VNA receiver converts the RF signal to a first IF frequency, which is further converted and detected in the IF processing path. In older analyzers, such as the HP-8510, this consisted of a synchronous analog second-converter which produced two DC outputs proportional to the real and imaginary portions of the RF voltage at the receiver input. These DC voltages were measured with DC ADCs that produced a digital representation of the real and imaginary values. More modern IF structures such as in the HP-8753 or HP-8720 used a separate stage of IF down-conversion, to bring the IF signal down to a frequency where an AC ADC could directly sample the waveform. The final IF frequency was set by the sampling rate of the ADC.

2.2.6.1 ADC Design

Now, most modern VNAs incorporate a high-speed ADC and perform direct sampling of the first IF signal. An example of a VNA digital-IF block diagram is shown in Figure 2.23. The IF signal is preconditioned with adjustable gain to optimize the signal-to-noise in the ADC. For some applications, it is useful to have a narrowband pre-filter before the ADC, so that the IF can be switched between a wideband IF and a narrowband response. An anti-alias filter is used just before the ADC, with a bandwidth of about one-third to one-fourth that of the ADC clock rate.

The field-programmable gate array (FPGA) that processes the ADC readings can be configured as a digital second converter of flexible IF frequency so the final digital IF frequency can be quite arbitrary. There are several modes of operation for the digital IF. For these high-speed ADCs, the raw ADC readings have very high bit rates. Some of the latest designs for VNAs have four channels of data, at 16 bits and 100 mega-samples per second to produce a data rate of 4.8 Gbits/second. Specialized conditioning of the signal and advanced digital signal processing (much of which is proprietary) can improve the performance of the IF ADCs to many more effective bits.

At these high data rates, the main CPU cannot process the data fast enough to keep up, so an FPGA is used to decimate and filter the signals before the processed data is sent to the main processor using shared DMA memory. The function of decimation and filtering is the basic data processing step of any digital IF; in this function, a measurement is performed by setting the source and receiver frequencies so that the first IF contains the signal of interest. The ADC samples the IF signal, typically with 2–4 times over-sampling, although it can be as much as 60 or 100 times over-sampled. A finite set of samples is processed by the FPGA to produce a final result that represents the real and imaginary part of the signal being measured. For example, if the digital IF is operating at 100 Msps, and the IF frequency is 10 MHz, and the IF filter is set to 100 kHz IF BW, then approximately 10 μsecs of data are captured, or approximately 1000 data samples. These 1000 samples are processed by a multiply-add chain

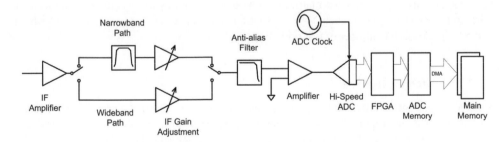

Figure 2.23   Digital IF block diagram.

in the FPGA to both filter the response and extract the real and imaginary values. In this way, the 1000 samples are reduced to two samples.

A second mode of operation for the digital IF is the "ADC capture" mode. In this mode, the FPGA does not process the data; rather, the data samples are simply captured into the local memory in the digital IF for a finite period of time. The entire ADC data stream is available for any further processing that might be beyond the algorithms available in the FPGA. Some modern VNAs have memory depths up to 4 Gbit allowing very deep memory captures. This mode of operation, while not typical, is useful for capturing anomalous effects such as transient or pulsed responses, as well as more complicated functions such as demodulation of IF signals.

2.2.7   Multiport Extensions

For a class of RF and microwave devices, the normal port count of two to four ports is not sufficient, and multiport measurements are required. There are two distinct classes of multiport test requirements that generate the need for two distinct RF architectures to support the measurements. RF switching test sets provide the basis for these multiport extensions.

The first class of devices requires multiple sets of two, three or four port measurements. As such, the native mode measurements of the VNA are sufficient, and all that is required is RF switching to route the VNA ports to the various port pairs of the DUT. One example of such a DUT is a satellite multichannel-diplexer (or multiplexer) which filters and separates signals from a common antenna path to each of several output channels, as shown in Figure 2.24. This unit has waveguide filters and interconnections to provide for the lowest possible loss.

This device requires two-port measurements for each path from the common port so a two-port VNA with one common port and one switch port can make all the required measurements. These are sometimes known as switching test sets or simple switch-trees.

The second class of devices requires a measurement from each port to every other port, and in general the response of any path depends upon the loading or match applied to every other port. A "Butler matrix" is a kind of signal dividing network used in phased-array radar systems, which has this attribute. An eight-port Butler has four inputs and four outputs, and the proper description is a 8×8 S-parameter matrix. To measure such a device, a switch matrix must be able to allow measuring every path of the device. Informally, these types of switch matrixes are called "full cross-bar switches" and implies that from the two ports of the VNA, any path of the DUT can be measured.

There is a further requirement on the Butler matrix; a full N-by-N port calibration measurement must be able to be performed to correct for the imperfect match at each port. This requires not only a full cross-bar matrix, but one that supports N-by-N calibration as well. A third style of test set allows such N-by-N S-parameters called an "extension test set" which extends or adds to the number of test ports from a VNA.

2.2.7.1   Switching Test Sets

Switching test sets contain only RF switches formed in a matrix to provide the needed measurement paths. Figure 2.25 shows the block diagram of a simple switch-tree test set. These test sets are typically constructed from either 1×2 RF switches or 1×4 to 1×6 RF switches. The 1×2 RF switches are sometimes used because some versions provide for an RF load on the unused ports. The 1×4 or 1×6 are typically mechanical switches and may

Figure 2.24 A satellite multiplexer with many outputs, Courtesy ComDev Ltd, with permission. © Copyright 2012 COM DEV Ltd. All rights reserved. Unauthorized duplication or distribution is strictly prohibited. Permission to use, copy, and distribute this image is subject to COM DEV Ltd.'s prior written consent.

not load the unused ports. If a multiport device has a path response between two ports that depends on the load match of a third port, the switch matrix must provide a load on the unused port. Larger switch configurations that have loads are often not available above 40 GHz, and so 1 × 2 matrix arrays are used. Electronic 1 × 2 switches are available over a wide range of frequencies, but there are few electronic switches with higher port counts, so electronically switched test sets are typically configured from 1 × 2 RF switches.

The simple switch matrixes of Figure 2.25 can be viewed as having a port 1 switch-set and port 2 switch-set, and any path from port 1 side to port 2 side can be measured, but no measurements are available between ports on the port 1 of the switch-set, nor between ports on the port 2 side. While there are 24 ports available in the test set, only 12 paths can be measured from any one of the 12 input ports. Thus, this simple switch-tree test set can support 144 paths, but a full 24 port device actually has 276 paths. There are 66 paths on VNA port 1 side that cannot be measured, and 66 paths on the VNA port 2 side that cannot be measured. To obtain a full matrix of paths, a so-called "full cross-bar" switch matrix is required.

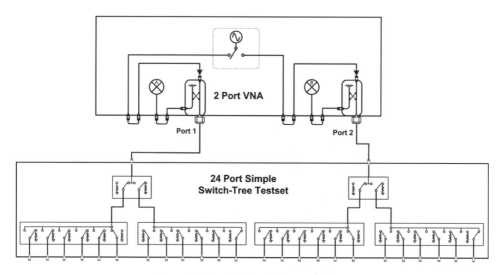

Figure 2.25   Simple switch tree test set.

To accomplish full cross-bar testing, a configuration of test set shown in Figure 2.26 is used. In the general configuration, sets of 1 × n switch trees are cross-connected to 1 × 2 switches at each port. This configuration provides for any path to be measured, but the unused ports are terminated back in the 1 × n switches which are terminated internally in a load. If the 1 × n switches are not internally terminated (rather, it is left open), then the 1 × 2 switch must provide a termination for an unused port. Figure 2.26 shows a full cross-bar switch constructed of a 1 × 2-port switch connecting to a pair of the 1 × n switches. With this configuration, every port that is not connected to the VNA is terminated in a switch load. However, it is difficult to use this type of switch matrix to perform full N-by-N calibrations as the exact value of the load termination of any port changes depends upon the switch settings of other ports.

For example, if test-set ports 1 and 6 are the active ports, port 2–5 are terminated in the 1 × 6 switch on the left, and if test-set port 5 is made active, then port 6 may be terminated in the 1 × 6 switch on the right. The fact that the termination of the port depends on the path selected makes calibration beyond the two ports selected more difficult.

Custom switching test sets might have a reduced number of paths, forming a combination of full cross-bar on some ports and simple switch trees on other ports. For high speed and reliability, solid state switching is preferred. Mechanical switches have almost no loss, but solid state switches can have considerable loss at microwave frequencies. This loss is after the directional coupler, and dramatically degrades the RF performance of the system. On the other hand, mechanical switches can have slight changes in return loss for each switch cycle, also leading to instabilities. Thus, this architecture of switches after the directional couplers of the VNA is a simple one but at a cost of substantially reduced stability and performance.

2.2.7.2   Extension Test Sets

To satisfy the requirement for making full N-by-N calibrated measurements, often referred to as "full N-port cal" measurements, a new test set design has been developed that includes

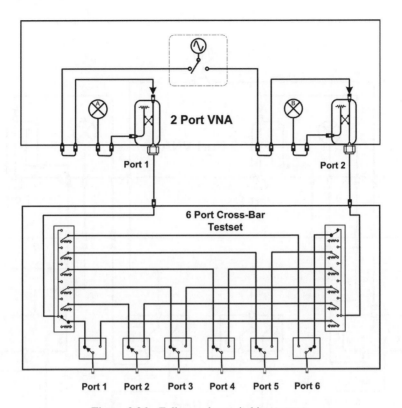

Figure 2.26   Full cross-bar switching test set.

both directional couplers and switches. The original implementation of this style of extension test sets was configured to supply two additional ports to a two-port VNA to create a four-port VNA for making the first balanced and differential measurements. The general idea of an extension test set is to essentially extend the source switch matrix of the VNA to more outputs through a source switch, and also extend the internal receivers to more ports through a receiver switch. This requires that an additional test-port coupler be provided for each additional port. Because the switching occurs behind the VNA directional couplers, they are still available as test ports; the ports on the test set extend the total number of ports available, hence the name extension test set. Figure 2.27 shows block diagrams for a simple two-port extension test set.

One key point of the block diagram is that the test set breaks into the source and receiver loops behind the test-port coupler. Since any number of switch paths can be supplied behind the test couplers, there is in theory no limit to the number of ports that can be used. Further, this block diagram allows additional test sets to be added so that the any number of test ports can be created by stacking extension test sets. Common configurations are four-port extension test sets for a four-port VNA to extend to a total of eight ports, ten-port extension test sets for a two-port VNA to achieve a total 12 ports, and 12-port extension test sets for a four-port VNA to achieve a total of 16 ports. Figure 2.28 shows a four-port VNA with two-port extension test sets to create a 12-port system.

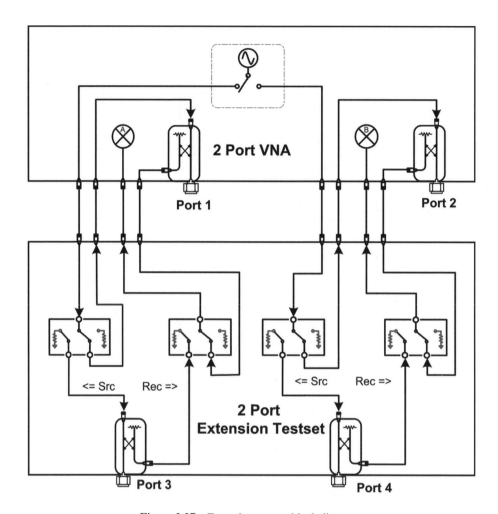

Figure 2.27   Extension test set block diagram.

The switches may be either mechanical switches or solid state switches. Because all the switching occurs behind the test-port couplers, the stability and performance of the measurements are much better than those of switching test sets, and loss in the switch, while it reduces the dynamic range, has no effect on stability of the measurements.

In some cases, an option may be provided to add a low-noise amplifier between the coupled port of the test coupler and the switch input. This improves performance as the gain of the LNA improves the dynamic range. Adding amplifiers in between the coupled arm and the switch also removes another source of error. In some cases, the source match of a port changes when the source and test port share the same VNA receiver: for example, ports 1 and 3 in Figure 2.27. This error is typically small, as the difference between the match of the VNA receiver and the match of the switch is small (on the order of −10 dB) and is further reduced by twice the coupling loss (32 dB) resulting in a typical source match error smaller than

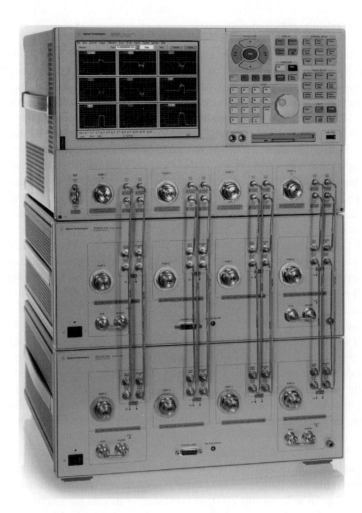

Figure 2.28 Twelve-port system using a four-port VNA and two extension test sets. Reproduced by permission of Agilent Technologies.

−40 dB. In most cases it has negligible effect, but in some measurements, particularly circulators or couplers, it can become significant and is not removed in calibration, so adding an amplifier ensures that the match presented to the coupled arm is constant. The test ports also change load characteristics depending upon whether they are terminated in a switch or the VNA internal load; however, the N-port calibration methods characterize both these states and fully correct for the difference.

Calibration is often a concern with multiport test systems. Traditional S-parameter calibrations require measurements between every path of a test system. However, new techniques have greatly reduced the total number of calibration steps to the point where a full N-by-N port S-parameter calibration can be achieved with a single one-port return loss calibration and N−1 thru measurements using the quick short open load thru (QSOLT) calibration. More details of these new calibration methods will be discussed in Chapter 3.

2.2.8   High Power Test Systems

Most VNAs have a maximum test port operating level on the order of 10–15 dBm, with a damage level on the order of +30 dBm. Beyond the operating level, the receiver will be in substantial compression, so the data is not valid. Many VNAs provide internal receiver attenuators that allow reducing the power to the receiver thus providing operation to much higher levels. The maximum input power to test-port couplers is often rated higher than the maximum level of other components behind the directional coupler, so that with proper padding and isolation, the VNA can be operated to levels as high as +43 dBm, depending on the model. Operation above these levels is possible but requires substantial external components including external couplers to ensure the power level at the VNA components is below the power damage level. Details of high power test configuration are shown in Chapter 6.

Another common practice is to add fixed attenuators between the DUT output and the VNA test port. This works well so long as the total attenuation between the test port and the DUT is less than 10 dB. Adding attenuation after the test-port coupler degrades the directivity by two times the attenuation (in dB), as will be shown at the end of Section 2.3.2. In practice, up to 10 dB of external attenuation can be added and compensated for with normal calibration techniques. If between 10 and 20 dB are added, the system becomes somewhat unstable, and for more than 20 dB of added attenuation, different techniques for calibration must be used, and the S22 measurements become unreliable.

For testing devices that require high power drives, it is common to add an amplifier to increase the power level normally available from the VNA. One method is to simply add a booster amplifier to the port 1 output and drive the DUT directly from the booster amplifier. This generally results in poor measurements of the DUT due to mismatch and gain errors in the booster amplifier. In this approach, it is common to add a booster amplifier, normalize the S21 trace, then add the DUT and measure the resulting gain relative to the normalized booster amplifier response. However, the normalization has errors due to mismatch between the booster amplifier and the load port. And the measurement has errors due to mismatch between the booster amplifier and the high-power DUT. Further, the input match or S11 of the DUT cannot be reliably measured because booster amplifier isolation eliminates the ability to measure a signal reflected from the DUT. A second error in gain measurements often occurs with this direct approach due to gain drift or gain compression of the booster amplifier.

A better systematic approach is to add the booster amplifier behind the test-port coupler, and use a second coupler as the reference-channel tap to generate a signal proportional to the booster amplifier output signal that can be routed to the reference channel. In this scenario, a directional coupler rather than a power splitter is typically used to provide lower loss after the booster amplifier. The output of the reference channel is directed through the test-port coupler of port 1 so that the S11 of the DUT can be accurately measured. In almost all cases, an accurate measurement of high drive-power devices requires a booster amplifier followed by a reference coupler. Chapter 6 provides a detailed discussion of high power amplifier measurements, including several alternative block diagram configurations to support various power levels.

2.3   VNA Measurement of Linear Microwave Parameters

In this section the fundamentals of making microwave measurements are discussed for a variety of parameters, along with the consequences of the practical limitations of the RF hardware as detailed in the previous section. This section discusses measurement methods, and the sources of errors and other complications to making microwave measurements related to test equipment limitations.

2.3.1 Linear Measurements Methods for S-Parameters

Linear measurements imply that the parameter being measured does not depend upon the level of the signal applied. For RF and microwave measurements, the principal linear parameters are S-parameters from which gain, match, impedance and isolation, among others, can be derived.

2.3.1.1 Signal Flow Graphs for VNA Hardware Configurations

A vector network analyzer is used to measure the S-parameters of a device by applying the source signal to the input of the DUT and measuring the response on the VNA receivers. With the source applied to port 1 of the DUT, the incident wave, a_{1M}, is measured from the reference channel and the test signals are measured at the input and output as b_{1M} and b_{2M}, but these do not represent the actual values of incident and scattered waves from the DUT. Errors in the VNA alter the source and receiver signals such that the measured values can vary significantly from the values at the DUT reference plane. There are a variety of errors associated with the S-parameter measurements but the principal ones can be defined in a signal flow diagram [2, 3] as shown in Figure 2.29. In conventional VNA measurements, each port of the DUT is

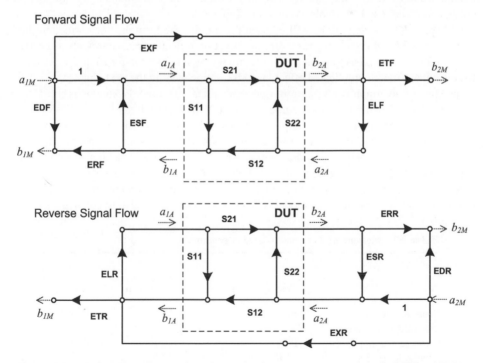

Figure 2.29   Signal Flow diagram for a forward and reverse measurements of a DUT and VNA.

Table 2.1 Systematic error terms in a VNA

EDF	Forward Directivity	EDR	Reverse Directivity
ESF	Forward Source Match	ESR	Reverse Source Match
ERF	Forward Reflection Tracking	ERR	Reverse Reflection Tracking
ELF	Forward Load Match	ELR	Reverse Load Match
ETF	Forward Transmission Tracking	ETR	Reverse Transmission Tracking
EXF	Forward Crosstalk	EXR	Reverse Crosstalk

stimulated by the source signal, and the other ports provide a nominally matched termination. For two-port devices, it's common to refer to stimulus at port 1 as the forward direction and port 2 as the reverse direction. In multiport systems, the F and R may be replaced with port numbers for the reflection port terms, such as ED1, ES1 and ER1, but the transmission terms must contain a port pair, such as ET21, or EL12. While it might seem odd that the load match is considered a transmission term, some VNA and multiport test systems present a different load impedance to the DUT depending upon the source port, and so the load match must be explicitly designated in terms of both the load port and the source port. It is standard practice to name the load port first and the source port second, so that ET21 is the tracking term that primarily affects the S21 parameter.

The traditional systematic errors in the measurement are shown, which can be characterized and removed from the measurements of a linear device. These error terms are listed in Table 2.1

Another way to view this is that the errors come in three types: Frequency Response, Mismatch, and Leakage. And the errors occur for each of the four S-parameters, so that they may be expressed in a 3×4 table as shown in Table 2.2.

In Chapter 3, these definitions will be used to simplify the formulas for correction. From this, it is implied that directivity has a similar effect on reflection measurements as crosstalk does on transmission measurements; a similar implication holds true for frequency response. Mismatch has a more complicated effect, as will become clear in the next chapter.

This signal flow graph configuration represents what is commonly known as the 12-term error model for S-parameter measurements. Most VNAs use this as the model for which the error correction algorithms are applied. In this signal flow model, the measurements are presumed to occur with a forward sweep of the VNA where the source is applied to port 1, followed by reverse sweep where the source is applied to port 2. Tracking terms represent relative losses (or how the receivers track each other) and match terms represent errors relative to the system Z_0. As described in Section 2.2.2.1, the source match error term refers to the ratio source match. Implicit in this model is that the port match at each port changes depending

Table 2.2 Measurements and associated error terms

Measurement	Error		
	Tracking Response	Mismatch	Leakage
Input Reflection	ERF	ESF	EDF
Forward Transmission	ETF	ELF	EXF
Reverse Transmission	ETR	ELR	EXR
Output Reflection	ERR	ESR	EDF

upon whether the source is active or not, as the internal switching of the VNA changes its impedance when the source is switched to a terminated state. This is represented in the naming where the port match is called the source match for the active port and the load match for the inactive port.

The crosstalk error term is largely ignored in modern VNAs, where the isolation between ports is greater than the noise floor of the system. In such a case, the crosstalk cannot be adequately characterized and thus is typically set to zero, reducing the 12-term error model to a 10-term model.

One subtle detail of the error terms is the unfortunate naming of the directivity error term, EDF (see Table 2.1), which in fact does not exactly represent the directivity of the test coupler, but rather the ratio of the leakage signal (from the isolation of the coupler) to the loss to the $a1$ receiver. This is essentially $b1/a1$ when measuring a perfect Z_0 load. In the case where the reflection tracking is 1, the directivity error term is identical to the directivity of the test system. But if the VNA setup causes the ERF term to be offset from 1 (or not 0 dB, for example different losses to the $a1$ and $b1$ receiver), then the $b1/a1$ leakage response to the load will change by that offset value, even though the coupler directivity is not changed (see Figure 1.30). For example, if an attenuator is added between the coupled arm of the port 1 test-port coupler and the test receiver, or between the $a1$ reference coupler and the main arm of the test-port coupler, the measured load response and the EDF term will change, as will the reflection tracking response ERF, but the system directivity is not affected. Thus the EDF error term does not represent the coupler or system directivity; rather, that can be computed as

$$SystemDirectivity = \frac{EDF}{ERF} \qquad (2.9)$$

If an external attenuator is added after the test-port coupler, between the coupler and the DUT, then the EDF term will not change, but the reflection tracking will (by twice the value of the attenuator), and so the effective directivity of the test-port coupler and the system will also be reduced by an amount equal to twice the added attenuation.

In Chapter 3, methods for characterizing these systematic errors, and solving the flow graph for the actual, or corrected, S-parameters will be described. It should be remembered that while the VNA itself may have well-matched ports, and with nearly ideal error terms, the errors in the flow diagram above often occur as a result of cabling, fixturing or wafer-probing of the DUT.

2.3.2 Power Measurements with a VNA

The signal flow diagram of Figure 2.29 is a simplified representation of the signal path that is suitable for S-parameters or other parameters that represent a ratio of a and b waves, where it is presumed that the VNA receivers measure the a_{1M} and b_{1M} directly. In the past, VNAs evaluated only ratio types of parameters. Modern VNAs can be calibrated to report power levels directly, and to control the source and receive powers in sophisticated ways, so that a more detailed signal flow diagram is required to understand the errors associated with parameters that are a direct measurement of the a and b waves at the DUT.

Power measurements of a DUT, strictly speaking, are not linear measurements of a component, but a combination of the DUT and the source signal. For example, the output power of an amplifier depends upon the input power, unless the amplifier is completely saturated. And if the DUT is linear, the power at any port can be computed by the S-parameters and knowledge of the input power. The measurement methods for making power measurements are very similar to making standard S-parameter measurements, and the errors in power measurements are quite closely related to the errors in S-parameter measurements. In most cases, power measurements are required to quantify the non-linear behavior of a DUT, such as the power level where the device changes from linear to non-linear response. Other power-related measurements include power-added efficiency, which measures both DC and RF power, and can compute the efficiency of converting the DC power to RF output signal.

Figure 2.30 shows a signal flow diagram that includes source and receiver error terms for a source that is active in the forward direction. Unlike the S-parameter error terms, error terms associated with the source loss and receiver loss do not have standardized names, so some are defined here and used consistently throughout this text.

The applied source signal (a_{1S}) at the input is measured by the reference receiver, which has its own loss and frequency response described by the reference transmission forward (RTF) error and represents the loss to the reference receiver. The loss of the source path to the test port of the DUT is represented by the source transmission forward (STF) error and represents the loss in the test set. The reflection test receiver (sometimes labeled the A receiver) measures the reflected power with an A-receiver transmission in the Forward flow graph (ATF); the transmission test receiver (sometimes labeled the B receiver) measures the transmitted power with a B-receiver transmission error in the Forward flow graph (BTF). Many manufacturers provide a factory correction of the receivers so that the loss and frequency response of the test and reference coupler are substantially removed. Note that the ATR may be equal to the ATF, but also might be different if there is some switching that occurs in the test set between the forward and reverse directions that changes the loss to the A coupler.

This particular representation of a signal flow graph will be useful in describing the source and receiver power corrections, as discussed in Chapter 3. These error terms are seldom mentioned in the current literature on VNAs and error correction, but aspects of them have been described in several papers. For example, the loss of the reference and test receiver is not explicitly articulated; in some manufacturers' error correction algorithms, the error terms

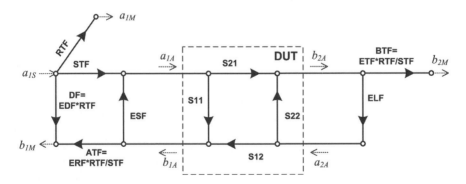

Figure 2.30   Signal flow diagram with source and receiver errors included.

are identified by type and port number and sometimes called the R receiver response tracking (R_{Tr}), or the A or B receiver response tracking (A_{Tr}, B_{Tr}). Other nomenclatures allow the use of the more formally correct, $a1$, $a2$, $b1$, $b2$, response tracking, and some allow the use of either naming convention.

This signal flow diagram separates the individual contributors to the S-parameter tracking terms, ERF and ETF, which can be computed from the receiver tracking terms, and the loss of the source path, as

$$ERF = \frac{STF \cdot ATF}{RTF}, \quad ETF = \frac{STF \cdot BTF}{RTF} \tag{2.10}$$

And with the configuration of the signal flow diagram separating the receiver response, the directivity error term is modified such that

$$EDF \equiv \left. \frac{b_{1M}}{a_{1M}} \right|_{\Gamma_{Load}=0} = \frac{DF}{RTF} \tag{2.11}$$

EDF should be recognized as the raw measurement of b_{1M}/a_{1M} with a perfect non-reflecting load at the test port. If the test-port coupler has coupling of ATF and directivity of DIR, then the isolation of the coupler is

$$CouplerIsolation = DIR \cdot ATF \tag{2.12}$$

EDF will change if the value of STF changes.

In Figure 2.30 the loss of the reference channel splitter or coupler is shown as RTF. In the past, the VNA was designed so that this loss was equal to the combination of STF and ATF for the reflection port, or STF and BTF for the transmission port; as such, losses of ATF and BTF were designed to have the same values; ATF and BTF will inherently take on the same values if identical couplers are used at ports 1 and 2. In this way, the ERF and ETF error terms would approach one, and the raw measured values of the receivers would more nearly match the actual values of the DUT. Modern VNAs sometimes add additional attenuation into the reference path to ensure that even at the highest source powers, the reference channel receiver will not be in compression. This allows extending the dynamic range of the analyzer when the DUT has high loss, described in Chapter 5.

Typically, the values of RTF and BTF are measured at the manufacturer's factory, and the raw values of $a1_M$, $b1_M$, $b2_M$ and $a2_M$ are mathematically adjusted to roughly compensate for this loss. This factory calibration allows raw measurements of a DUT to be close to the actual DUT values, and is helpful in setting up and optimizing a measurement before starting the calibration process.

Similarly, the loss of the source tracking term, STF, is compensated in the source settings so that a_{1S} is adjusted such that the power delivered to the test port is approximately correct. Often, a step attenuator is placed in the STF path, allowing the power at the test port to be substantially lowered while maintaining high power at the reference receiver, a_{1M}. In older analyzers, changing the step attenuator would drop the power into the test receivers while leaving the power in the reference receiver high, causing the apparent value of the S-parameters to change. Modern analyzers compensate the factory calibration by the nominal change in the attenuator, so that the displayed receiver power at the reference receiver is lowered by the attenuator value. Still, the attenuator loss is not identical to its nominal value, and the mismatch of the attenuator changes with the attenuator state, so that substantial ripple is often seen after calibrating in one state and changing to another state. Even these errors can be removed by some clever steps to characterize the attenuator response, as discussed in Chapter 9.

At the test port of a VNA with a factory correction, the source power, receiver power readings and S-parameters typically have less than 1 dB of error. Of course, if a cable is used, the loss of the cable is usually not compensated for in the VNA factory calibrations.

In some cases, additional internal step attenuators are added to the VNA block diagram between the coupled arm of the test-port coupler and the test receivers, which measure b_{1M} and b_{2M}. In these cases, the values of ATF and BTF may also be compensated in the VNA software by the value of the receiver attenuation, allowing the receiver attenuation state to be changed while the raw values of the DUT S-parameters remain relatively the same. In other cases, external attenuators are sometimes placed between the coupled arm of the test-port couplers and the receivers especially in high-power measurement situations, where the power absorbed by the external attenuator exceeds the internal attenuators' power handling. In these cases, the ATF or BTF error terms will incorporate the value of the attenuator offsets, as the factory calibration does not compensate for external components.

2.3.3   Other Measurement Limitations of the VNA

The systematic error terms of Table 2.1 are well known and many methods are used to essentially eliminate the effects of the S-parameter measurement results. However, other hardware limitations in a VNA cannot be so easily removed, and special care must be taken to diminish the effects of these limitations.

2.3.3.1   Noise Floor

The consideration of system noise floor is often not included in the specifications of measurement accuracy, on the assumption that noise effects can be eliminated with sufficient reduction in IF bandwidth or increase in averaging factor. While this is theoretically true, in many circumstances the increase in measurement time makes this impractical. In some situations, such as real-time tuning of microwave filters, the IF bandwidth must be increased to achieve a real-time update rate. The noise effects increase 10 dB for each $10\times$ increase in IF BW, so it is a simple matter to compute the effective noise floor at any IF BW given a noise floor at some IF BW, typically 10 Hz.

There exist two distinct noise effects in S-parameter measurements: noise floor and high-level trace noise. The noise floor is easily understood as the effect of added noise at the input of the receiver, due to the noise figure of the VNA receiver. The coupling factor of the test-port coupler reduces the measured signal further so that the effect of noise floor is more dominant. The effect of noise floor on a measurement can be determined by taking the RMS noise floor, converting it to an equivalent linear amplitude wave, and then adding it to the amplitude of the signal at the measured receiver.

The conversion to the linear b_2 noise is

$$b_{2\_Noise} = 10^{\frac{NoiseFloor_{dBm}}{20}} \tag{2.13}$$

Note that the raw measured noise floor on a VNA receiver will be the square root of the noise power, as the a and b waves are in units of square root of power. Often, the noise floor of a VNA is expressed as a dBc value relative to a 0 dB insertion loss measurement. Of course, for a constant noise power in the receiver, the relative noise floor will depend upon the source drive power.

The RMS trace noise apparent on an S-parameter trace can be computed by adding the RMS noise floor to the amplitude of the signal at the b receiver.

$$TraceNoise_{dB} = 20 \log_{10} \left[\frac{b_{2\_noise} + b_{2\_signal}}{b_{2\_signal}} \right] \tag{2.14}$$

when the noise floor is sufficiently below the measurement of interest. Of course, when the noise floor is above the measured value, the measurement becomes meaningless.

Take, for example, a filter with 80 dB of insertion loss (S21 $= -80$ dB), with a drive power from the source of 0 dBm, a VNA with an RMS noise floor of -127 dBm in a 10 Hz bandwidth. If it is measured using a 10 kHz IF bandwidth, as shown in Figure 2.31, the trace noise due to noise floor at any insertion loss can be computed.

The effective noise floor is 30 dB greater than the 10 Hz spec, for a level of -97 dBm. The measured b_2 noise would be

$$10^{\frac{-97}{20}} = 1.41 \cdot 10^{-5}$$

The output signal is

$$10^{\frac{[-80]}{20}} = 1 \cdot 10^{-4}$$

and the RMS trace noise level would then be

$$TraceNoise = 20 \log \left[\frac{\left(1.41 \cdot 10^{-5} + 1 \cdot 10^{-4} \right)}{1 \cdot 10^{-4}} \right] = 1.15 \, dB_{RMS}$$

This value is very close to the measured trace noise, shown as trace statistics computed near marker 1 on Figure 2.31 and displayed as SDEV $= 1.24$ dB (trace statistics measure the variation of signal of a trace, and in this case the computation is restricted to be a 5% region near the marker position). Thus one would see substantial noise on the filter stopband measurement. The RMS trace noise represents one standard deviation of noise. For this example, about 21 points are used to compute the trace noise near the marker. One would expect a peak-to-peak trace noise of about four standard deviations, worst case, or approximately 4.6 dB of peak-to-peak noise on a typical measurement. However, since noise can take on any value for any single instance, the RMS value is almost always used when describing noise-related values. As the S-parameter signal rises above the noise-floor of the VNA, the trace noise diminishes at a rate of about three times (in dB) for each 10 dB increase in signal level. But this 3-for-10 reduction doesn't continue at high signal levels.

A second cause of trace noise in a measurement is called "high-level" trace noise. At high signal levels, the noise from the source signal, typically due to phase noise of the source, can rise above the VNA noise floor and dominate the trace noise in the measurement. Further, if the source in the VNA has substantial internal amplification, the broadband noise floor from the source can dominate the phase noise far from the carrier. In this region, the trace noise

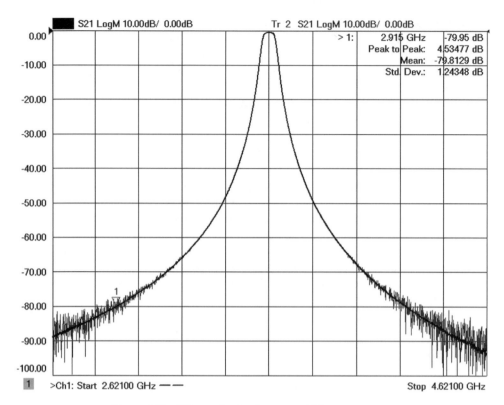

Figure 2.31   Effects of noise floor on an S21 measurement.

stays approximately the same as the S-parameter signal increases. Consider the trace noise on the skirt of a filter: when the signal through the filter is sufficiently high, the trace noise on the measurement decreases as the signal level rises above the noise floor, until the source phase-noise or pedestal noise, as it is sometimes called, becomes dominant. Above this level the trace noise stays constant as a dBc level even as the S-parameter loss diminishes. The problem of high-level trace noise is more commonly found on older VNAs where phase noise was typically worse than that of stand-alone signal sources due to the difficulty of integrating sources internally. The problem is also seen on more modern VNA systems at mm-wave frequency, where multipliers are used to increase the source frequency. With each two-times multiplication of frequency, the phase noise increases by 6 dB. These problems are typically seen only at high power levels because the use of attenuators for power level control reduces the source signal and the phase noise in the same manner.

Figure 2.32 shows the phase noise of a VNA source as it is increased to where the phase noise is higher than the noise floor. The memory trace, in light gray, is for a power level of −10 dBm. At this level, phase noise is below the receiver noise floor. The dark trace shows the phase noise when the source power is increased to +10 dBm. Here the phase noise is about 15 dB above the noise floor, and will limit the high level trace noise. This data was measured with a 10 kHz RBW, so the actual maximum phase noise is about −110 dBc/Hz at offsets below about 50 kHz.

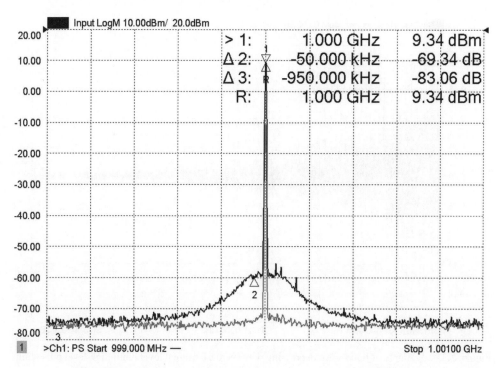

Figure 2.32   VNA source signal where phase noise rises above noise floor.

Figure 2.33 shows a plot of trace noise as a function of received power. In this normalized response, the trace noise limit is apparent in the high level region starting at about −10 dBm, where the trace noise no longer decreases directly as a function of increased signal level, indicated on the figure as the "hi-level trace noise" region.

2.3.4 Limitations Due to External Components

Often, the performance of external components used to connect from the VNA to the DUT present the largest contribution of errors to the measurement. These can come in a variety of configurations, and each has its own peculiarities that can affect measurements in different ways. The most common cause of external errors is due to cables and connectors.

Cables, connectors and adapters are ubiquitous when using VNAs to measure most devices. The quality, and particularly the stability, of the cable and connector can dramatically affect the quality of the measurement.

The first-order effect of cables is added loss and mismatch in a measurement. For short cables, the loss is not significant but the mismatch can add directly to the source match and directivity of the VNA to degrade performance. With error correction, the effect of mismatch can be substantially reduced (to the level of the calibration standards quality) if it is stable, but cable instability limits the repeatability of the cable mismatch and often is the dominant error in a return-loss measurement.

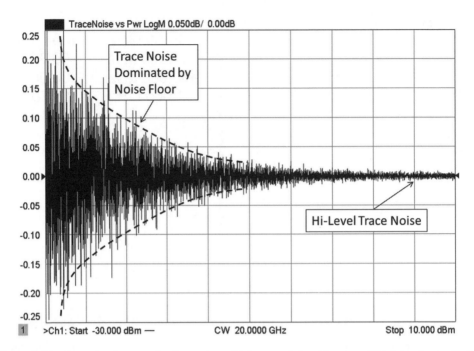

Figure 2.33 Example of trace noise decreasing with increased signal level, until high level noise limit is reached.

For transmission measurements, the effects of mismatch do directly affect calibration, though it is reduced to a small level by the quality of the calibration standards. Often, the major instability in a cable is the phase response versus frequency. Even if the amplitude of the cable is stable, if the phase response changes, the VNA error correction will become corrupted because of phase shift of the cable mismatch error. Methods for determining the quality of the cable and the effects of flexure will be described in Chapter 9.

2.4   Measurements Derived from S-Parameters

S-parameter measurements provide substantial information about the qualities of a DUT. In many cases, the transformation and formatting of these parameters is necessary to more readily understand intrinsic attributes of the DUT. Some of these transformations are graphical in nature, such as plotting on a Smith chart, and some are formatting such as group delay and SWR, and some are functional transformation such as time domain transforms. Some of the more important transformations are discussed below, with emphasis on some particularly interesting results.

2.4.1   The Smith Chart

The Smith chart is a visualization tool which every RF engineer should strive to master. It provides a very compact form for describing the match characteristics of a DUT, as well as being a useful tool for moving the match point of a device to a more desirable value. Invented by Philip Smith [4], it maps the normalized complex value of a termination impedance onto a circular chart, from which the impedance effects of adding lengths of transmission line onto the impedance termination are easily computed. The original intention for the use of a Smith chart was for the computing of impedances presented to a generator as lengths of transmission line were added to a load, and it was intended particularly for the use of telephone line impedance matching. Adding a length of transmission line changes the apparent termination impedance, Z_T, according to

$$Z_S = Z_0 \frac{(Z_0 + Z_T \tanh[(\alpha + j\beta)z])}{(Z_T + Z_0 \tanh[(\alpha + j\beta)z])} \tag{2.15}$$

where α and β are the real and imaginary propagation constants, and z is the distance from the load. This computation was tedious, in part because the argument of the hyperbolic tangent is complex, so a nomographic approach was desirable. A Smith chart solves by this mapping impedance to reflection coefficient (Γ), and plotting the return loss on a polar plot, as

$$\Gamma = \frac{(Z - Z_0)}{(Z + Z_0)} \tag{2.16}$$

The genius of the Smith chart is recognizing that rotating an impedance value through a length of transmission line is just the same as rotating the phase of the reflection coefficient value on the chart. The Smith chart maps the impedance onto the polar reflection coefficient plot, but with the graticule lines marked with circles of constant resistance and circles of constant reactance. As such, any return loss value can be plotted and the equivalent resistance and reactance can be determined immediately. To see the effect of adding some Z_0 transmission line, the impedance is simply rotated on the polar plot by the phase shift of the transmission line. If the line is lossy, the return loss is modified by the line loss (two-times the one-way loss of the line) and from this new position, the resistance and the reactance is directly read.

2.4.1.1 Series and Shunt Elements

The original intent of the Smith chart was to show S11 at a fixed frequency and use the chart to derive the change in impedance due to change in distance from the generator. But the use of the Smith chart in VNAs is different in that the display shows return-loss or S11 as a function of frequency, and the phase rotation displayed is due to phase shift in a transmission line or device caused by the increase in frequency. Various characteristics, such as capacitance, inductance, loss and delay can be directly inferred from the Smith chart trajectory displayed on a VNA, and it is often more informative than just the logmag plot or the phase plot individually. In many instances, the Smith chart is useful for determining the principal component characteristics of the DUT. Since, by most designs, the DUT should ideally be matched, the deviation from matched conditions is due to some parasitic series or shunt element. Series elements show up in a Smith chart trajectory as following a contour of constant resistance. Shunt elements are not intuitively deduced from a Smith chart, but can be deduced from an admittance chart (also called an inverse Smith chart) which follows the same conformal mapping of a Smith chart (impedance chart) but with the inverse of impedance (admittance) displayed as lines of constant conductance or susceptance.

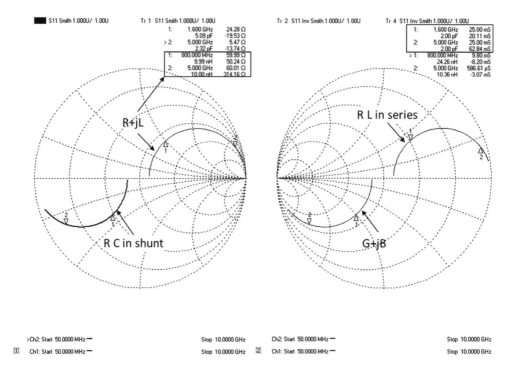

Figure 2.34   Impedance and admittance Smith charts.

For high-frequency measurements, shunt capacitance or series inductance are almost always the parasitic values that must be dealt with. Note that the parasitic effect of a series capacitance or a shunt inductance actually diminishes with frequency, with the capacitor becoming a short, and the inductor an open, and these elements typically cause only low-frequency degradation.

Figure 2.34 shows both an impedance plot (left) and admittance plot (right), each with two circuit elements: a 40 ohm load with a shunt capacitance (dark trace, left), and of a 60 ohm load with a series inductance.

Note that on the impedance chart, the highlighted marker values for the series inductive circuit display constant resistance and inductance, even as the frequency varies. The impedance plot of the shunt RC circuit (dark trace) does not show either constant resistance or constant capacitance. However, the same trace on the admittance chart (right) does show constant conductance and constant capacitance, whereas the series inductive LR circuit does not show constant value. From these charts, it is clear that the impedance chart allows determination of series elements easily as their trajectory follows constant resistance circles, and the admittance chart allows determination of shunt elements as their trajectory follows constant conductance circles.

In evaluating practical responses for input impedance, it is often the case that the series or shunt element is at the end of some short, or long, transmission line which wraps the response around the Smith chart. In this case, for useful information to be discerned, it is necessary to remove excess delay from the measurement. However, it is sometimes difficult to know the exact delay which must be removed. In such a case, it is possible to use two marker readouts

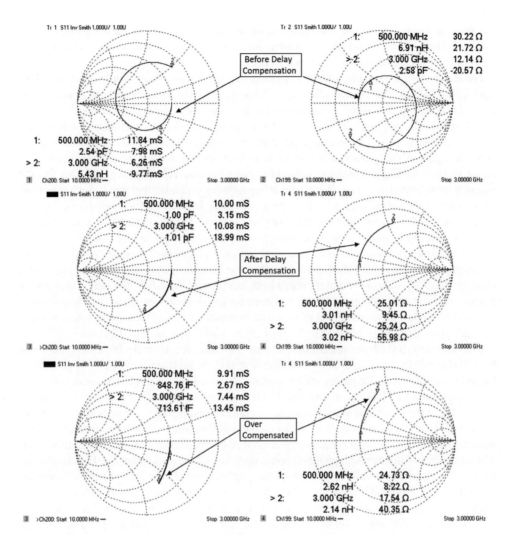

Figure 2.35 Left: Admittance chart, and Right: Smith chart, with wrapped phase (upper) and unwrapped phase (middle) and overcompensated (lower); for an inductor (right) and a capacitor (left).

of the VNA to attempt to determine the value of a parasitic element by means of unwrapping the phase of the response until the underlying value is determined. Figure 2.35 shows a set of responses where the DUT characteristic is delayed by a short length of transmission line, as might be found in a PCB fixture or on-wafer probe, followed by a series resistance and inductance (right, upper) or a shunt capacitance and conductance (left, upper). These are the two most common cases of parasitic characteristics.

The trajectories are distorted by a frequency dependent phase shift due to the delay of a portion of transmission line between the VNA reference port and the parasitic. For the inductive case, the elements are series elements, and the normal Smith chart, or impedance chart, is shown. The marker readout gives the resistance, in ohms, the reactance, in ohms, and the equivalent inductance. It is clear that the apparent value of the reactive element is not constant. Most VNAs include the equivalent inductance or capacitance value associated with the reactance for the marker position and frequency.

For the capacitive example, since it is a shunt-impedance, an inverse Smith chart, or admittance chart, is used. The value displayed for the real part of the admittance is the conductance, in millisiemens (or mS, inverse ohms), and for the imaginary part is the susceptance, also in millisiemens. The reactive part is converted to an equivalent shunt capacitance or inductance, determined by the sign of the imaginary part of the admittance. Again, it is clear that the apparent value of the shunt reactive element is not constant. In fact, both trajectories show an attribute of having a resonance, since they cross the real axis. However, the fact that the magnitude of reflection is not a minimum at the crossing indicates that this is not a true resonant structure, but rather a device whose phase response is distorted by a length or delay of a transmission line between the measurement plane and the discrete impedance or admittance.

It is reasonably simple to investigate the effect of removing the delay, by using two markers, spaced in frequency. By reading the value of the imaginary element of each marker while adding in electrical delay, the phase shift of the delay line can be removed and the resulting underlying element characteristics are revealed. When both marker readings show the same value for the reactive element, then the proper delay has been removed, as shown in the middle portion of Figure 2.35. In this case, the left plots give a capacitance of 1 pf in shunt with 100 ohms, and the right plots show an inductance of 3 nH in series with a resistance of 25 ohms.

The lower traces show the same measurement, but with even more electrical delay removed from the response. Electrical delay is a common scaling function in VNAs that provides a linear phase shift versus frequency for any particular trace. A related function is port-extension, which also provides a phase shift, but that shift is associated with the port of the analyzer, rather than with just the particular trace. With electrical delay scaling, only the trace which is active has the delay applied, and different traces of the same parameter can have different delays. With port extension, all traces that are associated with a particular port, for example S11 and S21 with port 1, will have their phase response modified by the port extension. Electrical delay applies the same phase shift regardless of the parameter type, but port extensions properly accounts for a two-times phase shift for reflection parameters in contrast with a one-times phase shift for transmission parameters. Therefore, it is perhaps better to use port extension to accommodate changes in reference plane, and reserve electrical delay when one wants to remove the linear phase shift of a particular parameter.

The delay or port extension is adjusted until the trace rotation is minimized, while still maintaining a trajectory that follows a clockwise rotation. Foster demonstrated that all real devices should have phase that increases with frequency causing the clockwise rotation, so the proper amount of delay to be removed can often be determined by looking at the rotation direction of the trace trajectory. This is demonstrated in the lower traces of Figure 2.35, where an additional 10% of the delay from the middle traces has been removed, making the response overcompensated, and causing reactive element value at the two markers to be different.

2.4.1.2 Impedance Transformation

One aspect of rotation on the Smith chart that is often misunderstood is that the rotation about the *center* of the chart for a transmission line delay only occurs if the transmission

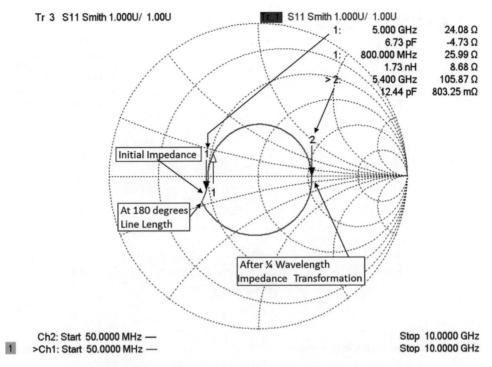

Tr 3 S11 Smith 1.000U/ 1.00U

S11 Smith 1.000U/ 1.00U

1:	5.000 GHz	24.08 Ω
	6.73 pF	-4.73 Ω
1:	800.000 MHz	25.99 Ω
	1.73 nH	8.68 Ω
>2:	5.400 GHz	105.87 Ω
	12.44 pF	803.25 mΩ

Initial Impedance

At 180 degrees
Line Length

After ¼ Wavelength
Impedance Transformation

Ch2: Start 50.0000 MHz —   Stop 10.0000 GHz
>Ch1: Start 50.0000 MHz —   Stop 10.0000 GHz

Figure 2.36   An impedance value rotated by 180°. 50 ohm line.

line impedance matches the reference impedance of the Smith chart. Consider the case of a termination consisting of a 25 ohm resistor to ground, shunted by a 3 pF capacitor, and evaluated from DC to 10 GHz. The impedance trajectory is shown in the light trace in Figure 2.36, and shows a small deviation from 25 ohms due to the shunt capacitance. The darker trace shows the same impedance, but at the end of a transmission line that has 180° of phase shift at 10 GHz. The value of the impedance trajectory centers on 50 ohms, and the value of the trace at 180° phase shift matches that exactly of zero phase shift. At the frequency where the phase shifts 90° due to the transmission line (5.4 GHz) plus the slight phase shift of the DUT, the impedance is nearly 100 ohms. This is a well-known aspect of quarter-wave (or 90°, or $\lambda/4$) transmission line transformers. If impedance of the line is Z_0, then the impedance at the end of a quarter-wave section is.

$$Z_{\lambda/4} = \frac{Z_0^2}{Z_T} \tag{2.17}$$

One consequence of this is that the maximum deviation of impedance due to a transmission line depends completely on the impedance of the transmission line. Figure 2.37 shows the Smith chart trajectories for the same termination, but this time with a 12.5 ohm line, a 25 ohm line and a 100 ohm line before the termination. Of course, at 180°, no transformation of impedance takes place and the impedance value at the end of the line is identical to that at the 0° phase shift. It is interesting to note that the smallest deviation of impedance is for the case

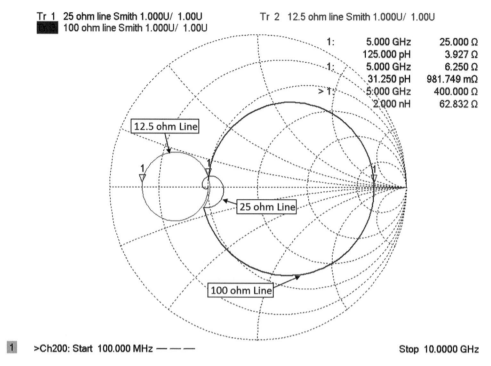

Figure 2.37 Twenty-five ohm termination proceeded by half-wavelength segments of 12.5 ohm, 25 ohm and 100 ohm lines.

where the line matches the impedance of the termination, rather than matching the system impedance as appears in Figure 2.36.

The other important aspect to note is that when the transmission line is of greater impedance than Z_L the resulting impedance will transform to a higher value, while when the transmission line is of lower impedance, the resulting impedance will be lower than Z_L.

2.4.2 Transforming S-Parameters to Other Impedances

While it is most common to define S-parameters in a 50 ohm impedance, or 75 ohms for cable-television applications, situations arise where it necessary to define an S-parameter matrix in other than 50 ohms, or to have it defined with 50 ohms on one port and a different impedance on another port. This requirement occurs for matching circuits, or impedance transformers, as well as for the use of waveguide adapters where it is common practice to define the terminal impedance as 1 ohm. Unfortunately, while S-parameter definitions don't prohibit different impedances on different ports, the most common data files for S-parameters, the so-called Touchstone™ or S2P file, provides for only a single impedance in its definition. (Recently a second revision of the S2P file format has been defined that allows different impedances on different ports, but it has not yet been widely implemented.) Thus, it is often necessary to transform S-parameters from one reference impedance to another. If the complete S-parameter matrix is available, then a matrix transformation [5] can be used to convert the impedance of by applying the following

$$[S'] = [X]^{-1} \, ([S] - [\Gamma]) \, ([I] - [\Gamma][S])^{-1} \, [X]$$

where

$$[X] = \begin{bmatrix} x_1 & 0 & 0 & 0 \\ 0 & x_2 & 0 & 0 \\ 0 & 0 & \ddots & 0 \\ 0 & 0 & 0 & x_n \end{bmatrix}, \quad x_n = 1 - \Gamma_n$$

(2.18)

$$[\Gamma] = \begin{bmatrix} \Gamma_1 & 0 & \cdots & 0 \\ 0 & \Gamma_2 & \cdots & 0 \\ \vdots & \vdots & \ddots & 0 \\ 0 & 0 & \cdots & \Gamma_n \end{bmatrix}, \quad \Gamma_n = \frac{Z'_n - Z_n}{Z'_n + Z_n}$$

This is a generalized formula, so that an impedance, $[Z_n]$, may be defined for any port of the original $[S]$ matrix and any other impedance $[Z'_n]$ may defined for any other port for the new $[S']$ matrix, but the two most common cases are where the transformation occurs for all impedances at the ports being equal, so that each element in the X matrix and Γ matrix are identical, or the two-port case where only one impedance is transformed, as when the S-parameters of a network are defined in two different impedances.

If the measurement system impedance is pure real, an alternative method for obtaining S-parameters at a different real impedance than the measurement system is to de-embed an ideal transformer at each port, with the turns ratio set to the square root of the impedance change. De-embedding methods are discussed in Chapter 9.

2.4.3   Concatenating Circuits and T-Parameters

In many instances, it is convenient to concatenate devices, and signal-flow charts provide a useful tool for understanding the interactions and determining the resulting S-parameter matrix. With appropriate transformations, the concatenation of S-parameter devices can be greatly simplified. One such transformation is from S-parameters to T-parameters, which also depend upon the wave functions but in a different relationship.

Figure 2.38 shows a concatenation of two devices, with a and b waves for each independently identified. Using normal signal-flow properties, the combined S-parameters of two devices is

$$\begin{bmatrix} S_{11} & S_{12} \\ S_{21} & S_{22} \end{bmatrix} = \begin{bmatrix} S_{11A} + \dfrac{S_{11B} \cdot S_{21A} \cdot S_{12A}}{(1 - S_{22A} \cdot S_{11B})} & \dfrac{S_{21A} \cdot S_{21B}}{(1 - S_{22A} \cdot S_{11B})} \\ \dfrac{S_{12A} \cdot S_{12B}}{(1 - S_{22A} \cdot S_{11B})} & S_{22B} + \dfrac{S_{22A} \cdot S_{21B} \cdot S_{12B}}{(1 - S_{22A} \cdot S_{11B})} \end{bmatrix}$$

(2.19)

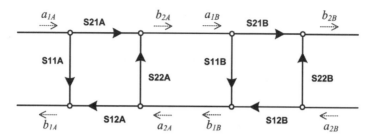

Figure 2.38 Concatenation of two devices.

However, signal-flow graph techniques get tedious for concatenating a long series of devices, and other transformations make this work easier and more programmatic.

The T-parameters [6] create a new functional relationship between input and output waves, with the independent variables being waves on the right, and the dependent variables being waves on the left

$$b_1 = T_{11}a_2 + T_{12}b_2$$
$$a_1 = T_{21}a_2 + T_{22}b_2 \tag{2.20}$$

or in the matrix form

$$\begin{bmatrix} b_1 \\ a_1 \end{bmatrix} = \begin{bmatrix} T_{11} & T_{12} \\ T_{21} & T_{22} \end{bmatrix} \begin{bmatrix} a_2 \\ b_2 \end{bmatrix} \tag{2.21}$$

From this, the T-matrix describing the first and second devices is

$$\begin{bmatrix} b_{1A} \\ a_{1A} \end{bmatrix} = \begin{bmatrix} T_{11A} & T_{12A} \\ T_{21A} & T_{22A} \end{bmatrix} \begin{bmatrix} a_{2A} \\ b_{2A} \end{bmatrix}, \quad \begin{bmatrix} b_{1B} \\ a_{1B} \end{bmatrix} = \begin{bmatrix} T_{11B} & T_{12B} \\ T_{21B} & T_{22B} \end{bmatrix} \begin{bmatrix} a_{2B} \\ b_{2B} \end{bmatrix} \tag{2.22}$$

And from inspection, one can recognize that the waves $a_{2A} = b_{1B}$ and $b_{2A} = a_{1B}$, so that the concatenation becomes the simple result:

$$\begin{bmatrix} b_{1A} \\ a_{1A} \end{bmatrix} = \begin{bmatrix} T_{11A} & T_{12A} \\ T_{21A} & T_{22A} \end{bmatrix} \begin{bmatrix} T_{11B} & T_{12B} \\ T_{21B} & T_{22B} \end{bmatrix} \begin{bmatrix} a_{2B} \\ b_{2B} \end{bmatrix} \tag{2.23}$$

or

$$\begin{bmatrix} b_{1A} \\ a_{1A} \end{bmatrix} = [T_A] \, [T_B] \begin{bmatrix} a_{2B} \\ b_{2B} \end{bmatrix} \tag{2.24}$$

Using this definition of T-parameters, the following conversions can be defined

$$\begin{bmatrix} T_{11} & T_{12} \\ T_{21} & T_{22} \end{bmatrix} = \frac{1}{S_{21}} \begin{bmatrix} -(S_{11}S_{22} - S_{21}S_{12}) & S_{11} \\ -S_{22} & 1 \end{bmatrix}, \quad \begin{bmatrix} S_{11} & S_{12} \\ S_{21} & S_{22} \end{bmatrix} = \frac{1}{T_{22}} \begin{bmatrix} T_{12} & (T_{11}T_{22} - T_{21}T_{12}) \\ 1 & -T_{21} \end{bmatrix} \tag{2.25}$$

Note that in this conversion, S21 always appears in the denominator. This can cause numerical difficulties in devices with transmission zeros, and sometimes causes de-embedding functions to fail. More robust de-embedding algorithms check for this condition and modify the method of concatenation in such a case.

Other definitions of T-parameter type relationships have been described, which exchange the position variables a_1 and b_1 on the dependent variable side, and the position of a_2 and b_2 on the independent variable side [7]. This version has similar properties, but care must be taken not to confuse the two methods as, of course, the resulting T-parameters are different. Another definition, which might seem more intuitive, would set the input terms a_1 and b_1 as the independent variable. Unfortunately, this has the undesirable effect of setting S12 in the denominator of the transformation parameter and thus gives difficulties when applied to unilateral gain devices such as amplifiers.

2.5   Modeling Circuits Using Y and Z Conversion

One common desire in evaluating the performance of a component is to model that component as an impedance composed of a resistive element with a single series or shunt reactive element, as demonstrated in Section 2.4.1.1. This desire was furthered by some built-in transformation functions on VNAs, first introduced with the HP8753A but now common on many models. The goal was to model a device in such a way that the S-parameters mapped to a single resistive and reactive element in the so-called Z-transform case (not to be confused with the discrete time Z-transform), or a single conductance and susceptance in the Y-transform case. These are quite simple models, and are represented in Figure 2.39.

2.5.1   Reflection Conversion

Reflection conversions are computed from the S11 trace, and are essentially the same values as presented by impedance or admittance readouts of the Smith chart markers. Thus, Z-reflection

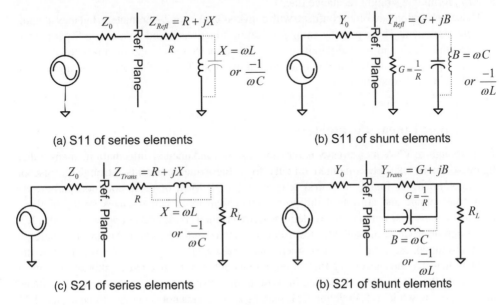

Figure 2.39   Y and Z conversion circuits.

conversion would be used with the circuit description from Figure 2.39a, and display the impedance in the real part of the result and the reactance in the imaginary part of the result. Y-reflection would be used with the circuit of Figure 2.39b, and display the conductance in the real part and the susceptance in the imaginary part of the result. The computations for these conversions are

$$Z_{Refl} = Z_0 \frac{(1 + S_{11})}{(1 - S_{11})} , \quad Y_{Refl} = \frac{(1 - S_{11})}{Z_0 (1 - S_{11})} \tag{2.26}$$

Typically, these conversions would be used on one-port devices and measurements. If it is used on a two-port device, one must remember that the load impedance will affect the measured value of the Z- or Y-reflected conversion.

2.5.2 Transmission Conversion

These reflection conversions are already well known as the models represented by the Smith chart, but a similar conversion can be performed for a simple transmission measurement. In this case, the circuits of Figure 2.39c and d are the reference circuits for these conversions. They are useful when analyzing the series element models such as coupling capacitors, and the models for series resistors and inductors. The underlying computation for the transmission conversions are

$$Z_{Trans} = Z_0 \frac{2 (1 - S_{21})}{S_{21}} , \quad Y_{Trans} = \frac{S_{21}}{2Z_0 (1 - S_{21})} \tag{2.27}$$

The Z-transmission conversion would be well suited to viewing the series resistance of a coupling capacitor. The Y-transmission would show the resistive value of a series-mounted SMT resistor with a shunt capacitance as a constant conductance with a reactance increasing as $2\pi f$, forming a straight reactance line.

These conversions are often confused with conversion to Y- or Z-parameters, but they are not, in general, related. These provide simple modeling functions based on a single S-parameter, whereas the Y-, Z- and related parameters provide a matrix result and require knowledge of all four S-parameters as well as the reference impedance. These other matrix parameters are described in the next section.

2.6 Other Linear Parameters

Even though a VNA measures S-parameters as its fundamental information, many other figures-of-merit may be computed directly from these measurements, through the use of the transformations found in several references [8, 9]. Most of these common parameters relate the voltage and current at the ports, rather than the a and b waves. Many of these transformations arise out of different definitions of terminal conditions as applied to Figure 1.2. These definitions arise out of DC or low frequency measurements, where it is an easy matter to short a terminal, meaning $Z_L = 0$, or open a terminal meaning $Z_L \to \infty$. An often confusing point is that it is not necessary that the terminals actually be opened or shorted, but most commonly the parameter is described in those terms. Just as it is most common to terminate the two-port network in Z_0 to define S21, making $a_2 = 0$, it is not necessary to do so, and S21 can be determined with any terminal impedance, as long as sufficient changes in *a1* and *a2* are made to solve Eq. (1.17) as shown in Eq. (1.21). Since the voltage and current relationships on the terminals of a DUT are easily determined from the S-parameters, many other linear parameters can be determined as well. Unless otherwise noted, these transformations apply to the simple case where the S-parameters are defined with a single, real-valued reference impedance.

2.6.1   Z-Parameters, or Open-Circuit Impedance Parameters

Z-parameters are one of the more commonly defined parameters, and often the first characterization parameter introduced in engineering courses on electrical circuit fundamentals.

The Z parameters are defined in terms of voltages and currents on the terminals as

$$V_1 = Z_{11} \cdot I_1 + Z_{12} \cdot I_2$$
$$V_2 = Z_{21} \cdot I_1 + Z_{22} \cdot I_2$$

(2.28)

where the V_N and I_N are defined in Figure 1.2. If we apply the condition of driving a voltage source into the first input terminal and opening the first output terminal, which forces I_2 to zero, and measure the input and output voltages, we can determine two of the parameters; similarly the other two parameters are determined by driving the output terminal, and opening the input terminal. Mathematically this can be stated as

$$Z_{11} = \frac{V_1}{I_1}\bigg|_{I_2=0} \qquad Z_{12} = \frac{V_1}{I_2}\bigg|_{I_1=0}$$

$$Z_{21} = \frac{V_2}{I_1}\bigg|_{I_2=0} \qquad Z_{22} = \frac{V_2}{I_2}\bigg|_{I_1=0}$$

(2.29)

But, the conditions for measurement of these parameters directly cannot be realized in RF and microwave systems for several key reasons:

1. When the ports of an RF circuit are left open, phase shift from the DUT reference plane, plus fringing capacitance from the center pin of the RF terminal to ground, reduces the impedance at high frequency, which makes the practical value of an open circuit deviate from the ideal.
2. The measuring equipment to sense V_1, I_1, V_2 and I_2 itself has a parasitic impedance to ground which also shunts some of the terminal current. At the driving port, this means that the measured current does not match the actual current into the DUT, and at the open port it means that while being measured, the output impedance does not match that of an open circuit.
3. For many active devices, the DUT is only conditionally stable and may oscillate if larger reflections are presented at the ports. Phase shift of the open circuit from the terminal port to the DUT active device can cause the reflection to take on almost any phase, and this ensures that, at some frequency, the reflection at the port will be such that the device will oscillate. This is perhaps the primary reason for S-parameters being used on active devices; they provide a consistent low-reflection load which in general prevents oscillations of the DUT.

Of course, Z-parameters are not restricted to just two ports, and the Z-parameters can be put in a matrix form as

$$
\begin{bmatrix} V_1 \\ \vdots \\ V_n \end{bmatrix} = \begin{bmatrix} Z_{11} & \cdots & Z_{1n} \\ \vdots & \ddots & \vdots \\ Z_{1n} & \cdots & Z_{nn} \end{bmatrix} \cdot \begin{bmatrix} I_1 \\ \vdots \\ I_n \end{bmatrix} \quad \text{or} \quad [V_n] = [Z] \cdot [I_n] \tag{2.30}
$$

where $[Z]$ is called the Z-matrix.

The Z-matrix and S-matrix can be computed from each other, provided the reference impedance is known for the S-parameters, and the same on each port:

$$
\begin{bmatrix} Z_{11} & Z_{12} \\ Z_{21} & Z_{22} \end{bmatrix} = \frac{Z_0}{\Delta S} \begin{bmatrix} (1+S_{11})(1-S_{22}) + S_{21}S_{12} & 2S_{12} \\ 2S_{21} & (1-S_{11})(1+S_{22}) + S_{21}S_{12} \end{bmatrix} \tag{2.31}
$$

$$
\text{where} \quad \Delta S = (1-S_{11})(1-S_{22}) - S_{21}S_{12}
$$

$$
\begin{bmatrix} S_{11} & S_{12} \\ S_{21} & S_{22} \end{bmatrix} = \frac{1}{\Delta Z} \begin{bmatrix} (Z_{11}-Z_0)(Z_{22}+Z_0) - Z_{21}Z_{12} & 2Z_{12}Z_0 \\ 2Z_{21}Z_0 & (Z_{11}+Z_0)(Z_{22}-Z_0) - Z_{21}Z_{12} \end{bmatrix} \tag{2.32}
$$

$$
\text{where} \quad \Delta Z = (Z_{11}+Z_0)(Z_{22}+Z_0) - Z_{21}Z_{12}
$$

An attribute of the Z-matrix is that if a DUT is lossless, the Z-matrix will contain only pure imaginary numbers; this is commonly found in filter design applications. If $Z_{21} = Z_{12}$ then the DUT is reciprocal, and if also $Z_{11} = Z_{22}$, the network is symmetrical. Note that in general $Z_{1n} \neq Z_{11}$ except for a one-port network. Z_{1n} represents the ratio of V_1 and I_1 for the DUT as it is terminated, normally in the system reference impedance Z_0, where Z_{11} is the ratio of V_1 and I_1, when all the other ports are open-circuited – a not very useful case in practice. Another important attribute of the Z-matrix is that its values do not depend upon the measurement system, unlike S-parameters whose values depend upon the reference impedance for each port, and whose values can change for the same network if these reference impedances change. Put another way, the S11 of a 50 ohm load will be quite different when measured in a 75 ohm reference impedance, but the Z-parameters will not change.

2.6.2 Y-Parameters, or Short-Circuit Admittance Parameters

Y-parameters are essentially an inverse of Z-parameters, and in fact the Y-matrix is the inverse of the Z-matrix. The definition of Y-parameters is from

$$
\begin{aligned}
I_1 &= Y_{11} \cdot V_1 + Y_{12} \cdot V_2 \\
I_2 &= Y_{21} \cdot V_1 + Y_{22} \cdot V_2
\end{aligned} \tag{2.33}
$$

And from this the common description of Y-parameters is

$$Y_{11} = \frac{I_1}{V_1}\bigg|_{V_2=0} \qquad Y_{12} = \frac{I_1}{V_2}\bigg|_{V_1=0}$$

$$Y_{21} = \frac{I_2}{V_1}\bigg|_{V_2=0} \qquad Y_{22} = \frac{I_2}{V_2}\bigg|_{V_1=0} \tag{2.34}$$

The Y-parameters can also be defined for more-than-two-port devices, and the matrix form is

$$\begin{bmatrix} I_1 \\ \vdots \\ I_n \end{bmatrix} = \begin{bmatrix} Y_{11} & \cdots & Y_{1n} \\ \vdots & \ddots & \vdots \\ Y_{1n} & \cdots & Y_{nn} \end{bmatrix} \cdot \begin{bmatrix} V_1 \\ \vdots \\ V_n \end{bmatrix} \qquad \text{or} \qquad [I_n] = [Y] \cdot [V_n] \tag{2.35}$$

where the Y-matrix refers to $[Y]$. The Y-matrix is related to the Z-matrix through its inverse

$$[Y] = [Z]^{-1} \tag{2.36}$$

The conversions between the S-matrix and the Y-matrix are

$$\begin{bmatrix} Y_{11} & Y_{12} \\ Y_{21} & Y_{22} \end{bmatrix} = \frac{Y_0}{\Delta_Y S} \begin{bmatrix} (1-S_{11})(1+S_{22}) + S_{21}S_{12} & -2S_{12} \\ -2S_{21} & (1+S_{11})(1-S_{22}) + S_{21}S_{12} \end{bmatrix} \tag{2.37}$$

$$\text{where} \quad \Delta_Y S = (1+S_{11})(1+S_{22}) - S_{21}S_{12}$$

$$\begin{bmatrix} S_{11} & S_{12} \\ S_{21} & S_{22} \end{bmatrix} = \frac{1}{\Delta Y} \begin{bmatrix} (Y_0 - Y_{11})(Y_0 + Y_{22}) + Y_{21}Y_{12} & -2Y_{12}Y_0 \\ -2Y_{21}Y_0 & (Y_0 + Y_{11})(Y_0 - Y_{22}) + Y_{21}Y_{12} \end{bmatrix} \tag{2.38}$$

$$\text{where} \quad \Delta Y = (Y_0 + Y_{11})(Y_0 + Y_{22}) - Y_{21}Y_{12}$$

2.6.3 ABCD Parameters

Just as the T-parameters provide for easy concatenation of devices using a and b waves, a similar matrix representation can be used when the terminal characteristics are defined in terms of voltage and current. These are sometimes called transfer parameters (reminiscent of the T-parameters) or chain parameters as networks can be chained together in a matrix multiplication manner.

The functional definition of ABCD parameters is found in at least two different forms, one of which is

$$V_1 = A \cdot V_2 - B \cdot I_2$$

$$I_1 = C \cdot V_2 - D \cdot I_2 \tag{2.39}$$

The second form replaces the minus sign with a plus sign with resulting changes in the derived values.

From (2.39) the values for ABCD parameters can be defined as

$$A = \frac{V_1}{V_2}\bigg|_{I_2=0} \qquad B = \frac{V_1}{-I_2}\bigg|_{V_2=0}$$

$$C = \frac{I_1}{V_2}\bigg|_{I_2=0} \qquad D = \frac{I_1}{-I_2}\bigg|_{V_2=0}$$

(2.40)

The transformations between the ABCD-matrix and the S-matrix are

$$\begin{bmatrix} A & B \\ C & D \end{bmatrix} = \frac{1}{2S_{21}} \begin{bmatrix} (1+S_{11})(1-S_{22}) + S_{21}S_{12} & Z_0\left[(1+S_{11})(1+S_{22}) - S_{21}S_{12}\right] \\ \dfrac{1}{Z_0}\left[(1-S_{11})(1-S_{22}) - S_{21}S_{12}\right] & (1-S_{11})(1+S_{22}) + S_{21}S_{12} \end{bmatrix}$$

(2.41)

$$\begin{bmatrix} S_{11} & S_{12} \\ S_{21} & S_{22} \end{bmatrix} = \frac{1}{\Delta} \begin{bmatrix} A + \dfrac{B}{Z_0} - CZ_0 - D & 2\,(AD - BC) \\ 2 & -A + \dfrac{B}{Z_0} - CZ_0 + D \end{bmatrix}$$

(2.42)

$$where \quad \Delta = A + \frac{B}{Z_0} + CZ_0 + D$$

2.6.4   H-Parameters or Hybrid Parameters

Due to its intrinsic transfer function, as a voltage-controlled current source, transistor performance has often been described using hybrid parameters. Their functional definition is

$$V_1 = H_{11} \cdot I_1 + H_{12} \cdot V_2$$

$$I_2 = H_{21} \cdot I_1 + H_{22} \cdot V_2$$

(2.43)

from which the definitions of individual H-parameters follow

$$H_{11} = \frac{V_1}{I_1}\bigg|_{V_2=0} \qquad H_{12} = \frac{V_1}{V_2}\bigg|_{I_1=0}$$

$$H_{21} = \frac{I_2}{I_1}\bigg|_{V_2=0} \qquad Y_{22} = \frac{I_2}{V_2}\bigg|_{I_1=0}$$

(2.44)

The H-matrix is most simply defined in terms of the other impedance matrixes as

$$[H] = \begin{bmatrix} \dfrac{1}{Y_{11}} & Z_{12} \\ \dfrac{1}{D} & Z_{11} \end{bmatrix} = \begin{bmatrix} Z_0 \dfrac{(1+S_{11})(1+S_{22}) - S_{21}S_{12}}{(1-S_{11})(1+S_{22}) + S_{21}S_{12}} & \dfrac{2 \cdot S_{12}}{(1-S_{11})(1+S_{22}) + S_{21}S_{12}} \\ \dfrac{-2 \cdot S_{21}}{(1-S_{11})(1+S_{22}) + S_{21}S_{12}} & \dfrac{1}{Z_0} \cdot \dfrac{(1-S_{11})(1-S_{22}) - S_{21}S_{12}}{(1-S_{11})(1+S_{22}) + S_{21}S_{12}} \end{bmatrix}$$

(2.45)

2.6.5   Complex Conversions and Non-Equal Reference Impedances

It is important to note that all the conversions described in the previous sections are valid only for the case of $Z_{01} = Z_{02} = Z_0$ and that Z_0 be pure real. The transformations for cases where the port impedances are not equal, and not real, have been computed and are available in papers by Marks and Williams [10] and, with using somewhat different wave definitions, by D.A. Frickey [11]. While the case of complex termination impedances is unusual, the case for differing reference impedances on ports is more common. Since the network elements do not change when the reference impedance of the system is changed, Y-, Z- and H-related parameters also do not change with reference impedance. S- and T- parameter values do change, and thus it is critical to know the reference impedance for each case.

References

1. Johnson, R.A. (1975) Understanding microwave power splitters. *Microwave Journal*, **12**, 40–51.
2. Fitzpatrick, J. (1978) Error models for system measurement. *Microwave Journal* vol. 21, pp. 63–66, May 1978 http://bit.ly/ICm9O4.
3. Rytting, D. (1996) Network Analyzer Error Models and Calibration Methods. RF 8 Microwave Measurements for Wireless Applications (ARFTG/NIST Short Course Notes).
4. Smith, P.H. (1944) An Improved Transmission Line Calculator, *Electronics*, vol. 17, 8-66–69, 1944.
5. Tippet, J.C. and Speciale, R.A. (1982) A rigorous technique for measuring the scattering matrix of a multiport device with a 2-port network analyzer. *IEEE Transactions on Microwave Theory and Techniques*, **30**(5), 661–666.
6. Agilent Application Note 154, http://cp.literature.agilent.com/litweb/pdf/5952-1087.pdf.
7. Mavaddat, R. (1996) *Network Scattering Parameters*, World Scientific, Singapore. Print.
8. Hong, J.-S. and Lancaster, M.J. (2001) *Microstrip Filters for RF/microwave Applications*, Wiley, New York.
9. Agilent Application Note AN-95-1, http://contact.tm.agilent.com/Agilent/tmo/an-95-1/index.html, original form can be found at http://cp.literature.agilent.com/litweb/pdf/5952-0918.pdf.
10. Marks, R.B. and Williams, D.F. (1992) A general waveguide circuit theory. *Journal of Research of the National Institute of Standards and Technology*, **97**, 533–561.
11. Frickey, D.A. (1994) Conversions between S, Z, Y, H, ABCD, and T parameters which are valid for complex source and load impedances. *IEEE Transactions on Microwave Theory and Techniques*, **42**(2), 205–211.

3

Calibration and Vector Error Correction

3.1 Introduction

The VNA is perhaps the most precise electronic instrument used in RF and microwave measurements. Modern VNAs can measure signals of high and low power with better precision than any other power sensor, and they can measure the gain across frequency of an electronic device with a performance traceable to measurements of physical dimensions. Of all the electronic measurement systems, the VNA derives the largest share of its performance and quality of measurement from error correction. However, when one speaks about calibration in reference to a VNA, it is in a different context from other instruments, and this is a common point of confusion. Also, VNAs provide correction for both magnitude and phase responses, which is commonly called vector error correction. Here, the term error correction should be understood to refer in general to vector correction methods.

Most other electronic test instruments are calibrated to their performance through a process of careful measurement with other electronic equipment of a higher quality, typically once a year. This process ensures that the equipment is meeting its specified performance, and in some instances the performance is measured and adjustments are made to optimize the performance. In the specific case of noise sources that have an excess noise ratio (ENR) data table, power sensors that have a flatness-versus-frequency coefficient table and other equipment that maintains correction arrays associated their performance the values for these corrections are updated when the equipment is sent in for its periodic calibration. In the case where an instrument cannot meet its performance specifications even after adjustment, repairs are made typically in the form of replacing defective modules in the instrument. For most electronic instruments, their performance pertains only to measurements made directly at the connector of the test equipment. This calibration process concludes with a reference sticker that states that the equipment is in calibration, with a recommended date for recertification.

Most VNAs go through a similar process, where the raw hardware performance characteristics are similarly characterized. But for VNAs the actual quality of the hardware, such as frequency flatness or coupler directivity, can be easily characterized in the field, and their effects as well as the effects of connectors, cables, fixtures and probes can be removed, to

Handbook of Microwave Component Measurements: With Advanced VNA Techniques, First Edition. Joel P. Dunsmore.
© 2012 John Wiley & Sons, Ltd. Published 2012 by John Wiley & Sons, Ltd.

yield a resulting measurement of the DUT itself that far surpasses the capability of the raw hardware. This process is commonly calibration, but a more proper term is error-correction. Calibration implies that measurements are made to characterize a performance parameter, and then adjustments are made to improve the actual performance. In contrast, traditional error correction in a VNA is strictly a post-processing function, where error correction algorithms are applied post-measurement to raw measured data to produce the corrected result.

The VNA error correction process consists of two steps: the first step, typically called calibration or VNA-cal, is to characterize known standards, such as an open/short/load, to determine the systematic VNA error terms. This might be formally called "error correction acquisition". The second step is measuring the DUT and applying the error correction algorithms to obtain a corrected result. This might be formally called "error correction application" or more informally just "correction". These processes are unrelated to the yearly calibration process, and are sometimes called user-calibration to indicate that it is done at the time of use, rather than at the instrument manufacturer's factory or service center.

Finally, some modern VNAs have had the error-correction acquisition process performed in the factory before being shipped, so that they have a built-in "factory cal" that can be applied to raw measurements even if a user does not perform the VNA-cal acquisition. In at least one example – the HP 8752A – a test port cable connected to port 2, was characterized at the factory and included in the factory calibration error terms. In most cases, a user VNA cal replaces the factory cal, and so the factory cal does not have any effect on the user calibration. Some exceptions to this are discussed in Section 3.13.7.

There have been many papers and publications discussing VNA calibration, and while not completely consistent in their terminology, common terms and symbols for some cases are recognized. Many of the terms come from the original implementations in the early HP VNAs, HP 8510 and HP 8753, which interestingly form both general types of measurement systems, the four-receiver VNA and the three-receiver VNA, respectively. Some advanced techniques utilize differing terminology in the original papers, but an attempt has been made here to maintain common terminology whenever possible. As with any subject of sufficient longevity, there are far more published papers on variations of VNA calibration than can be included even in a book of this nature. Interested readers are encouraged to review the references and bibliography for a more in-depth study of this area. Here, the theoretical treatment of VNA correction will be limited to the important results that are found in common practice today, and in the practical application of those methods to real-world problems faced by the practicing engineer.

3.2 Basic Error Correction for S-Parameters: Cal Application

Correction of systematic errors of S-parameter measurements in VNAs has been available for decades; the details of how the systematic errors affect the measurement results of a two-port S-parameter measurement are presented below.

There are two basic error models employed in VNAs, which rely on simultaneous measurements on either three VNA receivers, which is known as the 12-term model [1, 2], or four receivers, known as the eight-term model [3]. In modern VNAs, both are used, and it is a simple process to move between the two models. In fact, most VNAs represent the error terms strictly as the twelve-term model, but eight-term methods are often more convenient to use to determine the values of the error terms. Other models include many more effects, but these are not commonly used in practice [4].

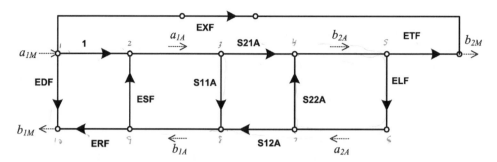

Figure 3.1   Forward error model for two-port measurement.

3.2.1   Twelve-Term Error Model

The 12-term error model actually consists of two separate six-term models, one in the forward direction and one in the reverse direction. This model was introduced in Section 2.3.1.1. In each case, three simultaneous or phase-coherent receiver measurements are required, including one incident wave and two scattered waves. Here the incident wave associated with the load port is assumed to be zero. Consider the forward error-model presented in Figure 3.1.

The measured values of a and b waves can be related to the actual values by computing the scattered waves from the incident waves, the error terms and the actual S-parameters, where here $a_2 = 0$, as ratios expressed as measured S-parameters. For the forward stimulation, this is

$$S_{11M} = \frac{b_{1M}}{a_{1M}} = EDF + \frac{ERF\left(S_{11A} + \dfrac{S_{21}ELF \cdot S_{12A}}{(1 - S_{22A} \cdot ELF)}\right)}{\left[1 - ESF \cdot \left(S_{11A} + \dfrac{S_{21}ELF \cdot S_{12A}}{(1 - S_{22A} \cdot ELF)}\right)\right]}$$

$$S_{21M} = \frac{b_{2M}}{a_{1M}} = \frac{(S_{21A} \cdot ETF)}{(1 - S_{11A} \cdot ESF) \cdot (1 - S_{22A} \cdot ELF) - ESF \cdot S_{21A} \cdot S_{12A} \cdot ELF} + EXF$$

$$(3.1)$$

An important point here is that the measured S-parameters depend upon all four actual DUT S-parameters. For the reverse stimulation, a similar analysis can be done for Figure 3.2.

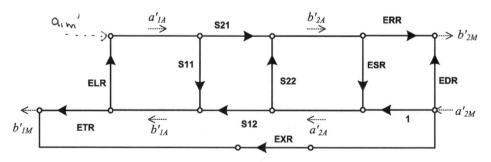

Figure 3.2   Signal flow diagram for reverse measurements.

This results in expressions for the measured reverse S-parameters S12 and S22:

$$S_{12M} = \frac{b'_{1M}}{a'_{2M}} = \frac{(S_{12A} \cdot ETR)}{(1 - S_{11A} \cdot ELR) \cdot (1 - S_{22A} \cdot ESR) - ESR \cdot S_{21A} \cdot S_{12A} \cdot ELR} + EXR$$

$$S_{22M} = \frac{b'_{2M}}{a'_{2M}} = EDR + \frac{ERR \left(S_{22A} + \dfrac{S_{21}ELR \cdot S_{12A}}{(1 - S_{11A} \cdot ELR)} \right)}{\left[1 - ESR \cdot \left(S_{22A} + \dfrac{S_{21}ELR \cdot S_{12A}}{(1 - S_{11A} \cdot ELR)} \right) \right]} \qquad (3.2)$$

Again, these reverse parameters depend upon the reverse error terms, and all four actual S-parameters. Between Eqs. (3.1) and (3.2), there are four equations, sufficient to solve for the four unknown actual-S-parameters, provided that the error terms are all known.

The solution to the actual S-parameters is

$$S_{11A} = \frac{S_{11N} \cdot (1 + S_{22N} \cdot ESR) - ELF \cdot S_{21N} \cdot S_{12N}}{(1 + S_{11N} \cdot ESF)(1 + S_{22N} \cdot ESR) - ELF \cdot ELR \cdot S_{21N} \cdot S_{12N}}$$

$$S_{21A} = \frac{S_{21N} \cdot (1 + S_{22N} \cdot [ESR - ELF])}{(1 + S_{11N} \cdot ESF)(1 + S_{22N} \cdot ESR) - ELF \cdot ELR \cdot S_{21N} \cdot S_{12N}}$$

$$S_{12A} = \frac{S_{12N} \cdot (1 + S_{11N} \cdot [ESF - ELR])}{(1 + S_{11N} \cdot ESF)(1 + S_{22N} \cdot ESR) - ELF \cdot ELR \cdot S_{21N} \cdot S_{12N}}$$

$$S_{22A} = \frac{S_{22N} \cdot (1 + S_{11N} \cdot ESF) - ELR \cdot S_{21N} \cdot S_{12N}}{(1 + S_{11N} \cdot ESF)(1 + S_{22N} \cdot ESR) - ELF \cdot ELR \cdot S_{21N} \cdot S_{12N}}$$

where a normalized S-parameter is defined as

$$S_{11N} = \frac{S_{11M} - EDF}{ERF} , \quad S_{21N} = \frac{S_{21M} - EXF}{ETF} ,$$

$$S_{12N} = \frac{S_{12M} - EXR}{ETR} , \quad S_{22N} = \frac{S_{22M} - EDR}{ERR} \qquad (3.3)$$

This is a different form than is given in other publications [5], in that it includes a normalized S-parameter which represents the measured S-parameter after subtracting the leakage terms and normalizing to the response tracking terms associated with the S-parameter. This form was anticipated in the description of error terms in Table 2.2. Here, directivity is recognized as the leakage term associated with reflection parameters, and crosstalk is the leakage associated with transmission terms.

This particular formatting of the solution gives good insight into some attributes of the error correction process. For example, if the VNA architecture is such that the forward source match and reverse load match at any test port is the same, then the equations for S21A and S12A are simplified to the normalized S-parameter divided by a common term in all equations that represent the mismatch loop equations.

The application of error correction to raw measurements is generally well known and well understood. The difficulty comes in determining the values of the error correction terms; the process, which here is called error correction acquisition or cal acquisition, is discussed in the section 3.3.

3.2.2   One-Port Error Model

For one-port devices, it is not necessary to complete a two-port calibration to obtain one-port reflection measurements. The one-port calibration application can be found from simplifying Eq. (3.3) to eliminate the two-port terms, such that the one-port correction utilizes only a single measured response and three error terms

$$S_{11A} = \frac{\left(\dfrac{S_{11M} - EDF}{ERF} \right)}{\left[1 + \left(\dfrac{S_{11M} - EDF}{ERF} \right) \cdot ESF \right]} = \frac{(S_{11M} - EDF)}{[ERF + (S_{11M} - EDF) \cdot ESF]} \tag{3.4}$$

This simplified version of the two-port response is sometimes called the Γ_1 correction as it provides the corrected reflection response of a two-port network in its currently terminated state. This is useful in computing some aspects of other corrections, such as the applied power to the port, which depends only on the match apparent at the port and not on the S11 of the DUT. As a reminder, the S11 of the DUT is the reflection when all the other ports are terminated in a matched load; the Γ_1 of the port represents the reflection or impedance of the DUT port with whatever termination is applied at the other ports.

3.2.3   Eight-Term Error Model

The eight-term error model is distinguished from the 12-term model in that it requires measurements at all four test receivers: two incident waves and two scattered waves. The error model is shown in Figure 3.3.

The eight-term error model has the advantage that the even though in practical implementation, the load match at each port changes depending upon whether the port is a source or a load, the eight-term model is not changed, as the change in load impedance is captured in changes in the incident waves at each port.

The eight-term error correction can be derived from the recognition that the measured S-parameters are a cascade of the input error box, the actual S-parameters and the output

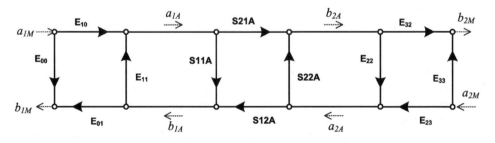

Figure 3.3   Eight-term error model, with four measured waves.

error box. The cascaded T-parameters (see Section 2.26) give the measured T-parameters of the DUT in terms of the actual T-parameters of the DUT and the input and output error-box T-parameters, as

$$\mathbf{T}_M = \mathbf{T}_X \cdot \mathbf{T}_{Act} \cdot \mathbf{T}_Y \tag{3.5}$$

where X and Y represent the port 1 and port 2 error boxes. From the definition of T-parameters, the relationship between measured values and actual values can be written as

$$
\begin{bmatrix}
\dfrac{(S_{21M}S_{12M} - S_{11M}S_{22M})}{S_{21M}} & \dfrac{S_{11M}}{S_{21M}} \\[2ex]
\dfrac{-S_{22M}}{S_{21M}} & \dfrac{1}{S_{21M}}
\end{bmatrix}
=
$$

$$
\dfrac{1}{E_{10}E_{32}}
\begin{bmatrix}
(E_{10}E_{01} - E_{00}E_{11})\ E_{00} \\[1ex]
-E_{11} \qquad\qquad 1
\end{bmatrix}
\cdot
\begin{bmatrix}
\dfrac{(S_{21A}S_{12A} - S_{11A}S_{22A})}{S_{21A}} & \dfrac{S_{11A}}{S_{21A}} \\[2ex]
\dfrac{-S_{22A}}{S_{21A}} & \dfrac{1}{S_{21A}}
\end{bmatrix}
\cdot
\begin{bmatrix}
(E_{32}E_{23} - E_{33}E_{22})\ E_{22} \\[1ex]
-E_{33} \qquad\qquad 1
\end{bmatrix}
$$

$$\tag{3.6}$$

and the actual values can be computed by inverting the input and output T-matrixes

$$\mathbf{T}_{Act} = \mathbf{T}_X^{-1} \cdot \mathbf{T}_M \cdot \mathbf{T}_Y^{-1} \tag{3.7}$$

with S-parameters determined by transforming the T-matrix back to the S-matrix. From the structure of Eq. (3.6) it is clear that the measured value of S21 must be non-zero; similarly the transmission tracking term, represented by $E_{10}E_{32}$, must be non-zero. Another aspect of the eight-term error model is recognition that only seven independent terms are needed to represent the model. The values E_{00}, E_{11}, E_{22} and E_{33} represent four of the independent values. The values of E_{10}, E_{01}, E_{23} and E_{32} always appear in the error correction Eq. (3.7) as products, $E_{10}E_{01}, E_{10}E_{32}$ and $E_{32}E_{23}$, thus the four terms represent only three independent values. A key aspect of the eight-term error model is that all four wave terms are utilized in its determination; the 12-term error model only requires measurements of three of the four waves (one incident and two scattered).

The eight-term error model is normally transformed into the 12-term error model before it is applied, by breaking it into two six-term models, with appropriate handling of change in load at port 2 during the forward sweep, and port 1 during the reverse sweep. From Figures 3.1 and 3.2, the one-port terms can be equated as

$$
\begin{aligned}
EDF &= E_{00}, & ERF &= E_{10}E_{01}, & ESF &= E_{11} \\
EDR &= E_{33}, & ERR &= E_{32}E_{23}, & ESR &= E_{22}
\end{aligned}
\tag{3.8}
$$

The other terms require a more complex evaluation. A key recognition here is that the effect caused by the port terminated as a load can be modeled as an additional error term at each port, which represents the ratio of incident to scattered waves at the measurement port. The terms in the forward and reverse direction can be defined as Γ_F and Γ_R, which are called "switch terms" and represent the change in match due to the source switch. These terms do not depend upon any external components or connections, and thus are entirely internal to the VNA. In most cases, these terms are very stable, and once determined in some manner, they can be utilized in other computations that rely on the eight-term model, replacing the measurement of the incident wave at the terminated port with a computation. A typical application is to use a 12-term model in a well-known connector type to find the switch terms, and then utilize the switch terms to determine the eight-term models in different connector types such as on-wafer probes. The switch terms are defined as

$$\Gamma_F = \left. \frac{a_{2M}}{b_{2M}} \right|_{Source \; at \; Port1} \tag{3.9}$$

$$\Gamma_R = \left. \frac{a_{1M}}{b_{1M}} \right|_{Source \; at \; Port \; 2} \tag{3.10}$$

From this the forward transmission tracking and load match terms can be determined as

$$ETF = \frac{E_{10}E_{32}}{1 - E_{33}\Gamma_F}, \quad ELF = E_{22} + \frac{E_{32}E_{23}\Gamma_F}{1 - E_{33}\Gamma_F} \tag{3.11}$$

and the reverse transmission tracking and load match terms can be determined as

$$ETR = \frac{E_{10}E_{32}}{1 - E_{00}\Gamma_R}, \quad ELR = E_{11} + \frac{E_{10}E_{01}\Gamma_R}{1 - E_{00}\Gamma_R} \tag{3.12}$$

Similarly, the eight-term error model, plus the switch terms, can be computed from the 12-term model as

$$E_{00} = EDF, \quad E_{11} = ESF$$
$$E_{33} = EDR, \quad E_{22} = ESR \tag{3.13}$$

$$\Gamma_F = \frac{E_{LF} - E_{SR}}{E_{RR} + E_{DR}\left(E_{LF} - E_{SR}\right)}, \quad \Gamma_R = \frac{E_{LR} - E_{SF}}{E_{RF} + E_{DF}\left(E_{LR} - E_{SF}\right)} \tag{3.14}$$

$$\frac{E_{23}}{E_{10}} = \frac{E_{TR}}{E_{RF} + E_{DF}\left(E_{LR} - E_{SF)}\right)} \tag{3.15}$$

$$E_{01}E_{10} = E_{RF}, \quad E_{32}E_{10} = \frac{E_{RR}E_{TF}}{E_{RR} + E_{DR}\left(E_{LF} - E_{SR}\right)} \tag{3.16}$$

In essence, the 12-term error model has an explicit description for both the source match at a port (when that port is the active source) and the load match at that same port (when it is not the active source). In general, these matches are not the same, and the difference is sometimes called the "switch-terms" of the VNA. The eight-term error model presumes that the source match and load match the same and employs an additional measurement to characterize the difference.

3.3   Determining Error Terms: Cal Acquisition for 12-Term Models

While the correction math is relatively straightforward, determining the error terms for corrections can be anything from trivially easy – for example using an electronic calibration module on a coaxial connection – to being extremely difficult – as in the case of measurements at cryogenic temperatures, at power level extremes or in cases with unusual DUT connections. In general, there are two distinct calibration methodologies based on either the 12-term error model, or the eight-term error model.

The measurement requirements for cal acquisition depend upon the number of required error terms. Essentially, one independent measurement is required for each error term. A device that has exactly-known properties is called a calibration standard or cal-std, and is used to generate one or more independent measurements. Depending upon the VNA architecture, more than one independent measurement may be acquired for each sweep, using a variety of measurement receivers to acquire the independent measurements on the same cal-std. For example, a one-port calibration standard (such as an open or a load) creates one independent reflection measurement. It may also be used to create independent transmission measurements, such as a measurement of crosstalk.

Calibration acquisition for the 12-term model requires six independent measurements for each direction, as there are no common error terms between the forward and reverse direction. The crosstalk measurements are most commonly omitted, as the crosstalk of modern VNAs is below the noise floor of the measurement system, except in special cases, and so any measurements would just be a measurement of noise, and impair the resulting correction rather than enhancing it. Thus the 12-term model becomes requires just 10 terms, and 10 independent measurements.

The standards for calibration come in two distinct types known as mechanical calibration standards and electronic calibration or Ecal$^{\text{TM}*}$ (see Section 3.4.6). Mechanical standards are physical representations of opens, shorts, loads and in some cases thrus (note, for purposes of clarity, the word "thru" is used to describe a calibration kit through-standard; this is historically how the standard appears in the VNA menu). These are usually sold together as a set, and form a mechanical calibration kit or Cal Kit as it is commonly called. Electronic calibration kits have built-in switchable standards that provide a similar function to the open/short/load standards, as well as a through state. More details on electronic calibration are described in Section 3.4.6.

The most common form of cal acquisition makes three measurements of one-port standards at each port, and two measurements of a known through standard, or "thru" as it is commonly called, in each direction, for 10 total measurements.

3.3.1   One-Port Error Terms

The one-port error terms are most commonly found using an open/short/load calibration. For ease of comprehension, consider first the case where the reflection standards provide the reflection of an ideal open $(\Gamma_{Open} = 1)$, ideal short $(\Gamma_{Short} = -1)$ and ideal load $(\Gamma_{Load} = 0)$. The measured S11 of each can be written in terms of the actual reflection coefficient and the error terms

$$S_{11M}^{Ideal\_Open} = EDF + \frac{ERF\,(1)}{[1 - ESF\,(1)]}$$

$$S_{11M}^{Ideal\_Short} = EDF + \frac{ERF\,(-1)}{[1 - ESF\,(-1)]} \tag{3.17}$$

$$S_{11M}^{Ideal\_Load} = EDF + \frac{ERF\,(0)}{[1 - ESF\,(0)]}$$

---

* ECal is a trademark of Agilent Technology, Ecal is used here to represent any electronic calibration method.

From which the one-port error terms are easily computed as

$$EDF = S_{11M}^{Ideal\_Load} \tag{3.18}$$

$$ESF = \frac{\left(S_{11M}^{Ideal\_Open} + S_{11M}^{Ideal\_Short} - 2EDF\right)}{\left(S_{11M}^{Ideal\_Open} - S_{11M}^{Ideal\_Short}\right)} \tag{3.19}$$

$$ERF = \frac{-2\left(S_{11M}^{Ideal\_Open} - EDF\right)\left(S_{11M}^{Ideal\_Short} - EDF\right)}{\left(S_{11M}^{Ideal\_Open} - S_{11M}^{Ideal\_Short}\right)} \tag{3.20}$$

Of course, these equations only apply to ideal opens, shorts and loads, but are useful in understanding the structure of the error terms. In particular, if one considers a system where the raw reflection tracking is 1, and evaluates the EDF and ESF terms graphically, one can show the relationships directly. In essence, the EDF term is just the load response. Figure 3.4 (left, upper) shows the load response as vector diagram, where the measured load is equal to the EDF term, and with the ideal load ($\Gamma_L = 0$) at the center of the chart, one finds $\Gamma_{LM} = EDF$, thus $\Gamma_{LM} - EDF = \Gamma_L = 0$. Figure 3.4 lower-left shows the construction of the measured short, and lower right the measured open. The upper right shows the construction of source match determination as Open + Short − 2EDF = 2 ESF. In practice, the ripple seen on a log magnitude (dB) trace of an open or a short represents the VSWR of the source match and the directivity of the port, and the mean value of the trace represents the reflection tracking.

In general, three known reflections are used, but they are not assumed ideal. There are models for each reflection standard built into the VNA firmware, and these models provide a way to modify the ideal standards to account for real-world errors such as loss or open-end fringing capacitance.

The generalized three-term solution [4] for the error terms is

$$\begin{bmatrix} EDF \\ ERF - EDF \cdot ESF \\ ESF \end{bmatrix} = \begin{bmatrix} 1 & \Gamma_{AO} & \Gamma_{AO} \cdot \Gamma_{MO} \\ 1 & \Gamma_{AS} & \Gamma_{AS} \cdot \Gamma_{MS} \\ 1 & \Gamma_{AL} & \Gamma_{AL} \cdot \Gamma_{ML} \end{bmatrix}^{-1} \begin{bmatrix} \Gamma_{MO} \\ \Gamma_{MS} \\ \Gamma_{ML} \end{bmatrix} \tag{3.21}$$

where Γ_{AO}, Γ_{AS}, Γ_{AL} are the presumed actual reflection coefficients of the three standards (essentially, the models of the standards), and Γ_{MO}, Γ_{MS}, Γ_{ML} are the measured values of the standards. While the labels imply open/short/load, it is understood that the standards can take on any values.

3.3.2   One-Port Standards

From Eq. (3.21), it is clear that any three known reflections can be used for one-port calibrations. Opens, shorts and loads are most commonly used as they provide very good separation of variables. While it's not possible to create standards that match the ideal assumptions of Eq. (3.17), at RF frequencies the standards are quite close to ideal. Most one-port standards are in

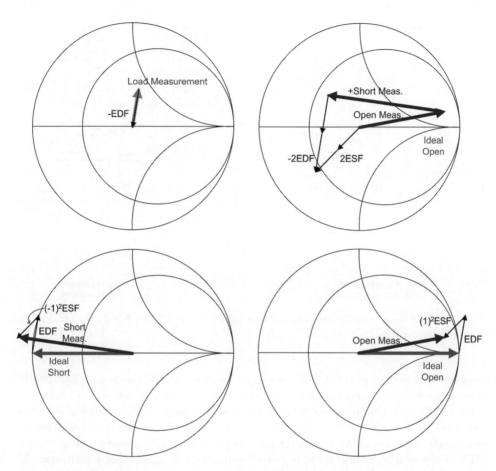

Figure 3.4 Determining the error terms graphically for open/short/load responses.

the form of a model for the standard and an input transmission line in the form of an offset delay and loss.

3.3.2.1 Open Standards

First things first: there is *no such thing* as an ideal open standard. All opens have fringing capacitance and virtually all opens have some offset length (certainly all commercial Cal Kits have only offset opens, even if they are not called that). Therefore you will *never* see a dot at the edge of the Smith chart at $Z = \infty$; it will *always* be an arc due to the offset open, as demonstrated in Figure 3.5.

This measurement is for a 7 mm connector, which is one of the only flush connectors commonly available. As a flush connector, it has no offset line. Every other coaxial connector type has an offset delay in both the open and the short. The offset delay produces greater than $360°$ of phase shift for connectors such as 3.5 mm. In the figure, the open fringing capacitance

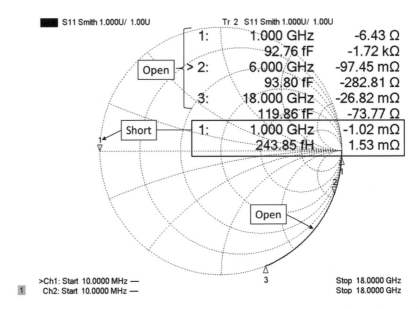

Figure 3.5 The correct re-measurement of an open and short after calibration; the arc is caused by the excess fringing capacitance on the open.

is not constant, as the model has increasing capacitance as frequency increases. The short has a very tiny amount of series inductance; from the model this amount is less than 1 pH.

The classic model for the open standard is shown in Figure 3.6. The principal deviations from ideal are due to a change in position of the open, rotated from the reference plane by a short length of transmission line, and fringing capacitance from the end of the open.

The fringing capacitance of the open end is illustrated by the small capacitors in Figure 3.7. In most instances, the male test port (female standard) uses a center-pin extender that produces a constant diameter to match the system impedance (Figure 3.7a). This produces phase shift associated with the length of the line. Fringing capacitance adds additional phase shift, and is typically modeled as a capacitance versus frequency polynomial in the form of

$$C(f) = C_0 + C_1 f + C_2 f^2 + C_3 f^3 \tag{3.22}$$

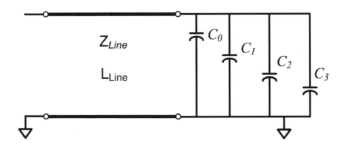

Figure 3.6 Model for an open circuit.

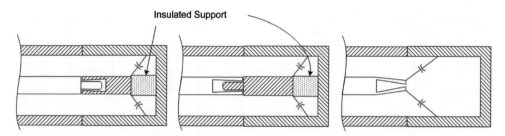

Figure 3.7 Physical construction of (a) female open, (b) male open with extender pin, (c) male open without extension.

The values for the capacitive coefficients are fitted to the performance of the open, and the actual extracted values may be different for RF versus microwave versions of the same standard. The polynomial fit can be adjusted to get better performance at low frequencies for the RF kit, or fitted to a broader overall response for a microwave kit. Newer techniques allow for essentially any arbitrary model of any standard, using so-called data-based methods.

Figure 3.7b shows a female test port with a male center-pin extender. The extenders are typically attached to the back of the standard with an insulated support made from Lexan or other plastic material. Figure 3.7c shows a female test port with a shielded open, but no center-pin extender. This is common on RF connectors for low-cost kits; for example, non-metrology type-N calibration kits typically don't have a center-pin extender for the female test port (male standard). It is still necessary to provide a shield or outer conductor to avoid radiation from the center pin. The fringing capacitance in this case is less certain than that of the solid cylinder that is formed by the center-pin extension. One reason for the uncertainty is that the center-pin female fingers may have a differing amount of gap depending upon construction and the size of male pins previously used. For the male test port (female standard), almost all calibration kits include a test port extender to keep the outer diameter of the male pin constant. On older Cal Kits, the male extension pin was a separate piece, and often lost from the Cal Kit; unfortunately, its use is critical for good calibrations.

A common problem seen with open standards is radiation of the open if the center pin is not shielded. For this reason, all precision calibration kits use a shield on the open circuit; but open circuits for fixtures, probes and adapters are sometimes used when a no-precision kit is available. Figure 3.8 shows two traces of a measurement of an SMA male open test port. The change in response is due to touching the outer coupling nut. The dip in the S11 response shows the frequency where the unshielded SMA-open is radiating.

3.3.2.2   Short Standards

Short standards are in some ways even more ideal than open standards, because they form nearly ideal reflections at the shorting plane. The typical model for a short standard is shown in Figure 3.9. The inductance model for a short circuit is in the form of an inductance versus frequency polynomial as

$$L(f) = L_0 + L_1 f + L_2 f^2 + L_3 f^3 \tag{3.23}$$

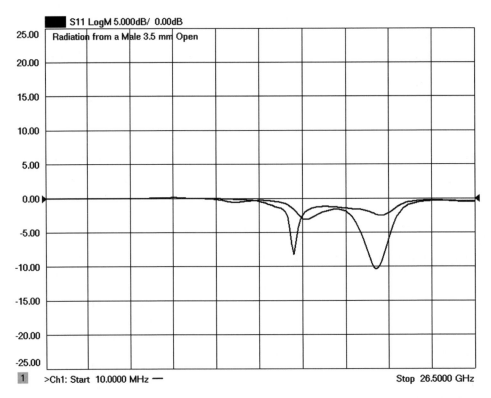

Figure 3.8 Variations in the open reflection coefficient due to radiation for an unshielded male SMA.

For older RF models, the values of all the inductance terms are zero, and the only non-ideal aspect is the transmission line model that provides the offset delay, but newer models have standardized on Eq. (3.23). Microwave models include inductance values, but they are very tiny indeed and the deviation from an ideal short is much smaller than that of an open circuit, as illustrated graphically in Figure 3.5.

The short circuit standards are the simplest of mechanical standards, and typically consist of just the center pin connected to ground, Figure 3.10. While there is no requirement for

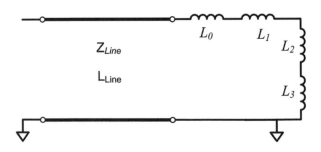

Figure 3.9 Model for a short standard.

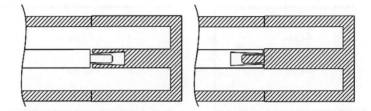

Figure 3.10 Short circuit standards (a) male test port, (b) female test port.

the length of the short to match the length of the open, it is best to make the length just slightly longer so that phase shift versus frequency of the short matches that of the open (which has some excess phase due to the fringing capacitance). Ideally, the opens and shorts should maintain 180° of phase separate throughout the frequency range of their operation. The short standard typically has the smallest residual errors. In a band-limited calibration, shorts of different delays, called offset shorts, may be used to provide all of the three calibration standards. Calibrations using offset shorts are often used in metrology applications where the diameter and length of the short circuit standard can be characterized with very fine precision, and the impedance and delay computed with minimal errors. These are particularly useful in high-frequency mm-wave applications as it is difficult to create a well-match fixed-load at high frequencies.

3.3.2.3 Load Standards: Fixed Loads

The load standard is usually the most difficult to produce, and errors associated with it may increase dramatically with frequency. The load element, as shown in Figure 3.11, is formed by terminating a coaxial standard into a resistive element, often a thin-film circuit with a patch of tantalum nitride designed to provide a constant impedance with frequency.

The typical model for a load element contains only a resistance and a delay line, as shown in Figure 3.12a. The value of the resistance can be set independently from the system Z_0, but it is typically set to be Z_0. The value of the transmission line impedance is also typically set to Z_0. An alternative model for a load element is a series R-L circuit as shown in Figure 3.12b. This is a common situation found in on-wafer load standards, and the determination of the value of inductance requires additional information during the calibration. A method called LRRM [6]

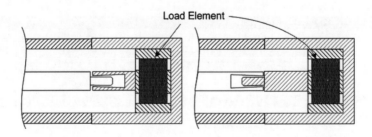

Figure 3.11 Load elements (a) male test port, (b) female test port.

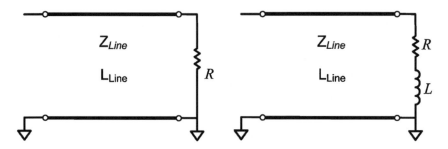

Figure 3.12   (a) Typical model for a load standard, (b) model for a load showing series inductance.

calibration provides this additional information by correlating the load response information across several frequencies to determine the values for the fixed R and L values. Since most VNAs do not provide model entries for the R-L combination, the effect of the inductance is generated by setting the impedance of the load delay line to a very high value (typically 500 ohms is the maximum allowed in many VNAs) and the length of the line is adjusted to give the equivalent phase shift of the effective inductance.

More details on modeling standards and determining parasitic values are discussed in Chapter 9.

3.3.2.4   Load Standards: Sliding Loads

The sliding load, which should more properly be called a sliding mismatch, is constructed from a length of precision airline followed by a moderately good termination, as show diagrammatically in Figure 3.13. The center conductor of the airline portion is typically created in such a way that it can slide into place while the outer conductor is not yet mated, to allow a bead-less connection. The load element is typically not a resistive element, but is more commonly a tapered bead of lossy material that essentially makes the airline look like a lossy element. It is designed to have an impedance that is not quite 50 ohms, normally in the range of 26–40 dB return loss. The far end of the airline is then clamped to the outer conductor at a precise connector gauge-length, so that any gap is only on the test port side. It is important to have this slight gap in the sliding load as all test ports must have a gap as well to avoid any

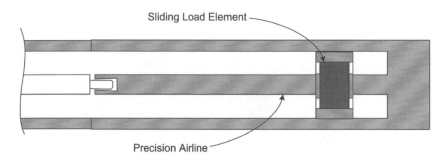

Figure 3.13   Representation of a sliding load.

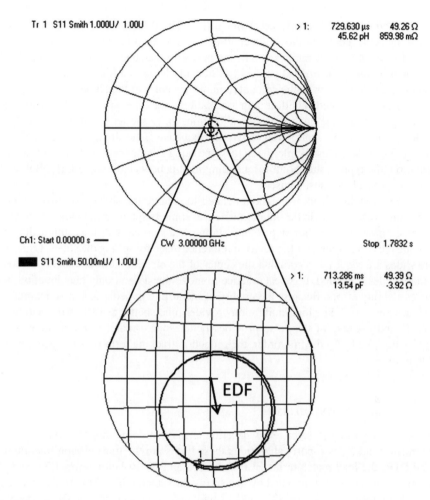

Figure 3.14 Smith chart measurement of a sliding load at a single frequency, while changing the slide position.

interference issues with the center pin connection; if the sliding load is pressed tightly against the test port center pin, it will provide a better, less reflective contact than the port will present to any DUT. The sliding load design allows the position of the center-conductor to be essential exactly at the reference plane. One might note that any DUT is likely to have a slightly greater gap depending upon its specific design, to avoid any interference with the test port. That is, both the test port and any DUT should have slightly recessed pins.

As the sliding load is moved, its apparent impedance rotates around the Smith chart, an example of which is shown at one frequency over the range of the slide, in Figure 3.14. The upper plot has the Smith chart at full scale, and the lower plot shows the locus of points for the sliding load at a smaller scale. In fact, the sliding load actually presents a spiral due to the finite loss of the transmission line. For the sliding load standard, several data acquisitions (typically five) are required, each at a different offset length for the slide. This forms a locus of points, the center of which is the impedance of the airline. The difference between this computed center and the actual center of the Smith chart determines the directivity error term, EDF.

The load element is deliberately set to be not perfectly matched, so that the diameter of the circle created by the locus of load points is sufficiently large that trace noise of the slide set position does not degrade from the computation of the center of the circle.

Some special care is needed with sliding load calibrations. The quality of the airline portion of the sliding load sets the quality of the calibration. Since the length of the line is limited, and the lossy element only works well above RF frequencies, the sliding load standard is only used above RF frequencies. While the actual operating frequency is determined in the Cal Kit definition, the typical start point for a sliding load is between 2 and 3 GHz. Below these frequencies, a fixed load must be used.

When using a sliding load, it is recommended to change the slide positions in non-even steps, to avoid creating a periodic situation where at some frequencies, the load response falls on the same phase as an earlier step. Many sliding loads come with logarithmically spaced marks along the outer conductor to provide a reference place for setting the slides. Usually, five or more slides are required to compute the center of the circle, and often the VNA code will allow more slides if desired. It is best practice to move the slide in only one direction, which minimizes stability errors. Because the quality of the airline impedance can be maintained at high frequencies, the sliding load calibration provides much better extraction of the directivity term, EDF, and because of this, also provides better source match determination, ESF, as implied by Eq. (3.20). To the first order, the reflection tracking term is not affected much by the EDF term.

3.3.3   Two-Port Error Terms

The two-port, or multiport, error terms describe terms that come into play on transmission measurements, and for two-port calibrations these terms are the transmission-tracking terms ETF and ETR, the load match terms ELF and ELR, and the isolation terms EXF and EXR. The transmission terms are computed by measuring thru standards and the isolation terms are computed from measurements of S21 and S12 with reflection standards on the test ports.

3.3.3.1   Isolation Standards

The isolation standard for the 12-term model is typically measured using load standards. From Eq. (3.1), the S21 and S12 measurements with loads at port 1 and port 2 have actual values of $S_{21LoadsA} = 0$, $S_{12LoadsA} = 0$, resulting in

$$S_{21LoadsM} = EXF, \quad S_{12LoadsM} = EXR \tag{3.24}$$

That is, the isolation or crosstalk error term is simply the S21 or S12 measurement with the load attached. Usually, substantial averaging must be used or the measurement is simply the noise floor of the VNA. In some cases, it may be advantageous to use the DUT (terminated by a load) to terminate the test port during an isolation calibration. In this way, any leakage from the reflection port (e.g., b1) due to mismatch of the DUT (as found in the stopband of a filter) may be characterized and removed.

3.3.3.2   Thru Standards – Flush Thru

The thru standard should most properly be called a "defined thru" standard, and represents a two-port standard for which all the S-parameters are known. There are two main forms of defined thru standards, sometimes called *flush thru* and *defined thru*.

The flush thru standard consists simply of the process of mating a pair of connectors together. For sexless connectors, such as the precision 7mm, the flush thru connection is always available if the physical test ports can be moved in such a way as to be directly connected. For sexed connectors, such as the 3.5 mm connector, the flush thru is available only for a port pair which includes one male and one female port. The S-parameters of the flush thru are also simply stated with $S_{21} = S_{12} = 1$, $S_{11} = S_{22} = 0$. From Eq. (3.1), the measured values of the flush thru become

$$S_{11FlushThruM} = EDF + \frac{ERF \cdot ELF}{(1 - ESF \cdot ELF)}, \quad S_{22FlushThruM} = EDR + \frac{ERR \cdot ELR}{(1 - ESR \cdot ELR)}$$

$$S_{21FlushThruM} = \frac{ETF}{(1 - ESF \cdot ELF)} + EXF, \quad S_{12FlushThruM} = \frac{ETR}{(1 - ESR \cdot ELR)} + EXR$$

$$(3.25)$$

from which the values of ETF and ELF can be determined, if the one-port terms and crosstalk terms have already been determined, as

$$ELF = \frac{(S_{11FlushThruM} - EDF)}{[ERF + ESF (S_{11FlushThruM} - EDF)]}, \quad ELR = \frac{(S_{22FlushThruM} - EDR)}{[ERR + ESR (S_{22FlushThruM} - EDR)]}$$

$$ETF = (S_{21FlushThruM} - EXF) (1 - ESF \cdot ELF), \quad ETR = (S_{12FlushThruM} - EXR) (1 - ESR \cdot ELR)$$

$$(3.26)$$

From this it is clear that the load match term is determined from the flush thru by applying a one-port error correction to the measured match of the flush thru. And the transmission tracking is similarly a measurement of the thru response, after subtracting the crosstalk and accounting for the source and load match interaction.

3.3.3.3   Thru Standards – Non-Insertable Thru

Whenever a pair of test ports uses the same-sex coaxial test-connector, such as for a DUT with SMA female connectors on each port – which is quite common in RF work – the calibration is known as the *non-insertable* case. The most typical case is a pair of male test ports which require a female-to-female thru. Unfortunately, built-in calibration kits for most VNAs have definitions only for the flush, male-to-female thru, and so do not have thru standards defined for male-to-male or female-to-female DUT port pairs. Failure to account for the delay of the thru (and to a lesser extent, the loss of the thru) is one of the most common causes of error in RF measurements. If a non-zero-length thru is used during calibration, the effective measured load match will be phase-shifted by the actual delay of the thru, and this can cause substantial error.

Figure 3.15 shows a measurement of a 15 cm airline where a 2 cm female-to-female thru was used in an 12-term calibration, and the delay and loss of the thru was ignored. The ripple

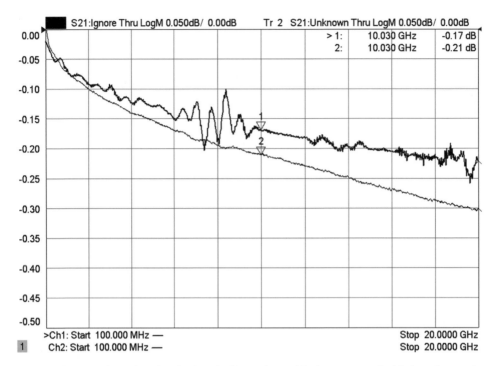

Figure 3.15   Error due to ignoring the length of a non-insertable thru, compared with the unknown thru calibration.

seen is due to the error in the measurement of the load match beating against the raw source match of the VNA; this error can cause the corrected or residual load match to be larger than the raw load match, by as much as 6 dB. The other, more insidious, error is the fact that the loss of the thru is not accounted for, so the loss of the DUT appears slightly less than it really is. In the case of a low-loss filter, this error allows a DUT to pass a test line limit even if the actual loss is out of spec.

In modern VNAs, this problem is seldom encountered as the unknown thru method (see Section 3.4.3) provides for highly accurate calibrations in the non-insertable case, as shown in the lower trace (the trace with marker 2 active) of Figure 3.15.

3.3.3.4   Swap-Equal Adapter

In the earliest days of modern VNAs, a common method for handling the non-insertable case was to add an additional male-to-female adapter to the test port. One could purchase a matched set of adapters which included a male-to-female, male-to-male and female-to-female thru adapter. These were designed to have very good return loss and equal electrical length. One performed the calibration by adding an additional male-to-female adapter on one of the ports, typically port 2, during the one-port calibration steps. Then during the thru step, the adapter was swapped for one which would mate with port 1, most commonly a female-to-female adapter. After the calibration was complete, the original male-to-female was replaced on the test port and fully calibrated measurements could be performed. While this was the typical way a calibration preceded, a better choice is to do the thru step of the calibration first, so that there is no need to disconnect and reconnect the male-to-female adapter between the one-port calibration and the thru measurement. This reduces the error due to connector repeatability. In modern VNAs the use of the unknown thru calibration eliminates the needed for swapping equal adapters.

3.3.3.5 Defined Thru

In most VNAs it is a very simply task to add a defined thru standard, set the connector sex to mate to the sex of the DUT connectors, and then complete the calibration using this defined thru standard instead of the flush thru standard. This is perhaps the simplest way of dealing with the simple non-insertable case.

The offset delay and offset loss must be determined for the connector, but the offset loss for a family of connectors is essentially the same for every standard, and is typically listed as part of the offset delay for the open or short standards in the calibration kit. For 7 mm connectors, it is about 700 Gohm/sec, and for 3.5 mm connectors it is 2000 Gohm/sec. The most common way for determining the offset loss is to compute it from the loss at 1 GHz, from the formula

$$L_{Offset(Gohms/sec)} = \left(\frac{\ln(10)}{10}\right) S_{21(dB\_loss)} \sqrt{f_{(GHz)}} \cdot \left(\frac{Z_{Offset}}{S_{21(delay)}}\right) \tag{3.27}$$

where the frequency is given in GHz, the loss of S21 in dB (as a positive number), and the resulting value is in Gohm/sec. This is the loss for a coaxial line; the loss for waveguide is computed using different models, and must be included particularly for mm-wave or higher frequencies.

For a measurement of an matched (50 ohm) offset line at 1 GHz, this simplifies to

$$L_{Gohms/sec} = \frac{11.5 \cdot S_{21(dB\_loss\_1GHz)}}{S_{21(delay)}} \tag{3.28}$$

For a defined thru, the math is essentially identical to (3.26), with the modification of replacing the one-port error terms with modified versions which include the de-embedded effects of the thru characteristics as described by Section 3.5. That is, the ELF and ETF terms are computed by measuring the defined thru, and applying Eq. (3.26) but first modifying the values of the EDF, ESF and ERF terms by way of de-embedding the actual defined thru S-parameter values from those error terms; a similar process applies to the reverse error terms. De-embedding techniques are discussed in detail in Chapter 9.

3.3.3.6 Adapter Removal Calibration

The non-insertable calibration case includes the common case of the same sex connector on each port of a DUT, but also includes the case of different connector families on each port. In such a case, an adapter (with different connector types on each end) must be used to connect between the ports, and it not easy to determine the delay of this adapter. This is also the case when making a measurement on a DUT with one waveguide and one coax port. To resolve this problem, a method for adapter removal has been developed that typically requires two two-port calibrations, one on either side of the adapter. From the error terms of the two

two-port calibrations, the characteristics of the adapter are computed. Forward reflection error terms from the port 1 side of the adapter of the first calibration (performed at the flush thru mating of the port 1 side) are combined with the forward reflection error terms from the second calibration performed at the port 2 side of the connector, to determine the S-parameters of the adapter as

$$S_{11} = \frac{(EDF_2 - EDF_1)}{[ERF_1 + ESF_1 \cdot (EDF_2 - EDF_1)]}$$

$$S_{21} = S_{12} = \frac{\sqrt{ERF_2 \cdot ERF_1}}{[ERF_1 + ESF_1 \cdot (EDF_2 - EDF_1)]} \qquad (3.29)$$

$$S_{22} = ESF_2 + \frac{ESF_1 \cdot ERF_2}{[ERF_1 + ESF_1 (EDF_2 - EDF_1)]}$$

where the subscript 1 refers to the error terms of the calibration on the port 1 side of the adapter and the subscript 2 refers to the error terms on the port 2 side of the adapter. In both cases, the error terms are the forward one-port terms and so this method for determining the S-parameters of an adapter could be used with only one-port calibrations just as well. In this computation, it is assumed that the adapter is reciprocal and passive. The computation of S21 contains a square root of a complex number. This will yield two values based on whether the positive or negative root is chosen. Put another way, the phase of the two results differs by 180°. For a single point computation, one cannot know the proper phase to choose. However, if one knows the approximate delay of the adapter, then the phase can be compared to the expected phase of the adapter for each frequency, and the phase value closest to the expected value is chosen. See Chapter 9 for more information about determining complex roots.

Once the S-parameters of the adapter are determined, it can be removed from the port 1 side of the second two-port cal using de-embedding as discussed in Chapter 9, to create a new 12-term calset that contains the one-port terms from the first cal for port 1, the one-port terms from the second calibration for port 2, and a modification of the transmission tracking and load match terms for both directions determined by the de-embedding math.

3.3.4   Twelve-Term to Eleven-Term Error Model

The 12-term error model shows 12 apparently independent error-terms. But other analysis [3] shows that there is an interdependency between the terms so that there are only 11 degrees of freedom; if we ignore crosstalk terms, there are only 9 degrees of freedom for the 10 error terms. This interdependency is represented by the relationship

$$ETF \cdot ETR = [ERR + EDR (ELF - ESR)] [ERF + EDF \cdot (ELR - ESF)] \qquad (3.30)$$

Though calibration acquisitions don't normally take advantage of this result, it implies that only 9 measurements are required to obtain the 10 error terms.

3.4   Determining Error Terms: Cal Acquisition for Eight-Term Models

The eight-term error models are determined through a different series of measurements, which rely more on transmission standards, and do not rely on the details of known reflection standards. As described in Section 3.2.3, the measurements require knowledge of both incident and scattered signals at each test port. Because of the nature of the eight-term models, there is not necessarily explicit computation of the one-port error terms, but rather they are computed as part of the solution to the two-port error model. In this section, the effect of crosstalk is ignored.

3.4.1   TRL Standards and Raw Measurements

The standards used in TRL calibrations are a thru, a reflect and a line. Another version of TRL is sometimes called LRL for line, reflect, line, but they are essentially the same, as most versions of TRL provide for non-zero length thru standards.

TRL calibrations are often described as the most accurate form of calibration. This is principally true in the case of connectorized coax and waveguide calibrations, but TRL may not provide the best calibration for in-fixture calibrations. The reason that TRL calibration is considered most accurate comes from the fact that the quality of calibration depends almost entirely on the precision of the knowledge of the line impedance, or in the case of waveguide, in the reflection of the line section.

TRL specifications only apply to metrology grade calibration standards, such as bead-less short airlines, which are seldom used in practice. These metrology kits are designed to be used with specialized insertable calibration lines, rather than normal thru lines. One can create a TRL calibration using a short high-quality adapter, but the impedance or S11 of the adapter will set the source and load match of the system (along with the quality of the open/short standards).

For in-fixture TRL calibrations, it is common to create the standards on separate PCBs; in such case differences in the coax to PCB launches (typically SMA) create errors in the calibration. Details of PCB calibration are discussed in Chapter 9. Important attributes of the TRL standards are described below.

While TRL calibrations are commonly called eight-term, it was shown earlier that there are only seven independent error terms. And since it is common to convert the eight-term model to the 12-term model for applying the error correction, two additional unknown error terms, Γ_F and Γ_R are required, making a total of nine unknown error-terms. Thus, at least nine independent equations are required to solve the TRL calibration problem.

3.4.1.1   Thru Standard

The thru standard is the simplest standard and consists simply of a flush thru. As such, its reflection is zero and its transmission is unity (1). Just as for defined thru calibrations in the 12-term model, the thru standard provides a very well-known standard. For coaxial connectors, the thru is established by simply making a male-to-female connection, excepting of course the sexless connectors such as 7 mm.

In the case of waveguide connectors, the thru standard is simply a mating of the two waveguide flanges. The loss and delay of this thru, like the flush thru of coaxial connectors, is zero.

In the case of on-wafer calibrations, it is not possible to make a direct connection from probe to probe, so the thru must have some length. In this case the calibration is more correctly called an LRL calibration. The LRL calibration has additional characteristics in that this first line might have some loss and some offset impedance. Normally, the impedance of the thru or the first line in LRL is presumed to be the system Z_0.

The reference plane is typically set at the center of the thru standard. In the case of a zero-length thru, this is typically the best choice. Most TRL calibration acquisition algorithms provide for setting the reference also at the reflect standard. This is a good choice if the thru standard is of substantial length, or if a flush short is used as the reflect standard.

During the thru standard measurement, the four raw S-parameters are measured, along with two additional parameters, which are sometimes called the *switch terms*. The switch terms are measurements of a2/b2 with the source at port 1, and measurements of a1/b1 with the switch at port 2. Some VNAs have limitations to the specific ratios that can be measured, so that the measurement of a2/b2 might be accomplished with a measurement of a2/a1 with the source at port 1, from which the a2/b2 measurement can be computed by realizing that the raw S21 measurement contains b2/a1, and that can be combined with the a2/a1 to obtain an a2/b2 measurement.

There are six raw measurements for the thru portion of the calibration:

$$S_{11ThruR} = E_{00} + \frac{E_{10}E_{01} \cdot ELF}{(1 - E_{11}ELF)}, \quad S_{22ThruR} = E_{33} + \frac{E_{23}E_{32} \cdot ELR}{(1 - ELR \cdot E_{22})}$$

$$S_{21ThruR} = \frac{E_{10}E_{32}}{(1 - E_{11}ELF)}, \quad S_{12ThruR} = \frac{E_{01}E_{23}}{(1 - ELR \cdot E_{22})} \tag{3.31}$$

$$\left.\frac{a_{1M}}{b_{1M}}\right|_{a_2\_active} = \Gamma_R, \quad \left.\frac{a_{2M}}{b_{2M}}\right|_{a_1\_active} = \Gamma_F$$

where

$$ELF = E_{22} + \frac{E_{32}E_{23}\Gamma_F}{1 - E_{33}\Gamma_F}, \quad ELR = E_{11} + \frac{E_{10}E_{01}\Gamma_R}{1 - E_{00}\Gamma_R}$$

Note that the two additional unknown terms, Γ_F and Γ_R, are explicitly measured as part of the thru-standard measurement. In fact, since these are related to the difference between the source and load match at the test port, and this difference is usually stable and does not change over time, so it is possible to acquire these values one-time for the VNA and not repeat this measurement. Some lower-cost versions of VNAs have three-receiver architectures that do not directly support the Γ_F and Γ_R measurements, but their values can be found using standard short open load thru (SOLT) techniques, where the difference between the source and load match is saved; this is often call the "delta match" calibration, and performing one on a three-receiver VNA allows it to then be used for any of the eight-term error model corrections.

From (3.31), there are nine error terms, and six independent equations. At least three more independent equations are required to solve for the error terms.

3.4.1.2   Line Standard

The line standard provides the key attributes to generate a good calibration. The quality of the impedance of the line standard sets the quality of the calibration. The one critical factor of the line standard is that its length must be different from the length of the thru, such that the phase shift through the thru is different from the phase shift of the line by at least $20°$ and not more than $160°$. While this is often given as a hard rule, in fact the range can be extended somewhat, at the risk of degradation due to noise and other errors in the measurement system. The difficulty comes from the necessity to have unambiguous results for the raw S-parameters of each standard. If the line standard were to be exactly $180°$, the raw measurements of the thru and the line would be identical (not including a small effect of loss) and there would not be sufficient independent measurements to determine the error terms.

In the case of metrology standards, the line standards are produced with very precise machining, often created in two parts as an airline that has no support beads. These airlines are measured using some of the most precise dimensional measurement methods, and the impedance is computed based on these measurements. Often, the computations include accounting for the skin effect of the line and even include effects such as the sag of the line's center conductor due to gravity effects. As such, the knowledge of the line's impedance is perhaps one of the most precisely know attributes in RF and microwave engineering.

In most cases, the line impedance is presumed to be the same as the system impedance, and errors in the line impedance become residual errors in directivity, source match and tracking. But for metrology applications, the precise value of the airline is entered into the calibration kit definition, and the TRL calibration algorithm is modified to use the line's actual impedance instead of the system impedance in determining the error terms.

In most cases, the loss of the line is defined in the calibration standard, but in some cases the loss is not well known. During the solution of the TRL acquisition, both the length of the line and the loss of the line can be determined, and this is sometimes called LRL autocharacterization. Auto characterization can be used when the thru and the line standards have the same offset loss and impedance.

A key area of concern for in-fixture or on-wafer calibration is variation between the impedance of the thru and line standards. Since both are typically manufactured in a photolithographic process, errors in creating the width of the line (due to over-etching, for example) or differences in the dielectric constant of the substrate material below the thru versus the line will create residual errors in the calibration.

The line standard measurements are

$$S_{11LineR} = E_{00} + \frac{E_{10}E_{01}S_{21UT}^2 \cdot ELF}{\left(1 - S_{21UT}^2 E_{11} \cdot ELF\right)}, \quad S_{22ThruR} = E_{33} + \frac{E_{23}E_{32}e^{-2\gamma L} \cdot ELR}{\left(1 - e^{-2\gamma L}ELR \cdot E_{22}\right)}$$

$$(3.32)$$

$$S_{21ThruR} = \frac{e^{-2\gamma L}E_{10}E_{32}}{(1 - e^{-2\gamma L}E_{11}ELF)}, \quad S_{12ThruR} = \frac{e^{-2\gamma L}E_{01}E_{23}}{(1 - e^{-2\gamma L}ELR \cdot E_{22})}$$

where

$$ELF = E_{22} + \frac{E_{32}E_{23}\Gamma_F}{1 - E_{33}\Gamma_F}, \quad ELR = E_{11} + \frac{E_{10}E_{01}\Gamma_R}{1 - E_{00}\Gamma_R}$$

where γL is represents the response due to the length of the line.

Here four more equations are added, but two more unknown values are also added, the delay and loss associated γL. Thus, the total number of independent equations has increased to 10, but the total unknowns have increased to 11.

3.4.1.3 Reflect Standard

The reflect standard is the simplest standard, and the only criterion is that it provide some non-zero reflection equally at each of the ports. While this sounds simple, it presents some problems in the case of coaxial connectors, due to the fact that the short standard for the male and female port must be different physical components. In such a case, the standards must be electrically identical. Thus, the reflect standards may be unknown, but they must be the same.

For probes, it is common to leave the port unterminated and simply lift the test probes. This provides an open reflection which is presumed to be the same on each port. While this may work for probes, it is not advised for coaxial connections, particularly for a male test port (female DUT) because the male pin and the nut may act as an antenna and radiate such that the reflection at some frequencies is no longer constant. This is one of the reasons that shorts are recommended for TRL reflect standards.

For waveguides, an open cannot be used for similar reasons. An open waveguide provides approximately 12 dB return loss due to radiation, and this value can change substantially due to re-reflections from the environment. For waveguides, shorts are also recommended as the reflect standard.

In some cases, the reflect standard is used to set the reference plane of the ports. The VNA calibration usually provides for a choice of using the center of the thru or the reflect standard to set the reference plane. In the case of the reflect standard, the phase of the standard must be approximately known, in a similar manner as the phase of the line standard. Often, it is defined to be zero delay and this is one case where an open is sometimes used. An example is a PCB TRL Cal Kit where the open is simply the PCB fixture with no DUT attached. The caution on avoiding effects of radiation of the open should be reiterated here.

The measured values for the reflect standard are

$$S_{11Refl} = E_{00} + \frac{E_{10}E_{01} \cdot \Gamma_{Refl}}{\left(1 - E_{11}\Gamma_{Refl}\right)} \ , \quad S_{22Refl} = E_{33} + \frac{E_{23}E_{32} \cdot \Gamma_{Refl}}{\left(1 - \Gamma_{Refl}E_{22}\right)} \qquad (3.33)$$

In Eq. (3.33), two more independent equations are available, but one more unknown is added: the reflection coefficient of the reflect standard.

Thus, there are now 12 unknowns and 12 independent equations, which can be solved to find the error terms. The solution is rather complex, but has been presented in several forums and is available in [7].

3.4.2 Special Cases for TRL Calibration

3.4.2.1 TRM Calibration

The thru reflect match (TRM) calibration is really a degenerate case of the TRL calibration, where the L could be considered as a lossy, infinite length line. The calibration match standard is the same as a load standard for the SOLT calibration of Section 3.3.2.3, and is typically considered to have a zero reflection. However, the mathematical expressions for the case of TRM can also allow an arbitrary match condition so long as the reflection coefficient is known.

TRM is, in fact, almost always used as part of the TRL calibration for low frequencies, with a common transition frequency being 2–3 GHz. The measured values for the match standard are

$$S_{11MatchF} = E_{00}, \quad S_{22MatchR} = E_{33} \tag{3.34}$$

It might seem that the TRM calibration eliminates the four measurements of the line, and replaces them with only two independent equations. But while the line standard provides four independent equations, it also adds two more unknowns; thus the match standard of TRM provides the same order of new independent equations as the line standard.

3.4.2.2   Other TRL Considerations

While TRL has widespread use, there are some special cases that require additional care to avoid poor results.

In waveguide calibration, the line standard loss is determined as part of the auto-characterize, but if the thru is not zero length, its loss and delay must be known. A special case of the TRL calibration provides for the dispersion of the thru, so that the phase shift error from waveguide dispersion is properly accounted for. In such a case, the Cal Kit must specify that the thru standard is waveguide. Further, it is standard practice to set the system Z_0 and the line Z_0 to 1 ohm for waveguide calibration. The concept of impedance is ill-defined in waveguides, and setting the value to 1 provides a common reference. Some older VNAs required both values to be set on the instrument, but newer VNAs provide these values as part of the calibration kit definition.

A similar situation occurs for microstrip lines, which are somewhat dispersive. Unfortunately, there are no means in most VNAs for providing for the dispersion of microstrip lines, so the best approach is to use the delay value at the center of the band of interest for the thru offset delay. The offset loss will be computed as part of the TRL self-calibration process. Some methods of LRL provide for a means to automatically characterize the loss and delay of the line standards, and so may be beneficial in this case.

For conditions where the line impedance varies greatly from the system impedance, special care must be used. An example of this scenario is a case commonly found when testing high power transistors, which often have a very low input or output impedance. The low impedance and high power output necessitates a wide trace. Thus, an impedance transformer, often in the form of a tapered line, is used to translate from 50 ohms to the transistor impedance. A TRL calibration of this very wide line can have some difficulties as the width of the line might support higher-order planar waveguide modes or orthogonal TEM modes, or even act as a patch antenna. In essence, the line becomes so wide that it appears to be another transmission line in shunt with the 50 ohm line.

Equations (3.31), (3.32) and (3.34) are sufficiently independent to be solved for the error terms; the solution is beyond the scope of this text but may be found in numerous references.

3.4.3   Unknown Thru or SOLR (Reciprocal Thru Calibration)

The unknown thru (UT) calibration, or short open load reciprocal (SOLR) calibration, as it is sometimes known, has become the preferred calibration method for most VNA measurement scenarios [8]. The UT calibration is also based on a combination of the eight-term error model and the 12-term error model. It requires the computation of the switch terms similar to the TRL calibration, but requires the same set of calibration standards as the 12-term calibration, and allows different degrees of freedom than either the SOLT or the TRL calibration.

3.4.3.1 Unknown Thru Standard

The thru standard for the UT calibration has only one requirement: it must be reciprocal in transmission, that is, S21 = S12. This requirement holds true for all passive devices excepting isolators or circulators. In practice, the loss of the UT must be sufficiently small to avoid numerical difficulties in the computation of the error terms. Different vendors of VNAs use different methodologies, with some allowing loss up to 40 dB and still maintaining good calibration integrity.

In the UT calibration, the error terms associated with the directivity, source match and reflection tracking are determined based on the same methods as a simple one-port calibration or the one-port portion of the 12-term calibration, as in Section 3.2.1. The raw measurements for the unknown thru are

$$S_{11UT\_R} = EDF + \frac{ERF \left(S_{11UT} + \frac{S_{21UT}^2 ELF}{(1 - S_{22UT} \cdot ELF)} \right)}{\left[1 - ESF \cdot \left(S_{11UT} + \frac{S_{21UT}^2 ELF}{(1 - S_{22UT} \cdot ELF)} \right) \right]}$$

$$S_{21UT\_R} = \frac{(S_{21UT} \cdot ETF)}{(1 - S_{11UT} \cdot ESF) \cdot (1 - S_{22UT} \cdot ELF) - ESF \cdot S_{21UT}^2 \cdot ELF}$$

$$S_{12UT\_R} = \frac{(S_{12A} \cdot ETR)}{(1 - S_{11UT} \cdot ELR) \cdot (1 - S_{22UT} \cdot ESR) - ESR \cdot S_{21UT}^2 \cdot ELR}$$

$$S_{22UT\_R} = EDR + \frac{ERR \left(S_{22UT} + \frac{S_{21UT}^2 ELR}{(1 - S_{11UT} \cdot ELR)} \right)}{\left[1 - ESR \cdot \left(S_{22UT} + \frac{S_{21UT}^2 ELR \cdot}{(1 - S_{11UT} \cdot ELR)} \right) \right]}$$

$$\left. \frac{a_{1M}}{b_{1M}} \right|_{a_2\_active} = \Gamma_R, \quad \left. \frac{a_{2M}}{b_{2M}} \right|_{a_1\_active} = \Gamma_F \qquad\qquad (3.35)$$

where

$$ELF = ESR + \frac{ERR \cdot \Gamma_F}{1 - EDR \cdot \Gamma_F}, \quad ELR = \overset{ESF}{E_{11}} + \frac{ERF \cdot \Gamma_R}{1 - EDF \cdot \Gamma_R}$$

where similarly to the TRL calibration, the values for Γ_F and Γ_R are determined by measurements of the a_1/b_1 and a_2/b_2 terms during the thru step of the calibration.

Equation (3.35) gives six independent equations, but also adds three additional unknowns: S_{11UT}, S_{22UT} and the product $S_{21UT} \cdot S_{12UT} = S_{21UT}^2$, realizing that the unknown thru must be reciprocal in transmission.

There are 12 total unknown, just as in the TRL case: the seven independent TRL errors, three unknown thru parameters, and Γ_F and Γ_R. These are solved in a simultaneous way to generate all the necessary error terms and give the values of the unknown thru as a bonus.

This calibration requires identical steps to the SOLT calibration, but does not require that the thru standard be defined. The applications for this greatly extend the flexibility of the SOLT calibration, allowing a vast array of thru devices to be used which can greatly simplify complex calibration tasks. The quality of the UT calibration is derived from the quality of the reflection standards. If there is substantial loss between the test port and the measurement plane, noise effects will degrade the measurement of the reflection standards, and thus the transmission measurements will be degraded to a greater extent than in the case of a defined thru calibration. Equation (3.35) provides the set of equations from which the unknown error terms are computed, the solution of which is beyond the scope of this book, but may be found in many references [8].

3.4.4   Applications of Unknown Thru Calibrations

Some common cases where unknown thru calibrations are beneficial are described below.

3.4.4.1   Non-Insertable Coaxial Calibration

For most components found in RF and microwave work, the connectors on each port are identical and typically female. The SMA female DUT connector is probably the most common connector used, perhaps followed by the type-N female. Most cables have male connectors, and thus many devices must be tested with male test port connectors on each port. As such, a female-to-female thru standard must be used during calibration. In the 12-term SOLT calibration, the attributes of this thru-adapter must be defined prior to calibration, as noted in Section 3.3.3.5, or determined through a second step of calibration, as described in Section 3.3.3.6. This is not necessary at all for the UT calibration. This flexibility can lead to many ingenious ways to reduce and reuse calibration steps to achieve complex calibration scenarios.

3.4.4.2   On-Wafer Calibrations

In some cases of on-wafer calibrations, the test probes are not aligned in a way that they allow a direct, straight-through line to connect between the probes. In on-wafer probing, it is common to refer to the probe positions as east and west when they are on the left and right side of the wafer, and north and south when they are located above and below the wafer. In cases where the device must be probed with a set of probes which do not line up, such as east and north, or south and west, the through-line on-wafer standard must contain a 90° bend, as shown in Figure 3.16. It is quite difficult to know exactly the impedance and length of the bend, and so utilizing an unknown thru calibration removes the necessity for providing a precision designed on-wafer thru standard. The calibration method requires a one-port calibration at each probe tip, but for the final step, any thru device may be used, including the DUT if it is reciprocal.

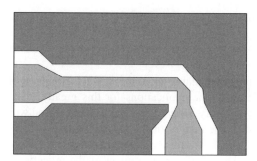

Figure 3.16   Using UT cal to provide a 90° on-wave calibration.

3.4.4.3   Fixed Port Calibration

Another common situation for the unknown thru calibration is a case where the test ports are fixed, and cannot be connected together. This often happens in waveguide-to-waveguide situations, where the waveguides are hard-mounted to a frame or bench, and cannot be easily moved. In the past, this provided a very difficult calibration situation and required moving the waveguides using extra waveguide bends and line sections to provide for a flush thru calibration, or characterizing a waveguide section that could be used as a defined thru standard. Even in the latter case, getting alignment of the defined waveguide thru to the fixed waveguide ports presented challenges.

With the unknown thru calibration, all that is needed is to perform a one-port calibration at each port and then provide any thru connection, including using a flexible cable with coax to waveguide adapters on each end. The only requirement is that the flexible cable, or other means to create a thru, are stable over the time it takes to acquire the six thru ratios (four S-parameters plus the two switch-terms). The one-port calibrations are typically performed using short/offset short/load.

Using a flexible unknown thru standard between the test ports has the potential to minimize significant errors that may be caused by using flexible test ports. In some cases, semi-rigid transmission lines can be used in the place of flexible cables, positioned so that they connect directly to the DUT. This eliminates the need to move them together to provide the normal defined thru connection step.

3.4.4.4   Switched Path Calibration

A common problem for calibration occurs when a switch matrix is added in front of the network analyzer ports (as shown in Figures 2.26 and 2.27), to provide the ability to measure a multiport device one port pair at a time. For an N-port device, there are $(N - 1)(N)/2$ possible paths, each one of which requires a two-port calibration. The unknown thru calibration can be used to measure a thru between one port pair, and use this same data it as an unknown thru for all the other port pairs. From this observation, it becomes clear that all that is needed to calibrate all the port pairs of a multiport test set is a one-port calibration on each port, and one measurement of an unknown thru on any single port pair. From this, the calibration of all the other port pairs can be determined.

3.4.5 QSOLT Calibration

The quick short open load thru (QSOLT) calibration is a different blend between the SOLR (unknown thru) calibration and the SOLT calibration [9]. As the name implies, it requires an open/short/load one-port calibration and a defined thru, but it is quick because the one-port calibration only needs to be performed on one of the test ports. In fact, any one-port calibration method can be used (such as offset short, or even Ecal) on just one of the test ports, and then a defined thru measurement is performed between ports 1 and 2. In this way, a full two-port calibration is easily performed on an insertable path using a Cal Kit or Ecal that has one sex of standards.

One may understand the underlying principle of the QSOLT calibration by realizing that, like the TRL calibration, there are seven unknown terms plus the switch terms. The switch terms can be found similarly to TRL during the thru measurement. The three one-port terms are determined during the one-port calibration, and there are four more equations that can be formed from measurements of the four S-parameters of the thru. Since the actual values of the thru are known, the four measured values yield four more independent equations, providing a means to compute the remaining four unknown error terms.

The QSOLT calibration is very convenient in cases where one-port calibrations are difficult. One use of the QSOLT calibration is in a multiport system. If a multiport DUT has N ports, all of the same sex, one can create an N+1 test system, with a flexible cable on the extra port that matches the DUT connector. A simple one-port calibration on this extra port, plus a thru connection to each of the other ports provides for a full N+1 port calibration. In this way none of the other ports need to move or even have calibration kits for them.

3.4.6 Electronic Calibration or Automatic Calibration

Electronic calibration (Ecal), or automatic calibration, was first introduced by HP in partnership with ATN, which together now form Agilent, for use with the HP 8510 in about 1995, and the term ECal (big C) is associated with Agilent. In this work, Ecal (small c) is used loosely to refer to any of the electronic calibration modules. Since this first introduction, the quality, capability and convenience of Ecal has been greatly enhanced, and now its use may exceed that of the mechanical calibration kits. Examples of some Ecals are shown in Figure 3.17.

Figure 3.17 Ecal modules are available in a variety of port configurations, connectors and frequencies. Reproduced by permission of Agilent Technologies.

The first electronic calibration modules consisted of a transmission line shunted by PIN diodes. If the diodes were reverse biased, the transmission line provided a thru connection from port 1 to port 2. If the diodes closest to the ports were forward biased, an offset short would be created with a small offset. If the diodes farther from the port were shorted, a short with a longer offset was created. The configuration allowed a variety of offsets to be created such that, for any frequency in the specified range, a good spread of reflection coefficients for three of the states could be obtained.

At the time of manufacture, the actual value of each of these reflection states are determined, as well as the S-parameters of the thru state, and these values are loaded into an on-board memory on the module. From these values, a 12-term error correction could be acquired using the formation of Eqs. (3.21) and (3.26) after de-embedding the thru S-parameters from the one-port calibration.

These older versions of electronic calibration could not be used at low frequencies due to the limitation of the length of the offset shorts used. More modern versions of electronic calibration use custom GaAs IC switches which provide an embedded nominal open, short, load and thru. These custom ICs may contain multiple short states in addition to the open state to ensure that a wide phase difference between standards is maintained over the entire frequency range. An example of these states over a small frequency range is shown in Figure 3.18.

The open state (marker 1) has a quite high reflection coefficient, very near the edge of the Smith chart. The short (marker 2), which is limited by series resistance in the solid state

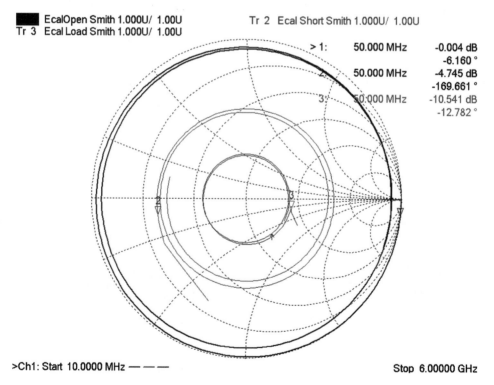

Figure 3.18   Measurement of the internal standards on an Ecal.

FET device, does not provide nearly as large a reflection, but is of opposite phase. The load (marker 3 in the figure) has the lowest reflection, but is not nearly as good as a mechanical load standard.

Because Ecal modules contain solid-state electronic switches, they are very repeatable and stable. Most have an internal heater which maintains a constant temperature within the unit, typically around 31°C, which ensures that it is at a constant temperature over the commonly specified range for VNAs of 20–26°C or sometimes 20–30°C. Because these are electronic switches, they can compress when driven with high power, so it is important to avoid driving the modules above their rated operational power, even though the damage level may be much higher.

3.4.6.1   Calibration Types for Electronic Calibration Modules

Ecal modules can support many of the calibration acquisitions described above, but the default type is unknown thru. Because each of the standards in the Ecal is a characterized device, and must be measured, the uncertainty of the standards is not the same for each standard. For the most part, reflection measurements have lower uncertainty than transmission measurements and so the reflection standards have less uncertainty in their characterization than the transmission standards. The UT calibration does not depend upon the characterization in the Ecal thru, but uses only the characterization data of the one-port standards. This method is referred to as "Ecal thru as unknown thru" as the Ecal unit is connected to both ports at the same time (to perform the one-port calibrations), but the thru connection does not rely upon the saved values for the thru standard

The three-receiver architecture of some lower-cost VNAs does not support unknown thru; in such a case only the SOLT with defined thru is used for Ecal. In other cases, where there is a great deal of loss after each test port, a defined thru may provide less error than an unknown thru when used as an enhanced-response calibration and thus might be preferred (see Chapter 6 for an example). Also, in the case where a QSOLT calibration is desired, the defined thru of the Ecal module might be utilized as well.

Another mode of operation for Ecal provides the ability to separate the one-port calibrations from the thru calibrations, such that a flush thru can be used instead of the Ecal thru, in cases where the test ports are insertable, or form a mated pair. In a related situation such as the fixed connector case, an Ecal may not be able to be connected to both test ports at the same time. In such a case, the Ecal can be used for the one-port calibrations, but another thru, such as a cable, can be used for the unknown thru step. This is simply called an "unknown thru" rather than "Ecal as unknown thru" and divides the normally single-step Ecal into three steps: Ecal at port 1, Ecal at port 2 and unknown thru.

Finally, Ecal modules can be used as calibration kits alongside mechanical calibration kits, where the Ecal ports don't match all the required DUT ports. An example case might be a coax-to-waveguide transition, where the Ecal might be used for one-port of a calibration on the coax side, a mechanical waveguide kit used on for the one-port calibration on the waveguide side and a waveguide-to-coax adapter used for the unknown thru step.

3.4.6.2   User Characterization of Ecal Modules

Another useful feature of Ecal modules is the ability to add adapters, cables or fixtures and characterize the Ecal with the adapter in place. This characterization consists of creating a two-port calibration (four-port in the case of four-port Ecal modules) in the desired connector type, then adding that connector to each port of the Ecal and running a characterization function. During the characterization, each internal standard of the Ecal is measured by the calibrated VNA in the new connector type. The values of these standards are then downloaded back into the Ecal, or saved on the hard-drive of the VNA, to be used later, perhaps in a production test. Current Ecals can support up to 12 user characterizations internally, and an unlimited number on the hard drive.

One convenient feature of user characterization is that it allows the Ecal to be extended beyond its normal frequency range. For example, the 3.5 mm Ecal is limited from the factory to 26.5 GHz frequency operation. However, the internal standards and the connector are substantially mode free until perhaps 33 GHz. It is possible to perform a user characterization of this Ecal using 2.92 mm connectors and a mechanical Cal Kit, such that the Ecal's useful frequency range can be extended. While this calibration is not guaranteed by the manufacturer, the quality can be traced through the mechanical calibration kit.

Finally, it is possible to embed an Ecal into a switching matrix, and perform a user characterization at the end of each of the switch-matrix ports, such that the user characterization can provide calibrations at each of the ports. Figure 3.19 shows an example of a custom calibration test set that has an integrated Ecal module, power sensor module and noise source module. The Ecal characterization is done at the test ports, and the loss to the common power sensor

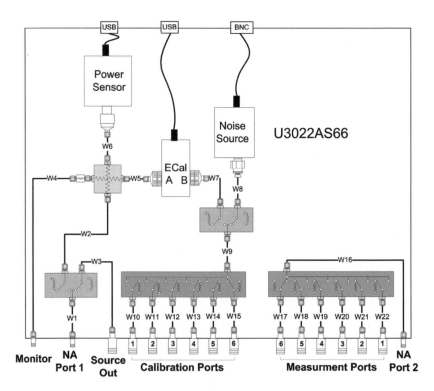

Figure 3.19 Custom multiport calibration test set including Ecal, noise figure and power calibration.

is determined from the difference in the reflection tracking term between the internal Ecal factory characterization and the external test port characterization, presuming that the power divider has equal loss. This kind of custom calibration test set allows multiple measurements, such as noise figure, power, IMD and S-parameters, to be calibrated with a single test port connection. Further, from the picture one can see that the test set includes both the calibration unit and a matrix switch to generate six test ports from the VNA port 2. This calibration system was designed to provide calibration for a 1×6 matrix test system used to test a 1×6 DUT.

3.5 Waveguide Calibrations

When the DUT ports are waveguides, the calibrations that are performed using waveguide calibration kits require some special care to obtain optimum results. The first difference is that a waveguide left with an open circuit does not have a well-defined reflection, thus open standards are not used. Instead, the Cal Kits contain what is called a quarter-wave shim, which represents a 90° phase shift in the center of the waveguide band. The quarter-wave shim over the nominal waveguide band has a phase shift from about 50–120°. In the case of using the shim as part of an offset short calibration, the phase shift at each frequency must be precisely calculated using the waveguide dispersion formula

$$\phi_f = \frac{360f}{c} \sqrt{1 - \left(\frac{f_c}{f}\right)^2} \quad \text{degrees/meter} \tag{3.36}$$

For example, the case of a Ka band waveguide (using WR-28, sometimes called R band), the waveguide band is 26.5–40 GHz, the center of the band is at 33.25 GHz, and the cutoff frequency for the waveguide is 21.081 GHz. A shim that is 90° at 33.25 GHz would have a length of 2.9168 mm, and have a phase shift of 56° at the lower band edge and 119° at the upper band edge.

The calibration process using a quarter-wave shim usually proceeds with a short, load, offset short (using the shim). This provides enough standards to perform the traditional one-port calibration. In the newest VNAs, an additional standard may be used, the offset load standard. In this case, the same shim that is used with the short is also used with the load. This should be used if the impedance and phase shift of the shim are known with less uncertainty than the impedance of the load. This is almost always the case with waveguide components, and so the offset load cal reduces the error caused by a non-ideal load down to the quality of the shim used.

Another aspect of waveguide calibration is that the TRL calibration is perhaps the best and simplest for a two-port system. It requires only a reflect standard (usually a short), a thru and a line. The quarter-wave shim makes an ideal line as it easily meets the criteria of providing a phase shift of 20–160°. In this way it is very simple to create a calibration kit, and use TRL to measure the exact phase shift of the quarter-wave shim.

For multiport waveguide calibration, the QSOLT calibration is well suited provided that the waveguide ports can be moved in such a way to directly mate. In this case, the one-port cal will use up to four standards: short, offset-short, load and offset-load. The remainder of the calibration is to connect a flush thru between ports, but this can be assigned in almost any order, and for N ports, only N−1 thrus are required. For example, consider a DUT with balanced waveguide input ports (e.g., 1 and 3) and balanced waveguide output ports (2 and 4). A multiport calibration can proceed by performing a one-port cal on port 1, then connecting it to port 2 to measure a 1–2 thru, followed by connection of port 2 to port 3, for a 2–3 thru, and finally connection port 3 to port 4 for a 3–4 thru. This order might minimize movement of the waveguide ports. While this minimizes the one-port calibrations required, the residual errors will be cascaded to the other ports. Another approach is to use a one-port calibration on each pair of ports, and calibrate the opposite port, then use unknown thru for the final port pair. For example, consider the case of ports 1 and 3 on the left and ports 2 and 4 on the right. A one-port calibration on ports 1 and 4 can then allow a QSOLT from 1 to 2, QSOLT from 3 to 4 and unknown thru from 1 to 3, 2 to 4, 1 to 4 or 1 to 3. Using a mixture of calibrations methods on multiport situations can provide a dramatic improvement in the efficiency of the calibration steps.

In the past, waveguide standards assumed that the loss of the waveguide was negligible. In most RF and microwave bands, that is a reasonable assumption, but in mm-wave and above, the loss of the calibration standards can produce a noticeable offset in the measured results. Recently, some VNAs have provided the capability to account for loss following a formula that computes loss based on offset delay, and the dimensions (width/height) of the waveguide. The computation follows the formula

$$
L_{Offset\_WG} = \frac{60\pi \ln(10)}{10} \frac{S_{21(dB)}}{S_{21(delay)}} \sqrt{\frac{f_c}{f}} \left[\frac{\sqrt{1 - \left(\frac{f_c}{f}\right)^2}}{1 + 2\left(\frac{h}{w}\right) \cdot \left(\frac{f_c}{f}\right)^2} \right] \tag{3.37}
$$

The use of some features on a VNA, such as port extension or electrical delay, also accounts for the dispersion of the waveguide in computing the effective physical length. Some capabilities, such as the time domain transform, require special consideration when applied to waveguide measurements

3.6   Calibration for Source Power

Traditionally, VNA measurements have been restricted to measurements of ratios of a and b waves. The value of the source power was not important, to the first order, as the ratios of waves that generate the S-parameters, as well as the S-parameters themselves, have always been considered as linear terms, that is, terms that don't depend upon the absolute power levels applied. But many devices are mildly to strongly non-linear, and their performance is specified at particular input powers.

Older VNAs, such as the HP 8510, used external sources, routed through the reflectometer test set. The source power setting of the source was substantially different from the incident power at the test port, sometimes by as much as 10–20 dB. Since then, almost all sources for VNAs are integrated, and a factor calibration ensures that the source power is at least nominally correct. But with the additional of cables, connectors, fixtures and switch matrixes, the actual loss to the DUT can be quite large, and must be accounted for if accurate source power is to be applied.

In the remaining discussion, the details are given for the forward condition, recognizing that the reverse measurements follow similarly.

3.6.1.1   Calibrating Source Power for Source Frequency Response

Starting with the HP 8720A and HP 8753D, the first VNAs with a built-in source and test set, the source power setting on the front panel referred to the incident source power at the DUT. These analyzers also had the ability to control and read a power meter connected via the GP-IB interface. They contained the first built-in firmware that performed source power calibration. The process for source power calibration was a point-by-point measurement and correction of the source power. The calibration process starts by connecting the power meter's sensor to the test port, setting the source to the first point of the frequency sweep data, and acquiring a power meter reading. The source power is then adjusted up or down, according to the offset in measured versus desired power. The offset is recorded and becomes the source calibration factor (SCF). Ideally, this is essentially the STF term in the signal flow graph shown in Figure 2.30. However, in practice, the source is non-linear, and can have errors in addition to the test set loss, so that the source calibration factor becomes

$$SCF = \Delta Src \cdot STF \tag{3.38}$$

Where ΔSrc accounts for any difference in the source setting value and the value incident to the test set. If there is no reflection at the test port

$$a_{1S} = \Delta Src \cdot a_{V_s}\big|_{\Gamma_{Load}=0} \tag{3.39}$$

where a_{V_s} represents the value of the VNA source power setting. Factors affecting ΔSrc are uncompensated loss between the source and test set, and non-linear response of the source output power to different source settings.

Figure 2.30 (also repeated in Figure 3.23) is enhanced below in Figure 3.20 to more closely represent the actual source situation, where reflections from the test port make their way back to the source and reflect off the power source match to affect the incident power. For the simplest case of source power calibration, assuming no reflection from the power meter measurement, the STF term is simply

$$STF = \frac{P_{Meas}}{a_{1S}}\bigg|_{\Gamma_{Load}=0} \tag{3.40}$$

where a_{1S} is the power incident to the test set, which can be monitored on the reference receiver. In general, the value of STF cannot be found directly without an independent measurement, but it is used in correction in a ratio with RTF. The source calibration factor can be found as

$$SCF = \frac{P_{Meas}}{a_{Vs}}\bigg|_{\Gamma_{Load}=0} = \frac{\Delta Src \cdot P_{Meas}}{a_{1S}}\bigg|_{\Gamma_{Load}=0} \tag{3.41}$$

Normally, the power offset is expressed in dB as

$$SCF_{dB} = P_{Meas(dB)} - a_{Vs(dB)}\big|_{\Gamma_{Load}=0} \tag{3.42}$$

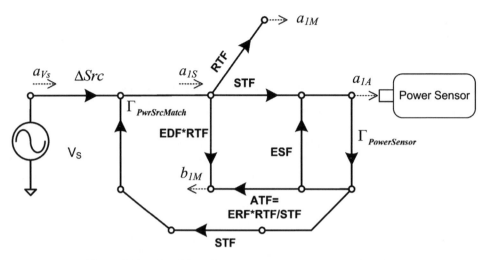

Figure 3.20   Signal flow diagram during source power calibration.

Since the ΔSrc changes with power level, it may be necessary to iterate the source power to achieve some particular value. For normal source power calibration, a cycle of measurement to find SCF is repeated by a user-specified factor, typically 3–10 times. Unlike S-parameter error correction, this is not a post-processed correction; rather, it is a calibration in the sense that the offset values are used to adjust the source settings before the data from the DUT is acquired. If the source is linear, the value for SCF is constant, but sometimes the source is not linear; that is, changing the source setting by 1 dB does not change the value of a_{Vs} by exactly 1 dB. Because of this, the value of ΔSrc takes up the non-linearity of the source, and is only precisely defined at a single power level.

3.6.1.2   Calibration for Power Sensor Mismatch

From the signal flow graph, it is clear that the match of the power meter meter can affect the power measured for a given source setting and thus the extraction of the values for STF and SCF. The value measured on the power meter is essentially a_{1A}, and is related to the VNA programmed source power, a_{1S}, as a function of STF, the source match ESF and the match of the power meter as

$$a_{1A} = \frac{a_{1S} \cdot STF}{1 - ESF \cdot \Gamma_1} = \frac{a_{1M}}{RTF} \frac{STF}{(1 - ESF \cdot \Gamma_1)} \tag{3.43}$$

where Γ_1 is the match of the power sensor. The power meter calibration factor includes a correction for mismatch loss of the sensor such that the power displayed on the power meter actually indicates the incident power to the sensor rather than the power absorbed by the sensor

$$P_{Meas} = \frac{P_{Absorbed} \cdot \Delta PM}{1 - |\Gamma_{Sensor}|^2} = a_{1A} \cdot \Delta PM \tag{3.44}$$

Here, ΔPM is the difference between the power applied and the power meter reading, usually caused by errors in the power meter calibration factor. From Eqs. (3.43) and (3.44) the match corrected value for the source test set loss, STF, with a power meter that has a non-zero match can be computed as

$$STF = \frac{P_{Meas}}{a_{1S}} (1 - ESF \cdot \Gamma_{PwrSensor})$$ (3.45)

In Eqs. (3.40) and (3.41), the values for source correction assume that the power sensor is well matched, but this is not always true, in which case the source calibration factor will be in error. Ignoring the second-order terms, the source power can be approximated as

$$a_{1S} \approx \frac{a_{V_S} \Delta Src}{1 - \dfrac{\left(\Gamma_{PwrSrcMatch} \cdot STF^2 \cdot \Gamma_1\right)}{\left(1 - ESF \cdot \Gamma_1^2\right)}}$$ (3.46)

where a_{V_S} is the power setting from the source and $\Gamma_{PwrSrcMatch}$ is the match associated with the source. Γ_1 is the return loss presented to port 1 by the DUT. The value of a_{1S} can be monitored at the reference receiver, and so while power level may change with mismatch, it can be precisely known. Note that, in general, one does not know the value for $\Gamma_{SrcPwrMatch}$, and this remains a source of uncertainty in the output power. Since the power meter is typically well matched, these errors can be quite small. Modern VNAs also measure the raw match of the power meter during the source power calibration. The value for STF is computed based on Eq. (3.43) where a_{1S} is the power incident to the test set. If the dB loss of STF is large (meaning that there is substantial loss between the reference coupler tap and the test port), the value of a_{1S} is nearly constant and does not change with changes in DUT match. This is the case when some source attenuation has been added, normally in order to provide a lower minimum power in the case of power sensitive devices. However, if the loss of STF is small, and the reflection of the load on port 1 is large, then the error described in Eq. (3.46) can be quite large. Figure 3.21 shows an example of the variation of incident power as the load match is varied through 360° for the case of $\Gamma_1 = 1, 0.5$ and 0.1 (0 dB, -6 dB, -20 dB). The upper plot shows the load contours for the three reflection states. The lower plot shows the incident power, a_{1A}, variation as the load varies. For the largest reflection, we can say that the peak-to-peak variation (1 dB in this case) is exactly the source VSWR, computed as match from

$$PowerSoureMatch = 20 \cdot \log_{10} \left(\frac{1 - 10^{\frac{p-p}{20}}}{1 + 10^{\frac{p-p}{20}}}\right) = 20 \cdot \log_{10} \left(\frac{1 - 10^{\frac{1}{20}}}{1 + 10^{\frac{1}{20}}}\right) = -24.8 dB$$

(3.47)

Thus, the power source match will cause the power of the a1 signal to vary as a result of the DUT loading. Advanced techniques will be described in Chapter 6 that can dramatically reduce this effect, using the a1 receiver as a leveling reference.

From this, three key points should be remembered:

1. The source power calibration provides the proper offset of the source setting to achieve the desired output power into a matched load.
2. Mismatch at port can affect the incident port power, a_{1A}, applied to the load.

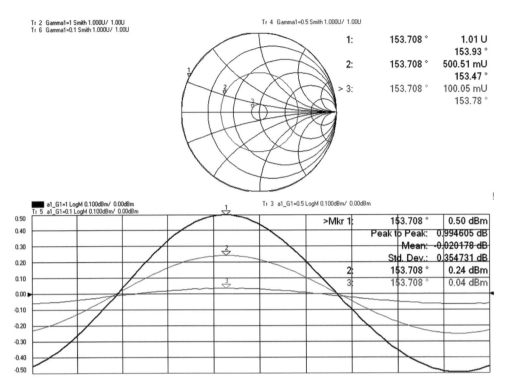

Figure 3.21   Upper: load reflection variation, lower: Variation in a1 due to mismatch on port 1.

3. The reference receiver measurement, a_{1M}, can be used to exactly monitor the incident power, provide that it is properly corrected.

Section 3.7 describes the proper methods for correcting the receivers to measure power.

3.6.1.3   Calibration for Source Power Linearity

The source power from a VNA can have, in addition to errors caused by frequency response, errors due to source power linearity. Linearity describes the accuracy of the source power output with respect to programmed changes in the source power level. Since a VNA receiver is typically more than an order of magnitude better in linearity (0.02 dB typical) than the source (0.2–0.5 dB typical) it is an easy measurement to determine the linearity of the source. Linearity is usually defined as error in the measured power relative to the set power, relative to the preset power. That is, the linearity error does not include source flatness level. The measurement can be performed by normalizing the reference channel power reading at the preset power level, over the frequency span of the analyzer, by using the data into memory and data/memory functions. Next, the source power is set to a new level (usually maximum or minimum specified leveled power) and the power level is read relative to the nominal offset. For example, if the preset power level is −5 dBm, and the measured power on the reference

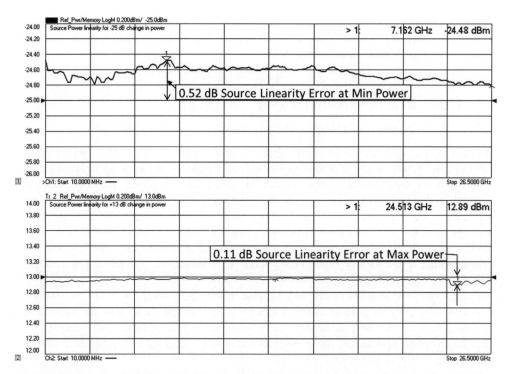

Figure 3.22 Results from a linearity error measurement for −25 dB (upper) and +13 dB (lower) changes in power.

receiver is −5.5, the trace is normalized to read 0 dB. If the power is now changed to −30 dBm, the reading should be at −25 dBc to the original power. Any error from this reference value is a linearity error, as shown in Figure 3.22, upper trace.

The lower trace shows a similar measurement but this time with a +13 dBm change in input power from the reference −5 dBm.

It is normally the lower power regions that have greater error because the ALC detector diodes are operating at much lower signal levels, and any DC drift or offset voltage will have a proportionately larger effect. The power level stability at lower power levels also tends to be worse than at higher power levels. At high power levels, the source harmonics can contribute to linearity errors. The example from Figure 3.22, from an Agilent N5242A PNA-X, shows very little error at high power as the harmonic performance of this VNA is very good, typically better than −60 dBc.

Normally, linearity errors are not corrected for unless the source power calibration is performed in a power sweep mode. Otherwise, only a single value of source power offset is saved for each frequency. This means that if a precise calibrated power is required, the only option available through calibration is to use the desired power for the calibration power. Measurements at other powers will be susceptible to the linearity error. However, advanced power control methods can remove even this error, as discussed below.

Finally, new methods of calibrating the reference receiver and using it as a leveling receiver instead of the internal ALC detector effectively eliminates the need for source power calibration altogether. The details of this capability are discussed in Chapter 6 under the topic of receiver leveling. The accuracy of this method of source power control depends entirely on the accuracy of the receiver calibration, discussed below.

3.7 Calibration for Receiver Power

3.7.1 Some Historical Perspective

The capability of a VNA to measure power has always been present in even the earliest VNAs. However, the power readings always contained the frequency response and roll-off of the VNA test-port coupler and the VNA first-converter frequency response. In VNAs composed of several individual instruments, such as the HP 8510A, no attempt was made to provide a well-known source power, nor was any calibration available to provide any accurate power readings from the VNA. In the VNAs with integrated sources and receivers, such as the HP8753A and HP8720A, the source power was defined at the test port, and the receivers were provided with a factory calibration (called "sampler-cal" or "mixer-cal") that roughly corrected for the frequency response of the VNA receivers.

These factory calibrations tended to be sparse in frequency (perhaps as few as 10 points across the frequency range) and therefore did not account for the fine-grain response of the receivers. As a result, the typical receiver flatness performance was on the order of ±0.5 dB. The complete error model for receiver measurements, repeated from Chapter 2 here for convenience, is shown in Figure 3.23.

Receiver calibrations were very basic, essentially consisting of simple response normalizations of the receiver to a 0 dBm reference power. The essential steps were:

1. Calibrate the source power to be exactly 0 dBm.
2. Connect the source to directly the receiver.
3. Normalize the receiver.

In the case of a reference receiver, step 2 consisted of simply leaving the power meter connected to the test port of the source so the match would be consistent.

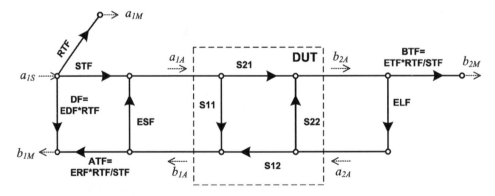

Figure 3.23 Signal flow error model for receiver power measurements.

This process has several drawbacks:

- Any error or drift in the source power calibration became error in the receiver calibration.
- Since only simple normalizations were available, the receiver calibration could only be performed at 0 dBm source power.
- Any mismatch between the source and the power sensor, or the source and the receiver, would cause an error.
- If any adapter was used to connect the source to the receiver, its loss and mismatch would be a direct error in the receiver calibration.

3.7.2 Modern Receiver Power Calibration

In more modern VNAs, the receiver response correction is enhanced by recognizing that during a source power calibration, the incident power is well-known. The reference receiver can be calibrated at the same time as the source power calibration occurs.

In the case where the VNA port is terminated in a well-matched load, the reference-receiver forward tracking term (RRF) is found by

$$RRF = \left. \frac{a_{1M}}{a_{1A}} \right|_{\Gamma_{Load}=0} = \frac{RTF}{STF} \tag{3.48}$$

When a power meter is used to measure the incident power, the mismatch is quite small, and so Eq. (3.48) can be approximated as

$$RRF \approx \frac{a_{1M}}{P_{Meas}} \tag{3.49}$$

For a power measurement at the transmission test receiver, sometimes called the B receiver, the B transmission forward tracking (BTF) is defined as

$$BTF = \frac{b_{2M}}{b_{2A}} \tag{3.50}$$

The normal method for finding BTF is to perform a source power calibration at port 1, connect port 1 to port 2 and assume that $a_{1A} = b_{2A}$. If the reference receiver has been calibrated, then BTF term can be approximated as

$$BTF \approx \left. \frac{b_{2M}}{\left(a_{1M} / RRF \right)} \right|_{S21\_Thru=1} \tag{3.51}$$

Here effects of the power sensor mismatch and port1/port2 mismatch and any loss in the thru connection are not accounted for. However, until very recently, this was the best estimate for receiver tracking that was available in commercial VNAs.

3.7.2.1 Calibration for Power Sensor Mismatch

In the past, a simple source power measurement was the normal way in which the reference receiver tracking was determined. But from signal flow graph, one can account for the match of the power meter as well as find the exact value of reference receiver tracking. From the signal flow graph, it is clear that the match of the power meter can affect the power measured for a given source setting and thus the extraction of the value for RRF. The value measured on the power meter is essentially a_{1A}, and is related to the VNA source power incident to the test set, a_{1S}, as a function of STF, RTF, the source match ESF, and the match of the power sensor as

$$a_{1A} = \frac{a_{1S} \cdot STF}{1 - ESF \cdot \Gamma_1} = \frac{a_{1M}}{RTF} \frac{STF}{(1 - ESF \cdot \Gamma_1)} \qquad (3.52)$$

where Γ_1 is the match of the power sensor. The power meter calibration factor includes a correction for mismatch loss of the sensor such that the power displayed on the power meter actually indicates the incident power to the sensor rather than the power absorbed by the sensor

$$P_{Meas} = \frac{P_{Absorbed} \cdot \Delta PM}{1 - |\Gamma_{Sensor}|^2} = a_{1A} \qquad (3.53)$$

Here, ΔPM is the difference between the power absorbed and the power meter reading, and is called the power meter calibration factor.

From the previous two equations, the reference receiver tracking term can be computed exactly in the presence of mismatch as

$$RRF = \frac{a_{1M}}{P_{Meas} \cdot (1 - ESF \cdot \Gamma_{Sensor})} \qquad (3.54)$$

Note that, with this method, if the reference receiver and the power meter measurement are recorded at the same time, it does not make any difference to the result what the actual value of the a1 source signal is. This removes the requirement for receiver calibration that an accurate source power calibration be performed first. In fact, this computation of receiver calibration can be done at the same time as the source power cal. A one-port error-corrected measurement of the power meter match can be acquired at the same time as the power sensor reading to determine Γ_{Sensor}

3.7.2.2 Response Correction for the Reference Receiver

In legacy VNAs and in most modern VNAs until very recently, receiver correction applied the receiver tracking term in a very simple way:

$$a_{1A\_RcrvCal} \equiv \frac{a_{1M}}{RRF} \qquad (3.55)$$

This simple response calibration does not provide an answer that exactly equal to the actual incident power due to the source match and DUT input match interaction.

To fully understand these effects, consider the block diagram shown in Figure 3.24. The directional coupler is used at the end of the test port to sample the incident power. The match at the end of the coupler is changed while monitoring the incident power on the test port channel and the power reported in the reference receiver. Differences in these power values represent uncompensated mismatch errors. For proper evaluation, one must also consider the loss of the coupler, as described in Section 2.2.2.2.

Figure 3.25 shows the effect of changing the match of the DUT at port 1, on the measured value of the actual incident power, a_{1A}, along with the a1 receiver power reading of the

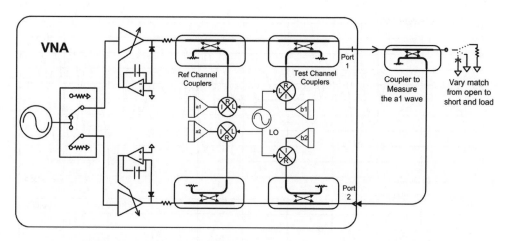

Figure 3.24   Block diagram for characterizing incident power mismatch.

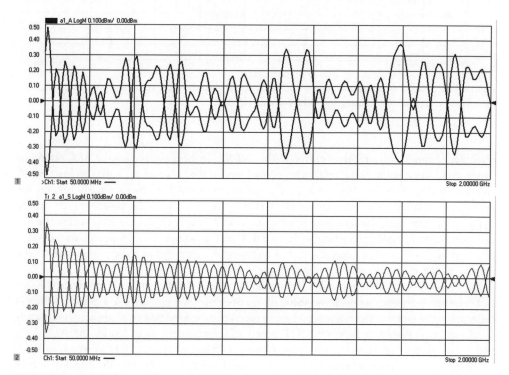

Figure 3.25   Ripple in incident power (a_{1A}) and measured source power (a_{1S}) due to DUT mismatch at port 1.

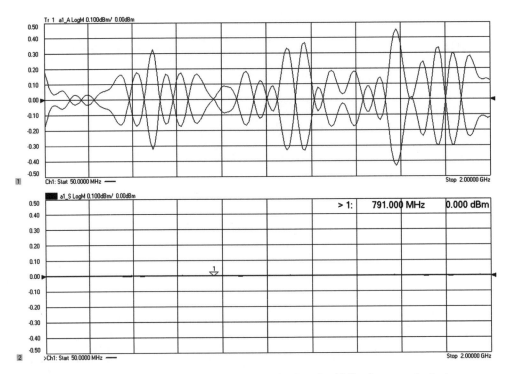

Figure 3.26   Ripple in the actual incident power (upper) when the ALC reference point is the same as the reference channel tap, causing a_{1S} to be constant (lower).

reference channel with a receiver correction, a_{1M}/RRF. In this case, the source ALC reference is not at the same point as the reference channel tap, so the value of a_{1S} varies as load match changes, and so does the value of a_{1M}. The reference channel reading properly reflects the portion of variation of a_{1A} due to the DUT port 1 mismatch with the power source match, $\Gamma_{PwrSrcMatch}$. However, the reference reading *does not* properly reflect the error due to the test set source match, ESF, interacting with the DUT port 1 mismatch, thus the measured value of $a_{1A} \neq a_{1S} \cdot STF$, and in general, a_{1A}, has larger ripple, although in some regions the source ripple can cancel the test port mismatch ripple.

In Figure 3.26, the system is modified so that the ALC reference and a_{1M} reference channel tap point are the same. As such the value of a_{1S} does not change with DUT match, which is reflected in the fact that the reference channel receiver measured power is perfectly flat (lower window). Any mismatch from the DUT port 1 to the source is detected by the ALC loop and compensated for. However, the power meter monitoring the incident power (the VNA calibrated test port in this example, upper trace) still has ripples indicating the effect of source match after the reference coupler is not being removed. Thus, the reference channel power with receiver calibration always reflects the error in source power due to the power source match, but does not reflect the error due to test set ratio source match, ESF, as described in Section 2.2.2.

3.7.3 Response Correction for the Transmission Test Receiver

For the testing power at port 2, the B test receiver response error correction becomes

$$b_{2A\_RcrvCal} \equiv \frac{b_{2M}}{BTF} \tag{3.56}$$

This version of receiver calibration only removes the requirement that the source power calibration be performed at 0 dBm. However, this receiver calibration accuracy is still limited by the source and receiver mismatch terms. Until very recently this was the only calibration method available for power measurements in a VNA.

Figure 3.27 shows the measurement of an amplifier's input and output power (lower window), as well as S21 (upper window). Also shown is the power gain computed as the output power divided by the input power (upper window, dark trace). The fact that there is substantial ripple on the input power is due to the mismatch of the input of the DUT. The fact that there is some ripple on the output power is a natural consequence of the ripple of the input power, as well as the error due to mismatch between the amplifier's output impedance and the load match of the test system. The power gain term in the upper window is simply the ratio of output power to input power, and accounts for the effect of the input power variation to show directly the variation of output power due only to mismatch of the port 2.

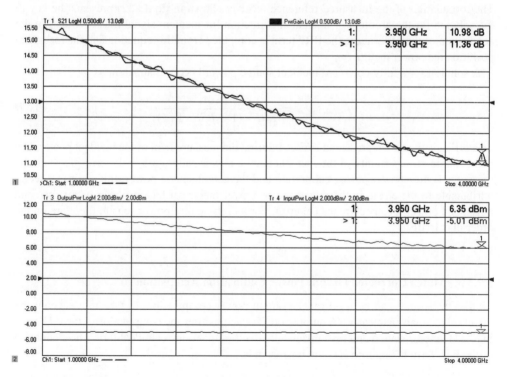

Figure 3.27   Power measurement of an amplifier after a receiver response calibration.

3.7.3.1   Enhanced Power Calibration with Match Correction

Power calibration in a VNA has traditionally been of lower quality than the S-parameter calibrations, and in many test systems, serious power measurements required the additional use of a power meter for only that purpose. Because power sensors have relatively good match, and can be directly connected to the output port of a DUT, they provided a reasonable if inconvenient solution to the problem of accurate power measurements. The VNA provides nearly perfect measurements of gain, which is the key characteristic of a component, and direct measurement of power was seldom part of the VNA test process.

However, starting around the year 2000, VNAs were being increasingly used to measure mixers and frequency converters. Since there did not exist good methods for gain measurements of converters based on ratio measurements, the use of unratioed power measurements became necessary. This resulted in the development of match corrected input and output power measurements for frequency converters. In late 2010, the full support of match corrected power measurements for non-converting devices was introduced in a commercial VNA in the Agilent PNA-X. The details of mixer measurements will be described more fully in Chapter 7, but is based on the enhanced power calibration techniques described below.

3.7.3.2   Match Corrected Incident Power Calibration

The formulation of the measured reference receiver shown in Eq. (3.52) presents the key to the solution for the problem of mismatch correction for incident signal measurement. In the acquisition of the calibration, the RRF term is computed from this equation and the measured power on the power meter.

A solution for a fully corrected incident source power becomes apparent as

$$a_{1A\_MatchCor} = \frac{a_{1M}}{RRF \cdot (1 - ESF \cdot \Gamma_{1M})} \tag{3.57}$$

where Γ_{1M} is the match presented at port 1 during the measurement of the incident power. The main source of uncertainty in the incident source power calibration is the error between the measured source power during calibration and the actual source power; this error is directly attributable to the errors in the power meter: power meter cal factor, power meter drift and noise, and power meter reference cal error. For a high-quality power sensor, these are on the order of 0.15 dB.

3.7.3.3   Match Corrected Output Power Calibration Acquisition

Similar techniques can be used for the output power computation, to take into account the effects of the load match at VNA output port. The acquisition and computation of the BTF term in (3.51) ignores source and load mismatch effects, and requires an ideal thru. The evaluation of the BTF term taking into account mismatch is

$$BTF = \frac{b_{2M}}{a_{1A}} \frac{[(1 - S_{11T}ESF)(1 - S_{22T}ELF) - (ESF \cdot ELF \cdot S_{21T} \cdot S_{12T})]}{S_{21T}} \tag{3.58}$$

Taking the case where the thru is well matched, so S11 = S22 = 0, this simplifies to

$$BTF = \frac{b_{2M}}{(a_{1M}/RRF)} \left. \frac{[1 - (ESF \cdot ELF \cdot S_{21T} \cdot S_{12T})]}{S_{21T}} \right|_{S_{11T}=S_{22T}=0}$$

$$= \frac{S_{21M} \cdot RRF \cdot [1 - (ESF \cdot ELF \cdot S_{21T} \cdot S_{12T})]}{S_{21T}} \qquad (3.59)$$

For a flush thru, this simplifies further to

$$BTF = S_{21FlushThruM} \cdot RRF \cdot [1 - (ESF \cdot ELF)] \qquad (3.60)$$

The computation of BTF can be performed directly, but it can also be computed without additional measurements if a two-port calibration has been performed, by recognizing that Eqs. (3.60) and (3.25) can be rewritten as

$$BTF = ETF \cdot RRF \qquad (3.61)$$

In fact, Eq. (3.61) holds true regardless of the calibration standards used for the full two-port calibration. Thus, if a reference receiver calibration and a full two-port calibration have been performed, any of the test receiver power calibration factors can be computed. Some manufacturers combine the source power calibration, reference receiver calibration and S-parameter calibration in a single guided power calibration function.

3.7.3.4 Match Corrected Output Power Calibration Application

Equation (3.61) provides a very good estimate of the B tracking error term, and will greatly improve the response correction given in Eq. (3.56), removing the effects of port 1 and port 2 mismatch during the acquisition of the error term. However, during measurement of a DUT, the mismatch caused by the S22 of the DUT and the load match of port 2 will not be properly compensated by the simple response calibration.

The match corrected power measurement is computed as

$$b_{2A\_MatchCor\_1-Port} = \frac{b_{2M}}{BTF} \cdot (1 - ELF \cdot \Gamma_2) \qquad (3.62)$$

where Γ_2 is output impedance of the DUT, presented at port 2 of the VNA. A simple one-port corrected S22 measurement is sufficient to determine the value of Γ_2, provided of course that the device is linear.

One might suspect that the output power of a DUT can be fully corrected if the input power to the DUT is known, by recognizing that the power into a matched load is simply the incident power times S21; and the match corrected incident power is found by Eq. (3.57) so that

$$b_{2A\_MatchCor\_Z_0} = a_{1A\_MatchCor} \cdot S_{21\_Full\_2-Port\_Cor} \qquad (3.63)$$

This is subtly different from the upper equation in that it will report the power as though the source and load were perfectly matched. Equation (3.62) gives the power incident from the DUT without altering it for the mismatch error at the input. So, while (3.63) gives the power that would have been measured if the source had been 50 ohms, Eq. (3.62) gives the power that is measured from the existing system's source match. Some advanced techniques described in Chapter 6 use the reference receiver to level the input power so that the effective source match is fully corrected for.

For non-linear devices, such as amplifiers in compression for example, more complex non-linear analysis is required for a complete solution, details of which are described in Chapter 6 under the topic of load pull and X-parameters.

3.7.3.5   Measuring Match Corrected Reflected Power

To complete the measurements of all powers, the match corrected reflected power can be found similarly to the match corrected output power by

$$b_{1A\_MatchCor} = a_{1A\_MatchCor} \cdot S_{11\_Cor} \tag{3.64}$$

where S_{11_Cor} is the one, two or N-port corrected reflection at port 1.

3.8   Devolved Calibrations

The calibrations described in earlier sections provided full correction for all the normal error terms found in a measurement. However, in some instances the full two-port correction will be inconvenient and in other cases will provide poor results depending upon the particular configurations of the measurement system. In these cases, a devolved, or lower-order, calibration may prove a better choice.

3.8.1   Response Calibrations

One instance where a full two-port calibration is impractical in a transmission measurement is where one-port calibration standards might not exist to provide for a full two-port calibration, and so a response calibration will provide a transmission correction even if some errors are not fully corrected for. Examples of this are some in-fixture measurements, measurements using new or unusual connectors and antenna measurements. For these cases, a response-only calibration may sometimes be performed.

The errors in a response calibration come from the source and load match interaction during the calibration, and their interactions with the DUT during the measurement. For response calibration, the errors can be minimized by improving the source and load match of the VNA system. Often, precision attenuators, which have very good match, are added to the test ports to improve the source and load match. Without mismatch correction, the errors in response calibration can be substantial. The measurement of a transmission thru follow Eq. (3.1), where the actual response includes effects of the S-parameters of the thru standard.

There is also a subtle detail when performing response calibrations on many VNAs. The response calibration is nearly the same as normalization. Normalization consists of measuring the thru, then using data-into-memory and data-over-memory to normalize the result to show 0 dB. The response calibration is like normalization, but it normalizes to a defined model, rather than magnitude 1 and phase zero, and using a memory trace is not required. For a "thru" response where the defined thru in the calibration kit is a flush thru (zero length, zero loss), the result is identical to normalization. For reflection measurements, if one does an open response calibration, the resulting trace will have a phase shift that starts at 0 at DC, and increases according to the model for the open circuit. A short response starts at 180° and follows the model of the short. Newer VNAs allow the choice of a pure normalization rather than normalizing to a calibration standard. But if there is some loss or delay defined for the thru, then the response calibration will differ from normalization by setting the loss and delay after calibration to that of the thru model. In this way, a non-zero length, or non-zero loss thru standard may be used for a response cal. Using response calibration allows the memory trace to be used for other things, such as comparison between different DUTs. Also, the response calibration will be interpolated if the span of the measurement is reduced or the number of points is changes; a data/memory trace will not adjust the memory values and will yield an erroneous answer if any stimulus values are changed.

Response calibrations are often used for "quick and dirty" calibration, but the quality of the calibration is frequently misunderstood. After the response calibration is complete, the response of the thru standard will appear to be a perfectly flat trace. This may give the impression that the calibration is of good quality, when in fact it simply means that the measurement after calibration matches the measurement before calibration. Figure 3.28, in the upper window, shows a measurement of a nearly ideal airline after calibration using a full two-port calibration and a response-only calibration.

The response shows substantial ripples, even though a thru response would show a flat line. This is because the errors in the response tracking exactly cancel the source and load match errors for the case where the DUT matches the thru, which occurs when re-measuring

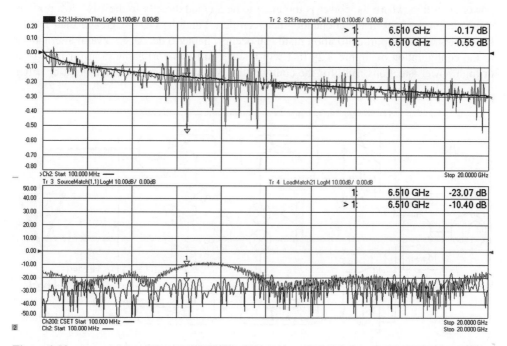

Figure 3.28 (upper) Measurement of an airline with a response calibration and a full two-port calibration; (lower) raw source and load match of the system.

the same thru. But measuring an airline, which also has nearly 0 dB insertion loss but a different insertion phase, generates a substantial ripple. Half the ripple is due to error during the calibration acquisition, and half is from error during the measurement. The source and load match raw error terms are shown in the lower window. It is very clear that the worst case ripple points occur where the combination of source match and load match are highest. At the position of the markers, the combination of source and load match is about −33 dB. The error due to this on the thru measurements is nearly equal to the mismatch error add to the thru, in the linear senses, then converting to dB and multiplying by two, or

$$S_{21Err} \approx 2 \cdot 20 \log_{10} \left(1 + 10^{\frac{ESF_{dB}+ELF_{dB}}{20}}\right) = 40 \cdot \log_{10} \left(1 + 10^{\frac{-33}{20}}\right) = .38 \, dB \qquad (3.65)$$

This exactly matches the error of the ripple at the marker point, which is the difference between the full two-port correction (0.17) and the response correction (0.55) or exactly 0.38 dB. Thus, from knowledge of system source and load match, the error in a response measurement is straightforward to compute.

3.8.2   Enhanced Response Calibration

Another reason for not using a full two-port calibration is when the configuration of the test system causes calibration acquisitions to have substantial error. A typical case of this is where a large loss is added in front of the port 2 test port. This case occurs when testing high power amplifiers where the power is too great to be applied directly to the VNA test port, as shown in Figure 3.29. If the attenuation between port 2 and the DUT is greater than 10 dB, difficulties with two-port calibration math can occur. Chapter 6 describes several alternatives to the configuration of Figure 3.29 that will allow a full two-port calibration for high gain or high power situations. But, for the configuration shown, a response calibration, or better

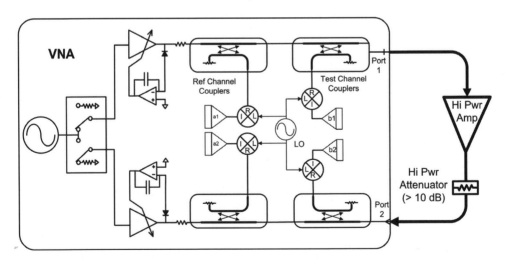

Figure 3.29   Using an external attenuator to reduce power to the VNA port 2.

yet the enhanced response calibration discussed below, may provide a better result than a full two-port calibration.

Another example where a full two-port calibration may have difficulties is when very long cables are used, which have substantial loss and degrade the ability to make reflection measurements at the ports. In such a case, the error terms associated with the port, such as directivity, load match or source match, can have very substantial errors. Further the loss causes the reflection tracking term to be very small and measured data becomes very noisy. The noise and loss in the reflection measurements will result in much noisier transmission measurements that can have very substantial errors. In the case of measuring a high gain amplifier, the reverse measurement, S12, can have substantial noise, or even be entirely noise, due to the high isolation of the DUT S12 and the loss due to attenuation of the test port. Thus if S22 and S12 have substantial errors and noise in their measurements, through the application of Eq. (3.3), they can cause noise in the corrected S21 result. Some specific examples are given in Chapter 6.

In these cases, doing a full two-port calibration can do more harm than good. Further, some test systems, particularly at high mm-wave or THz regions, do not support S-parameter measurements in both directions. For these cases, the enhanced response calibration (ERC) provides a useful addition to the error correction algorithms.

The acquisition of ERC proceeds identically to that of the forward full two-port calibration, but requires a defined thru calibration procedure. In the ERC acquisition, a one-port calibration is performed on port 1, and then a defined thru is measured. The best case is a flush thru, but for non-insertable devices, such as the very common SMA female-to-female case, a short female thru is used. The length of this thru is often ignored, but much better results are obtained if the delay and loss of the thru calibration standard are modified to account for the physical device being used. During the thru measurement, the load match of port 2 is also measured, so that the complete characterization of the forward path is accomplished. As such, the error terms for ERC are identical to the forward full two-port terms and there is no increase in uncertainty at this point.

During the measurement of the DUT, the raw input match and transmission are measured. The input match is corrected with a one-port calibration as in Eq. (3.4), and the transmission is corrected with a modified forward response calibration as

$$S_{21A} = \frac{(S_{21} - EXF)}{ETF \cdot \left(1 + \frac{(S_{11M} - EDF)}{ERF} \cdot ESF\right)} \tag{3.66}$$

which is identical to the result one would obtain by setting the forward load match term, ELF, to zero, in Eq. (3.3). Figure 3.30 shows the measurement of an airline with an enhanced response calibration (light grey trace, trace 1) and with a full two-port cal using the unknown thru method (dark trace, trace 2). The ripple in this measurement is much smaller than that of a response-only calibration (Figure 3.28). This is because half the ripple in Figure 3.28 is from errors in the response tracking term, and the rest of the ripple is caused about equally by the input mismatch and output mismatch, with some caused by the round trip mismatch of S21*S12*ELF*ESF. With the ERC, four of these five terms are compensated. Using the one-port calibration to compensate for Γ_{In} includes effects of the DUT and load as seen from port 1, so that only the term of S22*ELF is not compensated for. For an airline, S22 is very small.

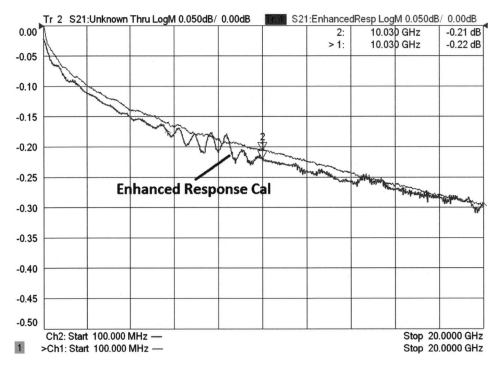

Figure 3.30 Measurement of an airline with enhanced response calibration, with actual airline performance also shown.

In the case of an amplifier measurement, where S12 is very small, the difference between a full two-port calibration and an enhanced response calibration is the effect of the load match term interacting with the amplifier S22.

When a large attenuator is added to port 2, in the case of measuring high power amplifiers, the load match error term is very small, if one uses a well matched attenuator. Thus, using the enhanced response calibration, which ignores the mismatch effect between the DUT S22 and the port 2 load match, can produce a smaller error effect than the noise in the S22 measurement for a full two-port. In this way, a lower-order calibration can provide less uncertainty and error than a full calibration.

3.9   Determining Residual Errors

3.9.1   Reflection Errors

After a calibration is complete, it is often desired to understand the quality of the resulting calibrated measurement. The error terms have been compensated for, but not perfectly, so inaccuracies in the acquisition of the error terms yield errors in the measurement.

The first test for calibration quality is to simply re-measure the calibration standards. This does *nothing* to tell one about the quality of the calibration, but it does show the repeatability and noise in the measurement system. If after reconnecting a standard and re-measuring, the results do not match exactly the model for that standard, then there is a stability or noise issue in the measurement which may dominate the determination of the residual errors.

3.9.1.1 Why is my Open and Short not a Dot?

Perhaps the most common misunderstanding about reflection measurements is the belief that the open standard should produce a single-point or dot at the $\Gamma = 1$ point (right hand of the Smith chart) and that the short standard should produce a single dot at the $\Gamma = -1$ point (left-hand side of the Smith chart). *This is almost always wrong!*

The model for an open is almost always an offset-open with an added fringing capacitance. This yields a phase response equal to the combination of the delay of the open plus the fringing capacitance. The model for a short is almost always an offset delay; at microwave frequencies it may include some small series inductance. For some connectors, it is possible to create a flush short that has almost no delay (virtually all connectors require some recession of the pin so that there is always the slightest gap to avoid damage due to interference at the center pin). The type-N connector has a built-in offset so that a flush short is not possible with type-N connectors.

One can verify that the calibration is repeatable and that the correct Cal Kit and model are used by looking at the Smith chart while displaying the impedance markers. Most VNAs provide a readout of effective capacitance and inductance along with the readout of the R + jB impedance values. The first step is to identify the offset delay of the standard. This can be done by opening up the VNA Cal Kit editor and selecting the appropriate standard. An example standard, a 3.5 mm open, is shown in Figure 3.31, with the offset delay shown as 29.243 psec.

When re-measuring the standard, port extensions can be used to add exactly the value of the delay, and then the marker readout will show the effective capacitance value, as shown in Figure 3.32. The plot shows the open and short results with the offset delay added, and a readout of the fringing capacitance of the open. At the marker frequency (10 GHz), it matches the expected value from the polynomial expression quite closely

$$
\begin{aligned}
C_F &= C_0 + C_1 \cdot f + C_2 \cdot f^2 + C_3 \cdot f^3 \\
&= 49.433 \cdot 10^{-15} - \left(310.13 \cdot 10^{-27}\right)\left(10 \cdot 10^9\right) \\
&\quad + \left(23.168 \cdot 10^{-36}\right)\left(10 \cdot 10^9\right)^2 - \left(0.15966 \cdot 10^{-45}\right)\left(10 \cdot 10^9\right)^3 \\
&= 48.48\text{fF}
\end{aligned}
\tag{3.67}
$$

which is very close to the first-order value, and the value displayed in the Smith chart. The difference is likely to be due to trace noise during the re-measurement. The short, after delay compensation (which is slightly different at 31.785 psec) forms an almost perfect dot as the parasitic inductance is very small. The difference in delay of the open and the short is intentionally set, such that the extra phase caused by the fringing capacitance of the open is almost perfectly compensated by the extra delay from slightly longer center conductor of the short so that the phase offset between the open and the short maintain almost 180° separation even at very high frequencies.

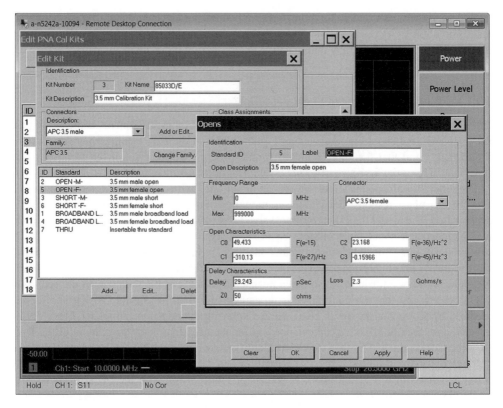

Figure 3.31   The values of a standard can be found in the kit model.

3.9.2   Using Airlines to Determine Residual Errors

Re-measuring the calibration standards does no good in determining calibration quality because the systematic errors behave in such a way that they will exactly cancel when the calibration standards are reapplied; but for any other DUT, the errors will not cancel. For example, if there is an error in the acquisition of the directivity error term, EDF, which is almost always caused by the load return loss being non-ideal (i.e., not equal to zero), that error will propagate through the extraction of the source match term and the reflection tracking term in such a way that when the short standard or open standard is re-measured, the error in source match and the error in load match exactly cancel for that particular standard measurement. However, if an offset short or offset open is measured, the errors add instead of canceling and will yield a large ripple, the peak deviation of which is essentially the combination of source match error and load error. If an offset load is measured, the error is essentially twice the residual directivity error. The best method for establishing the offset is to use a bead-less airline (see Section 1.9).

For these methods to be effective, the quality of the offset airline must be better than the quality of the calibration standards. Fortunately, airlines provide about the best possible impedance reference, and their values are very well known.

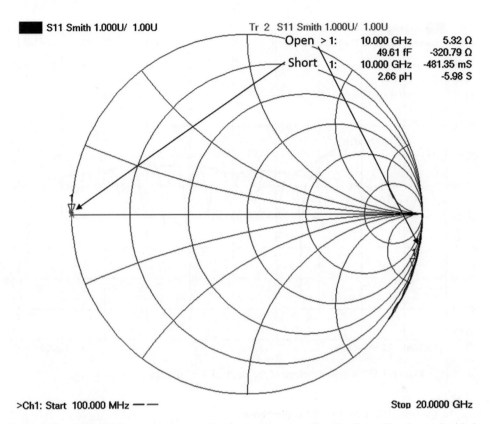

Figure 3.32 A Smith chart measurement of a short and an open directly after calibration, and with the appropriate offset delay.

3.9.2.1 Determining Directivity

Using an airline to determine the residual directivity can proceed in one of two ways. The first method relies on creating a ripple envelope, the peak value of which is twice the residual directivity.

The measurement of any load in the presence of residual errors is

$$S_{11A\_Load} = EDF_R + \frac{ERF_R \cdot \Gamma_L}{(1 - ESF_R \cdot \Gamma_L)} \approx EDR_R + ERF_R \cdot \Gamma_L \cdot (1 + ESF_R \cdot \Gamma_L)$$

$$= EDF_R + ERF_R \cdot \Gamma_L + ESF_R \cdot \Gamma_L^2$$

$$(3.68)$$

where the approximation assumes that the residual directivity and source match are small and the residual tracking is almost exactly one. For even moderately good loads, the measured S11 can be simplified to

$$S_{11A\_Load} = EDF + \Gamma_L \tag{3.69}$$

The residual directivity is the difference between the measured load and the actual load value. In the case where the model of the load is ideal, in that its reflection is presumed to be

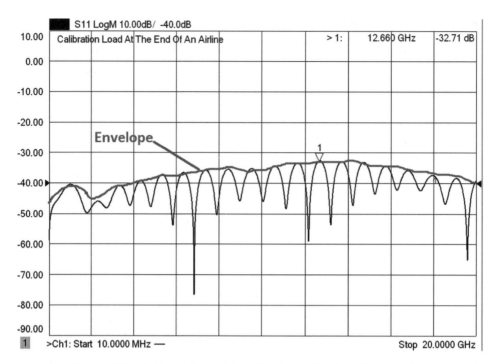

Figure 3.33   Ripple envelope of the calibration load at the end of an airline.

zero, the measured directivity term is defined as

$$EDF_M = S_{11M\_Load} \tag{3.70}$$

and the residual directivity is

$$EDF_R = EDF_M - EDF \tag{3.71}$$

Combining Eqs. (3.69), (3.70) and (3.71), the residual directivity is found as

$$EDF_R = \Gamma_L \tag{3.72}$$

Figure 3.33 is created by measuring the load used for calibration at test port 1, at the end of an airline. The ripple in the figure is caused by the load having different impedance from the airline, which should be at the system impedance. If this load is the load used for calibration, its return loss can be estimated from the airline measurements, and thus the residual directivity can also be estimated. At the frequencies where the length of the airline is a multiple of one-quarter wavelength, the impedance of the load is transformed to

$$Z_{L\_\lambda/4} = \frac{Z_0^2}{Z_L} \tag{3.73}$$

The calibration of the system sets the actual system reference to the load reference, so that the peaks of the load return loss represent twice the load's reflection coefficient (6 dB higher), referred to the desired system impedance. As an example, consider a case where the load has an impedance of 51 ohms, for a residual directivity of −40 dB. The effective Z_0 reference for the system for zero reflection after calibration is 51 ohms. But at $\lambda/4$ points, the impedance presented to the test port by the offset load is

$$Z_{L\_\lambda/4} = \frac{50^2}{51} = 49.02\ ohms \tag{3.74}$$

The reflection coefficient of the load offset by airline, when measured after calibration is

$$\rho = \left|\frac{Z - Z_{0A}}{Z + Z_{0A}}\right| = \left|\frac{49.02 - 51}{49.02 + 51}\right| = \frac{1.98}{100.02} = .0198\ or\ -34\,dB \tag{3.75}$$

where the actual reference impedance is that of the calibration load, or $Z_{0A} = 51$. This is 6 dB above the actual residual directivity value. Note that this method provides values in a simple way, but the estimate of the residual directivity is only at the peaks of the ripples. It is common to draw an envelope of the ripple pattern, presuming that the errors in the loads change slowly with frequency, as shown by the gray line in Figure 3.33. In this case, where the marker is set, the measured peak is −32.7 dB, thus the estimated directivity is approximately 38.7 dB. The source match and reflection tracking errors are negligible when measuring a load of even moderately low return loss. Of course, this method does not work if some characterized device, such as Ecal, is used for the calibration standards, or if a method such as TRL is used.

An alternative method makes use of the time domain transform to separate the reflection of the offset load from the directivity error term. For this method, one must have a sufficiently long line to separate the input error (directivity) from the reflection at the end of the airline. An example is shown in Figure 3.34, where the response of a measurement of an airline is

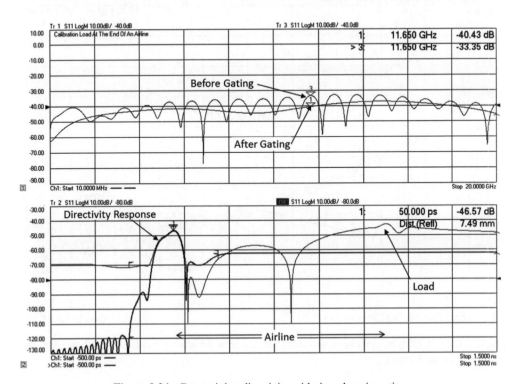

Figure 3.34   Determining directivity with time domain gating.

shown, after calibration, where the load element has about a 40 dB return loss. The upper plot, light trace, shows the S11 of the airline terminated in a load. The lower window shows the time domain response of the S11, along with the gated response around the input connector. The darker trace in the upper window shows the gated frequency response, which is the directivity.

3.9.2.2 Determining Source Match and Reflection Tracking Residual Errors

Using either an open or a short at the end of an airline creates a ripple pattern in the S11 trace. An example is shown in Figure 3.35 where both the open and short responses are shown. The light trace, labeled "Mismatch" shows the peak-to-peak response (in dB) of the error, after removing the effect of the loss of the airline. This is simply done by taking the open data divided by short data, and expressing the value in dB. The loss curve (trace 3 in the figure) is created by taking the square root of open data times the short data and matches very well to twice the through loss of the airline loss. If the loss of the airline is known, this can be compared to the measurement result, and from this errors in the reflection tracking can be inferred. However, the reflection tracking errors are typically quite small and are difficult to quantify with airline techniques.

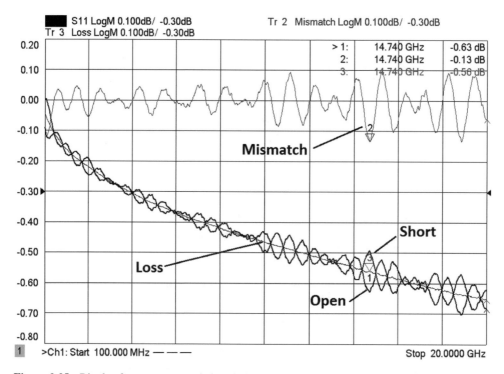

Figure 3.35 Ripples from an open and short help describe the source match in magnitude, with the loss and mismatch ripple.

The envelope of the ripple represents a combination of the reflection tracking error, the source match error and the load match error. The measurement of an open after calibration is

$$S_{11M\_open} \approx EDF_R + \frac{ERF_R}{[1 - ESF_R]} \approx EDF_R + ERF_R + ESF_R \qquad (3.76)$$

where here it is assumed that the open and short reflections have a magnitude of 1, and the residual error terms (represented by the subscript R) are relatively small; that is directivity and match are nearly zero, and reflection tracking is nearly one. From the signal flow diagram of Section 3.3.1, the source match is a combination of the errors in the open/short difference, and the residual directivity, so that

$$ESF_R \approx EDF_R + \frac{(\Delta O - \Delta S)}{2} \qquad (3.77)$$

where ΔO and ΔS are the differences between the model of the open and the actual open, and the model of the short and the actual short, respectively. In practice, the open and short are very well known, so the residual source match is almost entirely due to the residual directivity. Thus, the load response is the main source of error in the calibration kit. Errors in the load response will appear in the source match term for using SOL calibrations, forming a floor in the source match value. That is one reason why techniques such as TRL or even Ecal (which is characterized with a TRL traceable calibration) have better theoretical specifications than SOLT calibrations. A sliding load calibration also derives its directivity error from the quality of the airline used in the load, and so is close to the quality of a TRL calibration. In reviewing corrected system specifications, one will note that the source match is usually equal to or worse than the directivity. In some newer high-frequency calibration kits, such as the 1.85 mm or 1.0 mm, several standards are used to provide an over-determined solution; in such a case the simple rules for estimating residual error sources don't apply.

The source match error term can be estimated by knowing the directivity error, and the additional ripple must be due to a combination of reflection tracking and source match. An upper limit on the source match error can be determined if the entire mismatch error is assumed to be source match. Using both a short and an open helps to define the envelope of the ripple.

From the figure, the worst case ripple is 0.13 dB, which represents the VSWR of the source match. The equivalent source match in dB can be computed for the marker value of 0.13 dB as

$$ESF_{Resdiual\_dB} \approx 20 \cdot \log_{10} \left(\frac{1 - 10^{\frac{OS_{Ripple(dB)}}{20}}}{1 - 10^{\frac{OS_{Ripple(dB)}}{20}}} \right) = 20 \cdot \log_{10} \left(\frac{1 - 10^{\frac{0.13}{20}}}{1 + 10^{\frac{0.13}{20}}} \right) = -42dB$$

$$(3.78)$$

The value can be computed directly on the VNA if the equation editor function is used, and recognizing that the open data is in the Tr1 data and the short data is in the Tr1 memory. The trace values are stored internally in the linear form, and the results can be computed in terms

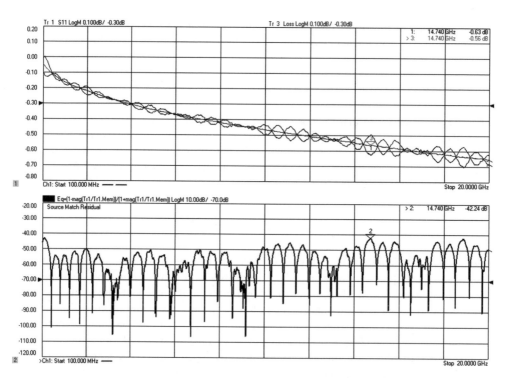

Figure 3.36   The computed residual source match shown in the lower plot.

of the linear source match; the equation to use is simply

$$ESF = \left(\frac{1 - \left| Tr1 / Tr1.Mem \right|}{1 + \left| Tr1 / Tr1.Mem \right|} \right)$$

(3.79)

This is shown in Figure 3.36 in the lower window. One might notice that the residual source match appears to rise at the low frequency portion of the trace. This is a measurement artifact arising from the fact that the airline impedance actually increases due to skin effect, and so its use to characterize source match is limited. Also, the lower trace represents the combination of source match and directivity errors, along with errors in the airline itself. Using the envelope of the peaks of the ripples gives a good conservative estimate of the source match. In this case, -42 dB is consistent with the previous measurements of directivity error.

3.9.2.3   Reflection Tracking Residuals

The reflection tracking, from Eq. (3.20), can be considered as the dB average of the open and short response or the geometric mean of the linear terms. The evaluation of the residual error can be computed by considering the case where the reflection tracking is defined as 1, and the actual open differs from the ideal open by ΔO, so the actual open is $(1 + \Delta O)$, and the short differs by ΔS, so the actual short is $-(1 + \Delta S)$. The derivation here follows the common practice when dealing with residuals of neglecting products of error terms (which are very small). It is instructive to go through the derivation for the error in ERF:

$$ERF_R = \frac{ERF_{Computed}}{ERF_{Actual}} = \frac{ERF_{Computed}}{1} = \frac{-2\left[(1+\Delta O) - EDF\right]\left[(-1-\Delta S) - EDF\right]}{(1+\Delta O)(-1-\Delta S)}$$

$$= \frac{-2\left[(1+\Delta O)(-1-\Delta S) - EDF \cdot (-1-\Delta S) - EDF \cdot (1+\Delta O) + \cancel{EDF^2}\right]}{1+\Delta O - (-1-\Delta S)}$$

$$= \frac{-\cancel{2}\left[(-1-\Delta O - \Delta S - \cancel{\Delta O \Delta S}) + \cancel{EDF} - \cancel{(\Delta S)} - \cancel{EDF} - \cancel{(\Delta O)}\right]}{\cancel{2}\left(1 + \frac{(\Delta O + \Delta S)}{2}\right)}$$

$$ERF_R = \frac{(1 + (\Delta O + \Delta S))}{\left(1 + \frac{(\Delta O + \Delta S)}{2}\right)} \cdot \frac{1 - \frac{(\Delta O + \Delta S)}{2}}{1 - \frac{(\Delta O + \Delta S)}{2}} = \frac{1 + \frac{(\Delta O + \Delta S)}{2} - \cancel{\frac{(\Delta O + \Delta S)^2}{2}}}{1 - \cancel{\left(\frac{(\Delta O + \Delta S)}{2}\right)^2}} = 1 + \left(\frac{\Delta O + \Delta S}{2}\right)$$

$$(3.80)$$

From this result, one can infer several attributes of the reflection tracking term, first among them being that for small residual errors, the error in the load does not contribute to the reflection tracking error. Also, it is the average of the open/short errors that contribute to the reflection tracking, so that if the short is a little long, and the open is a little short, the errors will cancel, leaving the reflection tracking with no error. It is only when the error in the open and the short are in the same direction that the reflection tracking picks up the error.

A similar analysis of the source match error due to open and short error yields the result that, in addition to a contribution of the load error, the source match error is proportional to the difference between the open error and the short error; thus, if the open is a little long and the short is a little short, the source match error is large. In a similar way, if the open and short are both a little long, the source match error is not affected.

In practice, the magnitude of the open and short, especially for mechanical standards, is well known, so the reflection tracking error is negligibly small. However, the phase of the open and short depend directly on the fact that their length matches the model for each, so that the phase error of the standards translates directly to a contribution to source math error, and can contribute to reflection tracking error. Examination of the residual errors specified for calibration kits will show that reflection tracking can almost always be ignored in considering the overall error contribution in reflection measurements.

In a normal situation, where the residual reflection tracking is nearly 1, the residual source match is typically a little worse than the residual directivity. But in some situations, where the coupler raw directivity is poor, such as when a lot of loss is added after the directional coupler, perhaps in the form of a long cable or attenuator pad, the residual directivity can be much worse than the residual source match. In a normal analysis of the error terms, one would find no difference with this test configuration, as only the load match of the calibration load is considered. But in the case of substantial loss after the coupler, the drift and noise in the raw directivity become the limiting factors in the directivity error term.

Consider the case of a large attenuator on port 1 which has been corrected with a careful low-noise calibration. Drift in the test-port coupler will give a large change in directivity, but almost no change in source match. In fact, the raw source match in this case might be quite good as well.

For the most part, uncertainty computations ignore drift in the system but for many test scenarios it is the dominant error. In these cases, the source match is often very good, as the attenuator or lossy cable presents a good match to the DUT. When such a situation occurs, using a devolved calibration such as enhanced response cal or even response-only calibration, with no mismatch correction, may provide a better result. This is due to a poor or unstable EDF term which in turn will yield a bad value for the DUT S11, and from that a bad value for the match correction.

3.9.2.4 Load Match Residual Error

The load match error term is measured as part of the transmission thru measurement. In the case where the thru standard is a flush thru, the error in the load term is essentially the same as the error in the directivity term. From Eq. (3.68), the source match and reflection tracking are both multiplied by the raw load match.

$$ELF_R = ELF_{Cal} - ELF_A = \left(EDF_R + ERF_R \cdot ELF_A + ESF_R \cdot ELF_A^2\right) - ELF_A$$

$$\approx \underline{ELF_A(ERF_R - 1)} + EDF_R + \underline{ESF_R \cdot ELF_A^2} \tag{3.81}$$

$$ELF_R \approx EDF_R$$

Since the raw load match is typically reasonably low, and the reflection tracking is typically very near one, the first product in the equation above can usually be neglected. The source match term, while typically larger than the directivity error term, is multiplied by the square of the load match term and can also be neglected. Thus, the residual load match is essentially equal to the residual directivity of the other (mating) test port. A quick example illustrates this: consider a VNA system with 15 dB raw load match (0.18 linear), 40 dB directivity (0.01 linear), 30 dB source match (0.032 linear), and 0.1 dB reflection tracking (1.012 linear). Converting all the terms to linear, Eqs. (3.81) becomes

$$ELF_R = (0.18)(1.012 - 1) + (0.01) - (0.032)^2(.18)$$

$$ELF_R = (0.0022) + 0.01 + (0.0002) \tag{3.82}$$

$$ELF_R \approx 0.01$$

Verifying the residual load match is quite similar to verifying the directivity, with a similar method. The directivity is verified by adding the calibration load to the end of an airline, and looking at the peak value. Alternatively, the time domain gated response of the airline plus load can be determined, as discussed in Section 3.9.2.1, to give the directivity. Once this is done, the airline can be connected to port 2, and the response can be viewed. Since the directivity error is known, the peaks of the response represent the sum of the directivity and the load match. Typically, this matches the result of using the same calibration load to terminate the airline because the load match and directivity error terms should be very similar. However, if the load port is connected to the VNA with a test port cable (the most common case) then cable flexure will add to the measured load match, and must be considered a part of it.

For the case where a non-flush-through is used during the calibration, such as a defined thru, then an additional error source occurs if the match of the defined thru is not zero, or if the delay and loss of the defined thru does not match the model in the calibration kit. In legacy VNAs, before the introduction of the unknown thru method, it was very common to have large errors in the load match, principally because the DUT had non-insertable connectors, and a thru adapter was used to connect the ports, but the Cal-Kit was not modified to add the delay and loss of the thru adapter. In these cases, the phase of the load match is incorrectly determined. The error effect of this depends upon the length of the adapter, but if it approaches one-quarter wavelength in the frequency of interest, the error due to the adapter is in fact up to 6 dB larger than if no port match correction was done at all. For the case of low-loss filter measurement, this was the most common reason for test engineers complaining about ripple in the transmission measurement. Fortunately, modern VNAs provide more techniques, such as Ecal and unknown thru, to avoid this issue. Some VNA calibration engines (for example, the PNA SmartCal^{TM}) require the user to enter the connector types and sexes for DUT ports, and will not allow a defined thru calibration if the Cal Kit does not have a thru adapter defined that matches the connectors of the DUT. In these cases, if a thru standard is desired to be used, one must modify the calibration kit to add a thru with the properly defined ports, most typically a female-to-female thru. The delay of the thru must be entered and can be reasonably estimated by a number of means including measuring the physical length, if it is an airline thru. Alternatively, using one-half of the S11 reflection delay measurement of the thru standard, after a one-port calibration, can be used, but one must consider the open fringing of the thru if it is not terminated (also a common practice) or account for the delay of the termination (short or open) if it is terminated.

To illustrate the measurement of the load match residual, consider the two cases shown in Figure 3.37. The upper trace shows a ripple in the S11 from a measurement of an airline, after a SOLT defined thru calibration (dark, marker 1). Here the thru adapter is ignored (the thru adapter in this case is one found in many calibration kits for the purpose of the "swap-equal-adapter", and is about 27 mm long). Also shown is a good unknown thru calibration (light, marker 3). The figure shows much worse S11 in the case of the defined thru, because the load match is not properly corrected for. The error represents a residual load match as bad as 30 dB compared to better than 40 dB for the unknown thru calibration. The lower window shows the equivalent S21 traces. The darker trace is from the defined thru calibration, ignoring the thru, and it has both lower loss (incorrectly showing lower loss because the loss of the thru adapter is not properly accounted for) and higher ripple, due to a poor load match acquisition.

A somewhat unexpected result is that in normal systems, where the raw terms are not too large, the residual errors for reflection depend *only* on the calibration kit and do not depend upon the quality of the test system. If the test system is stable, then a system with poor directivity and one with good directivity would yield the same results after correction. From this point of view, there is no benefit in trying to make a well-matched test system with good directivity. However, if there is any drift in the system (and there always will be) the residual performance will degrade more quickly on a system with poor raw performance.

3.9.2.5   Transmission Residual Errors

Transmission errors are evaluated in a manner quite similar to the load match evaluation. In this case a transmission measurement of the airline is used, and any ripple on the S21

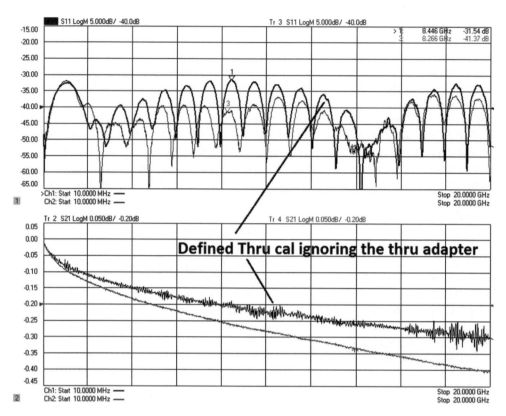

Figure 3.37 Measurement of a test port load match; upper is with a good calibration, lower is with an erroneous defined thru cal.

measurement is associated with transmission tracking error. The expected residual tracking error can be computed from Eq. (3.1), by considering the effects of the other residual values on the computation of the tracking error. For a flush thru calibration, the value of ETF is found from the measured S21 value and the values for the other error terms:

$$S_{21M\_Cal} = \frac{(S_{21A} \cdot ETF)}{(1 - S_{11A} \cdot ESF) \cdot (1 - S_{22A} \cdot ELF) - ESF \cdot S_{21A} \cdot S_{12A} \cdot ELF} + EXF$$

$$S_{21M\_Cal} = \frac{ETF}{1 - ESF \cdot ELF}\bigg|_{S_{11A}=S_{22A}=0,\, S_{21A}=S_{12A}=1} \qquad (3.83)$$

$$\therefore ETF = S_{21M\_Cal} \cdot (1 - ESF \cdot ELF)$$

The actual values for ESF and ELF are not used when computing ETF; rather the extracted values (or measured values) of the match terms are used in the computation. The distinction is that measured values are only estimates of the actual value. The residual tracking can be found taking the ratio of the computed tracking over the actual tracking

$$\Delta ETF = \frac{ETF_M}{ETF} = \frac{S_{21M\_Cal}\,(1 - ESF_M \cdot ELF_M)}{S_{21M\_Cal}\,(1 - ESF \cdot ELF)} = \frac{[1 - (ESF + ESF\Delta) \cdot (ELF + \Delta ELF)]}{(1 - ESF \cdot ELF)}$$

$$= \frac{\left(1 - ESF \cdot ELF - ESF \cdot \Delta ELF - ELF \cdot \Delta ESF - \cancel{\Delta ESF \cdot \Delta ELF}\right)}{(1 - ESF \cdot ELF)} \cdot \frac{(1 + ESF \cdot ELF)}{(1 + ESF \cdot ELF)}$$

$$= \frac{[1 - \cancel{(ESF \cdot ELF)^2} - (ESF \cdot \Delta ELF + ELF \cdot \Delta ESF) - \cancel{(ESF \cdot ELF)(ESF \cdot \Delta ELF + ELF \cdot \Delta ESF)}]}{1 - \cancel{(ESF \cdot ELF)^2}}$$

(3.84)

In the derivation, it is assumed that the values of the source and load match are substantially less than one, and the raw error terms are larger than their residual terms. In the simplification, higher-order terms are neglected where the residuals are considered to be on the order of the square of the associated error term. This means that the square of the product of error terms is a fourth-order term, and the product of an error term and a residual is a third-order term. Simplifying with these considerations yields the residual tracking term as

$$\Delta ETF = 1 - (ESF \cdot \Delta ELF + ELF \cdot \Delta ESF)$$ (3.85)

This equation is most significant and represents an important result in understanding the errors in a test system. Unlike the residual errors in the one-port terms, which depend only upon the quality of the calibration standards, the transmission tracking terms depend on both the quality of the calibration standards *and* the raw source and load match of the test system. The load match term has a similar dependency. This is in contrast with the result for one-port, where the raw system performance did not affect the residual error terms, ignoring drift and stability effects.

Because of this, modifying the test system to improve the match at the DUT will reduce the residual tracking error. This is typically accomplished by adding an attenuator pad to the test port cable, right at the DUT reference plane. While this will improve the transmission tracking, the loss after the directional coupler will cause any drift in the directivity of the system to be amplified by twice the attenuator's value. Further, source power is lost as well as receiver sensitivity. But, in many cases, a small attenuation (on the order of 3 or 6 dB) can reduce the residual errors in a transmission measurement without adversely affecting other measurements.

The importance of Eq. (3.85) can be seen if one looks at a manufacturer's data sheet for residual errors after calibration. In the case of mechanical calibration kits, the transmission tracking error nearly always computes to exactly the value predicted in Eq. (3.85) if the source and load match of the system, and the residuals of the Cal Kit, are used.

These are the important points to remember about airline ripple techniques:

1. They provide estimates of magnitude errors, but not estimates of the phase or delay errors. Some phase errors may be estimated as described in Section 3.12.
2. Imperfections in the airlines, and in particular in the end connectors of the airlines, will limit the level to which verification of residual errors can be determined.

3.10   Computing Measurement Uncertainties

Now that methods have been described to determine the residual errors of the measurement system, the overall uncertainty of any measurement can be determined. It's important to note that these errors apply to all RF measurements, whether they are on sources, spectrum analyzers, power meters or VNAs. With the exception of VNAs, the effect of source match and load match errors in the measurements cannot be removed with error correction (as the errors' effects depend on both the magnitude and phase of the interaction of the errors and the signals to be measured, and only the VNA measures the phase of both) and so the resulting uncertainty will depend on the raw match terms. For VNAs, the uncertainty depends upon both the raw match and the residual match, as will be discussed below.

In this discussion, the term "uncertainty" is sometimes used interchangeably with the term "measurement error", but strictly speaking the uncertainty is a derived value that bands the actual error. In many cases, several error terms are present so that at any particular frequency, some might cancel and the actual difference or error becomes small, even while the uncertainty remains as a computed limit-band of the errors.

For reflection, uncertainty is defined as a difference (in linear terms) between the reading and the true value. This may then be added to, and subtracted from, the measurement to come up with the upper and lower linear limits, each of which may then be converted to a dB uncertainty limit.

For transmissions terms, including source power, receiver power and S21, the uncertainty is defined as the ratio (in linear terms) between the reading and the true value. This can be converted to a dB error by taking $20 \log_{10}$ of the ratio.

3.10.1   Uncertainty in Reflection Measurements

The uncertainty of a reflection measure can be derived by comparing the S11 corrected with the actual error terms to that corrected with the estimated error terms. From (3.4), the difference between the actual S11 and the error corrected S11 can is defined as

$$S_{11A} - S_{11\_1PortCal} = S_{11A} - \frac{(S_{11M} - EDF_R)}{[ERF_R + (S_{11M} - EDF_R) \cdot (ESF_R)]} \tag{3.86}$$

From this it is clear that only the residual error terms play a part in the uncertainty of a one-port corrected measurement, and that the raw error terms play no part in the overall uncertainty. While true in the abstract, in fact the raw error terms do play a significant role in the overall uncertainty if there is any system drift (cable drift, coupler drift, receiver conversion loss drift) over the measurement period. These drifts can dominate the overall uncertainty if the raw error terms are large, even if the residuals are quite small.

3.10.2   Uncertainty in Source Power

For any system using a source, for example driving a complex modulated signal into a receiver from a vector signal generator, or driving a power amplifier from a VNA, the uncertainty of the source power depends upon the three errors: the source tracking error, STF; the source match, ESF; and the effective input match of the DUT, Γ_1. The uncertainty (in dB) can be computed directly from (3.43) in Section 3.6 as

$$\Delta SourcePower = \left|20\log_{10}\left(\frac{a_{1A}}{a_{1S}}\right)\right| = \left|20\log_{10}\left[\frac{STF \cdot \left[\left(a_{1S}/a_{1R}\right) \cdot SourceLinearity\right]}{1 - |ESF \cdot \Gamma_1|}\right]\right|$$

(3.87)

Where a_{1S} is the source power setting and a_{1A} is the actual source power applied to the DUT, and a_{1R} is the reference power for the source amplitude specification or measurement. Source linearity represents the error that occurs when the source power is changed for a calibration or reference value and is often represented as a dB error for particular dB change in value, and must be converted to linear terms for use in the equation above. Here the absolute value of the product of $ESF \cdot \Gamma_1$ is used to give worst case uncertainty, and the absolute value is taken for the whole equation because uncertainty is traditionally expressed as an absolute value.

In the case of a complex modulated source, the modulation produces power over a range of frequencies and the match of the DUT and the source may vary over the that range of frequencies. If the signal modulation is narrow, the source match and load match are likely slowly moving and the resulting source power error occurs similarly to entire modulation envelop, yielding only an error in the modulation power. However, if the signal is wideband, then the variation of the source power due to mismatch can cause amplitude and phase uncertainties in the wideband signal.

The error term STF is essentially the error in the source power, when it drives into a perfect 50 ohm matched load. For signal sources, this error can be caused by amplitude offset error and by source flatness and source linearity errors. If a power meter is used to calibrate the source power, then the STF becomes the residual source tracking error as defined in Eq. (3.45).

These same errors apply to the source power for a DUT driven by a VNA source. In the case of the VNA, the ESF term is not strictly correct as that is usually associated with the ratio source match, where the error depends upon the power source match as defined in Section 3.6. However, in the case of using enhanced power calibration with receiver leveling, the error terms in Eq. (3.87) become residual error terms. In general, source power errors don't affect any of the ratio measurements (e.g., S-parameters) in a VNA measurement, but certainly will affect gain measurements when using a signal source and other receiver such as a power meter or spectrum analyzer.

3.10.3   Uncertainty in Measuring Power (Receiver Uncertainty)

Just the analysis of Section 3.6 provided the basis for computing source uncertainty, so too can the analysis of receiver calibration in Section 3.7 be used to compute the uncertainty of receiver measurements. Similar to source uncertainty, the receiver uncertainty computation applies equally to all types of receivers including power meters, spectrum analyzers and VNA receivers. Also, for all receivers except VNA receivers, the uncertainty in power readings depend upon the raw output match of the DUT, Γ_2, the raw input match of the measuring receivers, ELF, and the receiver tracking, BTF. In power meters the receiver tracking is a combination of the reference calibration error, the calibration factor (or cal factor) accuracy, and the power meter linearity. For a spectrum analyzer, it depends upon the amplitude flatness calibration (which in itself may depend upon the match and accuracy of the calibration source) and SA linearity. The uncertainty of a measuring receiver can be computed from (3.62) as

$$\Delta RcvrPower = \left| 20 \log_{10} \left(\frac{b_{2A}}{b_{2M}} \right) \right| = \left| 20 \log_{10} \left\{ \frac{(1 - |ELF \cdot \Gamma_2|)}{BTF} \left[\left(b_{2M} / b_{2R} \right) \cdot R_{DA} \right] \right\} \right|$$

(3.88)

where b_{2A} is the actual power at the receiver (in this case, the b2 receiver), b_{2M} is the measured power, R_{DA} is the receiver dynamic accuracy and b_{2R} is the reference power for the receiver dynamic accuracy. In practice, receiver dynamic accuracy is given as a dB error for some dB change in power, and so must be converted to linear form to be used in the equation above. In practice, for modern VNAs, the receiver dynamic accuracy can be very small; for example, the Agilent PNA-X has less than 0.01 dB error over 80 dB of power change, and so generally be neglected; a typical power sensor has an error of 0.004 dB over a 10 dB range, plus a range-error offset for each range that is traversed.

For power meter and SA measurements, the errors are all in terms of raw or actual system errors. For VNA measurements, the BTF term is always a residual term, after error correction, and is computed as in (3.60). For response-only corrections, the ELF term is an actual (raw) load match; for match corrected power measurements, the residual value of the ELF term should be used.

3.11   S21 or Transmission Uncertainty

While there are many contributors to S21 uncertainty, it is dominated by the source and load match, both raw and residual, of the test system. The uncertainty of an S21 measurement can be derived from the results of Eq. (3.1) as

$$\Delta S_{21} = \left| 20 \log_{10} \left(\frac{S_{21M}}{S_{21A}} \right) \cdot \left[\left(\frac{b_{2Cal}}{b_{2M}} \right) \cdot R_{DA} \right] \right|$$

$$= \left| 20 \log_{10} \left[\frac{ETF}{(1 - |S_{11A} \cdot ESF|) \cdot (1 - |S_{22A} \cdot ELF|) - |ESF \cdot S_{21A} \cdot S_{12A} \cdot ELF|} \right] \cdot \left[\left(\frac{b_{2Cal}}{b_{2M}} \right) \cdot R_{DA} \right] \right|$$

(3.89)

where b_{2M} and b_{2Cal} are the power levels at the test receiver during measurement and calibration, respectively, and R_{DA} is the receiver dynamic accuracy. If this is an uncorrected S21 measurement, or a gain measurement using a signal source and power sensor or SA as a receiver, then the error terms are the raw error terms. If this is a VNA measurement, and a calibration has been performed, then the error terms are residual error terms. In many cases, a response calibration is performed, ignoring mismatch, so the transmission tracking term becomes a residual term, but the other match errors that remain are the raw match terms.

The source and load match terms are easily understood, but the transmission tracking error term must be derived from the thru measurement.

The flow graph for an ideal flush thru is shown in Figure 3.38. From this flow graph, the ETF error term can be derived from the measured value of b_{2M}/a_{1M} as

$$\frac{b_{2M}}{a_{1M}} = \frac{ETF}{(1 - ESF \cdot ELF)}$$

(3.90)

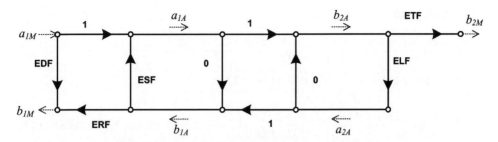

Figure 3.38   Signal flow graph for a thru measurement.

However, response calibrations use a simple normalization so that the value derived for ETF, that is, the estimated tracking term $E\hat{T}F$ is simply b_{2M}/a_{1M}, and the residual error in the tracking term is

$$\Delta ETF_{\text{RespCal}} = \left| 20\log_{10}\left(\frac{E\hat{T}F}{ETF}\right) \right| = \left| 20\log_{10}\frac{1}{(1 - |ESF \cdot ELF|)} \right| \qquad (3.91)$$

In the case of a full two-port VNA measurement, the residual tracking error must be computed taking into account the corrections performed on the VNA. The estimated tracking term for the full calibration case is

$$E\hat{T}F = \frac{b_{2M}}{a_{1M}}\left(1 - E\hat{S}F \cdot E\hat{L}F\right) \qquad (3.92)$$

where $E\hat{S}F$, $E\hat{L}F$ are the estimates (measured values) of the source and load match. The residual tracking error is computed as in (3.91) but applying Eq. (3.92) for the estimated tracking to yield

$$\Delta ETF_{\text{2PortCal}} = \left| 20\log_{10}\left(\frac{E\hat{T}F}{ETF}\right) \right| = \left| 20\log_{10}\left(\frac{1 - \left|E\hat{S}F \cdot E\hat{L}F\right|}{1 - |ESF \cdot ELF|}\right) \right| \qquad (3.93)$$

The estimated source and load match can be expressed in terms of the actual source and load match and a residual source and load match (ΔESF, ΔELF respectively) as

$$\begin{aligned} E\hat{S}F &= ESF + \Delta ESF \\ E\hat{L}F &= ELF + \Delta ELF \end{aligned} \qquad (3.94)$$

Taking the inner term of Eq. (3.93) and expressing it only in terms of raw and residual source and load match one obtains

$$\left(\frac{1 - \left|E\hat{S}F \cdot E\hat{L}F\right|}{1 - |ESF \cdot ELF|}\right) = \left(\frac{1 - |ESF \cdot ELF + ESF \cdot \Delta ELF + \Delta ESF \cdot ELF + \Delta ESF \cdot \Delta ELF|}{1 - |ESF \cdot ELF|}\right)$$

$$(3.95)$$

To simplify this, products that are much smaller than other products are neglected; in general, residual terms are more than an order of magnitude smaller than raw terms so that

(3.95) simplifies to

$$
\left(\frac{1 - \left|ESF \cdot ELF + ESF \cdot \Delta ELF + \Delta ESF \cdot ELF + \cancel{\Delta ESF \cdot \Delta ELF}\right|}{1 - |ESF \cdot ELF|}\right) \approx
$$

$$
\left(\frac{1 - |ESF \cdot ELF| - |ESF \cdot \Delta ELF + \Delta ESF \cdot ELF|}{1 - |ESF \cdot ELF|}\right) =
$$

$$
\left(\frac{1 - |ESF \cdot ELF|}{1 - |ESF \cdot ELF|} - \frac{|ESF \cdot \Delta ELF + \Delta ESF \cdot ELF|}{1 - |ESF \cdot ELF|}\right) \approx (1 - |ESF \cdot \Delta ELF + \Delta ESF \cdot ELF|)
$$

$$(3.96)$$

Recognizing in the last equation that $1 - |ESF \cdot ELF|$ is nearly one for most normal systems, the transmission tracking uncertainty becomes

$$
\Delta ETF_{2PortCal} = \left|20 \log_{10} (1 - |ESF \cdot \Delta ELF + \Delta ESF \cdot ELF|)\right| \qquad (3.97)
$$

From this it is clear that the transmission tracking error relies on both raw and residual terms. Note that this is a different result from that of uncertainty in reflection measurements, which is only dependent on the residual errors. Following a similar argument, the S21 uncertainty for a two-port calibrated measurement can determined. In this case, the uncertainty is computed for a low-loss device, so that receiver dynamic accuracy doesn't contribute much error. To compute the uncertainty for a two-port calibration, the measured values for the error terms are substituted for the actual error terms

$$
\Delta S_{21\_2PortCal} = \frac{S_{21A}}{S_{21Corr}} = \left[\frac{\dfrac{ETF}{(1 - |S_{11A} \cdot ESF|) \cdot (1 - |S_{22A} \cdot ELF|) - |ESF \cdot S_{21A} \cdot S_{12A} \cdot ELF|}}{\dfrac{E\hat{T}F}{\left(1 - \left|S_{11A} \cdot E\hat{S}F\right|\right) \cdot \left(1 - \left|S_{22A} \cdot E\hat{L}F\right|\right) - \left|E\hat{S}F \cdot S_{21A} \cdot S_{12A} \cdot E\hat{L}F\right|}}\right]
$$

$$(3.98)$$

This can be further simplified for the case of a matched device where S11 and S22 are negligible (a common assumption as attenuators are often used as reference versification devices) to become (in linear form)

$$
\Delta S_{21\_2PortCal} = \frac{S_{21A}}{S_{21Corr}} = \left[\frac{\dfrac{ETF}{1 - |ESF \cdot S_{21A} \cdot S_{12A} \cdot ELF|}}{\dfrac{E\hat{T}F}{1 - \left|E\hat{S}F \cdot S_{21A} \cdot S_{12A} \cdot E\hat{L}F\right|}}\right]
$$

$$
= \frac{ETF}{E\hat{T}F} \cdot \frac{(1 - |(ESF + \Delta ESF)(ELF + \Delta ELF) S_{21A} \cdot S_{12A}|)}{(1 - |ESF \cdot S_{21A} \cdot S_{12A} \cdot ELF|)}
$$

$$
= \Delta ETF \frac{1 - \left|S_{21A} \cdot S_{12A} \cdot \left(ESF \cdot ELF + ESF \cdot \Delta ELF + \Delta ESF \cdot ELF + \cancel{\Delta ESF \cdot \Delta ELF}\right)\right|}{(1 - |ESF \cdot S_{21A} \cdot S_{12A} \cdot ELF|)}
$$

$$
\approx \Delta ETF \left[1 - \frac{|S_{21A} \cdot S_{12A} (ESF \cdot \Delta ELF + \Delta ESF \cdot ELF)|}{\left(1 - |ESF \cdot \cancel{S_{21A} \cdot S_{12A} \cdot ELF}|\right)}\right]
$$

$$
= \Delta ETF (1 - |S_{21A} \cdot S_{12A} (ESF \cdot \Delta ELF + \Delta ESF \cdot ELF)|)
$$

$$(3.99)$$

For the case of a low-loss device, where S21 and S12 are nearly one, this simplifies to

$$\Delta S_{21\_2port} = \Delta ETF \left(1 - |ESF \cdot \Delta ELF + \Delta ESF \cdot ELF|\right) = \Delta ETF^2$$

$$\Delta S_{21\_2port(dB)} = 20 \log_{10}(\Delta ETF^2) = 2 \cdot \Delta ETF_{dB}$$

(3.100)

Just like the ETF term, the uncertainty of an S21 measurement is primarily due to the source and load match, raw and residual.

An intuitive way of thinking about this is to recognize that when one does error correction, the mismatch terms are subtracted from the flow graph equations through their characterization. But the characterization of the source and load match is not perfect, and some residual error is left. Thus, when a reflection off the load occurs, the reflection is compensated for by the estimated load match, but the residual load match is free to reflect off the load and re-reflect off the source. Since this residual re-reflection is unknown by the error correction math, it is not corrected for by the source match term, and thus the error becomes the product of the residual load match times the raw source match. Similarly, when there is a reflection off the load match that re-reflects off the source match, it is corrected for by the estimated source match, but there is a remaining portion of the reflection that is not compensated for, and it is equal to the raw load match term times the residual source match.

These residuals occur in two stages: first there are the residuals during the acquisition of the error terms, yielding the residual tracking term; this is the calibration residual. Next there is the residual due to the measurement of the DUT, and for low-loss DUTs, it is identical to the calibration residual. In fact, if the phase of the DUT matches the phase of the calibration thru, these residuals will cancel and there will be no error in the S21 measurement. That is why re-measuring the thru standard tells one nothing about the quality of the calibration. However, measuring an airline of a different length causes phasing of the residual terms, and the envelope gives a good estimate of the uncertainty, as was demonstrated previously in Figure 3.37.

3.11.1.1   General Uncertainty Equation for S21

For completeness, in the case where S11 and S21 are not zero, their residual effect can be added as a simple input and output mismatch error. The loop term is ignored as it contains a product of residuals, and will be quite small. Thus, the general uncertainty of a two-port calibrated S21 measurement is

$$\Delta S_{21\_2Port(dB)} = 20 \log_{10} \left[\frac{1 + (|ESF \cdot \Delta ELF| + |\Delta ESF \cdot ELF|)(1 + |S_{21A} \cdot S_{12A}|)}{(1 - |\Delta ESF \cdot S_{11A}|)(1 - |\Delta ELF \cdot S_{22A}|)} \right]$$

(3.101)

A final note on S21 uncertainties: it is clear from Eq. (3.101) that the uncertainty of an S21 measurement depends upon the actual characteristics of the DUT and not only on the test system. This is an uncomfortable result for some engineers, who would like to have a simple number for the uncertainty of a measurement. In fact, one could compute a worst case uncertainty by taking all the S-parameters to equal 1 (not physically realizable) and find that this would give a limit to S21 uncertainty for any device. In fact, this can be extended further to any arbitrary test system by presuming the raw source and load match as unity, so that the S21 uncertainty limit for any device on any test system is determined only by the residual source and load match. For example, an arbitrary test system calibrated with 40 dB residual source and load match would have a worst case limit of S21 uncertainty of about 0.5 dB.

3.12   Errors in Phase

The error computations given thus far have been vector in nature, so magnitude and phase errors could both be computed. However, a simplified method for computing phase errors given an amplitude error can be determined by assuming that the error term represents a locus of points around the true value, as seen in a vector plot. The magnitude error occurs when the error adds or subtracts in phase with the measured value. The phase error occurs when an error term adds in quadrature with the measured value, as shown in Figure 3.39.

From this construction, one can see that the phase of the error is the arctan of the ratio of magnitude of the error and the signal, or

$$
\Delta dB = 20 \log \left(\frac{Signal + Error}{Signal} \right), \quad Error = Signal \cdot 10^{\frac{\Delta dB}{20}} - Signal
$$

$$
\Delta \phi_{\deg} = \frac{180}{\pi} \cdot \arcsin \left(\frac{Error}{Signal} \right) = \frac{180}{\pi} \cdot \arcsin \left(10^{\frac{\Delta dB}{20}} - 1 \right)
$$

(3.102)

For small errors, the ratio of phase error to dB error approaches a constant value of 6.6 degrees/dB; the analysis to demonstrate this is shown below in Eq. (3.103)

$$
\frac{\Delta \phi_{\deg}}{\Delta dB} = \frac{180}{\pi} \cdot \frac{\arcsin \left(10^{\frac{\Delta dB}{20}} - 1 \right)}{\Delta dB} \approx \frac{180}{\pi} \left(\frac{10^{\frac{\Delta dB}{20}} - 1}{\Delta dB} \right) \Bigg|_{\arcsin(x) \approx x \, for \, small \, x}
$$

$$
= \frac{180}{\pi} \cdot \lim_{\Delta dB \to 0} \left(\frac{10^{\frac{\Delta dB}{20}} - 1}{\Delta dB} \right)
$$

$$
= \frac{180}{\pi} \frac{\lim\limits_{\Delta dB \to 0} \dfrac{d}{d \Delta dB} \left(10^{\frac{\Delta dB}{20}} - 1 \right)}{\lim\limits_{\Delta dB \to 0} \dfrac{d}{d \Delta dB} (\Delta dB)} = \frac{180}{\pi} \frac{\lim\limits_{\Delta dB \to 0} \left(\dfrac{10^{\frac{\Delta dB}{20}} \log_e 10}{20} \right)}{1}
$$

(3.103)

$since \lim\limits_{x \to 0} (10^x) = 1,$

$$
\frac{\Delta \phi_{\deg}}{\Delta dB} = \frac{9}{\pi} \log_e (10) = 6.6 \,, \, or \quad \Delta \phi_{\deg} = 6.6 \cdot \Delta dB
$$

This useful result applies to errors from unknown signals, from mismatch, from calibration errors, from noise or from any other situation where an error signal adds in a vector manner with the desired signal. It allows one to predict with confidence the error or ripple in phase due to some error signal by observing only the dB level of the error or ripple. Inspection of the

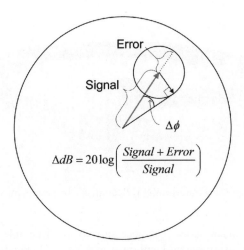

Figure 3.39   Phase error as a result of an error signal.

calibration residual presented in many data sheets will show that the phase errors in tracking are approximately 6.6 times the dB error, as an example.

3.13   Practical Calibration Limitations

While the details of calibration and error correction have been thoroughly derived and described, the results presented so far apply to an idealized situation that considers only the systematic errors, that is ones that can be characterized and removed. This is the common situation found in most discussions of microwave metrology. These conditions include attributes such as calibrating at the optimum power level, using very narrow low-noise IF bandwidths, making measurements only at exactly the same frequencies as calibrated, using only the test ports of the VNA without cables or adapters and, in particular, only specifying performance when using a particular connector with particular high-performance calibration kits.

In the real-world experience of the RF and microwave engineer, it is often necessary to diverge from these idealized conditions for reasons of measurement speed, device limitations or other practical applications. In this section, the effects of these real-world conditions are discussed. Some of the errors are random errors, such as noise, stability and connector repeatability, which must be treated in a statistical manner. Others errors are related to systematic errors, and while they are not compensated for, error bounds might reasonably be assigned to them.

3.13.1   Cable Flexure

Cables are a bane to the existence of the RF and microwave engineer. By way of a simile, it has been said that cables are like dogs; either they are bad, they've been bad or they are going to be bad, and when they're good, they only stay good with great care.

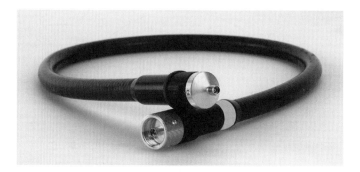

Figure 3.40   Metrology grade cable for a VNA. Reproduced by permission of Agilent Technologies.

In almost all measurement cases, it is necessary to use a flexible or semi-flexible cable to connect between the measurement instrument and the DUT. For S-parameter and related measurements, the cables add mismatch, group delay and loss to the measured results. These additions to the systematic errors are assessed during the measurement calibration process, and for the most part are removed. However, in some cases, the cable effects remain the dominant source of error in measurements due to drift and flexure in the RF cable. The length and loss of the cable contribute directly to the size of the error.

Metrology-grade cables (Figure 3.40) have thick outer jackets that restrict the cable bending radius to try to avoid kinking or other damage to the cable.

In very old VNA systems, the RF cables had specifications for phase tracking because of limited calibration capabilities. In all modern VNAs, since integrated error correction has been added, the phase matching of cables is entirely unnecessary. What is critically important is the cable stability, for both magnitude and phase, and in both return loss and transmission. Often semi-rigid cables are used as a lower-cost alternative and give very good stability performance. Flexible braid cables are used for lower cost and more flexibility in connecting to DUT interfaces.

The loss of the cable has the effect of degrading the directivity stability of the system. Because the loss occurs after the directional coupler, the raw directivity (which is the directivity error term divided by the reflection tracking) is a combination of the coupler directivity minus twice the cable loss. If the VNA has a coupler directivity of 30 dB, followed by a cable with a loss of 12 dB, the effective directivity will be only 6 dB. In this sense, directivity is the difference between a full reflection (open or short) and a perfect load. The stability of the cable match is also degraded by the loss. Adding a fixed attenuator at the end of the test cable improves the match of the cable, but further reduces the effective directivity and stability.

Some details of characterizing system and cable stability are discussed in Chapter 9.

3.13.2   Changing Power after Calibration

The source or port power setting is one of the stimulus settings associated with the calibration state. In older VNA systems, the calibration would display an indication that the stimulus settings had changed; this was an on-screen annunciator "C?". This annunciator led many users to believe that the calibration was questionable, but for the most part this was not the case. Later VNAs use the annunciator "$C\Delta$" to indicate a change to the calibration settings. The idea behind identifying a change in power as a possible condition that would degrade the calibration comes about due to the receiver linearity of the VNA, and the notion that changing the power after calibration would cause the receiver to be operated in a different portion of its linearity curve and add a dynamic accuracy error to calibration. However, for almost all conditions where the power is changed, this is not the case.

The first thing to note is that the receivers for the test signal will be at a different power level than the calibration power level for any device that is not exactly 0 dB S21 gain. The common reason for changing power level is to avoid overdriving an amplifier during test. The usual way to go about calibrating is to set the source power to the desired test power level, and perform the two-port calibration at that level. However, for an amplifier test, that level may be very low, and so the calibration will have substantial noise due to low drive power. When the amplifier is connected, the power to the test receiver is changed by the amplifier gain, so the dynamic accuracy error occurs on the output receiver. The noise in the calibration may be removed with averaging and IF bandwidth reduction, but this often leads to unacceptably long calibration and measurement times.

Consider instead if the calibration is done at a higher power where the noise is less of a problem. After calibration, the power is lowered. The reference channel receiver sees the change in power level and the dynamic accuracy error of the reference receiver adds to the calibration error. But for most VNA systems, the reference receiver is padded down 5–10 dB lower than the test receiver, just to ensure that it stays out of compression and so its dynamic accuracy performance is better than the test receiver. Further, when the amplifier is connected, its gain will bring the signal at the transmission test receiver up closer to the level that it was during calibration, so that the dynamic accuracy error of the test receiver is reduced or eliminated. From this analysis, it is clear that changing the power level has essentially no effect on the calibration accuracy provided the signal levels remain within the linear-gain range of the DUT.

For a low-loss device such as a filter, where the test channel power is nearly equal to the calibration power, there is a possibility for reduced accuracy due to dynamic accuracy error as the test and reference receiver will both see a different power level if the source power is changed. But there is seldom reason to change the power level when measuring low-loss passive devices.

The final reason for changing power level is to investigate the non-linear behavior of a device. The DUT is often measured at a low power state, then the power level is increased, and the change in S21 is recorded as device compression. In these cases, the dynamic accuracy error will be introduced into one of the two power measurements. It is often the case that the calibration will be performed at the lower power level, and then the power level raised for measurement, but this is exactly the opposite of best practice, as the lower power level will likely mean higher noise and a poor calibration. Experience shows that the dynamic accuracy error for a change in power is almost always less than the error due to increased noise in the calibration. Another idea is to calibrate at each power level, but again the error due to noise in the low power level will likely overwhelm the dynamic accuracy.

This is especially true on more modern VNAs, where improved methods have yielded nearly perfect dynamic accuracy, especially at lower power levels. Figure 3.41 shows a measurement of S21 for a thru connection, where the test receiver is padded so that it is measuring in a very linear region, and essentially characterizes the linearity and noise of the reference receiver as

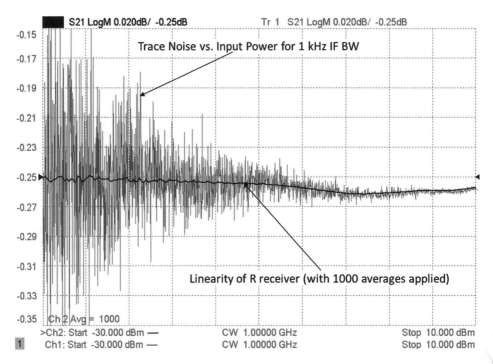

Figure 3.41   Noise and dynamic accuracy error versus drive.

the power level is changed There are two traces, both taken at a normal IF bandwidth of 1 kHz, and one taken with a very large number of averages (1000). These demonstrate the relative effects of noise and dynamic accuracy on the reference channel receiver. For this example, an Agilent PNA-X was used. The specified compression at maximum power is 0.1 dB, but it is clear that the actual receiver compression is much less, on the order of 0.01 dB (note the scale is at 0.02 dB/div). Even at 1000 averages, at lower power, trace noise is nearly the same as the compression.

This is from the reference receiver. The test receiver will normally have more compression at high power, as it is not padded down as the reference receiver is. From these measurements, it is clear that if one calibrates at a power where the receiver is not in compression – normally 10 dB below the maximum specified power – and then moves to any other power, the error in calibration due to both noise and dynamic accuracy is minimized. In fact, the rule for modern network analyzers can be changed from "don't change power after calibration" to "calibrate at the power that gives the best result, and then move to any other power level". This applies as long as the source attenuator is not changed. What happens if the attenuator is changed? Please carry on the to the next section.

3.13.3   Compensating for Step Attenuator Changes in Step Attenuators

Many VNAs included integrated step attenuators between the reference and the test couplers. These allow extending the power range of the VNA to much lower levels. The placement between the reference and test couplers means that changes in the power due to the step attenuator are not seen by the reference receiver, and so the calibration appears to be invalidated if the step attenuator is changed. This placement is intentional, to allow the reference channel receiver to maintain a relatively high signal level, and thus lower noise, even when the signal to the DUT is set to a very low level.

Many VNAs compensate for nominal changes to the test set source attenuator by modifying the values displayed on the reference receiver by the nominal value of the attenuator change. In this way, the readings of the reference channel always represent the source power applied to the DUT. However, the step attenuator is not perfect, and so it often has between 0.25–0.5 dB of variation from the nominal value. Further, the step attenuator has different port matches at different attenuator states, with the largest change being from the 0 dB state to any other state. In the 0 dB state, the source and load match are determined by quality of the components behind the step attenuator, such as the directivity of the reference channel coupler. For any attenuator step greater than 0 dB, the loss of the attenuator effectively isolates the source match from the test port, meaning that the error from changing the step attenuator between non-zero states will be much smaller. Figure 3.42, upper window, shows the S11 trace after calibrating

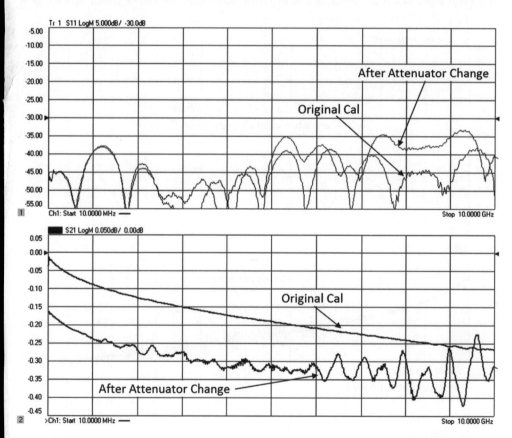

Figure 3.42   (upper) Error in S11 changing the attenuator difference for 5 dB and 10 dB; (lower) error in S21 due to attenuator change, measuring an airline.

with the attenuator in the 5 dB state (memory), and changing to the 10 dB state (data). The lower window shows the same result, but for the S21 response. The DUT is an airline, the ripple of which highlights the quality of the calibration.

Also, the match of the attenuator is better than the source match of the system behind it, so a best practice is to set the step attenuator to at least the first position (5 or 10 dB) before calibration, unless the source power available is not sufficient for the tests to be performed. If the initial calibration is performed with the source attenuators at each port set to the first position, lowering it to other positions will have smaller effect on the mismatch.

There is one simple way to pre-characterize the attenuator states and remove most of the attenuator switch effect from the measurement. After a two-port calibration, leave the thru standard in place, or put a thru standard from port 1 to 2. If it is a non-insertable DUT, use a non-insertable thru and measure the thru's four S-parameters and save this as an S2P file. Then de-embed the thru from port 2 (here one must reverse the sense of S11 and S22 in the S2P file of the thru, if the VNA requires port 1 of the de-embedding file to face test port 2). After de-embedding, the result should be a perfectly flat S21 and very low S11. Switch the attenuator to each of its states, and record the state as an S2P file. This represents the set of S-parameters that is the difference or offset in the original attenuator setting and the new

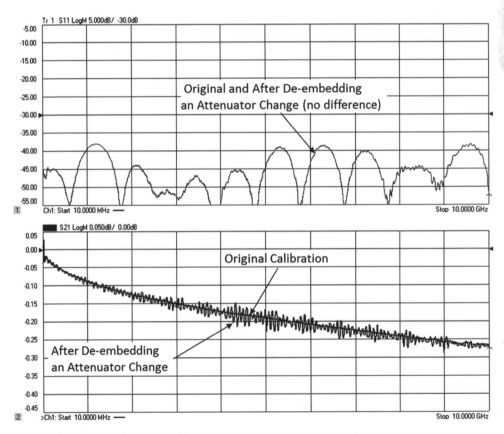

Figure 3.43   Attenuator offset applied as a de-embedding after the attenuator change.

setting. De-embedding these S-parameters from the original calibration will remove most of the error from the attenuator switch. To measure a DUT, turn off the de-embedding on port 2 (as the thru is not used anymore), and turn on the de-embedding for the appropriate attenuator state S2P file on port 1, and measure the DUT with the new attenuator setting.

This removes most but not all the errors associated with the attenuator change. When the attenuator is switched, both the source match and the load match of the test port change, but in slightly different ways. The S2P file of the switch state that is measured captures the change in load match, but not the change in source match. For the most part this is a small error, and on the order of 0.05 dB or less if the original calibration is performed in a non-zero attenuator state, and the new state is also not the zero attenuator state. The residual error even in the zero state is on the order of 0.10 dB, which is quite good enough for most applications.

When the de-embedding method, as described above, is applied to the measurement, the S11 trace is improved dramatically such that it is indistinguishable from the 5 dB state, and all of the offset for the S21 trace is also removed, as shown in Figure 3.43. The memory trace shows the data of the 5 dB state, and the data trace shows the result after changing attenuation and de-embedding the attenuator offset. The fine grain ripple on S21 is a result of differences between the source and load match at port 1, and simple de-embedding cannot compensate for that, but the worst case error here is less than 0.05 dB.

Changing the attenuator after calibration is seldom recommended, but it can actually produce less error than calibrating at a very low power level, where noise dominates. Using the de-embedding technique can reduce the error to a negligible amount which is especially useful when drive signals need to be at a very low level, for example, below −60 dBm.

3.13.4   Connector Repeatability

Connector repeatability is closely related to the problem of cable flexure, in that it is a non-repeatable error and cannot be removed with a normal error correction process. Unlike cables, however, the location of the connectors and their associated mismatches is well known. The use of time domain gating, as discussed in Section 3.9.2.1, can be used to remove the residual connector repeatability error provided the connector is far enough from the DUT so that the time gate resolution can fully isolate it.

While classically connector repeatability referred to only the change that occurs when a connector is de-mated and re-mated, it can also be applied to the situation commonly found in using PCB fixtures, where different standards are realized in different PCB layouts of each standard on a calibration PCB. Each standard, opens, shorts, loads and thrus, are PCB traces with a separate coax-to-microstrip launch, typically an SMA connector. Even when identical connectors are used, difference in mounting or soldering can create differences in match between the connectors. This has exactly the same effect as connector repeatability where a connector is de-mated and re-mated for each step of the calibration.

Good connectors typically have repeatability an order of magnitude better than their mismatch, or greater. Precision connectors using slot-less contacts, such as 3.5 mm, have repeatability on the order of 65 dB or so. Commercial grade type-N connectors can have repeatability as low as 40 dB. PCB coax connectors have connector-to-connector repeatability on the order of 30 dB up to about low microwave frequencies of 3 GHz or so, and 20 dB up to 20 GHz. An odd reality for PCB connectors is that the repeatability between connectors can be worse than the connector return loss. Consider if one connector is low in impedance by a few ohms, and another is high, the difference in the match is greater than the mismatch of either connector. If this case exists, it is actually more detrimental to calibrate on the PCB than to ignore the connector effect. The connector-to-connector repeatability can be determined on PCB standards using time domain gating techniques, and is discussed in Chapter 9.

3.13.5   Noise Effects

Noise effects have been somewhat discussed already in the context of setting and changing source power after calibration. Noise comes in two forms that affect calibration: noise floor and high-level trace noise. At lower power levels, they are the same thing, and the trace noise apparent on the measurement trace can be entirely attributed to the contribution of the noise floor. The low-level noise floor will affect the measurement trace noise as

$$N_{Trace(dB)} = 20 \log \left(10^{\frac{Signal_{(dB)}}{20}} + 10^{\frac{N_{Floor(dB)}}{20}} \right) - Signal_{(dB)} \tag{3.104}$$

For a -60 dB signal measured on a system with a -100 dB noise floor, the trace noise will be about 0.1 dB. The noise floor of the VNA is set by the IF bandwidth used. In the case of calibration, if source power is set very low, say -60 dBm, then some standards such as the load standard will be right at the noise floor, causing inaccuracies in the determination of the error terms.

From the power sweep trace of Figure 3.41, it is clear that the trace noise increases with lower power level; it does so at the rate of 3 times more noise for each 10 dB change in power level relative to the noise floor. One simple way to lower the noise floor of a measurement is to lower the IF bandwidth. The noise floor goes down 10 dB for each 10 times decrease of the IF bandwidth and trace noise goes down 3 times for each 10 times change in IF bandwidth. Unfortunately, changing the IF bandwidth lowers the sweep speed in the same ratio. In some VNAs, the source is swept at wider bandwidths and automatically changes to stepped sweep at lower bandwidths. This sometimes causes errors due to the delay of the DUT (this effect is discussed in detail in Chapter 5). This effect can be avoided if calibration is always performed in a stepped mode; many VNAs have an explicit stepped mode frequency sweep selection.

In some cases where the highest speeds are needed, such as filter tuning, the same effect of noise reduction can be accomplished using sweep-to-sweep averaging. Using averaging maintains the same sweep dynamic, but averages a number of responses to generate the final measurement response. Usually, sweep averaging displays all the intermediate average results, and unless the sweeps are stopped after the max average value is reached, the average value will improve somewhat during subsequent sweeps even after the max average number is reached. This is because the sweep averaging function operates like a two-tap infinite-impulse-response (IIR) filter. Rather than accumulate N sweeps of data and average all of them together, the averaging function in most VNAs follows the formula

$$A_N = \left(\frac{N-1}{N} \right) Data_{old} + \left(\frac{1}{N} \right) Data_{new} \tag{3.105}$$

Thus, on the $N + 1$ or even $N + 100$ sample, data from the first reading will be a small portion of the result.

Once the signal-to-noise gets to 80 dB, the noise floor effect on the trace noise is less than 0.001 dB. However, at high signal levels one may find that the trace noise does not further diminish. That is because at high signal levels, the noise on the signal may come from the source phase noise rather than the receiver noise floor. Once one is in the high-level trace noise region, raising the power to a higher level has no effect on the level of the trace noise. For older VNAs, which often had relatively poor trace noise, the transition point from low-level to high-level noise was between -30 to -20 dBm power at the receiver. For more modern VNAs, which have phase noise comparable to a signal generator, the high-level trace noise may not become apparent until 0 to $+10$ dBm. In Figure 3.41, the effect of power level and noise floor on the trace noise is very apparent in the light gray trace. Here the test receiver was padded down to maintain good compression so the receiver power is 35 dB below the x-axis source power displayed. However, at the high power end it is difficult to ascertain the properties of the noise signal.

One may more readily see the effects of high level noise by taking the noise trace of Figure 3.41, which is a measurement of a thru, and computing the added noise by the formula

$$(S_{21} - 1) \cdot a_1 = \left(\frac{b_2 + N_{Added}}{a_1} - \frac{b_2}{a_1} \bigg|_{Thru} \right) \cdot a_1 = N_{Added} \qquad (3.106)$$

This will show the effective noise floor for all power, and it is clear from the lower plot in Figure 3.44 that the noise power added is flat over much of the x-axis range, but then increases with power of the source and rises above the receiver noise floor for source powers above -5 dBm. This was measured on a PNA-X, which has rather good phase noise. And the absolute level of the noise floor depends upon the IF BW used (here a wide bandwidth is used). Older VNAs such as the HP 8753 or HP820 show this effect at power levels as much as 20 dB lower. This is most likely due to the source phase noise rising above the receiver noise floor when the source has sufficiently high gain, that is, at the higher powers.

3.13.6 Drift: Short-Term and Long-Term

A common question from VNA users is "How long is my calibration good for?" And a common answer is "Until it goes bad!" Most VNA manufacturers are silent on the question, as the answer depends strongly on the particular conditions of the VNA in use. If a VNA is used in a temperature-controlled environment, with very stable test port cables, a calibration can last many days or weeks. In almost all cases, the drift of the VNA itself is less than the drift of the interconnect cables and the repeatability of the test connectors.

The local environment has a strong influence on the stability of a test system. If the temperature varies between day and night (a common occurrence in an office building) then the expansion and contraction of components, particularly cables inside and outside the VNA, can degrade the quality of the calibration, and small ripples will start to appear in the calibrated measurements.

Even in a temperature-controlled lab, the heating and cooling systems can cause many degrees of localized temperature changes, and care must be taken to isolate the air flow of the heating and cooling systems from the test equipment.

Short-term drift can occur over several minutes of measurement time and can be related to aspects such as relaxation in test port cables (a phenomena whereby after flexing, a cable takes

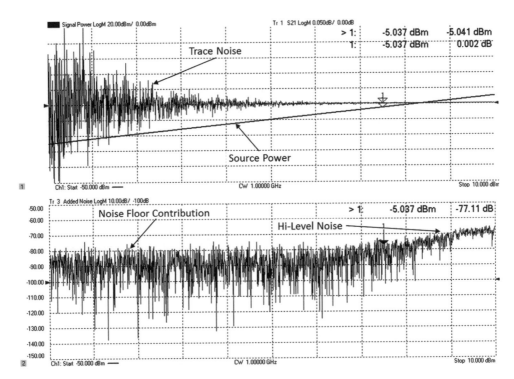

Figure 3.44   Noise added in a signal trace vs source power.

some time to return to a previously quiescent state), and slowly varying response due to subtle heating and cooling caused by external environmental factors, as well as slight internal heating in the VNA itself. This internal heating can occur if a VNAs internal construction is composed of several modules which turn on and off as the VNA sweeps across bands of frequency or from one port to another. For the very best of metrology measurements, these drifts can be avoided by always ensuring that the swept data is taken with the same dynamics. For example, to ensure that the delay on a retrace from one sweep to another is consistent, it is sometimes best to take groups of sweeps rather than single sweeps. Single sweeps can have arbitrary delay between a first and second sweep for forward and reverse sweeps. But a group of sweeps have the same dynamic on the second and subsequent sweeps. These effects are very small indeed, and only need to be attended to in the cases where the highest metrology is required.

3.13.7 Interpolation of Error Terms

Whenever calibration topics are discussed, it is almost always assumed that the VNA will be set to exactly the desired frequencies before the calibration is performed, and the measurements will be made only at those frequencies. Older analyzers such as the HP 8510A would turn off the calibration if the frequencies were changed at all. Starting with the HP 8753A, interpolation of error terms was provided which allowed one to change the number of points, and start or stop frequencies, as long as the final frequency set was within range of the original calibration. Later, the Wiltron 360 and the HP 8510 introduced the concept of "zoom-cal" which allowed changing the start and stop frequencies but would reset all the frequencies to land exactly on the original calibration points, reducing the overall number of points. This provided a way to zoom the display although no more data resolution was provided.

Interpolation of error terms has often been controversial, and even experts at the same measurement company have not always agreed on its utility, but if the limitations are well understood, interpolation of error terms can be set up to generate reasonably small errors, often less than the specified uncertainty of the VNA. Lowering the span or changing the number of points slightly will cause the VNA to interpolate the error terms, and one can look for any changes in the VNA trace after interpolation to see if errors occur.

Error terms such a transmission tracking and reflection tracking have slowly varying functions and are usually more amenable to interpolation than match terms. Match terms often have responses that vary dramatically with frequency, because they are composed of mismatch elements that are spaced some distance apart, such as mismatch at each end of a cable. Since the functions vary quickly with frequency, it is more difficult to interpolate between data points. While simple interpolation of complex numbers interpolates the real and imaginary parts separately, a preferred method is to interpolate the magnitude and phase separately. Note that mismatch elements separated by transmission lines form circles on the Smith chart. Recognizing this, some VNAs utilize circular interpolation to generate an improved interpolation result. Circular interpolation uses three points to define a circle, and then computes the interpolated result by linearly interpolating between the angles of the two points surrounding the targeted point, based on frequency spacing. Figure 3.45 (upper) shows an example of the

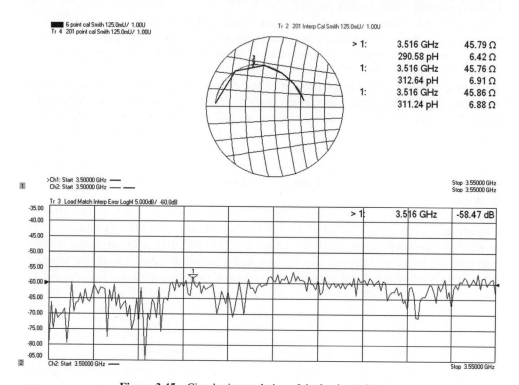

Figure 3.45 Circular interpolation of the load match term.

circular interpolation and a linear interpolation of load match of the same points, along with a calibration done on the exact points. In this case, calibration was performed over a 50 MHz span with six points (10 MHz spacing), then interpolated to 201 points. A second calibration was performed at 201 points. The six-point cal indicates the trajectory that linear interpolation would apply, while the other smoother traces are the 201-point circular interpolation and the 201-point calibration (nearly indistinguishable on the Smith chart). The lower window shows the difference between the interpolated load match and the 201-point measured load match. Over most of the band, the interpolation error is less than −55 dB, which is on the same order as the error in the calibration kit.

Of course, if the frequency point density is not sufficient to capture three points of a circle on the Smith chart, the interpolation will result in large errors. One can estimate the number of points required by determining the lengths associated with the mismatch error terms. Usually, this will be the somewhat longer than the length of the test port cable. For good interpolation the point density should be set to something on the order of 25 points for each wavelength, so that the phase shift between points is less than 15°. For a typical test situation, where the test port cable is approximately 1 meter, a point spacing of about 5 MHz usually gives very good results.

An example of interpolation results is shown in Figure 3.46, where a calibration was performed with point spacing of 10 MHz, over a 1–10 GHz frequency range. After the calibration the span was changed to 10 MHz to 1 GHz with a 1 MHz point spacing. The resulting S11 and S21 plots of an airline measurement show the error due to interpolation artifacts. The error from interpolating is less than 0.035 dB over the entire trace for S21, and less than −45 dB residual match for S11.

Pay particular attention to the points at low frequencies. One issue with interpolation is the assumption that the error term functions are smooth. But many VNAs operate in octave bands of frequency, due to source stimulus or LO implementations, and often at these band breaks there can be a step discontinuity of the error term. At low frequencies there are many band breaks. Furthering the problem at low frequencies is the coupler roll-off, which makes the tracking responses change very rapidly near the bottom of the VNA frequency range. Left alone, these steps would cause large errors in interpolation. For that reason, some VNAs incorporate a factory-generated receiver calibration, the purpose of which is to correct the raw receiver response so that it provides approximately correct values and no step discontinuities. Having this receiver calibration in place means the error terms based on receiver measurements will not have step discontinuities either, but should maintain a smooth response.

One common point of confusion is whether the DUT characteristics have any effect on whether interpolation can be used or its quality: they do not. The DUT characteristics have no effect on the interpolation: even very narrow-band DUTs, DUTs with long delay or DUTs with complex frequency response may be measured with interpolation. It is not the measured data which is interpolated, but only the error terms. If the error terms are smooth over the range of interest, interpolation will work quite well regardless of the DUT response.

3.13.8   Calibration Quality: Electronic vs Mechanical Kits

From a specification or theoretical standpoint, the best TRL mechanical calibration kits provide the highest quality of calibration. Next are Ecal standards using the best Ecal modules,

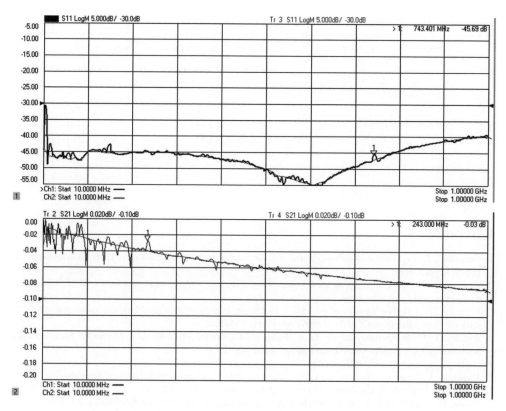

Figure 3.46 Interpolation results with various point spacing.

followed by SOLT with sliding loads. Fixed load SOLT calibrations typically have the poorest performance.

In the earliest versions of Ecal, the module stability and calibration methods made the calibration quality worse than the SOLT calibrations, but with unknown thru calibrations, modern Ecal modules perform better than any but the best metrology TRL Cal Kits.

In fact, if RF cables are used, the error from the cable flexure will undoubtedly cause sufficient errors in the TRL calibration to degrade its quality below that of the Ecal. If one includes the likelihood of human handling error and added connector repeatability for mechanical calibrations, then there is little question that, in practice, Ecal modules almost always provide superior calibration to mechanical calibration kit methods. In some cases, it is best to modify the default Ecal method to use separate unknown thru (rather than the Ecal as unknown thru) to minimize the cable flexure after cal. For low or moderate loss DUT, using the DUT itself as the unknown thru can produce the best results. This gives the Ecal quality to the one-port error terms, and improves the stability of the thru measurement.

Table 3.1 shows the S21 uncertainty (worst case, and RSS with coverage factor of two) for various calibration kits used with an Agilent PNA-X VNA. Choosing a different VNA with different source and load match will change the value of the uncertainty, but the relationship between the relative quality of the calibration with each Cal Kit will remain the same.

Table 3.1   S21 uncertainty vs calibration kit

Uncertainty computed at 20 GHz with Agilent N5242A option 423 VNA, 3.5 mm		
Kit Type	Worst Case S21 Uncertainty (dB)	RSS Uncertainty (dB)
85052C TRL short line	0.047	0.029
N4291B Ecal	0.081	0.057
85052B SOLT Sliding Load	0.134	0.103
85052D SOLT Fixed Load	0.192	0.166

Reference

1. Fitzpatrick, J. (1978) Error models for system measurement. *Microwave Journal* vol. 21, pp. 63–66, May 1978 http://bit.ly/ICm9O4.
2. Rytting, D. (1980) An analysis of vector measurement accuracy enhancement techniques. RF & Microwave Symposium and Exhibition, 1980.
3. Marks, R.B. (1997) Formulations of the basic vector network analyzer error model including switch-terms. 50th ARFTG Conference Digest-Fall, Dec. 1997, vol. 32, pp. 115–126.
4. Rytting, D. (1996) Network Analyzer Error Models and Calibration Methods. RF 8: Microwave Measurements for Wireless Applications (ARFTG/NIST Short Course Notes).
5. Agilent Application Note 154, http://cp.literature.agilent.com/litweb/pdf/5952-1087.pdf.
6. Davidson, A., Jones, K., and Strid, E. (1990) LRM and LRRM calibrations with automatic determination of load inductance. 36th ARFTG Conference Digest – Fall, Nov. 1990, vol. 18, pp. 57–63.
7. Engen, G.F. and Hoer, C.A. (1979) thru-Reflect-Line: An improved technique for calibrating the dual six-port automatic network analyzer. *IEEE Transactions on Microwave Theory and Techniques*, **27**(12), 987–993.
8. Ferrero, A. and Pisani, U. (1992) Two-port network analyzer calibration using an unknown 'thru'. *Microwave and Guided Wave Letters, IEEE*, **2**(12), 505–507.
9. Ferrero, A. and Pisani, U. (1991) QSOLT A new fast calibration algorithm for two port S parameter measurements. eighth ARFTG Cant Dig., San Diego, CA, Dec. 5–6, 1991.

4

Time Domain Transforms

4.1 Introduction

For most engineers, their introduction to electrical circuits is based on the time domain response of the signals and circuits. The first experiments in college-level electrical engineering courses are measuring sine wave signals on an oscilloscope. The first lab exercise using electrical networks is often finding the Bode plot response of a filter by measuring the output sine wave from the filter on an oscilloscope while varying the frequency. The better labs include measuring the phase from the sine wave zero crossing. From this, the concept of the frequency response of a network is often developed, with the frequency domain being demonstrated to be the Fourier transform of the time domain impulse response, and from then on, the electrical engineer works and thinks in the frequency domain. The penultimate instrument for frequency domain measurements is the VNA, and it provides unrivaled accuracy in obtaining the frequency domain response. For the new RF or microwave engineer, the frequency domain response becomes second nature, and the time domain response of networks is sometimes considered an anachronistic approach, useful as a learning tool and then forgotten.

However, for the microwave engineer working on distributed circuits, components separated by cables, transmission lines or waveguides, the time domain (TD) response provides unique insight into the attributes of the circuit as well as methods for improving measurement results by removing artifacts caused by test fixturing and equipment that is separated in time from the device to be measured. The details for using time domain in a variety of situations are discussed later in this chapter, but first it seems appropriate to put down in one place sufficient details so that the time domain response shown on a VNA is clearly understood.

First, a frequently asked question is "Why doesn't the VNA time domain transform match the result I get if I run an FFT on the data?" It should be understood that the TD response is not at all the same as the FFT (or more correctly, inverse FFT) of the frequency response, and has many subtleties that can cause confusion to the casual observer. To that end, the next few sections give the exact mathematical details of the time domain transform used in many modern VNAs. The VNA time domain transform is developed here from the definition of the Fourier transform, by applying the various limits and compensations that are necessary when used in the VNA. This is by far the most rigorous use of mathematics in this book; the author apologizes in advance for copious use of integrals.

Handbook of Microwave Component Measurements: With Advanced VNA Techniques, First Edition. Joel P. Dunsmore.
© 2012 John Wiley & Sons, Ltd. Published 2012 by John Wiley & Sons, Ltd.

4.2 The Fourier Transform

While a network is mathematically characterized by its transfer function, the frequency response of a network provides the physically measurable response of a network, utilizing sinusoidal signals as the stimulus, and measuring the response as magnitude and phase changes in the stimulus signals. Fourier analysis is ideally suited to represent the physical response, and can provide for useful analysis of a network. However, measurement systems are limited to measuring finite frequency points over specified bandwidths, so any interpretation of the measurements must include these limitations. This chapter provides some important details of Fourier analysis as applied to device measurements using VNAs. Since the data is measured in the frequency domain, the transformation we are most interested in is the inverse Fourier transform (IFT), generating the time domain response from the frequency domain data. Most statements about the IFT have corollaries in the forward transform.

4.2.1 The Continuous Fourier Transform

The Fourier transform can be interpreted as a Laplace transform with the special case of $s = j\omega$. Many of the significant theorems of the Fourier transform are quite similar to their Laplace counterparts, and those that are particularly useful are presented here. When data is measured in the frequency domain, an inverse Fourier transform is used to determine the time domain response of the DUT, such as a filter or transmission line. If the data represents the frequency response of the filter, then the inverse transform represents the impulse response of the filter. Since the Fourier transform plays such a key role in the VNA time domain transform, it is appropriate to review some of its details, as well as standardize the nomenclature. The Fourier transform pair (forward and inverse) is defined as

$$\mathbf{F}(f(t)) = F(\omega) = \int_{-\infty}^{\infty} f(t)\, e^{-j\omega t}\, dt \tag{4.1}$$

$$\mathbf{F}^{-1}(F(\omega)) = f(t) = \frac{1}{2\pi} \cdot \int_{-\infty}^{\infty} F(\omega)\, e^{j\omega t} d\omega \tag{4.2}$$

applied to analytic functions $f(t)$ and $F(\omega)$ over all time and all frequency, respectively [1]. The careful reader will note that nomenclature used in the forward transform by electrical engineers differs slightly from the commonly defined transform where the value for frequency used [2] is $\omega = 2\pi s$.

4.2.2 Even and Odd Functions and the Fourier Transform

Functions are even if $F(\omega) = F(-\omega)$, and are odd if $F(\omega) = -F(-\omega)$. All functions can be represented as a sum of an even function and an odd function. Evenness, oddness and other types of symmetry can simplify calculating transforms, and is often assumed for cases of some transforms. A function $f(t) = e(t) + o(t)$ has the Fourier transform

$$F(\omega) = 2\int_{-0}^{\infty} e(t) \cdot \cos(\omega t)\, dt - 2j \int_{-0}^{\infty} o(t) \cdot \sin(\omega t)\, dt \tag{4.3}$$

where $e(t)$ and $o(t)$ are even and odd functions, respectively. From this result, many Fourier transform relationships can be deduced. For modeling physical functions, a key relationship is that for a pure-real time function, f(t), the Fourier transform must be in the form

$$\mathbf{F}[f(t)] = E(\omega) + jO(\omega) \tag{4.4}$$

that is, the Fourier transform of a pure-real time function has an even real part and an odd imaginary part.

4.2.2.1   Hermitian Functions

The transform of functions such as those described in Eq. (4.4), which have an even real part, and an odd imaginary part are called hermitian. This can also be written as $F(\omega) = F^*(-\omega)$. Functions which are real and symmetric (even) – that is, which exist from negative to positive infinity, with the response in negative time equal to that positive time – have pure real transforms, but do not exist in nature. Time functions that represent real networks are pure-real and non-symmetric (that is $F(t)$ for $t < 0$ is 0) and must have hermitian transforms. Note that all physically realizable networks have non-symmetric real impulse responses, due to causality (the output response of a network to an impulse at time zero must be zero for all negative time), and thus must have hermitian Fourier transforms. A consequence of this is if one knows the frequency response in the positive frequency domain, one also knows the negative frequency response, since $F(\omega) = F^*(-\omega)$.

4.2.3   Modulation (Shift) Theorem

Many filter derivations and communication analyses are based on low-pass to band-pass transformations. This represents a shift in frequency. It is sometimes useful to use a similar transformation in time domain analysis, if the DUT is band limited, such as a filter. The shift or modulation theorem can be derived from the definition of Fourier transform:

$$if\ \mathbf{F}^{-1}(F(\omega)) = f(t), \quad then \quad \mathbf{F}^{-1}(F(\omega + \Delta\omega)) = f(t)\,e^{-j\Delta\omega t} \tag{4.5}$$

Note that the resultant time function is in general a complex function, so pure shift in frequency is not physically realizable. To transform a low-pass prototype to a realizable band-pass filter, one must replicate a positive shifted response and a negative shifted response. Thus, if $H_{LP}(\omega)$ is a low-pass filter's frequency response, and

$$H_{BP}(\omega) = H_{LP}(\omega + \omega_0) + H_{LP}(\omega - \omega_0) \tag{4.6}$$

is the band-pass filter frequency response, then the inverse transform of this is

$$h_{BP}(t) = h_{LP}(t)\,e^{-j\omega_0 t} + h_{LP}(t)\,e^{+j\omega_0 t} \tag{4.7}$$

Expanding the complex exponential, we find that

$$h_{BP}(t) = h_{LP}(t)\cos(\omega_0 t) - j\,\cancel{h_{LP}(t)\sin(\omega_0 t)} + h_{LP}(t)\cos(\omega_0 t) + j\,\cancel{h_{LP}(t)\sin(\omega_0 t)} \tag{4.8}$$

with the result

$$h_{BP}(t) = 2 \cdot h_{LP}(t)\cos(\omega_0 t) \tag{4.9}$$

The sum of these two shifts results in the imaginary terms canceling. The real portions add, and the result is that if $h_{LP}(t)$ is the low-pass prototype time (or impulse) response, the correct band-pass impulse response of the band-pass filter will be a cosine wave at the center frequency of the band-pass, with an envelope of two times the low-pass prototype's impulse response. However, this band-pass time response is *not the same* as the response obtained from the band-pass mode of a network analyzer time domain transform; this will be discussed later in some detail.

4.3   The Discrete Fourier Transform

Since measured frequency response of networks consists of discrete data, it is appropriate to discuss the discrete version of the inverse Fourier transform to determine the associated time response. The inverse discrete Fourier transform, which is defined only at discrete time points, for a discrete frequency data set, is

$$f(\tau) = \sum_{n=0}^{N-1} F(\nu)\, e^{j2\pi(\nu/N)\tau} \tag{4.10}$$

where (ν/N) is analogous to frequency in samples per cycle, τ is the discrete time increment, and $F(\nu)$ is the discrete frequency data set. The inverse fast Fourier transform (IFFT) is a very efficient way to compute $f(\tau)$ over the entire discrete time set. It might appear that the conversion of VNA frequency domain data to the time domain can be simply accomplished with an IFFT for computational efficiency. However, the IFFT limitations on the flexibility of the data (time) output can hide important effects that occur between calculated time samples, as described below. Further, much more conditioning is done in the VNA transform to enhance its applicability to practical problems.

4.3.1   FFT (Fast Fourier Transform) and IFFT (Inverse Fast Fourier Transform)

The FFT and IFFT are well-known algorithms for calculating the Fourier transform pair of a discrete data set as described in (4.10). If the discrete data set is generated from a sampled data set of a frequency response, and the data is sufficiently sampled as describe below, then the IFFT generates the time response of the network associated with the sampled data. FFTs and IFFTs have the attribute of greatly reducing the number of computations needed to compute a Fourier transform, but are limited in the data that is used and presented. One common limitation on FFT/IFFT transforms is that the sampled data and transformed data must have the same number of points. Some transforms also require that the number of points be in the form of 2^n, and all IFFTs distribute this limited number of points across the entire range of the time transform, which is the inverse of the frequency spacing of the VNA data.

4.3.1.1   Fine Structure Response

If an IFFT is applied to a frequency response, the resulting time response must have the same number of points, and the time intervals must evenly span the time period. A consequence

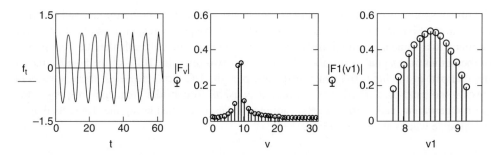

Figure 4.1 (left) Cosine of frequency 8.5 Hz, (middle) FFT of the waveform in the left plot, (right) Fourier transform of the waveform around the frequency 8.5 Hz.

of this is that fine-grain time response is not necessarily evident in the IFFT data, and this fine-grain response is where much of the intuitive understanding of the network can be found. The IFFT is equivalent to the analytic IFT, sampled evenly over a time period with the number of sample points equaling the number of frequency response points. Thus, any time domain response information that is present between these points is not evident in the IFFT data. This is also true of the FFT of a time-sampled signal.

It is illustrative to use a familiar example to demonstrate this fact. Take a time function shown in Figure 4.1 consisting of a cosine signal of known frequency such as

$$V(t) = 1 \cdot \cos(8.5t) \tag{4.11}$$

If several cycles of the time signal are sampled at higher than the twice the highest frequency, it is sufficiently sampled to avoid aliasing. One might naively assume that the FFT of this time signal should return the original frequency of the time waveform. However, if the frequency of the signal is not synchronous with the sampling, the FFT does not have an output "bucket" at the frequency of the cosine, and the FFT appears to have two main output signals as the spectrum of the sampled time waveform shows in Figure 4.1(middle), neither of which is the correct amplitude of 0.5 based on the time function. Thus the fine grain nature of the signal is not revealed by the FFT. A Fourier transform can be performed at discrete frequencies over the range of the two largest-valued FFT outputs, using the same time data set, from which the correct magnitude of the original frequency of the signal is revealed, Figure 4.1(right). In fact, since the time data is a finite set of discrete sampled points, the frequency response must be a periodic and continuous function with an infinite response to represent the transitions at the start and end of the data set. The FFT is exactly a sampled version of this continuous frequency function. The non-zero values of the FFT for all the other frequencies is a consequence of taking the time data over a finite time, essentially turning the sine wave signal into a pulsed sine wave, with values assumed to be zero before and after the sampled data. This has the effect of spreading the signal over several frequency buckets. Reducing this effect is a key attribute of the VNA time domain transform.

In order to get faster computation speeds, FFTs are often used instead of direct calculations of the discrete Fourier transform (DFT). However, many commercially available signal analysis tools take further shortcuts in calculating the FFT. One common shortcut is to assume that the time response is real. From this the frequency response must be hermitian, and therefore only half the FFT need be calculated to obtain the full frequency response. With an IFFT, it is common to assume a hermitian frequency response input, and only calculate the real portion of the output time signal. Thus, the IFFT is simply two times the IFFT of the positive half of the real part of the input frequency response, plus the DC term. However, there are several cases – the band-pass transforms – where it is useful to consider frequency responses that are not hermitian, and in these cases, care must be used in considering the shortcuts that are permissible when calculating IFFTs.

4.3.2 Discrete Fourier Transforms

The fine structure of a time response can be determined if an inverse discrete Fourier transform (IDFT) is used, in which the time axis can be arbitrarily small. If a small time-spacing were used for the FFT, an extremely large number of frequency points would have to be used as the input, greatly slowing measurement time to generate the frequency response terms. However, the DFT takes considerably longer to calculate than the FFT, and therefore is also not satisfactory where real-time transformation is needed. Fortunately, if the transform is needed over a relatively small portion of the time response, faster methods of calculation are available that provide finer resolution in time, without sacrificing speed.

Most faster-transform methods require equally spaced time points. The time spacing will be the time span divided by the number of points minus one. However, there are no restrictions on the start or stop times. If the start and stop times are chosen to be $t_{start} = 0$ and $t_{stop} = 2\pi/\Delta\omega = 1/\Delta\omega$, and the number of times points equals the number of frequency points, then the DFT will return the same values as the IFFT.

4.4 Fourier Transform (Analytic) vs VNA Time Domain Transform

The limitations of the IFFT, as applied to microwave measurements, required other techniques for analyzing these networks. The time domain transform of VNA measurements was first introduced in 1974 [3], and has been widely used since its real-time commercial introduction with the HP 8510A (1984), which allowed increased accuracy and real-time gating [4,5]. This VNA provided the capability to calculate the time domain response of the frequency domain data, using a form of the inverse Fourier transform. However, there were several modifications that are important to note, which causes the time domain response of the VNA to be different from the actual inverse Fourier transform of the frequency response of a network, that is, different from the impulse response of the network being measured. These differences come from the mode of the VNA transform (low-pass step, low-pass impulse or band-pass impulse), data windowing and truncation, window renormalization and data gating. For much of this time, the principal use was in low-pass impulse mode for fault location, and much has been written about the interpretation of the low-pass step-mode time domain response. Recently, the time domain response has been applied to solving the problem of filter tuning, using the band-pass mode [6]

A rigorous analysis comparing this time domain mode to the analytically derived impulse response may be obtained by applying appropriate functions to the analytical frequency response until the inverse Fourier transform of this modified response exactly matches the

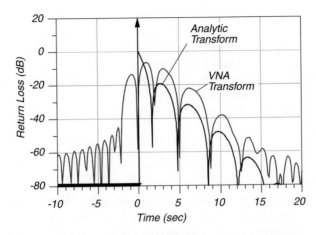

Figure 4.2   Analytically derived impulse reflection response vs VNA time domain response for a three-pole Butterworth filter.

VNA time domain response. Each of these functions applied to the frequency response can be evaluated in the time domain, and their associated time domain effects can be individually determined. This approach differs from [3] in that Hines and Stinehelfer develop the time domain response from assuming a periodic time function, the Fourier transform of which reproduces the measured frequency response. However, in the treatment presented below, a continuous analytic frequency response is assumed, and modifications are applied to account for discrete frequency sampling and windowing to directly obtain the VNA time domain transform in terms of the original frequency response and these modifications.

4.4.1   Defining the Fourier Transform

The IFT of a function provides directly the impulse response of that network, and provides the same time domain result as driving the network with an impulse, $\delta(t)$, and determining the time response. Figure 4.2 shows the analytically derived transform of S11 of a three-pole Butterworth filter (meaning that the reflection frequency response is calculated using standard network theory, and the inverse Fourier transform from Eqn. (4.2) is calculated to get the time response), along with a VNA time domain transform of the same function. Clearly they are not the same though they have some similarities in their structure. The differences will be reconciled in the following sections by describing the way in which each aspect of the VNA measurement must be accounted for with the appropriate mathematical transformation to achieve the same result as the IFT.

4.4.2   Effects of Discrete Sampling

The Fourier transform operates on continuous functions, while the VNA time domain transform must operate on measured (discrete) data. One approach is to assume that the measured data is a sampled version of a continuous analytic frequency response. Since the data applied to the time domain transform is discrete, the time domain transform must differ from the analytically calculated IFT of the network, but an equivalent discrete representation of an analytic function can be obtained by a mathematical representation of the sampling process. Note that such a time function would be identical to one determined in [3] but this approach is more readily applicable to the problem of comparing the VNA time domain transform to the analytic impulse response of a network.

A frequency sampling function can be represented as $III(\omega)$, which is defined as,

$$III(\omega) = \Delta\omega \cdot \sum_{n=-\infty}^{\infty} \delta(\omega - n\,\Delta\omega) \tag{4.12}$$

and can be visualized as a collection of delta functions with $\Delta\omega$ spacing. The effect of discrete data in the measured frequency response can be analyzed by forming a sampled function composed of the analytic frequency response multiplied by the sampling function, such that its value is zero between measured points, and the scaling factor of the delta function at each frequency is the measured value of the frequency response. The IFT of the sampled function, $f_S(t)$, can now be represented analytically by multiplying the original frequency response function by a sampling function and taking the IFT of the result:

$$f_S(t) = \mathbf{F}^{-1}(F_S(\omega)) = \frac{1}{2\pi} \cdot \int_{-\infty}^{\infty} F(\omega) \cdot \Delta\omega \cdot \sum_{n=-\infty}^{\infty} \delta(\omega - n\,\Delta\omega) \cdot e^{j\omega t} d\omega \tag{4.13}$$

or, through the operation of the integral on the delta function:

$$f_S(t) = \mathbf{F}^{-1}(F_s(\omega)) = \frac{1}{2\pi} \cdot \sum_{n=-\infty}^{\infty} F(n\Delta\omega) \cdot \Delta\omega \cdot e^{jn\,\Delta\omega t} \tag{4.14}$$

This operation can also be understood by noting that multiplication of two functions in the frequency domain is the same as convolving the inverse transforms of functions in the time domain. Convolving a function by a delta function returns the original function, at the origin of the delta function. Thus, the inverse transform of the sampling function returns another sampling function,

$$III(t) = \frac{1}{\Delta\omega} \cdot \sum_{n=-\infty}^{\infty} \delta\left(t - n\frac{1}{\Delta\omega}\right) \tag{4.15}$$

Sampling in the frequency domain is the same as convolving the original time domain response by the sampling function $III(1/\Delta\omega)$. Therefore the transform of an analytic function can be related to the transform of the discrete sampled version by convolving the inverse impulse response of the original function with the sampling function of Eq. (4.15). The effect of discrete data sampling can be seen to create images of the original function (sometimes called aliases) spaced at the inverse of the sampling spacing. The time range of $\pm\pi/\Delta\omega = \pm1/2\Delta f$ is referred to as the alias-free range of the inverse transform for sampled data. Many commercial products display a maximum range of $\pm1/\Delta f$. If the impulse response of the original function does not tend to zero by $\pm1/2\Delta f$, then the appearance of the transformed sampled-function in the alias-free range will be distorted by effects from previous and subsequent images. Figure 4.3 (upper) shows a frequency plot of the sinc^2 function, $(\sin(x)/x)^2$, the transform of

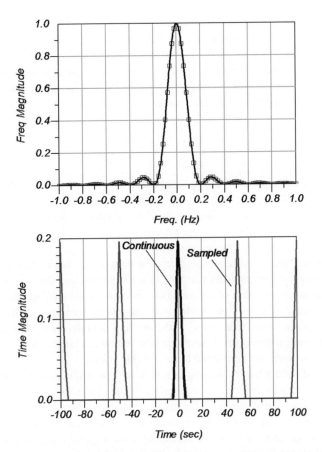

Figure 4.3 Sinc-squared frequency response continuous and sampled with a sampling frequency of 0.02 Hz; time domain response continuous and showing repeated time responses due to sampling (lower).

which is well known analytically to be a triangle pulse, along with the sampled data points. The lower plot shows the IFT of the continuous function, and also the IFT of the sampled function.

The inverse transform of a sampled frequency response must have an infinitely repetitive (periodic) time response. Even if the frequency response is discrete, the time response may still be continuous. Only if the frequency response is discrete and periodic will the time response be discrete. Since any **real** sampled frequency response must be sampled over a finite frequency span (and thus cannot be periodic), the time response associated with any measured frequency response will be continuous and periodic. That is, every time domain response displayed on a VNA represents a periodic time function.

4.4.3   Effects of Truncated Frequency

Another consequence of taking a transform of measured data is that the frequency response must be truncated, rather than extend to plus and minus infinity. That is, all VNAs are limited in their range of response measurements, and the sample frequency data will not have an infinite response. For transmission responses, this does not present much of a problem, as the responses of most networks act as filters and become arbitrarily small at high frequencies; the high frequency contribution to the inverse Fourier integral is negligible. However, for reflection responses, the value of the response remains large at high frequencies. In fact, these responses are not strictly Fourier transformable, as they do not satisfy the equation

$$\int_{-\infty}^{\infty} |f(\omega)| \, d\omega < \infty \tag{4.16}$$

However, most reflection functions can be represented with the help of the generalized function, $\delta(t)$. But, if the response is truncated, and the response data is finite, then the Fourier transform of the data strictly exists. In fact, if functions are derived from an accurately specified physical quantity it is a sufficient condition for the existence of a transform. Or put another way, if a thing is real, its time domain response must exist.

Truncation of the frequency response data of a network is mathematically equivalent to multiplying the data by a rectangular window. In the time domain, this can be represented as convolving the impulse response of the network with a $\sin x/x$ function, which is the inverse transform of the rectangular window. In this way, the inverse transform of truncated data will always have a response with "side lobes" if the original data does not go to and remain zero some time before truncation occurs. These side lobes can be so large as to obscure the impulse response, and much work has been done to reduce this effect.

For the most part, side lobes, or ringing as it is sometimes called, can be controlled through the appropriate use of windowing. Taking the IFT of the product of the original function and a rectangular window can represent the effect of truncated data in the VNA time domain transform. Referring to Eq. (4.13), this truncation is equivalent to redefining the limits of the integral to be the endpoints of the measured data. Figure 4.4 shows an example of a one-pole filter response (upper, gray trace) with the analytic function $F(s) = 1/(s+1)$, or $F(\omega) = 1/(1 + j\omega)$ where $s = j\omega$, along with its truncated frequency response (upper, black trace). Then Figure 4.4 (middle) shows the time domain response of the truncation function, which is the IFT of a rectangular window, which in turn is a $\sin x/x$ function. The filter response has an analytic time response of $f(t) = e^{-t} \cdot U(t)$ (where $U(t)$ is the unit step function) as shown in Figure 4.4 (lower, gray trace). The truncation effect on the analytic time response can be obtained by convolving the IFT of the original function with the $\sin x/x$ function, and this is shown in Figure 4.4 (lower, black trace). From this, the original transform is almost not recognized due to the distortion of the side lobes caused by the truncation effect.

For a sampled data set, over the range of $\omega = -N\Delta\omega$ to $+N\Delta\omega$, the IFT becomes

$$f_s(t) = \frac{\Delta\omega}{2\pi} \cdot \sum_{n=-N}^{N} F(n\Delta\omega) \cdot e^{jn\Delta\omega t} \tag{4.17}$$

Equation (4.17) might be called the sampled inverse Fourier transform. Note the similarity to the inverse discrete Fourier transform of Eq. (4.10). The sampled inverse Fourier Transform of Eq. (4.17) can be used to calculate the inverse transform for any particular time t, so there are no limits on the time span or point spacing in time, as there is with the FFT.

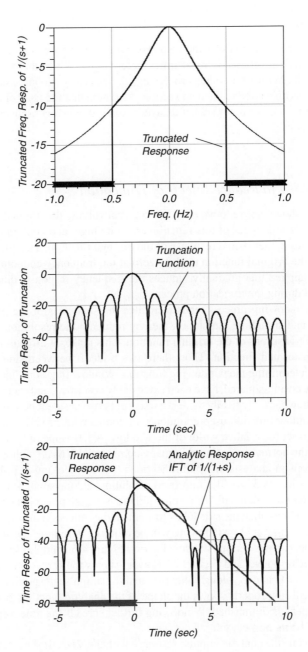

Figure 4.4   One-pole filter frequency response with and without truncation.

4.4.3.1   Causality

A consequence of truncation is that the response appears to be non-causal, that is, the side lobes for the impulse response occur at times before zero. While this is not desired, it is a mathematical fact. With some other processing, the causality violation can be reduced to acceptable levels, as discussed in the next section. Also note that the peak of the function does *not* occur at time t = 0, but is somewhat delayed. Again this is a consequence of truncating the frequency response.

4.4.4   *Windowing to Reduce Effects of Truncation*

Data truncation is shown above to have the effect of convolving the original transform with a $\sin x/x$ function. The side lobes of this function are quite high, and continue for a substantial extent, often obscuring the desired response of the original function. The effects of truncation are minimized if the original function tends to zero at the frequency endpoints. A windowing function may be applied that gradually reduces the frequency response, thus controlling the side lobes created during the truncation process.

However, the windowing process tends to reduce the sharpness of the original response, spreading pulses and stretching out slopes, thereby reducing the resolution of the transform and distorting the transitions of the original function. This makes it difficult to assess the true nature of the transformed function. Thus, there is a tradeoff between side lobe height and resolution when determining the windowing function. Window functions, including Hanning, Hamming, cosine, cosine-squared, have been extensively described, and each window function has benefits and drawbacks; typically there is a tradeoff between side lobe suppression and loss of rise time; a window function used commonly in commercial products uses a Kaiser-beta (KB or β) value to set the relative width of the window. A KB value of 0 gives no window, a KB value of 6 is the normal values used in many commercial VNAs, with a maximum value of 12. For time domain analysis of small reflections (such as looking at in-line connectors), using KB values as low as 3 provides improved resolution and the small reflections mean that side lobe values don't interfere with the results. If there is a large reflection (such as an open or short at the end of a line), then the side lobes from the large reflection may mask the reflection of interest. In such a case, more windowing will be required to remove the side lobe effect.

Figure 4.5 (upper) shows various window factors, Figure 4.5 (middle) shows these applied to a one-pole filter response, and Figure 4.5 (lower) shows the time response of windowed functions for KB = 0 and 6, along with the analytic impulse response. This windowed version of the transform properly shows the shape of the analytic function, but the rise-time is extended and the peak is diminished. Windowing further spreads the time response, adding to the appearance of being non-causal.

In order to reconcile the analytic impulse response with the VNA time domain transform, the effects of finite frequency, sampling and windowing on the analytic IFT can be mathematically represented below as f_{SW} (for sampled, windowed)

$$f_{SW}(t) = \frac{\Delta\omega}{2\pi} \cdot \sum_{n=-N}^{N} F(n\Delta\omega) \cdot W(n\Delta\omega) \cdot e^{jn\Delta\omega t} \tag{4.18}$$

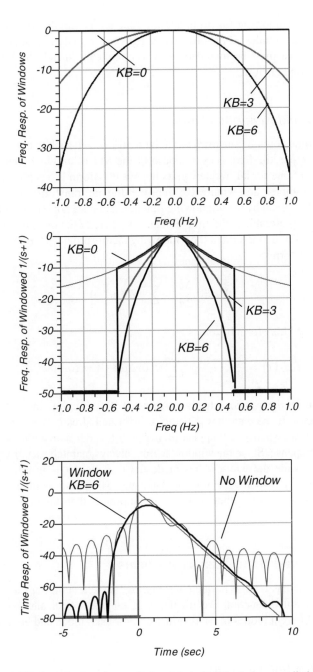

Figure 4.5 (upper) Windows for beta factors 0, 3 and 6; (middle) windows applied to a one-pole filter; (lower) time response of windowed trace.

where $W(\omega)$ is the windowing function, and the function is sampled over $\omega = -N\Delta\omega$ to $+N\Delta\omega$. This response includes all the obvious changes to the analytic function but there is one final modification that must be included so as to completely match the VNA time domain transform, as described below.

4.4.5 Scaling and Renormalization

The value of the time domain transform has to be renormalized such that it retains its physical meaning. For example, the frequency response of the S11 of an ideal open circuit, with no delay, has a value of one for all frequency; its inverse transform is a delta function. However, when the data is sampled and windowed, the time domain transform of the response of an open circuit will be spread by the windowing function and does not return an impulse of unity height. It would be preferable if the time domain response of the open circuit had a value of unity at time $t = 0$. Taking the sum of the windowing factors provides the correct scaling factor for subsequent transforms:

$$W_0 = \frac{\Delta\omega}{2\pi} \cdot \sum_{n=-N}^{N} W(n\Delta\omega) \tag{4.19}$$

and the renormalized transform becomes

$$f_{VNA}(t) = \frac{1}{W_0} \cdot \frac{\Delta\omega}{2\pi} \cdot \sum_{n=-N}^{N} F(n\Delta\omega) \cdot W(n\Delta\omega) \cdot e^{jn\Delta\omega t} \tag{4.20}$$

Note that this scales the transform to always return 0 dB for a unit frequency input, regardless of windowing factor. If the data that is being transformed already tends to zero at the band edges, the windowed response will appear higher after this normalization, when compared to an analytic time response. Since the window scaling always maintains a unity peak amplitude, regardless of how wide the window has made the response, it is in effect amplifying the DC and low frequency responses. For some data, such as a low-pass filter response, this can result in a windowed response that is higher in amplitude than the corresponding analytic impulse response.

4.5 Low-Pass and Band-Pass Transforms

Since measured data has a finite frequency sampling, some assumptions are made about the behavior of the sampled function. Vector network analyzers offer alternative assumptions, which yield two different modes of transformation: low-pass mode and band-pass mode.

4.5.1 Low-Pass Impulse Mode

The assumption for low-pass impulse mode is that the underlying frequency response is that of a real network. As such, the frequency response is hermitian and the time domain response is pure real. Also, it is assumed that the network response becomes asymptotic at low frequencies, that is, the response at low frequencies is roughly constant, and the frequency response beyond the measured frequency range contains no important information about the network. In other words, everything of interest occurs over the frequency of measurement. The data points must be linearly spaced over the range of $\omega = n\Delta\omega$ from $n = 1$ to N. Thus, the frequencies must be harmonically related. For this transform, the windowing function is centered at $\omega = 0$, and extends to the max frequency $\omega = N \cdot \Delta\omega$. From this, it follows that the complex sum in Eq. (4.20) becomes

$$f_{LP}(t) = \frac{\Delta\omega}{2\pi}F(0)\frac{W(0)}{W_0} + \frac{1}{W_0}\frac{\Delta\omega}{2\pi}2 \cdot \text{Re}\left[\sum_{n=1}^{N}F(n\Delta\omega) \cdot W(n\Delta\omega) \cdot e^{jn\Delta\omega \cdot t}\right] \quad (4.21)$$

Given a hermitian function, the imaginary parts of the negative and positive transform cancel, and the real parts double, so that only the real portion is computed. Further, it is clear that a value must be determined for $F(0)$, which is done with DC extrapolation. From Eq. (4.21), it can be seen that the time domain transform consists of sums of sines and cosines, and that the highest frequency measurement point determines the highest frequency element. Thus, the rise time is determined by the maximum slope of the highest frequency measured. The transform will repeat itself at intervals determined by the frequency step value, which is the same as the lowest frequency point.

4.5.2   DC Extrapolation

In addition to being limited in upper frequency response, measurement equipment is limited to its minimum frequency response. However, the Fourier transform includes effects of the DC value on the frequency response. Since VNAs do not commonly measure the DC response, DC extrapolation is used. Some analysis programs allow DC to be directly entered. DC extrapolation requires the assumption that the network response approaches DC asymptotically, and different algorithms are used by different applications. Consequences of DC extrapolation are discussed below.

4.5.3   Low-Pass Step Mode

Up to now, the discussion has focused on the impulse response of a network. The step response of a network can be useful in directly determining the network characteristics, particularly in the case of concatenated transmission lines, and evokes the normal mode of operation of a time domain reflectometer (TDR), which employs a stepped DC stimulus. The unit step function $U(t)$ is defined as

$$U(t) = \begin{cases} 0 \text{ for } t < 0 \\ \frac{1}{2} \text{ for } t = 0 \\ 1 \text{ for } t > 0 \end{cases} \quad (4.22)$$

and from this its Fourier transform may be determined as

$$\mathbf{F}[U(t)] = \pi\,\delta(\omega) - j\,\frac{1}{\omega} \quad (4.23)$$

The time domain step response may be found by multiplying the Fourier transform of the unit step-function, Eq. (4.23), by the frequency response, $F(\omega)$, of a network and taking the inverse transform

$$f_{Step}(t) = \frac{1}{2\pi} \int_{-\infty}^{\infty} F(\omega) \cdot \left(\pi \, \delta(\omega) - j \frac{1}{\omega} \right) e^{j\omega t} d\omega = \frac{F(0)}{2} - \frac{j}{2\pi} \int_{-\infty}^{\infty} \frac{F(\omega)}{\omega} e^{j\omega t} d\omega$$

(4.24)

Taking the derivative of the step response yields the desired impulse response of the network.

The low-pass mode of the VNA time domain transform has two forms: low-pass impulse, which is defined by Eq. (4.21), and the low-pass step, which is essentially the integral of the low-pass impulse response, with respect to time and with some particular choice for the constant of integration. The step response of the VNA should retain the property that its derivative is the VNA time domain impulse response, and since the frequency-domain sampling function creates a periodic time domain response, with a period of $1/\Delta f$, the step response should retain this aspect of the periodicity, and low-pass data will be valid between $t = 0$ to $t = 1/\Delta f$

Figure 4.6 shows the step response stimulus (labeled "VNA Unit Step Resp") that meets the properties of having a periodic impulse response for its derivative. This response differs from the square-wave response described by Hines and Stinehelfer [3], and from the plot, it is obvious that this function cannot have a Fourier transform. However, it may be written as the sum of two functions, the first one being periodic (labeled "Periodic Portion") and the second being a ramp function (labeled "Ramp Portion").

The time step response can be determined from the network function and the unit step stimulus by applying the appropriate Fourier transform to the periodic portion, and some appropriate Laplace transform to the ramp portion. From Eqs. (4.21) and (4.23) the step

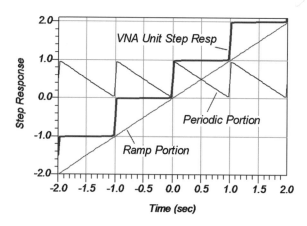

Figure 4.6 VNA unit step response comprised of a periodic portion (which is Fourier transformable) and a ramp portion.

response for a sampled, truncated, windowed function can be proposed to be

$$f_{Step}(t) = \left\{ \frac{F(0)}{2} + \frac{\Delta\omega}{2\pi} \cdot 2 \cdot \text{Re} \left[\sum_{n=1}^{N} \frac{F(n\Delta\omega) \cdot W(n\Delta\omega)}{jn\Delta\omega} \cdot e^{jn\Delta\omega \cdot t} \right] \right\} + F(0) \cdot \frac{\Delta\omega}{2\pi} \cdot t + C$$

(4.25)

Differentiation of Eq. (4.25) clearly results in Eq. (4.18). The second term in the equation (the ramp portion) is needed if the impulse response contains a DC term. The final constant of integration is chosen to give the proper response value of the transform at time t = 0.

Thus the step response can be obtained by taking the inverse Fourier transform of the frequency response divided by j times the step frequency, and adding to this a linear time ramp. The time domain step response is only available in the low-pass mode.

4.5.4 Band-Pass Mode

The band-pass mode provides an alternative method of time domain transform that may be used when the low-pass mode assumption of harmonically related frequencies cannot be met. This might occur, for example, in the measurement of a network that is band-pass or high-pass filtered. The output of a VNA measurement is typically an odd-numbered set of points, linearly spaced in the form of $\omega = \omega_c + n\Delta\omega$ from $n = -N/2$ to $N/2$, and ω_c is the center frequency of the data. The inverse Fourier transform is calculated only on the data points measured, rather than presuming the negative frequency response to be the conjugate of the measured data. That is, the band-pass mode does not presume a hermitian frequency function, and uses data as though the frequency response was positive only (of course, this cannot represent a real device, but it is very useful). Windowing is applied, where the center for the windowing function is the center frequency of the data set. In contrast, the center of the windowing function in low-pass mode is centered on the DC term, or the first point of the data set. The inverse band-pass transform is defined by

$$f_{BP}(t) = \frac{1}{W_0} \frac{\Delta\omega}{2\pi} \cdot \sum_{n=-N/2}^{N/2} F_{BP}(\omega_C + n\Delta\omega) \cdot W(n\Delta\omega) \cdot e^{j(\omega_c + n\Delta\omega)\,t}$$

(4.26)

This is an important difference between the VNA band-pass mode and that described by Hines and Stinehelfer [3] which results in a pure-real time domain response. In contrast, the VNA band-pass response results in a complex time domain response, and this choice of transform is key to the useful application of the band-pass mode response, one example being the application to filter tuning. To illustrate the band-pass transform mode, consider the frequency function of a band-pass filter. The frequency response tends to zero away from the center frequency, so the windowing function will have little effect on the transform. If the frequency response F_{BP} represents a band-pass version of a low-pass prototype response [7], such that $F_{BP}(\omega) = F_{LP}(\omega - \omega_C)$ and thus $F_{BP}(\omega_C) = F_{LP}(0)$, the relationship between the time domain band-pass transform of the band-pass filter and the low-pass prototype's frequency response can be established as

$$f_{BP}(t) = \frac{e^{j(\omega_c)t}}{W_0} \frac{\Delta\omega}{2\pi} \cdot \sum_{n=-N/2}^{N/2} F_{LP}(n\Delta\omega) \cdot W(n\Delta\omega) \cdot e^{j(n\Delta\omega)t}$$

(4.27)

or in terms of the low-pass time domain response

$$f_{BP}(t) = e^{j\omega_c \cdot t} \cdot f_{LP}(t) \tag{4.28}$$

From this it follows that the band-pass time domain mode always returns a complex time domain response. This effect is due to removing the assumption that the frequency response contains negative frequency elements. The magnitude response of the band-pass transform is the same as the low-pass prototype

$$|f_{BP}(t)| = |f_{LP}(t)| \tag{4.29}$$

Thus, the band-pass mode response of the time domain transform is quite different from the analytic impulse response of the network. Consider a network, such as a filter, that has a low-pass response $f_{LP}(t)$. If this filter is used as a prototype for a band-pass filter, and is shifted to create a band-pass response, the band-pass filter will have an analytic impulse response of

$$f_{Imp}(t) = 2f_{LP}(t) \cdot \cos(\omega_C \cdot t) \tag{4.30}$$

which *is* pure real as would be expected of an analytic transform of a real network. So, the band-pass mode transform has, in addition to the windowing, sampling and frequency truncation effects, an effect due to the data being taken as though the network had a single-sided (positive frequency only) response. Also, since the windowing function is centered on the center frequency of the transform, it forces the function to zero at the lowest as well as the highest frequency; there is no point in extrapolating the DC term.

One consequence of the band-pass transform is that the resolution is half that of the low-pass transform. This can be seen from Eq. (4.26), which shows that the maximum frequency in the complex exponential is one-half of the frequency span (since the data ranges from $n = -N/2$ to $N/2$). The alias-free range for this transform remains the same as the range of the low-pass transform.

With this introduction to the time domain transformations used in VNAs, the concept of time-gated measurements can be better understood.

4.6 Time Domain Gating

Time domain gating refers to the process of selecting a region of interest in a portion of the time domain, removing unwanted responses and displaying the result in the frequency domain. Gating can be thought of as multiplying the time domain response by a mathematical function with a value of one, over the region of interest, and zero outside this region. The gated time domain function can then be forward transformed to display the frequency response without the effect of the other responses in time. The gating effects, however, are somewhat subtle in their response and there are consequences of the gating function that are not readily apparent.

In practice, the gating is not a "brick wall" function. This is because a sharp transition in the gate function causes a similar sharp transition in the gated time function. As such, the frequency response will have ringing associated with the sharp transition (as the frequency response is limited to the measured data region). To avoid this ringing, the gating function is windowed in the frequency domain before being transformed to the time domain. For a rectangular time-gating function centered at $t = 0$, the Fourier transform can be calculated analytically, with the result that the gate frequency response will have a $\sin(\omega)/\omega$ or $sinc(\omega)$

function. The width of the sinc main lobe is inversely proportional to the width of the time gate. If the center of the gate time is not at $t = 0$, the resulting Fourier transform produces a response that corresponds to the sinc function multiplied by a complex exponential factor, namely sinc $(\omega) \cdot e^{j\omega t_c}$. This is windowed in the frequency domain which sets the maximum gate transition slope (think of it as the gate rise time) in the time domain, and reduces gate side lobes. The gate side lobes cause a phenomenon sometimes called "gate-leakage", which is observed when a large reflection signal is gated-out near a small reflection signal. The gate leakage can still produce a significant response that will produce an erroneous gated response. The gate function is then transformed to the time domain and multiplied by the time domain response to display the gated time domain response. If the gate frequency response is desired, the gated time domain can be transformed back to the frequency domain. In practice, an alternative computation is used whereby the time-gated frequency response is computed by convolving the gate frequency response by the measured frequency response. This method reduces the number of transforms required, makes for faster processing, and this convolution interpretation of gating allows a more intuitive understanding of one of the subtle gating effects, described below.

4.6.1   Gating Loss and Renormalization

A curious effect of the gating function occurs at the endpoints of the time-gated frequency-domain response if applied as described above: these endpoint regions are lower by 6 dB, as though the gating had caused some loss at the frequency endpoints. In fact, the gating is a kind of filtering function and the loss is real. The 6 dB offset can be understood by comparing the center point and last point of a gated frequency-response of a unit function $F(\omega) = 1$, that is, a frequency response that is constant (such as a thru line). The time domain response will approach a delta function, $f(t) = \delta(t)$, but the frequency response is just a flat line. This function is the dark solid trace in Figure 4.7.

In the convolution process, the gated value at any frequency ω_1 can be determined by multiplying $F(\omega)$ by the frequency-reversed version of the gating frequency function centered at that frequency, ω_1, and integrating (summing since it is discrete) the product, this being the

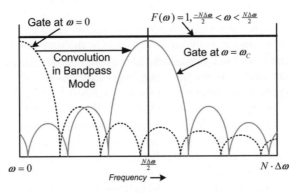

Figure 4.7   Convolution of the frequency gate response.

definition of convolution:

$$F_g(\omega_1) = \sum_{n=-N}^{N} F(n \cdot \Delta\omega) \cdot G(\omega_1 - n \cdot \Delta\omega) \tag{4.31}$$

For the center frequency point in band-pass mode (or zero frequency, DC, in low-pass mode), the gated frequency-response is a sinc function with the response centered at $\omega = \omega_C$, multiplied by the frequency response (which is just 1), and appears as the light gray trace in the figure above. The integration of the light gray trace is the gated response at that frequency. For the case of the first (or last) frequency point, the sinc function is centered on the first (or last) data point, and half the gate function is multiplied by zero (for frequencies outside of the original response), and thus does not contribute to the sum, as shown by the dashed trace in the figure. So the first data point will be one-half the value of the center point, or 6 dB lower; this can be seen graphically in Figure 4.7 as the area under the dashed line is one-half the area under the light gray line. Unfortunately, this creates the result that any gating will distort the first and last few points (last points only in low-pass mode) of the gated frequency response. This span of the distorted data is related to the span (in frequency) of the gate; for wider time gates, the distortion appears over fewer of the points at the edges of the gated frequency response.

The VNA time domain compensates for this roll-off through a post-gate renormalization. The post-gate renormalization is determined by creating a frequency response that is unit magnitude. A pre-gate window is applied to this unit response that is the same as the pre-transform window applied to the normal frequency response data. This unity-magnitude frequency response is convolved with the gate frequency response, to generate the final normalizing frequency response. The time-gated frequency response is divided by this function to remove the roll-off effects of the time gating. This normalizing function works perfectly for a unit time response at the center of the gate. If the gate is not symmetric around the time function, there will be some errors in the gated response when compared to the original frequency response.

It is instructional to view the actual gate shape in the time domain, which can be done using a function not normally available in commercial VNAs. The gate shape may be generated by creating a delta-like frequency response ($F(0) = 1; F(\omega) = 0$ for $\omega \neq 0$, but with some spreading due to windowing), applying gating and transforming the result to the time domain to see the actual gate shape. This is useful in understanding how the gate shape affects the gated response.

Figure 4.8 (upper) shows the gating function for various gate center times. Figure 4.8 (middle) shows a unit frequency response ($F(\omega) = 1$) in the time domain, with two of the gates applied the extremes of the gate offset times. Note that the peak of the time domain response is nearly unchanged as the time-gate fully encompasses the impulse at all three center times, but there is some difference in the side lobes for the shifted gates. Figure 4.8 (lower) shows the frequency response after gating. Here there is a substantial difference in the response at high frequency for the different gate center times. It is clear that normalization is optimal when the gate is centered on the response being gated.

The gated time-response may be viewed in the time domain by taking the IFT and displaying the result. In fact, it is almost always required to first view the time domain response to assign proper gating start and stop values: the transform function is turned on and the resultant time domain response is displayed and the gate start and stop are set. Next gating is turned on.

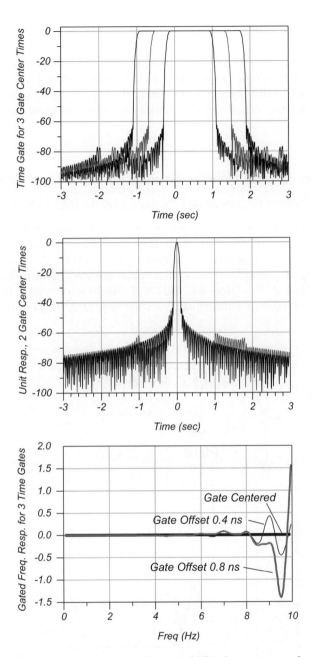

Figure 4.8 (upper) Time gates at three center times; (middle) time response of gated unit response, with the first and last gate center; (lower) gated frequency response of three gate centers, showing normalization error at the band edge (only positive frequency shown).

Finally, the transform is turned off, and the time-gated frequency-response is shown in the frequency display.

A study of the time domain response of several examples of composite responses of several component elements will show how time gating can be used to separate the responses in time, and display the individual frequency responses of the component elements, but with some distortion due to masking effects [8]. From this a method is developed that compensates for these effects.

4.7 Examples of Time Domain Transforms of Various Networks

4.7.1 Time Domain Response of Changes in Line Impedance

For ladder networks – that is, networks that consist of series-connected elements – the time domain transform provides very good insight into the nature of the discontinuities by which the frequency response is generated. As a first example, consider the network in Figure 4.9 (upper), consisting of a short length of Z_0 line followed by a $Z_0/2$ impedance line segment terminated in another Z_0 line. Note that there will also be re-reflections if the timescale is extended. There are two main reflections from the impedance steps at the beginning and

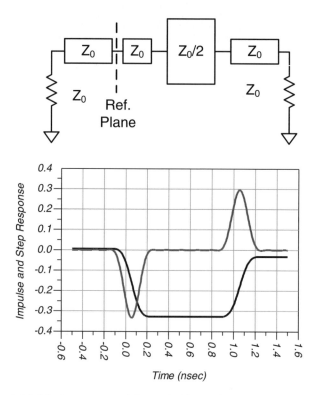

Figure 4.9 (upper) Model of concatenated lines of different impedances, (lower) step (black) and impulse (gray) response of the lines in time.

end of the $Z_0/2$ line segment. The impedance value of a discontinuity caused by a step in impedance of a transmission line can be directly related to the time domain step response, which shows reflections as a function of time. The reflections are relative reflection coefficient, so for a 50 ohm reference impedance, a 1% reflection relates to approximately 1 ohm change in impedance, as

$$\Gamma = \frac{Z - Z_0}{Z + Z_0}, \text{ and for } Z \cong 50, \Gamma(\%) \cong \Delta Z, \text{ where } \Delta Z = Z - Z_0 \qquad (4.32)$$

Care must be used in this interpretation, as other factors such as loss in the transmission line, changes in line impedance and previous reflections can affect the apparent reflection being investigated. For the lines in Figure 4.9, the step in impedance is apparent for the reflection being investigated and the step in impedance is quite large, but the reflection coefficient of each step is the same, $|\Gamma_1| = |\Gamma_2| = 0.33$. However, the second apparent reflection coefficient, computed by the difference between the level of the $Z_0/2$ line and the final Z_0 line after the second transition, $\hat{\Gamma}_2$, is only 0.30, as shown in Figure 4.9 (lower). Also, the impulse response shows a similar "masking" effect in the second reflection response. This masking effect is a consequence of the response of the network and illustrates why the simple concept that a time domain traces shows the impedance along a transmission line is not completely valid. In fact, the TDR shows the reflected voltage along the line, and the offset associated with masking can be determined from first principles, as discussed in the next section. Several more measurement examples using time domain and time gating are discussed in Chapter 5 on measurement of passive devices.

4.7.2   Time Domain Response of Discrete Discontinuities

As a second example, concatenated transmission lines with discrete discontinuities between sections are evaluated with a time domain transform of the frequency response, and the values of the various discontinuities are individually determined. Figure 4.10 (upper) shows a schematic of a Z_0 reference, followed by a first capacitive discontinuity, and followed by a Z_0 line then a second identical capacitive discontinuity terminated in a Z_0 load.

The time domain response of this network is shown in Figure 4.10 (lower). This is the low-pass step response, which shows capacitive discontinuities as negative dips in the time domain. The reflections of the discontinuities repeat at the spacing of the discontinuities, and these repetitive reflections should ideally continue on, at diminishing levels, for infinite time (though actually they get added to all the aliased responses). Also, note that even though the responses are caused by identical discontinuities, the response of the second discontinuity appears smaller than the first. The second response is somewhat masked by the first, though by a different amount than in the example of Figure 4.9, indicating a different masking mechanism.

4.7.3   Time Domain Responses of Various Circuits

The time domain responses of many elements are well known, the low-pass mode in particular can be used to identify the type and relative value of these discontinuities. This is a particularly useful aspect in evaluating unwanted discontinuities in connectors, cables and transmission line structures. A catalog of useful responses is shown in Table 4.1.

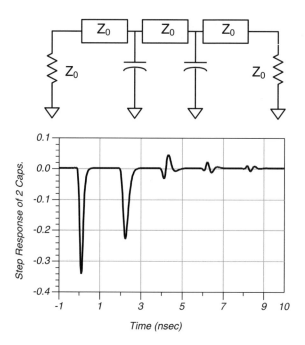

Figure 4.10   (upper) Model of two capacitive discontinuities, (lower) step response of the S11 of two discontinuities.

4.8   The Effects of Masking and Gating on Measurement Accuracy

The concept of time gating above refers to mathematically removing a portion of the time domain response, and viewing the result in the frequency domain. The intent is to remove the effects of unwanted reflections, say from connectors and transitions, leaving the desired response of the device being measured. This should improve the quality of the response; that is, the gated response should more closely resemble the device response as if it were measured with no other reflections. However, the effects of previous reflections can have an effect on the time-gated measurement. Previous work reported on the compensating for the effect of loss [9] but ignored the effect of previous reflections. Others have proposed an error associated with previous reflections but have not provided for compensation methods, or for errors associated with change in impedance. These effects are mathematically described below, along with new compensation methods, and with an uncertainty analysis on the time-gated frequency response, applied to several particular examples.

4.8.1   Compensation for Changes in Line Impedance

For the lines in Figure 4.9, the apparent reflection coefficient of the second transition, Γ_2, is only about 90% of the actual value. The normal time domain transform does not present the true reflection coefficient at that particular point. To understand this, consider that at the interface of the first reflection, the reflection coefficient, is calculated as defined in (4.32).

Table 4.1 Catalog of time domain responses

Element	Step response	Impulse response
Open	Unity reflection	Unity reflection
Short	Unity reflection, −180 °	Unity reflection, −180 °
Resistor $R > Z_0$	Positive level shift	Positive peak
Resistor $R < Z_0$	Negative level shift	Negative peak
Inductor	Positive peak	Positive then negative peaks
Capacitor	Negative peak	Negative then positive peaks

However, the signal that continues down the transmission line structure is changed by the transmission coefficient defined as [10]

$$T_1 = \frac{2 \cdot Z_1}{Z_1 + Z_0} \tag{4.33}$$

where Z_0 is the input line, and Z_1 is the second section of line. The reflection apparent at the input due to the second step in impedance, Γ_2, is further changed by a second (reverse) transmission coefficient, T_2 as defined by

$$T_2 = \frac{2 \cdot Z_0}{Z_1 + Z_0} \tag{4.34}$$

The total apparent reflection, $\widehat{\Gamma}_2$, due to the second step is now computed as

$$\widehat{\Gamma}_2 = \Gamma_2 \cdot T_1 \cdot T_2 = \frac{(Z_0 - Z_1) \cdot (4Z_1 Z_0)}{(Z_1 + Z_0)^3} \tag{4.35}$$

or, for the example, $\widehat{\Gamma}_2 = +0.30$, which precisely matches the measured value in Figure 4.9

Further, for the case of a response following a change in line impedance, where the first line impedance is not the reference impedance, two compensations are required. First, the reflection response must be compensated by dividing the apparent response by the transmission coefficient term product, $T_1 \cdot T_2$ to produce a reflection response relative to the line, $S'_{11} = S_{11}/(T_1 \cdot T_2)$ (derived from $\hat{\Gamma}_2$ as shown above). The second compensation is renormalizing the response by the impedance of the line just before the desired response. The frequency response assumes a reference impedance of the system impedance, typically 50 ohms. The renormalization consists of converting the reflection response to an effective impedance Z_{eff}, using the line impedance Z_{line} just before the desired response (S'_{11}), as the reference impedance. This is then reconverted from the resulting effective impedance back to effective reflection response $(S_{11(eff)})$ using the system impedance [11]:

$$Z_{eff} = Z_{line} \cdot \frac{1 + S'_{11}}{1 - S'_{11}}, \quad S_{11(eff)} = \frac{Z_{eff} - Z_0}{Z_{eff} + Z_0} \tag{4.36}$$

4.8.2 Compensation for Discrete Discontinuities

Figure 4.10 shows the time domain response of two capacitive discontinuities. The second discontinuity, which is caused by an identical element in the circuit, has a different time domain response from the first element. The most noticeable aspect is that the magnitude of the response is smaller, which is consistent with the first example. However, in this case, there is no change in reference impedance to account for the difference. Instead, the first reflection removes some of the energy from the forward (incident) wave, such that there is less energy available at the second discontinuity. A similar effect occurred in the first example, and was accounted for by the transmission coefficients. For a localized discontinuity, with the same impedance on each side, the effect on the transmitted wave must be determined in a different manner.

From power conservation, the magnitude of the voltage wave incident on the second reflection, $|V_2^+|$, (assuming the first reflection is lossless) is

$$|V_2^+| = |V_1^+| \cdot \sqrt{1 - |\Gamma_1|^2} \tag{4.37}$$

where V_1^+ is the incident voltage wave and Γ_1 is the first reflection. The magnitude of the reflected voltage, $|V_2^-|$, from the second reflection is

$$|V_2^-| = |V_2^+| \cdot \Gamma_2 = |V_1^+| \cdot \left(\sqrt{1 - |\Gamma_1|^2}\right) \cdot \Gamma_2 \tag{4.38}$$

The signal V_2^- reflects again off Γ_1 with a portion transmitted, V_3^- (the portion of the signal from Γ_2 that is actually measured at the input port), which is reduced in the same manner as Eq. (4.37) to yield the effective value of the second reflection as

$$\left|\hat{\Gamma}_2\right| = \frac{|V_3^-|}{|V_1^+|} = (1 - |\Gamma_1|^2) \cdot \Gamma_2 \tag{4.39}$$

This result only applies to the magnitude of the reflection, as the power conservation argument does not apply to the phase of the transmitted signal, and while consistent with the results described in other publications [8], it goes further to provide a means to remove the effects of the first discontinuity.

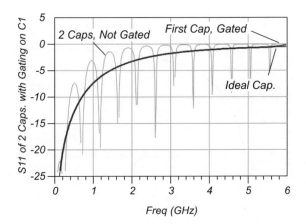

Figure 4.11 S11 response of two capacitive discontinuities (light gray) and gating around the first cap (dark gray); also shown is the S11 of just the first cap (black).

4.8.3 Time Domain Gating

4.8.3.1 Gating the First of Two Discontinuities

The effectiveness of gating can be evaluated using the circuit from Figure 4.10. Figure 4.11 shows the original frequency response in light gray, with the characteristic ripple pattern found from two discontinuities separated by a length of line. The thin black trace is the result of computing the ideal S11 of just a single capacitive discontinuity, terminated in Z_0. Gating around the first capacitive discontinuity yields a response (Figure 4.11, dark gray) nearly identical to the frequency response calculated for only the first discontinuity. The difference is seen only at the high frequency of the response, most likely due to the errors in the renormalization, as described in Figure 4.8 (lower). Clearly, gating about a first discontinuity terminated in Z_0 is very effective in removing effects of other elements. However, if the gate is applied to the second discontinuity, the response is not similar

4.8.3.2 Gating the Second of Two Discontinuities

The time-gated response of the second discontinuity is quite different from the underlying response as shown in Figure 4.12, (thin dark trace, labeled "2nd Cap, Gated, No Comp."). The frequency response of the gated measurement of the second discontinuity may be compensated by taking the gated response of the first discontinuity, and using Eq. (4.39) creating a compensation for Γ_2 as

$$\Gamma_2 = \frac{\left|\hat{\Gamma}_2\right|}{(1 - \left|\hat{\Gamma}_1\right|^2)} \tag{4.40}$$

where $\hat{\Gamma}_1$ is the gated result of the first capacitor, and $\hat{\Gamma}_2$ is the gated result of the second capacitor. This compensation has been applied in Figure 4.12, with the result showing remarkably

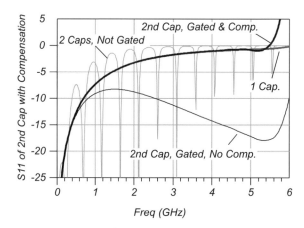

Figure 4.12   S11 of two capacitive discontinuities (light gray) not gated, S11 response of one capacitive discontinuity (dark gray), S11 of second discontinuity gated (thin black) with no compensation, and S11 of second discontinuity, gated and compensated (thick black).

good compensation over most of the frequency range (thick black trace, labeled "2nd Cap, Gated & Comp."). However, the band edge response deviates because of the normalization does not completely compensate for the error due to the gate, as described in Figure 4.8. This error was apparent as a slight upturn of the S11 trace of the first capacitor, after gating. The effect of this slight upturn on $\hat{\Gamma}_1$ moves it closer to 1, and the denominator of Eq. (4.40) moves closer to zero, causing the apparent value of Γ_2 to increase more rapidly at the band edge. Increasing the gate window function will reduce the edge effect, or alternatively, the frequency response can be performed over a wider span, and the upper 10% can be disregarded as a gating effect.

Also shown in Figure 4.12 is the ungated S11 measurement of two discontinuities (light gray), and, for reference, the S11 of just a single discontinuity (dark gray, labeled "1 Cap."). This shows the effectiveness of the compensation method, where there is substantial deviation from the ideal in the uncompensated gated response of the second capacitive discontinuity.

Also note that this is a very large first discontinuity, having a frequency response return loss value of nearly 0 dB over much of the frequency range. A more normal situation is where the first discontinuity, often a connector interface from coax to PCB, has a much smaller reflection, and so the compensation method works even better.

Most VNAs do not have multiple section compensation, so it is up to the user to create the compensation function. However, many VNAs have equation editor functions that can allow real-time computation of these compensations.

4.8.3.3   Compensation for a Combination of Discontinuities and Line Impedance Changes

Many practical applications of time-gated measurements include effects of both discrete discontinuities and impedance steps. For example, the measurement of a connector at the far end of a cable is affected by the cable's near end connector, and the impedance of the cable.

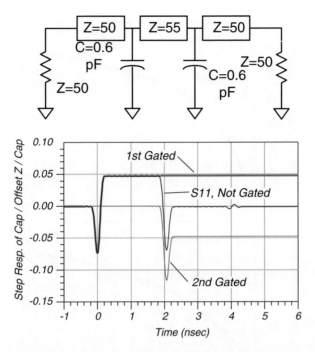

Figure 4.13 (upper) Circuit with two capacitive discontinuities, and an offset impedance line; (lower) time domain response of the circuit (black) with first cap after gating (wide gray) and gating around only the second cap (thin gray).

Figure 4.13 (upper) shows a circuit diagram and (lower) time domain response of a 55 ohm cable with a 20 dB return-loss (at 1 GHz) input and output mismatch due to capacitive loading, terminated in 50 ohms. In this example, the discrete reflections are much smaller than those shown in the previous example and more in line with a typical application. Here, the cable itself has a 5% reflection due to impedance error, and each discontinuity is approximately 10% at 1 GHz.

Figure 4.14 (upper and lower) shows the time-gated frequency response of the first and second discontinuity as a result of gating about each respectively. In the case of gating the first discontinuity, there is only an impedance step to be accounted for.

A normalization of the first case can be performed by recognizing in the step response that the value of the step after the gate represents an offset impedance in the termination. The effective reflection results in

$$\hat{\Gamma}_1 = \Gamma_1 + \cdot \frac{\Gamma_2}{(1 - \Gamma_1 \cdot \Gamma_2)} \tag{4.41}$$

which can be reduced to

$$\Gamma_1 = \hat{\Gamma}_1 - \Gamma_2 - \frac{\Gamma_1(\Gamma_2)^2}{(1 - \Gamma_1 \cdot \Gamma_2)} \approx \hat{\Gamma}_1 - \Gamma_2 \tag{4.42}$$

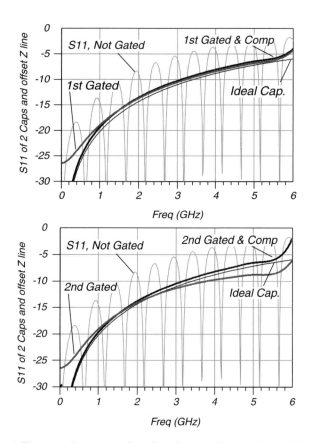

Figure 4.14 (upper) Time-gated response of the first discontinuity (thick gray, "1st Gated"), and with compensation (thick black, "1st Gated & Comp"), along with original S11 (light gray) and ideal single capacitive discontinuity (thin black); (lower) similar to the upper trace but with gating and compensation applied to the second discontinuity.

Here Γ_2 is the reflection coefficient of the 55 ohm cable that forms the termination after gating, and there is assumed to be no delay before Γ_1. Figure 4.14 (upper) shows the first discontinuity, after gating and gating plus compensation. If there is delay before the first discontinuity, the phase of Γ_2 must be shifted to account for the delay.

The masking of the second discontinuity is more difficult to account for. There are three effects: first, masking due to signal loss in the first discontinuity as described by Eq. (4.39); second, by the change in reference impedance as described by (4.35); finally, by the change in terminating impedance, relative to the modified reference impedance, as described by Eq. (4.42). Thus three compensations are required: first, the effective reflection after the first discontinuity is found by applying the first compensating equation; second, the effect of the impedance transformation is compensated by applying Eq. (4.35) to the result of the first equation, which in this case is a small effect; finally, compensation is made for the step in impedance at the termination. Also, as this compensation includes an additive element associated with the terminating impedance, phase becomes important. Both Γ_2 and Γ_3 are phase shifted by the delay of line Z_1, which may be determined directly from the time domain response. Thus, the effective impedance $\widehat{\Gamma}_2$ is

$$\widehat{\Gamma}_2 = \frac{\Gamma_2 \cdot (4Z_1 Z_0)}{(Z_1 + Z_0)^2} \cdot (1 - |\Gamma_1|^2) \cdot e^{j\omega \cdot 2\tau_d(Z_1)} + \Gamma_3 \cdot e^{j\omega \cdot 2\tau_d(Z_1)} \qquad (4.43)$$

where $\widehat{\Gamma}_2$ is the gated response of the second discontinuity, Γ_2 is the reflection of only the second discontinuity, Γ_1 is the (gated) first discontinuity, Z_1 is the 55 ohm line, Γ_3 is the reflection from Z_1 to the terminating impedance, Z_0, and $\tau_d(Z_1)$ is the delay corresponding to the length of the line Z_1. The compensation is determined by solving Eq. (4.43) for Γ_2.

Figure 4.14 (lower) shows the results of gating the second discontinuity, and applying the compensations as described in (4.42) and (4.43). Also shown is the frequency response of a single (ideal) discontinuity associated with the first capacitance as though it were on a matched line, and the original response of the two discontinuities with the offset impedance line in between. In Figure 4.14, for both the (upper) and (lower) plots, there is remarkable improvement in the gated measurement, especially at low frequencies, when evaluated against the ideal response. In this example, the compensated result is quite sensitive to the delay selected for the 55 ohm line. This delay was determined by choosing the delay displayed at the peak of the second discontinuity. These compensations are appropriate for single discontinuities that are nearly lossless and non-distributed, as might be found from a coax to PC-board adapter.

4.8.4   Estimating an Uncertainty Due to Masking

The proposed compensations described above may generate some error, in part due to lack of consideration of loss in the network, and due to inability to totally separate responses. In some cases it may not be necessary to actually compensate the network for gated response, but rather to establish an estimate of uncertainty associated with the response that is gated out. This uncertainty can be derived from Eq. (4.39), for a second gated response on matched line. Additional uncertainty will come from non-matched lines leading up to the reflection of interest, and following the reflection of interest. The magnitude of the uncertainties can be determined in a manner similar to (4.35) and (4.43) with the resultant total uncertainty after gating determined as

$$\Delta\Gamma_2 = \frac{\Gamma_{2G}}{1 - |\Gamma_1|^2} \left(|\Gamma_{Z1Z0}|^2 + |\Gamma_1|^2 \left| \frac{(4Z_1 Z_0)}{(Z_1 + Z_0)^2} \right| \right) + \Gamma_{Z2Z1} \qquad (4.44)$$

where Γ_1 is the first (gated out) discontinuity and Z_1 and Z_2 are the lines before and after the desired reflection Γ_2, respectively, and Z_0 is the system impedance, and Γ_{Z1Z0} and Γ_{Z2Z1} are the reflection between lines Z_1 and Z_0 and Z_2 and Z_1 respectively, and Γ_{2G} is the value of the gated response.

4.9   Conclusions

The exact relationship of time domain transform used in VNAs to the analytic impulse response of measured networks has been described in detail with mathematical rigor, and details of the time-gating function have been presented. From this background, the effects of time gating on measured results have been explored, and a method is given for compensating for undesired masking effects, yielding superior time-gated measurement results. Additionally, the uncertainty of time-gated measurements has been quantified, and qualitative errors due to the subtle effects of gate properties and renormalizations have been presented. More examples of the use of time domain and gating will be shown in subsequent chapters, as it is applied to specific component measurements.

References

1. Glass, C. (1976) *Linear Systems with Applications and Discrete Analysis*, West Publishing Co., St. Paul.
2. Bracewell, R.N. (1986) *The Fourier Transform and Its Applications*, 2nd edn, Revised, McGraw-Hill, New York.
3. Hines, M. and Stinehelfer, H. (1974) Time-domain oscillographic microwave network analysis using frequency-domain data. *IEEE Transactions on Microwave Theory and Techniques*, MTT-22(3), 276–282.
4. Sharrit, D. Vector network analyzer with integral processor. US Patent No. 4,703,433.
5. Rytting, D. (1984) Let time-domain provide additional insight into network behaviour. Hewlett-Packard RF & Microwave Measurement Symposium and Exhibition, April 1984.
6. Dunsmore, J. (1999) Advanced filter tuning using time-domain transforms. Proceedings of the 29th European Microwave Conference, 5–7 Oct. 1999, Munich, vol. 2, pp. 72–75.
7. Blinchikoff , H.J. and Zverev, A.I. (1976) *Filtering in the Time and Frequency Domains*, John-Wiley & Sons, New York.
8. Lu, K. and Brazil, T. (1993) A Systematic Error Analysis of HP8510 Time-Domain Gating Techniques with Experimental Verification. IEEE MTT-S Digest, p. 1259.
9. Bilik, V. and Bezek, J. (1998) Improved cable correction method in antenna installation measurements. *Electronic Letters*, 34(17), 1637.
10. Pozar, D.M. (1990) *Microwave Engineering*, Addison-Wesley, Reading, MA. Print.
11. Dunsmore, J.P. (2004) The time-domain response of coupled-resonator filters with applications to tuning. PhD Thesis, University of Leeds.

5

Measuring Linear Passive Devices

Linear devices are at the same time both some of the simplest of microwave components, yet also some of the most difficult to characterize because the quality of measurements must be very good, especially for low-loss devices. Filters, transmission lines, couplers and isolators, to name a few, are all designed to be as nearly ideal as possible, and any error in the VNA measurements can lead to production test failures and loss of yield. In this chapter, some particular details of the real-world problems and solutions to measuring these devices are presented.

5.1 Transmission Lines, Cables and Connectors

Cable and connectors are perhaps the simplest of RF components, usually having very low loss and good match. However, making measurements on devices with such ideal performance requires careful calibration and measurement techniques so that the error in measurement doesn't overwhelm the devices' performance.

For short, low-loss lines, the main difficulty is dealing with the mismatch of the input and output connection, and properly calibrating the source and load mismatch of the VNA. These devices might be interconnect cables with integrated connectors, or might represent transmission line traces on PCBs. If the device includes connectors of standard type, then any mismatch associated with the connector is appropriately assumed into the measurement. On the other hand, if a device represents some transmission line structure on a PCB, and the connectors are only used as a fixture to allow connection to the VNA, then the effects of the connectors should properly be removed from the overall measurement.

5.1.1 Calibration for Low Loss Devices with Connectors

If one reviews published specifications for most calibration kits, one often sees that TRL based mechanical calibration provide the best performance, followed by nearly identical performance

---

Handbook of Microwave Component Measurements: With Advanced VNA Techniques, First Edition. Joel P. Dunsmore.

from sliding load SOLT calibrations and Ecal[1] modules. However, in practice, Ecal almost always provides better calibrations because the quality of TRL and mechanical standards used in practice is not of the same quality as the metrology versions. However, Ecal modules are calibrated to a very high quality, and retain this high quality in any measurement application. Regardless of which type of calibration kit is used, the best calibration method is almost always using the SOLR or "unknown-thru" calibration, unless the device to be tested is an insertable device, that is, it has one male and one female connector of the same type.

For the system setup for low-loss devices, where even a small amount of trace noise can be significant, it is best to use a moderate IF bandwidth and sweep-to-sweep averaging to reduce trace noise. While reducing the IF bandwidth to a very narrow value will reduce noise, it will also increase the sweep time, and very slow sweep times lead to instability in the calibration due to system drift which might occur over a long sweep time. One cause of system drift is a slowly varying temperature in a measurement lab, due to cycling of the air conditioning. Even if the lab is ostensibly held to $\pm 1°C$, the instantaneous air temperature around test systems and cables can vary by several degrees in a short time. If the sweep rate is slow relative to this, there will be drift in the measurement during the sweep. If, however, sweep averaging is used on a faster sweep, the drift effect will average out over several sweeps. Before starting a measurement, evaluate the effect of trace noise. The simplest way to do this is to set up the frequency span and power and turn on about 10 sweep averages. After the sweep average completes, normalize the trace using the "data into memory" and "data/memory" functions, and turn the averaging off. One can use the trace statistics functions (often found under the analysis section of the VNA user interface) to determine the trace noise, which is usually expressed as a standard deviation or peak-to-peak. However, because of its noise-like nature, the peak-to-peak on any particular sweep can sometimes have a single high noise point even if the average noise is low, so typically the mean-to-peak noise is specified as two or three times the standard deviation noise and the peak-to-peak is four to six times. Using averaging to store the memory trace eliminates double counting the noise because the deviations from the memory trace will add to the deviations of the data trace and this gives a higher value for trace noise than would normally appear in the measurement. Alternatively, for slow sweeps, one can simply normalize the trace with no averaging, and look at a new trace with data/memory, and recognize that the standard deviation noise will be approximately 1.41 times the actual noise due to the noise of the normalization memory trace. This should be done before any calibrations are performed.

One can reduce the trace noise by either decreasing the IF bandwidth, increasing the averaging or increasing the power – see Figure 5.1. A rule of thumb for improvement is that the standard deviation trace noise will improve three times for every ten times change in IF bandwidth or averaging, and for every 10 dB increase in source power. One caution with changing power: if the power is changed after calibration, there will be the additional error of the power range or source attenuator value changes. This sometimes occurs if the power ranges are set to automatic.

Calibration power for low-loss devices should be low enough to ensure that the test receiver is not in any compression, on the order of 10 dB below the specified 0.1 dB compression point. On most VNAs, this is in the range of -10 to 0 dBm. At high enough power, the trace

---

[1] Ecal is a registered trademark for Agilent Technologies. Its use here should be understood to refer to both the Agilent implementation specifically, and also the more general category of electronic calibration modules.

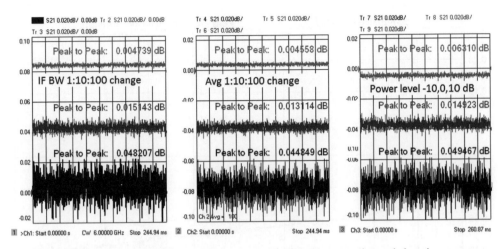

Figure 5.1 Measured trace noise with changes in IF BW, trace averaging and changing power.

noise does not improve further due to high-level noise caused by the phase noise of the source (see Section 3.13.5).

For the very best measurements, it is helpful to improve the source and load match of the VNA. This can be done by adding a well-matched precision attenuator to the source and/or load ports. While this reduces the power (thus increasing the noise) it typically improves the raw source and load match by two-times the attenuator value. The tradeoff is that it will degrade the raw directivity by the same amount. If the test system is very stable, it is a good tradeoff; if the test system is not stable or the pad is at the end of a cable, it may degrade the S11 measurement even as it improves the transmission measurement. This will also improve the corrected load match. This is particularly important if one is unsure of the quality of the calibration kit. An attenuator of 6 dB on port 2 provides a good tradeoff between increased noise and better match calibration.

Figure 5.2 shows two S21 measurements of a low-loss airline, first connected directly to the VNA (light trace), and then connected to a VNA with a 6 dB pad on port 1 and a 10 dB pad on port 2 (dark trace). The improvement in S21 ripple is most likely due to the improved raw match. Also shown are two S11 traces under the same conditions. Note, since the S11 calibration accuracy depends almost entirely on the quality of the one-port standards, and not on the quality of the raw test port, the improvement in port 2 raw load match has little effect on S11.

5.1.2 Measuring Electrically Long Devices

Electrically long devices such as long cables or SAW filters (which have substantial delay) can cause some particular problems in characterization, especially with older VNAs. Many VNAs have both a swept frequency mode (where the source is continuously swept across frequency band) and a stepped frequency mode (where the source is stepped to discrete frequencies across a frequency band) [1]. The swept mode has the advantage of having no dwell time (or frequency settling time) at each point and therefore can be much faster in the fastest sweep

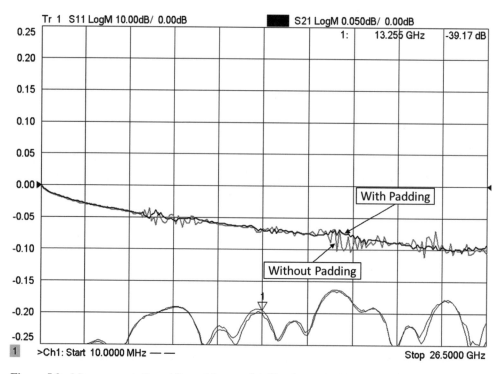

Figure 5.2 Measurement of an airline with normal calibration and with additional 6 dB pad on input and 10 dB on the output.

modes. In one sense, the swept frequency mode provides a better characterization of the DUT in that it does not skip any frequencies, but rather ramps through all frequencies. If there is some particular frequency that has a dropout or resonance mode, it will be indicated by the swept-frequency measurement. However, in stepped mode, if the dropout occurs at a frequency that is not one of the stepped frequencies, there will be no indication on the trace.

However, for electrically long devices such as cables, even the swept mode can cause difficulties, due to an effect known as "IF delay".

5.1.2.1 IF Delay

IF delay occurs when the source and receiver of a VNA are continuously swept (as opposed to stepped) and there is a long delay from the source to the receiver. Figure 5.3 illustrates the problem with measuring long cables. If the sweep generator is moving at a constant rate of change, $\Delta f / \Delta t$, the delay of the DUT will delay the measurement of the signal from the source while the receiver moves in frequency, causing an apparent shift of the signal in the IF of the receiver channel. This shift causes an apparent drop in signal which results in more than expected loss. The amount of shift is computed as

$$IF_{Shift} = \left(\frac{\Delta f}{\Delta t} \right) \cdot DUT\_delay \tag{5.1}$$

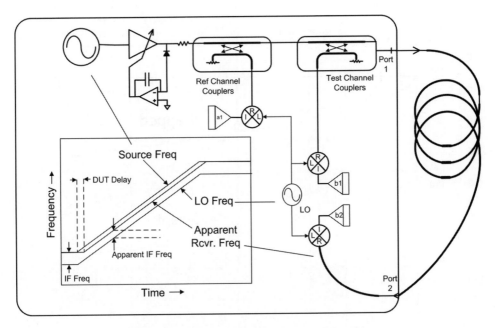

Figure 5.3   Illustration of IF delay for a long cable.

This IF delay effect occurs differently at different bands of the VNA as the sweep rate changes depending upon the source and phase lock architecture; for example, if the fundamental oscillator has a maximum rate of 300 GHz/sec, then in the doubled band, it can have a rate of 600 GHz/sec and show twice the IF delay effect. As an example, consider a system that sweeps at 37 GHz/sec, which translates into about 270 msec sweep on a 10 GHz span. This corresponds approximately to a 3 kHz IF BW with an 801 point sweep. Suppose the group delay of the DUT is 10 nsec (approximately the group delay of 3 m cable), then the IF shift would be

$$IF_{Shift} = \frac{10 \text{ GHz}}{.267 \text{ sec}} \cdot 10 \text{ nsec} = 0.37 \text{ kHz} \tag{5.2}$$

This corresponds to about 10% of the IF bandwidth, giving approximately 0.3 dB of error in the measurement. For a low-loss device, this is unacceptable.

The effect of IF delay is shown in Figure 5.4. Two measurements are made, one in stepped mode and one in swept mode. The calibration is done in the same mode as the measurement. The IF delay amplitude offset is somewhat compensated for by calibration, but for a DUT with sufficiently long delay, the IF delay offset will become apparent.

In Figure 5.4, the measurement parameters are adjusted to get the fastest possible sweep to illustrate the issue – in this case a sweep with only 51 points and with a 300 Hz IF BW. For such a low point count, the sweep rate can be very high, and the narrow IF BW means that any apparent frequency shift due to delay will cause a drop in amplitude. In the figure, the swept trace shows four discrete steps, each one corresponding to a different band break for the instrument, an HP 8753. This VNA used a sampler with a high harmonic order (low VCO

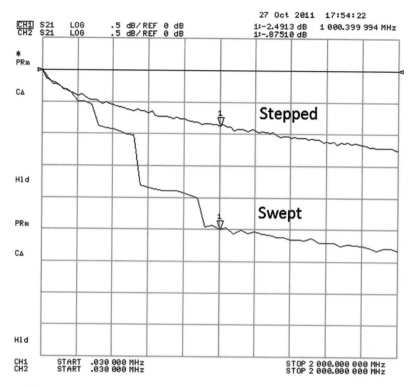

Figure 5.4 Comparing stepped with swept mode on a 3 m cable insertion loss measurement.

frequency) and so it has many bands. If the number of points is changed to 201, a more normal value, or the IF BW increased to a wider value, the difference between swept and stepped becomes much smaller.

IF delay effects also occur when the calibration is done in one sweep mode and the measurement is performed in another sweep mode. This can occur if the calibration is done in a different IF bandwidth from the measurement. Sometimes, the IF bandwidth is reduced to lower the trace noise of the measurement, and sometimes the calibration is performed at a lower IF bandwidth to reduce the noise effect of the calibration, and the IF bandwidth is widened to speed up the measurement trace. Either case can cause an IF delay shift as demonstrated in Figure 5.5.

The IF delay becomes apparent in wideband measurement traces as a drop in amplitude (or shift in phase) at the VNA's discrete band breaks. Because of some of the phase locking methods, the first point or first few points in a band may be stepped. The VNA determines the fastest method for sweeping and this sometimes means using stepped points at the band edges. These stepped points become obvious as spikes in the trace, since they are stepped, and have no error relative to the calibration. They are apparent in the figure where the markers are posted. At these points, the band break occurs but there is no change in sweep rate since these first points are stepped. Typically, the sweep rate increases at high bands because multipliers are used on the fundamental oscillator, giving the staircase appearance. However, much effort

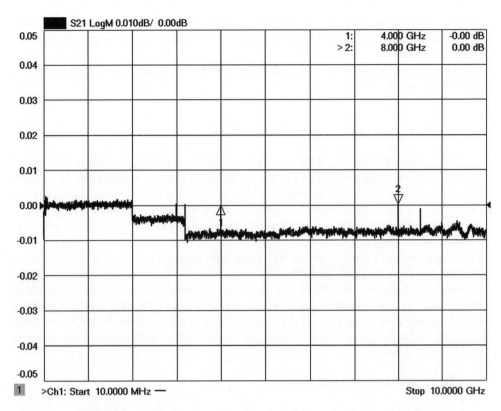

Figure 5.5 Calibrating in a stepped mode and measuring in a swept mode.

goes into programming the sweep rate so that the steps are minimized. In the figure, the worst error due to swept-versus-stepped is about 0.01 dB.

Typically, swept-mode is used when real-time response is needed for a tuning application, as in tuning a filter. For historic reasons, many VNAs default to swept mode. In the HP 8510, swept mode is sometimes known as "lock and roll" as the frequency is only phased locked at the first point, and the rest of the sweep is performed in an open-loop mode. The HP 8510 provides a software switch to change from swept to stepped-mode. One drawback in the swept mode is that the analyzer sweep accuracy may not be adequate to determine band edges in filters and similar components, so for the HP 8510, it was common to tune the filters in swept mode then change to stepped mode for a final measurement. In the HP 8720, PNA and ENA families, there is also a switch, but unlike the HP 8510, the HP 8720 maintains phase lock throughout the sweep. Some analyzers, such as the R&S ZVA do not have a swept mode at all, and provide only stepped sweep.

The HP 8753, which is used extensively in filter tests, does not have an explicit mode for switching between swept and stepped mode; it switches into a stepped sweep in three conditions: (1) if the IF BW is less than 30 Hz, (2) if the sweep time is greater than 15 msec per point, (3) if the source power calibration is activated. These conditions also apply to the HP 8720.

For the Agilent PNA series, stepped sweep mode is automatically enabled for IF bandwidths of 1 kHz and below, or for any of the advanced application channels or whenever source power calibration is enabled.

Realistically, one should always calibrate and measure electrically long devices with a stepped-sweep mode.

5.1.3 Attenuation Measurements

Attenuation measurements of cables are relatively straightforward: essentially S21 versus frequency. As described above, any cable measurements should be performed in stepped sweep mode. For very long cables, where the loss is large, the IF BW should be set narrow enough to reduce trace noise. Figure 5.6 shows the attenuation of a 1 m cable used as a typical test-port cable. These cables are chosen for their high quality and stability. Markers show the attenuation of the cable at various frequencies.

One major source of error in measuring cables is the input and output connectors used to mate from the VNA to the cable. If the cable is an assembly including connectors, then the only issue is to ensure a good calibration at the connector interface. However, it is common to measure a reel of cable that does not have connectors (such as a CATV cable) or the connectors are used only for test purposes before shipping the cable. For long cables, the loss of the cable diminishes the mismatch effects, and input and output match correction of a full two-port

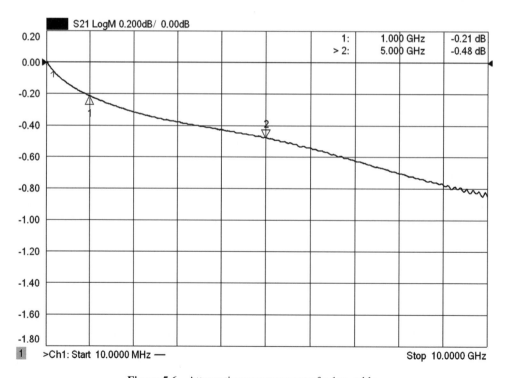

Figure 5.6   Attenuation measurement of a 1 m cable.

calibration compensate somewhat for the mismatch loss. For shorter cables, the mismatch between connectors can add significant ripple to the cable measurement.

5.1.3.1 Connector Compensation Using Port Matching

There are two techniques that may be used to reduce the effect of test connectors on the measurement of a long cable: connector compensation and time domain gating

Connector compensation involves creating a simple model of the input connector, typically a series inductive element and a shunt capacitive element. A compensating model is created using a port matching function in the VNA, to compensate for the input connector. A common method is to generate a time domain display of the S11 of the cable, and adjust the series and shunt elements until the input mismatch is minimized. An example of this is shown for the measurement of a cable with poor input and output connectors, in Figure 5.7, with the S21 frequency response shown in the upper left corner, and the S11 and S22 frequency response shown in the upper right. The time domain response for S11 and S22 is shown in the lower plot. In this case, the low-pass time domain mode is used, so that the type of discontinuity, capacitive or inductive, is readily apparent. For S11, the first response shows a negative dip indicating a capacitive discontinuity in the connector. For the S22 trace, the first response shows a larger, positive peak, indicating a larger inductive discontinuity.

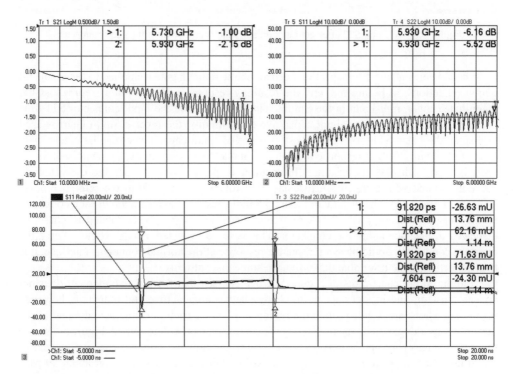

Figure 5.7 Cable measurement with poor connecters, frequency domain (upper) and time domain response (lower).

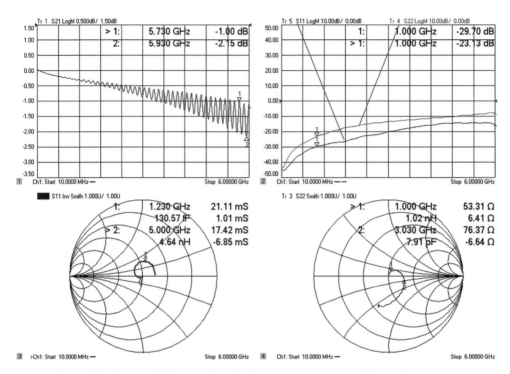

Figure 5.8   S11 and S22 after gating, indicating the vector error of the each connector.

To help determine the compensation needed, a time domain gate is applied to the input and output reflections, in Figure 5.8, and the resulting response is shown on as return loss log magnitude (upper, right) and in the Smith chart (lower). Since the input connector S11 time domain response indicated a shunt capacitive discontinuity, the inverse Smith chart is used to display the gated S11, so that the best estimate of shunt capacitance may be determined. The output connector S22 response indicated an inductive discontinuity, so the normal Smith chart is used to display its response. From these responses, one can see the characteristic curve associated with a mismatch that has been rotated in phase from the reference plane (refer to Figure 2.35). The display in Figure 5.8 indicates that the mismatch in each connector is not located precisely at the VNA reference plane, but is somewhat rotated in phase, or displaced in distance along the connector. This is to be expected as one would assume the interface connector should have a reasonable response at the normal connector interface; it is the connection to the raw cable that will have some discontinuity.

From the Smith chart plots of Figure 5.8, one can see that the reactive element does not appear constant, and in fact appears to change between capacitive and inductive; however, if port extensions are used to rotate the reference plane of each port, the actual value of the discontinuity, along with its location along the connector, can be determined. Figure 5.9 shows the result of applying port extension to each port, individually, until the reactive portion of the Smith chart response is as constant as possible. For port 1, this occurs at about 49 psec of extension, which is adjusted until the difference in the reactive capacitance value between the markers at 1 GHz and 5 GHz is minimized. For port 2, the value is about 40 psec of delay where the difference in the reactive inductance value is minimized. In this case, the port 2 discontinuity is more ideally inductive (has a more constant value) than the capacitive discontinuity of port 1.

From these Smith chart displays, an average value for the discontinuity is determined as a starting point for using the port matching feature of the VNA.

The port matching function in a VNA allows one to add a simulated or virtual matching-network at the input or output port of the DUT as illustrated in Figure 5.10. With the port extension used, the reference position for the added network is rotated so that the network's effect is at the desired point along the connector. Figure 5.10 also shows the dialog used for adding port matching of −0.220 pF of shunt capacitive (negative so that it compensates for the positive capacitance shown in Figure 5.9) to port 1. A similar simulated network adds −1.1 nH of series inductance to port 2. With port matching on, the Smith chart traces are nearly fully compensated and appear close to the center of the Smith chart. The log-magnitude traces of the time-gated return loss in the upper-right corner of Figure 5.10 show that the effective input match has been improved to better than 30 dB. Finally, the upper-left plot shows the initial S21 response (including the ripple and loss effects of the bad connectors) and the S21 response with port matching. Clearly, the port matching removes very nearly all the ill effects of the poor connectors.

As a final comparison, the same cable is measured using very good connectors, so that the input and output return loss are very small. The resultant S21 trace is slightly less lossy

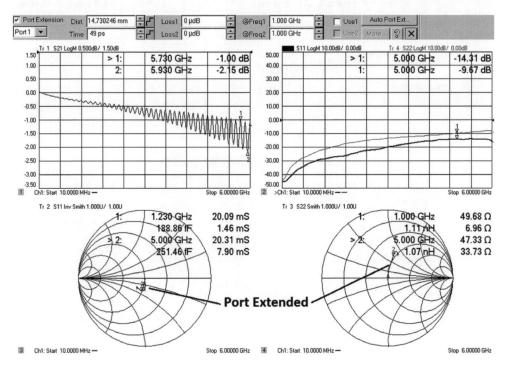

Figure 5.9   Port extension is applied to each port to determine the location and value of each discontinuity.

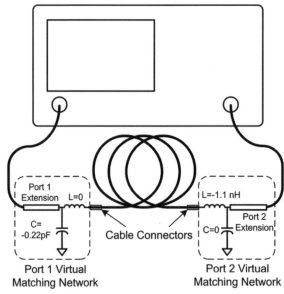

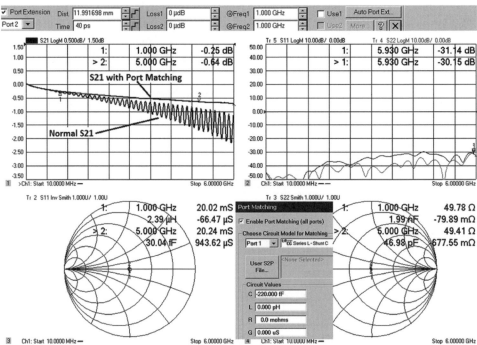

Figure 5.10   Port matching adds the negative of the reactive element for cable compensation.

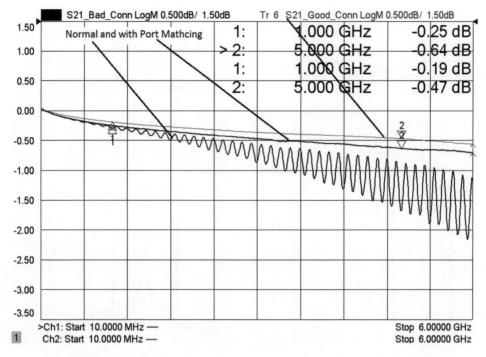

Figure 5.11 Comparing the cable S21 with compensation and with good connectors.

than the measurement with bad connectors and compensation with port matching, as shown in Figure 5.11.

The difference in loss could be due to some slight non-compensation since the capacitive discontinuity was not single-valued (it ranged from −0.188 to −0.251 pF), but more likely the cause is due to resistive or radiation loss in the poor connectors. The fact that there exists a discontinuity that yields a poor return loss also suggests that the connector might have some radiation or other lossy response. However, when comparing the measurements of the uncompensated S21 with poor connectors to the compensated value, one can see remarkable improvement, with results nearly as good as using a higher-quality connection.

Connector compensation using port matching is quite robust, and does not depend upon any attributes of the DUT, but it does require that the connector discontinuity be simple if simple port matching is to be used. In the next section, a method that does not require any knowledge of the discontinuity to remove its effect is described.

5.1.3.2 Connector Compensation Using Time Domain Gating

Time domain gating involves using the VNA time domain function to remove the effect of the input and output mismatches from the S21 measurement. However, it is very common to mistakenly apply the same gating settings that may have been determined from a reflection (S11) measurement to the transmission (S21) and this can yield completely erroneous results. To be very clear: if one uses time domain to remove the effect of reflection (either input or

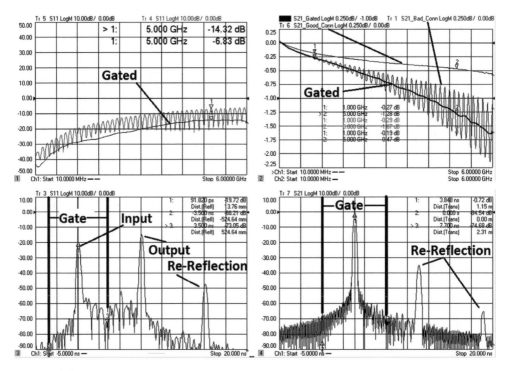

Figure 5.12 A cable with input reflections showing S11 (right) and S21 (left) with frequency response (upper) and time domain response (lower).

output) in an S11 measurement, the same gating settings *cannot* be used with a transmission measurement. To fully understand the time domain response, and reflection effects on a transmission measurement, consider the sample cable measurement shown in Figure 5.12.

In this figure the same cable and connectors are used as in Figure 5.7; the S11 traces are on the left, and S21 traces are on the right. The upper portion shows the frequency response and the lower portion shows the time domain response. In the S11 time domain (lower, left) there are three distinct peaks representing the reflection from the first test connector (marked "Input"), the second test connector (marked "Output") and a reflection representing a double reflection from the output, back off the input and off the output again (marked "re-Reflection"). The scale is 2.5 nsec per division and the reference (0 time) is at the second division. A time domain gate is placed around the first reflection, and the gated response is shown in the upper S11 frequency response. This represents the energy reflected from the first input test connector. While these connectors are not very good, they are quite typical of some common test port cases, in this case having a reflection coefficient ranging up to −14 dB for the input connector (at high frequency) and −10 dB for the output connector (not shown in the figure). The time domain response peak represents, in a way, an average reflection over the frequency range of 20 dB for the input connector and 18 dB for the output connector; the output connector appears better in the S11 measurement than it actually is due to the loss of the cable. The gate to separate the input reflection from the output reflection is set centered on the input reflection, and the span of the gate is set so that the stop is exactly between the two reflections. Not coincidentally, this time will be equal to the time domain transmission time, or group delay of the cable, and represents placing the stop gate at exactly halfway into the cable. It is not required to gate so far into the cable and some experiment can determine when the gate span is sufficiently wide to account for the entirety of the input connector reflection.

The S21 time domain transmission (TDT) response is shown in the lower right portion of Figure 5.12. It shows three response peaks, but these represent the transmission response as a function of time. The first, main response shows the time domain transmission through the cable, and the peak is located at a time that represents the group delay of the cable, in this case, 3.848 nsec. The other smaller peaks represent the arrival of the re-reflected signals from the input and output test connectors. Note that if only one of these connectors causes a reflection there will be no re-reflection in the S21 response, and the only effect on S21 will be the loss of energy represented by the gated signal of the S11 response. However, with a poor input and output connector, the S21 response has substantial ripple on it. This ripple is shown in the upper right portion of Figure 5.12. Also shown is the time-gated response of S21. In this example, it is possible to apply time domain gating to the S21 response, as the re-reflections caused by poor test connectors are easily separated in the time domain response. But note that the gating factor for S21 is not at all related to the S11 gate center. In fact, it must be set so that it is centered on the transmission response peak, which is at the group delay of the cable. The gate-span must be small enough to eliminate the re-reflection from the output and input connectors. Generally these will occur at three times the delay of the cable, so setting the span at two times the cable delay is a reasonable starting point. In fact, a second higher-order re-reflection can also be seen at five times the delay of the cable, which represents the transit time of the main lobe of the TDT response, plus two-times the transit time for the first reflection and two-times more the transit time for the second reflection for the total of five transit times.

The time-gated frequency response is shown in the upper trace, along with the ungated response which shows the effects of all the re-reflections. It is clear for a low-loss cable that these re-reflections dominate the response. What might not be so clear is that the energy loss from the reflections also dominates the response so that the S21 measured response of this cable section, which should not include the mismatch effects of the test connectors, is nearly completely dominated by the test connectors' response. Fortunately, the compensation techniques described in Section 4.8.2 can be applied to this cable measurement with good effect.

The energy lost from the reflection at the input port, and output port, can be described as

$$Loss_{Input} = \sqrt{1 - |S_{11\_Gated}|^2}$$
$$Loss_{Output} = \sqrt{1 - |S_{22\_Gated}|^2}$$

(5.3)

where S_{11_Gated} represents the gated frequency response of the S11 trace. The gated response of S21 is essentially missing this energy, but can be compensated by dividing the gated response by the product of these losses:

$$S_{21\_Compensated} = \frac{S_{21\_Gated}}{\left(\sqrt{1 - |S_{11\_Gated}|^2}\right)\left(\sqrt{1 - |S_{22\_Gated}|^2}\right)}$$

(5.4)

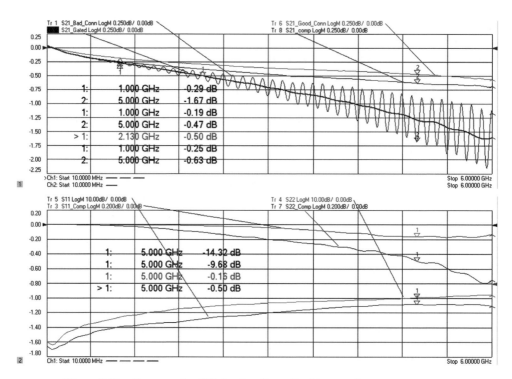

Figure 5.13 Compensating S21 for the effect of bad input and output connectors.

While not shown in Figure 5.12, the S22 gated response is determined in a similar manner as the S11 response. These results of these computations are shown in Figure 5.13.

The upper plot shows the S21 response with no gating or compensation showing the ripple (marked "S21Bad_Conn") and with gating applied (marked "S21_Gated", which passes through the middle for the ungated response). Also shown is a trace with the gating and compensation applied (marked "S21_Comp") and the ideal response for this cable (measured with good connectors). Clearly, the gated-compensated S21 is a much closer match to the expected result than the original S21 measurement, or the gated-only S21 measurement (no compensation) from the upper-right plot of Figure 5.12.

The lower portion of Figure 5.13 shows the gated S11 and S22 measurements, on a scale of 10 dB per division, and the computed compensation for each from Eq. (5.3), marked "S11_Comp" and "S22_Comp" in the figure, and scaled at 0.05 dB per division. The equation editor feature of the VNA was used to compute the compensation traces. It was also used to apply these individual compensations to the S21 trace utilizing Eq. (5.4).

A small note on practical application of the equation editor: for this VNA, the equation editor math can be applied to either S-parameter results or trace results. In this case, since Eq. (5.3) must be applied to the gated response of S11 and S22, a pair of traces displaying the gated results must first be created (Tr5 and Tr4 in the figure above). Next, the equation editor function is applied to these traces to generate Tr3 and Tr7 above. Finally, the S21 trace (the "Gated S21" trace from Figure 5.12, Tr8) is compensated in the equation editor using the equation

$$S_{21\_Comp} = \frac{S_{21}}{(Tr3 \cdot Tr7)} \tag{5.5}$$

This trace also has the S21 transmission time domain gate applied.

Since time domain is a linear function, it would be equally valid to create a gated S21 trace first, and then another equation editor trace taking the gated response of the S21 divided by the compensation for S11 and S22. Care must be used when applying time gating to equations to ensure that the gate is applied in the proper operational order. In the case of Eq. (5.5), the values of Tr3 and Tr7 are simple scalar numbers due to the absolute value function in Eq. (5.3), and do not affect the phase response of the result, and thus can be included either inside or outside the Fourier transform, as

$$\mathbf{F}\left(\frac{S_{21\_Gated}}{Tr3 \cdot Tr7}\right) = \frac{1}{Tr3 \cdot Tr7}\mathbf{F}\left(S_{21\_Gated}\right) \tag{5.6}$$

Note that, in this case, we refer to the forward Fourier transform as we are in a sense converting back from the gated time-response to the frequency domain.

Finally, here again as in Figure 5.11, the compensated S21 trace is slightly lower than the S21 trace of the cable with good connectors. This is likely due to the uncompensated resistive insertion loss and radiation of the poor connectors, just as in the example of Figure 5.11. It is reassuring that in both methods of connector compensation, the resulting S21 traces agree within 0.01 dB!

This second method of connector compensation does not rely on obtaining any models or de-embedding of the test connectors. Some care should be taken if applying this technique to more complex transmission structures, as a prerequisite for this type of compensation is that the entire S21 TDT response must be able to be identified in the S21 time domain transform, separately from the input and output reflections. In the case of more complex structures, re-reflections in the DUT itself might appear in the time domain response near the test-connector re-reflections and render the use of gating impractical. In such cases, using port matching may yield a better result.

5.1.3.3   Attenuation Measurements on Very Long Cables

One problem that is particular to cable measurements is the occasional need to measure very long cables in situ. This might be an example of a cable between two buildings, or a cable in some other structure such as a ship or an airframe, or up to the top of a radio tower. There are two use cases for long cable measurements: one is when both ends of the cable are accessible, although perhaps with some difficulty, and the other is that only one end of the cable is accessible. Options for the first case are described here; options for the second case are described in Section 5.1.3.5. Another use for long cables is in the measurement of a DUT that is not itself large, but is situated far from the VNA. This might occur in a satellite test

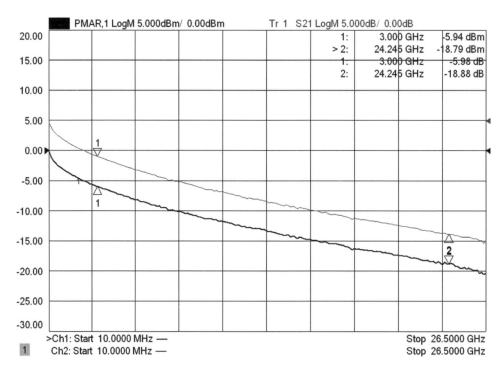

Figure 5.14 Using a power meter as a receiver at the end of a long cable, compared to S21, offset one division.

scenario where the satellite is in a thermal vacuum (TVAC) chamber. A long test-port cable may be required to connect through the chamber to the satellite.

The difficulty with measuring very long cables is that the input and output must be widely separated. One method that can be employed is to use a remote power sensor and a source to measure the cable loss. This is a scalar technique that can provide reasonable results as long as the loss to the power sensor is not too great. Power sensors are broadband detectors and have dynamic range limited to about 50 dB for the thermistor type sensors, and 90 dB for diode sensors. Many power meters can be controlled and triggered by LAN or USB control, so it is possible to remotely locate the head. Some systems provide source and power meter control directly, such as the Agilent PNA, which can use a power-meter-as-receiver (PMAR) mode to directly display a synchronized trace of the power meter reading at triggered source frequencies. Such a measurement is shown in Figure 5.14, where the dark trace is the power meter response and the light trace is the traditional S21, for a 10 m cable. The reference for the traces is intentionally offset by one division for clarity but from the marker readings, one can see that they have nearly identical values.

The noise floor of the power meter does affect the reading at higher frequencies, as well as the lack of match correction, so the worst-case error using a power meter in this way exceeds 0.4 dB. The power meter based trace in the figure above shows more ripple at high frequency than a measurement the same cable using a full two-port calibration.

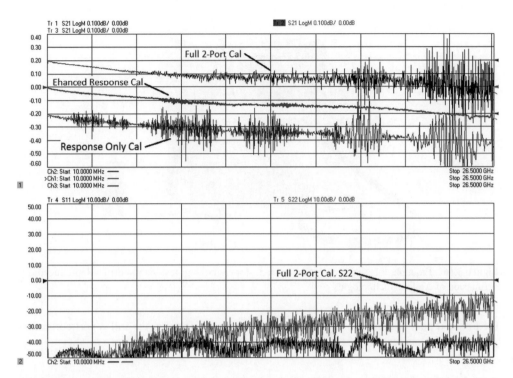

Figure 5.15 Measurement at the end of a long test-port cable comparing full two-port calibration with enhanced response calibration, after movement of the test-port cable.

However, in many cases the phase or delay response of the cable is required which means that a full vector measurement must be made. The most common method is to position the VNA at one end of the cable run, and then use a long test-port cable to connect to the other end. This presents a particular problem in that the stability and loss of the test-port cable will severely limit the overall response. For very long cables, the mismatch can dominate the measurement result of a full two-port calibration as the output match correction at the end of long test-port cable is very poor. One method to remove this output match correction error is to use response-only (S21 normalization) or enhanced response calibration (ERC) that removes the input mismatch from the S21 measurement. Figure 5.15 shows a comparison of a measurement of a short, low-loss line using a 10 m test-port cable after flexing the test-port cable following calibration. Shown are the S21 traces using a full two-port correction, a response correction and an enhanced response correction, each one separated by two grid divisions for clarity. In the lower window the S11 and S22 traces are shown from the full two-port calibration. The full two-port correction shows errors due to the poor measurement quality of the S22 trace. The response-only correction shows errors due to mismatch at the input and output. The enhanced response correction shows the best overall measurement of the cable under test, as the input match is corrected for but the output mismatch correction is not performed. Since the actual value of the DUT S22 is essentially hidden from the VNA by

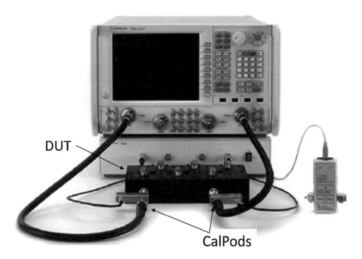

Figure 5.16 Example of a re-correction system for removing test cable drift. Reproduced by permission of Agilent Technologies.

the loss and poor match stability of the test-port cable, performing the full 2-port correction actually yields poorer results.

5.1.3.4   In-Situ Calibration and CalPods

Recently, a new system and method has been developed that addresses the stability issue. The cause of the stability issue is the fact that the test-port cable can have loss and delay errors due to flexing and changes in loss or mismatch. This new system makes use of an in-situ calibration method that can reduce or eliminate the instability of a long test-port cable. The heart of the system is an electronically controlled module that provides three discrete reflection standards, nominally an open, short and load, as well as a thru state, connected directly in-line with the DUT. An example of the system is shown in Figure 5.16.

To use the modules for re-correction, a calibration is performed at the output end of the re-correction module. Immediately after calibration, the standards of the re-correction module are measured and recorded. This is known as the initialization process, and it is critical that the initialization measurements are performed before any drift in the test-port cable occurs. It is not even necessary to do the initialization using the long test-port cable; in fact, a short cable or no cable at all can be used for the initial calibration and initialization. Next, the output end of the in-situ calibration module is connected to the DUT. This can be using the same cable as the initial calibration, or a different, longer cable. The three standards are re-measured and a re-correction array is computed that represents the difference between the initialization and the re-correction measurement. Any drift or change in loss or match of the test-port cable is captured in this difference array. The mathematics involved is described in Chapter 9, as a difference array.

The quality of re-correction is shown in Figure 5.17 where the light traces show the poor S21 and S11 measurements after drift in the test-port cable. In this case, the test-port cable is a 10 m high-quality cable typically used in satellite test systems. The dark traces show the result

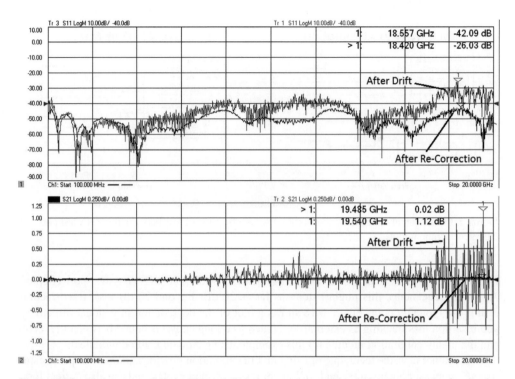

Figure 5.17 Light traces: drift in S21 and S11 due to a long cable; dark traces, result after in-situ re-correction (utilizing Agilent CalPod).

after an in-situ re-correction. In fact, this method of re-correction is so beneficial for removing cable drift that it is useful in very-low-loss, low-reflection measurements where slight cable drift can degrade performance. In most cases, it is actually more accurate than a calibration with a normal Ecal or mechanical standards, as simply the act of disconnecting the Ecal and connecting the DUT can cause a change in the test-port cable response that is great enough to benefit from re-correction.

The only limitation is the assumption that the change in loss is reciprocal, and for the proper computation of the change in phase, there must be less than 180° phase change between the initialization and the re-correction between each measurement point. For normal differences due to an unstable cable, this is a very minimal restriction. However, if the initial calibration is made with a very short cable, and the measurements are made with long cables, then the phase change for the difference can be quite large; in such a case a large number of points are required to ensure that the phase response is captured without any aliasing. It is often desirable to do the initial calibration with shorter, lower-loss cables to reduce the error and noise in the initial calibration. The re-correction quality is limited by the noise of the initialization data and the repeatability of the reflection states of the in-situ calibration module. These errors after re-correction can typically be kept below 0.02 dB insertion loss repeatability and −50 dB return loss repeatability for in-situ calibration modules.

The greater loss between the test-port coupler and the re-correction module, the more the recorrected performance is degraded. With commercially available modules, the degradation

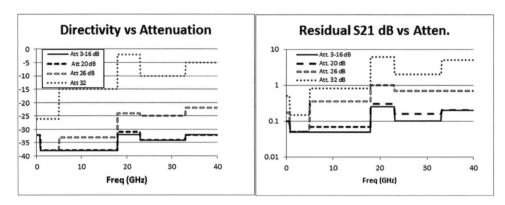

Figure 5.18 Residual directivity and insertion loss as a function of loss between the VNA test port and the re-correction module.

is minimal with less than 16 dB loss between the module and the VNA test-port coupler. At lower frequencies, this might extend as far as 20 dB. Beyond 26 dB loss, re-correction results are typically poor. Figure 5.18 shows a plot of the degradation in residual directivity and insertion loss tracking as external-loss (such as a long cable) is added between the VNA and the in-situ calibration module (in this example, using Agilent CalPod modules).

While rather new, the technique of using in-situ calibration provides dramatic improvement in making measurements at the ends of long or lossy test-port cables, and can also correct for a variety of other instabilities include switch repeatability in the case that a multiport switching network is placed between the VNA test port and the DUT.

5.1.3.5   Time Domain Responses, and One-Way Measurements

Occasionally, a cable is located in such a way that only one end is accessible to test equipment and it is still desired to measure the frequency response (magnitude and phase). A classic example is a cable in an airplane wing connected to an antenna element. In such a case, one may still be able to characterize the cable using time domain reflection techniques.

To use reflection techniques, it is necessary to provide a reflection at the far end of the cable. This can be accomplished by disconnecting it from its terminating element, or in the case of a cable to an antenna, putting a reflector close to the antenna to provide a total reflection into a cable. As an example, consider a planar antenna embedding in the structure of a wing; this could be wrapped in a conductor (aluminum foil) to provide a total reflection.

A simple way to approach this is to simply look at S11 of the cable and determine that the one-way loss is simply the square root of S11 (or one-half the dB value). However, if the cable has much loss, and any reflections, this method will yield poor results. Typically, such a cable will have a substantial reflection at the input connector.

In Figure 5.19, the dark trace shows the S21 of a cable, and the lighter trace shows the square root of S11. At lower frequencies, where the cable loss is small, this gives a good estimate of the transmission loss. However, at higher frequencies, where the loss is greater and the reflection from the input connector is also greater, the loss determined by the square

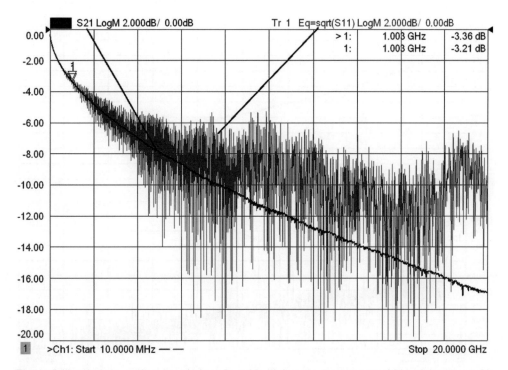

Figure 5.19 Dark trace, the transmission of a cable; light trace, square root of S11 of the same cable terminated in a short.

root of S11 deviates substantially from the transmission response. This is due to the input connector's reflection adding to, and subtracting from, the two-way loss of the cable and effectively masking the cable's loss.

One can use the techniques similar to those of Section 5.1.3.2 to improve this one-way measurement of a cable's attenuation by taking advantage of time domain gating [2]. Figure 5.20 shows in the upper window both the normal S21 measurement of the cable, and the gated time domain transform of the S11 of the shorted cable.

In the lower window is the time domain response with the transmission response shown with gates around it. With even limited knowledge of the cable's length, it is easy to determine the location of the shorted end of the cable. The peak on the time domain is lower than expected for a short due to the average (across frequency) two-way loss of the cable; but also, a close inspection of the impulse-like response associated with this transmission peak shows some spreading (or dispersion) at the base of the time domain response for this transmission impulse. This widening of the transmission response is due to an effect similar to windowing, and implies that there is a non-constant frequency response. Thus, it contains information on the magnitude and phase response of the two-way transmission. By placing a gate around this reflection and gating out the other reflections (principally from the input connector, and a discrete reflection in the cable) the loss of the cable is isolated from the other reflections.

The traces in the upper window are the square root of the gated frequency response of S11, along with the previously measured S21; they are nearly indistinguishable. In comparing it

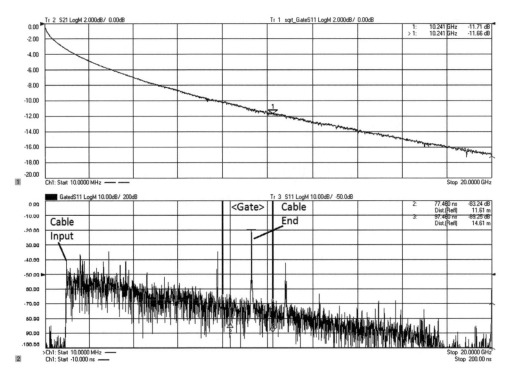

Figure 5.20 Lower window: time domain response of the shorted cable, with gate set about the short; upper, frequency response of the cable showing S21, and time-gated S11.

to Figure 5.19, the improvement in the quality of the estimate of S21 is remarkable. Even at this small scale of 2 dB per division, this is nearly identical to the S21 measured using standard techniques, and demonstrates the usefulness of this method. Further if one or a few large reflections are identified in the time domain response, their masking effect can be compensated for just as described in Eq. (5.3). The lower plot shows the time domain response with the cable input and cable end identified, along with the gate location. It's also clear that there is a substantial reflection in the cable at about 75 ns, likely from some damage in the cable. The re-reflection from this off the short at the cable end is clearly seen at about 110 nsec.

5.1.4   Return Loss Measurements

Return loss measurements on cables and connectors are relatively straightforward for cables with common connectors, measured as an assembly, and for the connectors themselves when both ends of the connector are of a type for which a calibration kit exists. However, return loss measurements on bare cable, very long cables, and on connectors that do not have common calibration kits can be quite problematic.

Return loss measurements on cables and connectors are almost always two-port measurements, where the cable or connector requires a termination, typically in the VNA port 2. If the DUT has a connector to which a high-quality metrology-grade load can be mated, then often the best return loss measurement will result from a simple one-port calibration and terminating the DUT with the high-quality load. This is especially true in older, legacy VNAs that lack the newer calibration methods and where the DUT is non-insertable, that is, when it does not have a mating male and female connector on each port. For older VNAs, the best calibrations could only be performed with flush (zero-length) mating connectors, that is, male on one port and female on the other. If the DUT had the same sex connector on each port, or connectors of different types on each port, then performing a proper full two-port cal was not practicable, unless a well-matched, and well-known defined thru standard was available. In many cases, the error from using a non-zero-length thru, and ignoring the length and match effect, can be greater than the return loss to be measured.

Later versions of legacy analyzers, such as the HP 8510 and HP 8753, incorporate an adapter removal method that allows mixed connector calibration by performing two two-port calibrations, one on each side of the single adapter that mated to each port. This calibration process required both calibrations to be performed, and then a new calibration set was extracted from the two calibrations.

For high-quality calibrated measurements on most cables and connectors, some type of non-insertable calibration is really required. A common DUT is a cable with male connectors on each end, and a common error is to perform a normal SOLT calibration with female open/short/load standards, and then use a non-insertable male-to-male adapter as the thru standard. This will cause an error in the load match and insertion loss, as described in Section 3.3.3.3.

For almost all cables and connectors, using the unknown thru calibration will provide the best results. In fact, surprisingly, the best calibration is likely to occur if the DUT itself is used as the unknown thru. If the DUT has relatively low loss, for example less than 10 dB, it will provide a sufficiently good unknown thru. If the unknown thru step is the last step of the calibration, then there will be no test-port cable flexure after the calibration to add error to the measurement. Most engineers assume that using the small thru out of the calibration kit will provide the best calibration, but for reasonably small loss cable and connectors, the DUT itself will, in practice, be the best choice.

This concept applies to cables or connectors that have the same connectors or mixed-series connectors, provided that a calibration kit or Ecal is available for the each connector type.

As an example, a short length of formed semi-rigid cable is tested by calibrating with an Ecal used as an unknown thru, and another calibration using the DUT cable itself as the unknown thru, with the results shown in Figure 5.21. In this case, a slight flex of the test-port cable results in slightly more loss measured using the Ecal thru compared to using the DUT as an unknown thru. In this case the difference is less than 0.02 dB.

5.1.4.1 Measuring In-Line Cable Connectors

For some industries such as the cable-TV infrastructure, large hardline cables are used for their very low loss. These cables are connected using an in-line splice or connector. A similar situation occurs for type-F cables used in the home, where the center wire of the cable provides the center pin for a type-F interface, or where the in-line adapter is a type-F female-to-female adapter. In these cases, which can be classified as in-line connectors, the quality of the adapter is difficult to judge in isolation as its effects are only apparent when used between two cables.

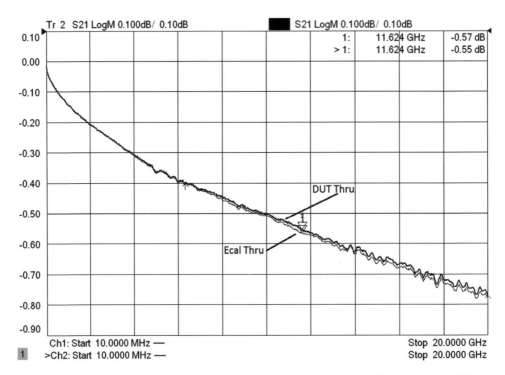

Figure 5.21 Comparing measurements of a short piece of formed semi-rigid cable using an ECal as unknown thru, and the DUT as an unknown thru.

As such, measurement methods for in-line connectors have been developed that allow them to be measured in-situ.

The difficulty with measuring in-line connectors is that they must be connected to cables, and these cables must then be connected to the VNA through a test connector, which is most likely an adapter as well (transitions from the cable type to the VNA connector type). In many cases, the return loss of these test connectors is worse than the in-line connector to be measured. These in-line connectors are often used as cable splices and have no intrinsic connection type of their own. They are designed to make the spliced transitions as cleanly as possible.

Normal return loss measurements of in-line connectors are almost always dominated by the reflections of the test connectors and other defects in the cable, thus requiring more sophisticated techniques. An in-line test configuration is shown in Figure 5.22. The improved methods turn again to time domain and gating techniques to remove unwanted reflections. Similar to the test methods for cables, the in-line connector test method relies on the ability of a time domain transform to separate the effects of the test connectors from the in-line DUT connector [3].

Figure 5.23 shows a frequency response of an in-line connector, which is inserted in between two approximately 2 m cables; the cable length is chosen to allow good separation of the responses in the time domain. The upper plot shows the S11 and S21 response of the overall system of input test connectors, input line, DUT in-line connector, output line and output connector. Also shown in the upper plot is the trace of just the connector (marked "Connector Only") derived from a simulation model.

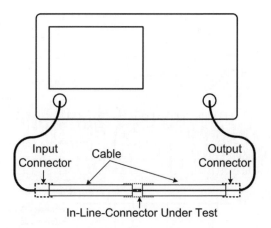

Figure 5.22   Configuration for in-line connector test.

The lower plot shows the overall time domain response from which the in-line connector response can be discerned in the middle. Knowledge of the approximate position and length (delay) of the in-line connectors helps in setting the gates and in recognizing the in-line connector separately from other defects in the cable (marked as "Cable defect"). One thing to note from the time domain response in Figure 5.23, before gating, is that both the input and output connectors of the test line have significantly higher return loss responses than the in-line

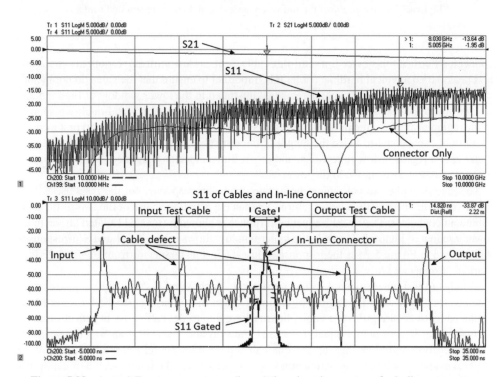

Figure 5.23   (upper) Frequency response; (lower) time domain response, for in-line connector.

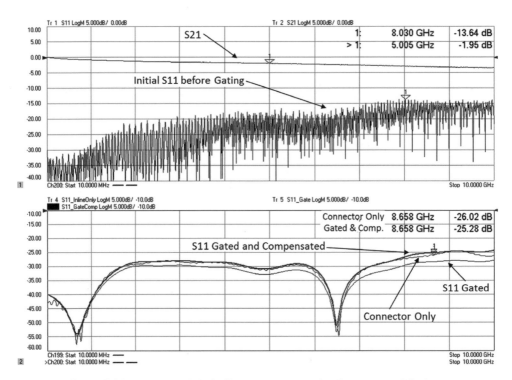

Figure 5.24   Response of the in-line connector, gated and compensated for loss.

connector. Further, this device is a semi-flex cable, so it has less uniform response (more return loss peaks) than a typical hardline cable. It is important to ensure that the gates do not include cable reflections from other cable defects (as illustrated in the lower window of Figure 5.23), so care should be taken to ensure that the test cables have reasonably good-quality responses in the region near where the in-line connector is applied. A common practice is to first measure a length of cable equal to twice the length of the test cable to check the quality. This cable is then cut at a region of low reflections, and the in-line connector is applied. The time domain gated response is also shown in the lower plot.

The gate is applied to determine the in-line connector response, but the response will be lower (showing incorrectly a better return loss than expected), because of the loss of the input cable and connector. The gated frequency response is shown as the "S11 Gated" trace in the lower window of Figure 5.24, while the upper window shows ungated S21 and S11 responses. Also shown in Figure 5.24 is the response of just the in-line connector from Figure 5.23 (still marked "Connector Only").

Finally, a third trace (marked "S11 Gated and Compensated") shows the response of the gated S11, compensated for the loss of the input cable. In this case, since the DUT connector is placed halfway along the test cable, the masking due to the cable loss in reflection at the halfway point is very nearly equal to the S21 transmission loss, since an S11 measurement will travel forth and back through the one-half input cable. Therefore, the S21 trace of the upper window is used in the equation editor to compensate for the input loss. Gating may also be applied to the transmission measurement to remove the extra ripples from the input and output connectors (see Section 5.1.3.2). The "S11 Gated and Compensated" is computed as

$$S_{11\_GateComp} = \frac{S_{11\_Gate}}{S_{21\_Gate}} \tag{5.7}$$

If the DUT is exactly in the middle of the test cable, the gate center-time value for transmission and reflection are the same. This final response is almost identical to the ideal in-line "Connector Only" response, proving the quality of this method. At very high frequencies, near the band edge, time domain gating effects will cause an up-tick in the response. As with all applications of gating, it is important to discount the last 5–10% of the frequency response due to gating edge effects, as described in Chapter 4. Therefore, over-sweeping the required frequency range by at least 10% is recommended.

Finally, it is important to note that this method for the most part only applies to return loss measurements. The insertion loss is difficult to ascertain in this method, and it is perhaps better to determine the insertion loss by doing a true insertion test. If the S21 of the test cable is measured before the in-line connector is added, then the difference can be determined when the in-line connector is added to look for added loss. In practice, these connectors are nearly lossless and so the in-line S21 loss is very nearly

$$S_{21\_Inline} = \sqrt{1 - \left| S_{11\_GateComp} \right|^2} \tag{5.8}$$

This technique for testing in-line connectors and removing the effects of the input and output test connectors can equally well be applied to other embedded component measurements such as SMT components on a PCB. In such a case, creating a fixture with long input and output lines will greatly help in separating the effects in the time domain.

5.1.4.2 Structural Return Loss

Structural return loss (SRL) differs from normal return loss in that it measures reflections relative to the average impedance of the cable rather than to some reference impedance [4]. The main reason for making structural return loss measurements is to look for minute, periodic defects in long reels of very-low-loss cable, typically used for cable TV mainline installations, as described in Section 1.8. For many of these systems, there was available some small amount of impedance adjustment in the system so the absolute value of the cable impedance was not as critical, but variation in impedance or return loss caused substantial problems. As such, the SRL was defined relative to the cable average impedance, and the average impedance was specified within some range, typically within plus or minus one or two ohms of the cable's nominal impedance.

The difficulty in SRL measurements arises from the fact that for very long, low-loss cables, very small but periodic impedance deviations can add up and create very narrow but very high return loss spikes. These are often caused by a defect in some part of the manufacturing, often from some rotating spindle that is slightly out of round or has some other defect on it.

Before the use of modern VNAs, the structural return loss was measured using a variable impedance bridge, illustrated in Figure 5.25. This bridge had an adjustable impedance factor (typically using a variable resistor in one leg) and also included a variable capacitor at the

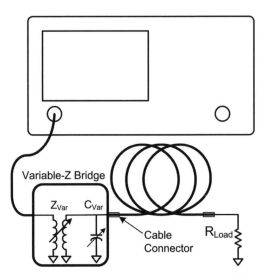

Figure 5.25   Measuring a cable with a variable impedance bridge.

input that canceled another fixed one inside the bridge. Thus, this adjustable capacitor could add or subtract a small capacitance, to compensate for the defects in the test connector.

Theoretically, one should also apply a variable impedance load as well, to match the far end of the cable to its impedance, but in practice the length and loss of the cable make this unnecessary.

Modern VNAs take a fixed bridge measurement, and use computation to simulate the effects of the variable impedance bridge, using port matching similar to Figure 5.10, with the addition that the impedance of the port can be varied as well as the matching elements.

To investigate how this response appears on a VNA, a circuit created with a 10 mm long, approximately 1 ohm deviation of a cable impedance, every 1 meter is simulated and the data loaded into a VNA using the built-in S2P file reader. The simulation also modeled a non-ideal input and output connector, and made small variations on the size and position of each discontinuity to emulate real-world conditions. A 3200 point sweep and a 201 point sweep were both created in the simulator. The return loss and insertion loss of the 201 point sweep are shown in the upper window of Figure 5.26.

The lower window of Figure 5.26 shows the resulting SRL after applying the effects of a virtual variable impedance bridge. The first step is to turn on the trace statistics and look at the mean return loss. While monitoring this, the port matching function is used to add either some capacitive or inductive compensation to "lay the trace down" as it is known in the cable test industry. In this case about −1.1 nH of inductance provides the lowest mean value. Next, some small port extension, the maximum length of which is equivalent to the length of the input connector, is added to further reduce the mean value. Finally the port impedance transformation is used to both find the average impedance and reduce the mean return loss to its lowest possible value. The impedance at which the return loss is lowest is called the average impedance of the cable. For the cable in Figure 5.26, this value is 76.7 ohms.

Finally, the peak of the S11 trace is observed to determine the worst case structural return loss; in this case it appears to be −37.03 dB. For these types of cables, a typical specification

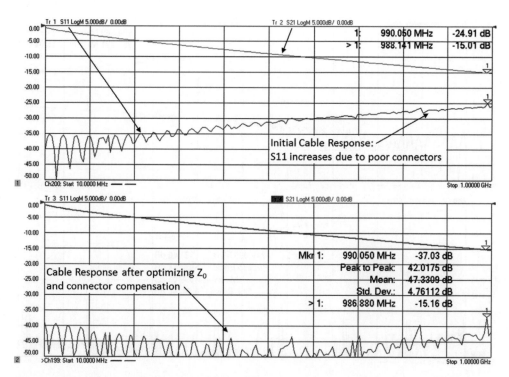

Figure 5.26 Return loss and insertion loss of a long cable; upper is normal return loss and lower is applying a virtual variable impedance bridge.

is −32 dB SRL, so it would appear that this cable would pass. However, from a fundamental analysis of the cable, one can determine that the resolution needed to see any and all peaks in the SRL is much smaller than the 201 point sweep above provides.

The cable in question has a length of approximately 500 m, and a velocity factor of about 0.9. From this, the frequency at which this represents one-half wavelength can be determined as

$$\frac{\lambda}{2} = \frac{V \cdot c}{2 \Delta f} = 500 \text{ m} \quad \therefore \quad \Delta f = \frac{0.9c}{2 \cdot 500} = 270 \text{ kHz} \tag{5.9}$$

For the 1 GHz sweep shown, around 3700 points would be required to ensure that a measurement point occurs every half wavelength. The loss of the cable diminishes the summing of periodic discontinuities, and they will not be spaced precisely periodically, so a somewhat lower number of points may be used. If the measurement is repeated with 3200 points, a different picture of the SRL response appears, as shown in Figure 5.27. Now the very narrow responses usually associated with SRL problems become clear. The adjustments are made to the input connector compensation and the cable impedance, with nearly the same results, although the best match for cable impedance was 76.6 ohms. In fact, this matches very nearly with the expected value of 76.5 ohms for this cable from the simulation.

Of course the most important result is the actual value of SRL, in this case −31.4 dB. This cable would be just out of specification due the periodic disturbance. The periodic impedance error here was 0.9 ohms, over a 10 mm region, occurring approximately every 1 m. This period

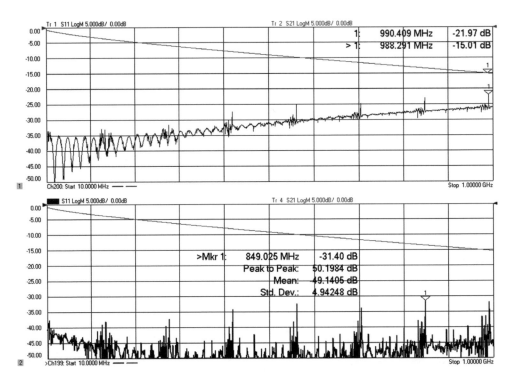

Figure 5.27   SRL measurement at 3201 points; upper trace is before connector compensation.

corresponds to a frequency of about 140 MHz given the velocity factor of the cable, and this is exactly the repetition frequency where the SRL response is seen in Figure 5.27.

Using a variable bridge with older VNAs, similar results may be achieved with the exception of connector compensation. A very consistent problem with the older variable bridge was difficulty in getting the trace to "lay-down". This is because the electrical delay between the variable capacitance in the bridge and the mismatch of the test connector meant that perfect compensation was never possible, and resulted in many cases of failed cables, or repeated re-measurements with different test connectors. With the advent of modern compensation techniques, the physical variable-impedance bridge has been almost entirely replaced. The first implementation of this technique was found in the HP8711 VNA with its option-100 SRL measurement code. However, these techniques can be applied with nearly any modern VNA.

5.1.4.3   Cable Impedance

While the measurement of the average impedance of a long cable was described in the previous section as part of the SRL discussion, the techniques used are not appropriate for shorter cables where the far-end load will have a more dramatic effect on the result. For shorter cables, the measurement of impedance becomes somewhat problematic as it is seldom realized that impedance of cable is a two-dimensional characteristic. In Chapter 1, the details of transmission lines are discussed and it is clear that any physical transmission line (one with loss) has an impedance that varies with frequency. Further, any real cable may have perturbations along the length so that the cable impedance varies as a function of distance along the cable as well. So any discussion of cable impedance should really be defined at a particular point in the cable and at a particular frequency. However, almost no one specifies cable impedance in that way and almost everyone desires to have a simple, single number when referring to a cable.

Thus, cable impedance is often defined as an average over frequency with variation only along the distance of the cable or even as just as an average impedance over both frequency and distance [5]. One traditional way of measuring a cable's impedance is using time domain reflectometry, and a VNA provides a modern version of this method. To illustrate the measurement technique, consider a cable with a stepped impedance every 10 cm, with a less than ideal input connector. An S2P file of a cable simulated with these attributes loaded into a VNA is shown in Figure 5.28. The upper plot shows the normal response, and the lower plot shows the response after applying connector compensation, as described in Section 5.1.3.1.

Note the difference in the value of marker 2, which represents about a 0.04 ohm difference in the apparent value, due to the effect of the less than ideal input connector. Also note that the frequency response is flattened with the connector compensation in the lower plot. The time domain response clearly shows the impedance steps in the cable. The values of marker 1 are shown in reflection coefficient. Some VNAs allow the marker format to be set differently

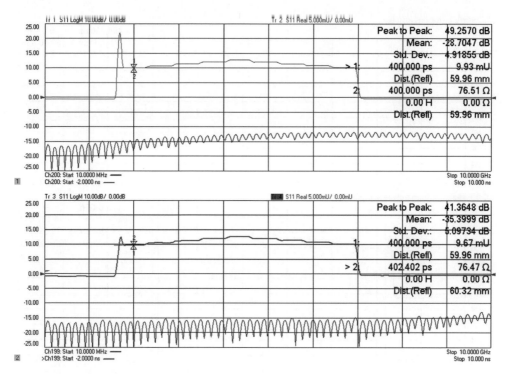

Figure 5.28 Frequency and time domain of a cable with stepped impedances; upper is normal and lower is with connector compensation.

from the trace format; here marker 2 is set to show R + jB, which provides the impedance in the case of a time domain trace, which is always pure real in a low-pass transform. From these plots, the impedance as a function of delay down the cable can be directly determined, by moving the marker along the line.

Using the built-in equation editor feature, one can also compute the impedance as a function of delay down the cable directly by converting the reflection coefficient to impedance. The conversion is simply

$$Z = Z_0 \frac{1 + S_{11}}{1 - S_{11}} \tag{5.10}$$

and is applied to the upper time domain trace, as shown in Figure 5.29. One anomaly of using such an equation editor function is that the time domain transform occurs on the raw data before the equation editor math is applied. Since converting to impedance is not a linear function, the conversion must be performed after the transform is performed. Thus, the transform is performed in the upper window (trace 4) and then the equation is applied in the lower window. An unfortunate consequence is that the x-axis retains its frequency label rather than having the time label. Trace statistics can be applied to a segment of this converted measurement, to produce a mean value. In the case of the lower window, the mean is computed between the region between marker 1 and marker 2, showing a mean value of 76.68 ohms. This is very close the value expected from the design used in the cable simulation.

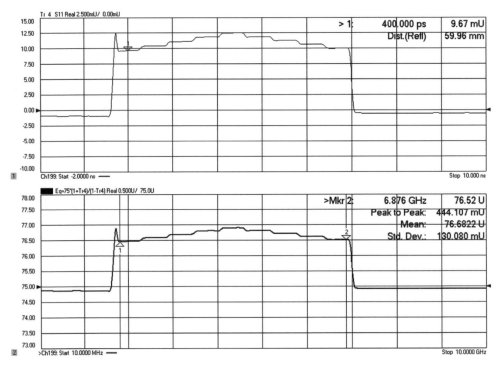

Figure 5.29 Cable impedance as a function of delay down the cable; upper shows linear reflection coefficient; lower has impedance in ohms as the y-scale.

One caution, however, with using the time domain transform: because the DC value is not directly measured in the VNA, it must be extrapolated. The DC value represents the value of the transform at some distant time, and should be equivalent to the termination value plus the DC loss of the cable. This value is set as described in Chapter 4. However, both alias effects (which occur if the time domain response does not return to zero inside the alias free range) and extrapolation error effects can cause the apparent impedance at the time just before zero to be different from the expected value, which is usually the system reference impedance. In such a case, one should measure the apparent reference impedance on the time domain trace and accommodate for any difference in the measurement results.

5.1.5 Cable Length and Delay

A common question when measuring a cable is to know what is its length or delay. Usually this entails measuring the group delay of the cable, although time domain can also be used for a similar measurement. For long cables, a key issue is under-sampling the phase response, which means that the frequency spacing is such that more than 180° phase change occurs between measurements points. Figure 5.30 shows the (admittedly complicated looking) result of measuring a 10 m cable with four different choices of number of points over the same frequency span. Since the length of the cable is already known, it is straightforward to compute

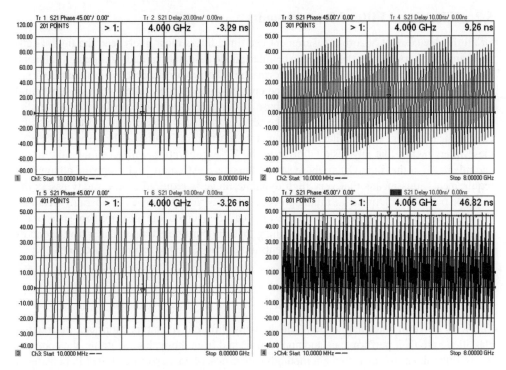

Figure 5.30 Examples of changing the phase sampling for a long cable; only the lower-right plot is correct.

the frequency spacing needed using the formula from (5.9), which yields about 10.5 MHz for a cable with a 70% velocity factor. In the first case shown in the upper-left window, an 8 GHz span is measured with the default 201 points for a frequency spacing of about 40 MHz, and it is clear that the response is under-sampled. In each window, the phase response and group delay are shown and in the upper left the phase looks abnormal and the group delay is negative. The upper-right window shows the result when the number of points is increased to 301 points. In this window, the phase looks somewhat normal and the delay looks positive, but the delay number does not correspond to the expected delay of a 10 m cable, which should be around 47 nsec, with a presumed velocity factor of 0.7. The 9.26 nsec delay must be wrong, but the under-sampling is such that the phase slope is negative and the delay is positive. The lower left window shows the response for 401 points. Here the delay is again negative, Finally, in the lower right window shows the response for 801 points. Here the delay is positive again, and the delay number is near the expected value of 47 nsec. The frequency spacing here is just under 10 MHz per point, and so we would expect this result to not be aliased.

The plots on the left illustrate caution in increasing the number of points: if the number of points is simply doubled, and the phase response is under-sampled sufficiently, doubling the points can result in the exact same incorrect delay indication. Therefore it is best to increase the number of points in non-uniform increments to illuminate the phase response with various spacings.

5.2   Filters and Filter Measurements

Of all the devices that absorb measurement test time, filters most likely lead the field. High-performance filters, such as those used in cellular-phone base stations and satellite multiplexers, must be tuned to achieve the desired performance. The product tolerances in manufacturing filters do not come close to being controlled enough to create these filters without substantial tuning. The tuning process on a complex base station filter might take up to an hour, and the tuning on a satellite multiplexer might take several days or weeks. During all this time, the careful measurement of the filter's S-parameters must be maintained. The very low loss of these filters means that any residual error in the VNA calibration will accentuate the ripple and mismatch of the filter. Often, these same filters have extremely high stopband attenuation requirements; this means that very high dynamic range is required to accurately assess the isolation of the filter. During the tuning process, which typically involves an operator manually turning a tuning screw while watching a result on a VNA screen, the measurement speed must be fast enough to provide for real-time tuning, usually considered ten updates per second or faster. Specific aspects of filter measurements are discussed in the following sections.

5.2.1   Filter Classes and Difficulties

The wide range of filter applications yields many different filter types that can be roughly classified as to their performance and application, and from these classifications, the common needs, difficulties and measurement methods can be described. A sorting of the filter classes is shown in Table 5.1. These can be roughly sorted into three main classes: fully tunable filters, trimmable (one-way adjustable) filters, filters fixed by design.

Table 5.1 Filter classification by application type

Filter Classes/ Applications	Technology	Loss (dB)	Match	Complexity/ Tuning	Other Attributes
Base station transmitter (Tx)	Silver-plated, larger air-coupled resonators	0.1–1	20–26 dB	Complex/tune resonators, couplings	Orders from 6 to 20, deep notches near passband
Base station receiver (Rx)	Silver-plated, larger air-coupled resonators	0.1–1	20–26	Complex/ tune resonators, couplings	May include LNA integrated into filter, very high Tx isolation
Base station duplexer	Silver-plated, larger air-coupled resonators	0.1–1	20–26 dB	Complex/tune resonators, couplings	three-port combination of Tx and Rx
Satellite multiplexer	Highest quality silver plating, tuned resonators and couplers	0.1–1	20–26 dB	Very complex/tune resonators, coupling, multiplexer	High port count (up to 20 or more), multistage adjustment
Handset	Ceramic coupled resonator	1–3	10–15 dB	Low/one-way	Low cost
RF subsystem	Microstrip coupled line on PCB	0.5–3	10–20 dB	Simple/no tuning	Often integrated into RF PCBs
RF subsystem	LTCC on ceramic	0.5–3	10–20 dB	Simple/no tuning	Often part of RF system on chip
Channel	SAW	1–10	5–15 dB	Moderate/no tuning	Very narrow, very high order

Tunable filters often have the most strenuous requirements of loss, isolation and match, which is why they must be made tunable to achieve the difficult specifications. These are used in high-power applications where any loss must be minimized, and in low-noise applications where a receiver's noise figure is directly degraded by filter loss. These also have high isolation requirements to ensure that in a transceiver application, the high transmitter (Tx) power is isolated from the low-noise receiver (Rx) path. These filters are often found in a combined form as a diplexer.

5.2.2   Duplexer and Diplexers

One common question is "What's the difference between a duplexer and a diplexer?" While there is no official answer, one can understand the difference in nomenclature by appreciating the difference in application. A diplexer is a filter that combines signals of two different frequency ranges at two different ports, into a single combined port, and isolates each port from the other. A multiplexer is a diplexer that combines more than two ports into a single port. Most commonly, a diplexer is used to channelize a receiver or a transmitter.

A duplexer allows a transmitter and receiver to operate at the same time using the same antenna. So, a diplexer is used as a duplexer when it combines a transmitter and receiver such as in a base station.

5.2.3 Measuring Tunable High-Performance Filters

Making measurements on tunable filters means setting up a VNA in a way that allows high speed updates while maintaining the desired performance. Typically, this is a passband only measurement, so high isolation is not generally needed during the tuning process. In such a case, wide IF bandwidths can be used. However, many VNAs change sweep type (from stepped to swept source) depending upon the IF bandwidth, with wide bandwidths using swept source by default. The fast sweep times needed for filter tuning are accomplished in some VNAs using continuous swept frequency (as contrasted with stepped frequency sweeps) and this can in some cases cause IF delay problems, as described in Section 5.1.2.1. For the most part, these effects only occur with very narrow filters which by nature must have very long delays. Typically this is only seen in crystal filters or SAW filters. Thus the IF delay errors seldom affect tunable filters.

In measuring tunable filters, there are many choices to be made in the configuration of the test and calibration that can affect both the quality and speed of the measurement; some of the common attributes are described below.

- **IF bandwidth:** This is the single most important setting affecting the tradeoff between speed of measurement and the noise in the results. For most tuning measurements, the IF bandwidth can be set quite high, but with some limits. At very wide bandwidths, other overheads such as computing and displaying the trace results, and band switch time and retrace time will be a larger percentage of the overall sweep cycle time. At some very wide bandwidths, the data-taking time is swamped by the other overhead times, and increasing the IF bandwidth further does not result in any substantial increase in overall cycle speed. A typical value for tuning is between 10 kHz and 100 kHz IF BW. Finally, the frequency resolution is limited by the IF BW, in that the frequency response of a filter is "smeared" across the IF BW. One cannot use a wide IF BW on a narrowband filter. A narrow IF bandwidth might be required to accurately assess the corner frequency of a filter.
- **Number of points:** As with IF BW, the number of points for a sweep directly impacts the cycle time unless the number of points is very small, where other overhead will swamp out any improvement from further lowering the number of points. For many applications, the number of points is set by required resolution at the band edges, but the interpolation of the trace response is quite good in modern VNAs, and marker or limit line values are computed off the interpolated result. Usually, this interpolation is good enough to allow reducing the number of points to a reasonable value over the bandwidth. A typical value for most tuning applications is 201 or 401 points.
- **Sweep mode:** As discussed in the section on IF delay, changing from swept to stepped mode may have a substantial effect on the sweep rate. However, modern VNAs have very fast synthesizers, so that for IF BWs below about 10 kHz, the effect is quite small and stepped mode may be used without a dramatic increase in cycle time. Some VNAs provide two forms of step sweep: standard and fast step. In fast step, some of the settling times

associated with the source ALC loop and the individual receivers used in a non-ratio mode are decreased or eliminated. For almost all ratioed measurements, such as S-parameters or gain, the errors from settling are removed by the ratio process so there is no ill effect in reducing or eliminating these settling wait times. But for cases where absolute power control is important, such as amplifier test, the normal stepped mode should be used to avoid power settling issues.

- **Calibration type:** For most high-performance tunable filters, a full two-port calibration is necessary to remove the effects of the test system load match from the return loss measurement. Older legacy VNAs, such as the 8753, provided a special mode that would update the reverse (S22) sweep only occasionally, at user specified interval, such as every 10 forward sweeps. However, the improved dynamic range of modern VNAs allows low-noise measurements at wide IF bandwidths, so that even full two-port calibrations can proceed with near real-time speed.

5.2.3.1   Filters with Very Low Loss and Well Matched In-Band

Measuring filters with very low loss which are well-matched for the in-band portion, requires the same techniques discussed in Section 5.1.1. Careful attention to using good calibration techniques and good cables are required. Because filters have long delay for their physical size, they can have a lot of phase change in the passband and this can cause extra ripples in the response if the VNA system source and load match are not characterized properly. In particular, since most filters are non-insertable (they have either different connector families on each port or the same type and sex on each port), it is very important to use either the unknown-thru calibration method or, for older VNAs, an adapter removal calibration method. Using the traditional SOLT calibration and ignoring the thru delay is a common cause of poor calibration that can make filter tuning difficult due to poor load match correction.

5.2.3.2   Measuring Filter Return Loss

The return loss of a filter is almost always used as the principal way to tune a filter, even though the key specification of a filter is S21 insertion loss and isolation. However, the return loss is more sensitive to tuning variations, and obtaining a good return loss provides better system performance and nearly always guarantees a good insertion loss result.

When testing for return loss it is common to use a limit line to set a visual pass/fail criterion. Another convenient technique is to place a marker at the passband edges to see the actual values while tuning. A third, less well known but very convenient technique, is to use a feature found in some VNAs to have a marker track the worst case (maximum) return loss value in the passband. Often, a technician is instructed to tune for "best possible match" even after the filter passes the minimum specification. Using a technique of tracking the worst case point makes this quite convenient to look for the best case. This test scenario is set up in Figure 5.31. Marker 1 is set to track the maximum value of return loss within a narrow region of the passband. Markers 2 and 3 track the edges of the filter.

In trace 2, marker 1 is set to track the minimum value of S21 over the same region as in the return loss trace; it's common to show both passband insertion loss and return loss at the same time when tuning filters. While tuning this filter, these marker values will update on

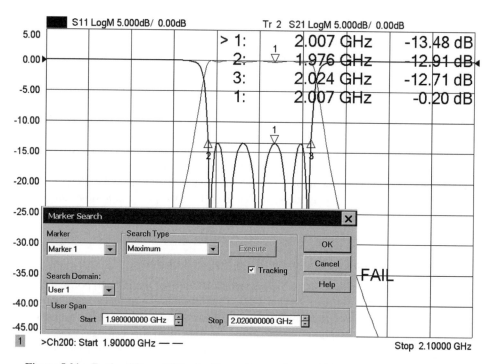

Figure 5.31   Testing S21 and S11 of a filter, using marker tracking to find the worst case S11.

every sweep and provide a convenient way of tracking the filter's performance. The limit-line indicator will change from fail to pass when the filter is tuned below the limit-line; usually failed regions of the test trace are highlighted in a red color.

5.2.4   Measuring Transmission Response

The transmission response of filters represents a measurement of the fundamental purpose of a filter. The response has two key attributes: the passband insertion loss and the stopband isolation. Band-pass filters have both an upper and lower stopband, though typically the stopband requirements are much more stringent on one side of the filter than the other, especially in filters used for duplexers in communications systems.

5.2.4.1   Pass-Band Measurement

Measuring a filter in the passband is very similar to measuring low-loss cables and connectors, and many of the details described in Section 5.1.1 should be followed for low-loss filters. One attribute of filters that distinguishes them from other devices is that the match in the passband is not, ideally, zero, but rather the reflection is determined by the number of sections, the ripple in the passband and the passband isolation. In most cases, the filter response is a tradeoff between accepting poorer return loss in exchange for sharper cutoff frequency. Thus, while most filters are tuned for a good return loss, even an ideal filter (one that operates precisely

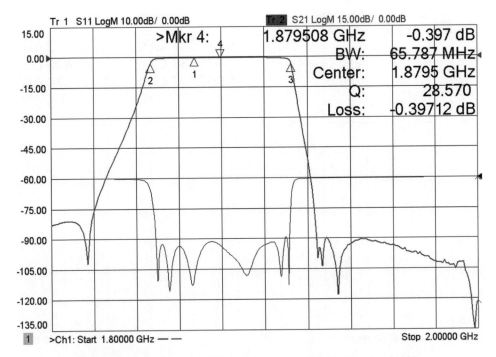

Figure 5.32 Using the marker search function to find a filter bandwidth.

as designed) will not have zero reflection across the passband. In contrast, a good cable or connector will have nearly zero reflection coefficient.

If a filter is nearly lossless, then the ripple in the transmission is directly related to the peaks in the return loss through the well-known equation

$$|S_{21}|^2 \leq \left(1 - |S_{11}|^2\right) \tag{5.11}$$

On most VNAs, the bandwidth of the filter can be automatically determined using marker search functions, as shown in Figure 5.32. Most such functions allow the bandwidth level, typically −3 dB, to be specified. For an equal-ripple filter, the bandwidth should be defined as the ripple level. The search function finds the maximum value, and computes the bandwidth from that level.

After finding the bandwidth, markers are placed on the upper and lower corner frequencies (markers 2 and 3 above), at the center (marker 4) where the loss of the filter is reported, and at the position of the maximum value (marker 1).

5.2.4.2   Excess Loss

Though not a typical specification, a useful figure of merit is the excess loss of a filter, defined as

$$L_{Excess\_dB} = 20 \log_{10} \frac{S_{21}}{\sqrt{1 - |S_{11}|^2}} \tag{5.12}$$

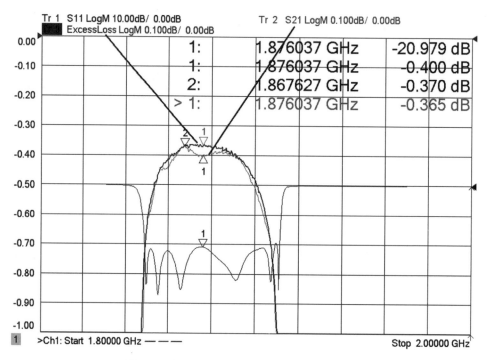

Figure 5.33   S11, S21 and excess loss of a filter.

This represents the energy absorbed by the filter (or any other passive device), in excess of the mismatch loss. Figure 5.33 shows the S21 and S11 of a filter and using the equation editor, the excess loss.

This loss is mostly independent of filter tuning and represents the best case insertion loss of an ideally tuned filter. It is sometimes useful to see the excess loss when tuning a filter to understand whether the desired insertion loss is achievable with tuning. In some cases, degradation of manufacturing processes, such as poor plating or bad soldering, can lead to an unexpected increase in excess loss, creating a filter that cannot be tuned for low-pass insertion loss, even if the return loss is properly tuned. In such a case, it is useful to find this out before a substantial time has been wasted in some fine-tuning process.

Because even ideal filters have reflections in the passband, the transmission response is very sensitive to the VNA port mismatch. This can lead to larger mismatch error in the filter response as the additional uncertainty is equal to S11 times the residual source match and S22 times the residual load match. Thus, for filters with very tight tolerances, it is very important to use good calibration techniques.

5.2.4.3   Limit Testing for Transmission

Transmission responses are usually tested against a minimum limit line value, but in some VNAs, the limit testing is only performed on discrete data points, and the actual limits might lie in between data points. For example, if a filter is tested across a 60 MHz span, with 201

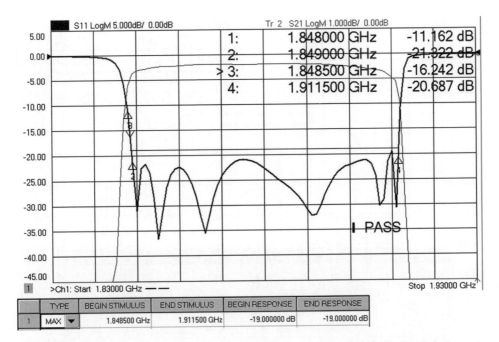

Figure 5.34   Limit testing when the measurement point does not equal the limit edges.

points, the point spacing will be every 300 kHz. If a limit line is centered on the filter center frequency and set to have a span of 50 MHz, then the limit edge points will lie exactly 25 MHz above and below the filter, but the discrete data points of the filter will be at a span of 49.8 MHz and 50.4 MHz, just inside and just outside the limit. If the point just inside the limit passes, and the point just outside fails, the limit will pass, even though the displayed traces may cross the limit line. In such a case, the number of points must be increased or segmented sweeps used to ensure that the limit test occurs exactly on the measurement point. Figure 5.34 shows an example where the S11 measurement points do not match the edges of the limits thereby causing a pass indication when the filter actually fails the true limit criteria.

In particular, markers 1, 2 and 4 are positioned on discrete measurement points, where marker 3 is interpolated between markers 1 and 2. It is clear that the trace breaks the limit line, but the limit test passes as there are no measured points outside the limit. Doubling the number of points so that marker 3 lies on a measured point will result in the limit test failing.

5.2.4.4   Evaluating Ripple Using Statistics

One key figure of merit in filters used for communications systems is the amplitude deviation or ripple in the passband; this is sometimes called filter flatness. This ripple is usually caused by the reflection of the filter, and is one of the design parameters. In many cases, the ripple of the passband will be acceptable if the return loss is properly tuned, but in some cases the return loss specification is less important than the filter flatness. The passband ripple is easy to characterize using the trace statistics function and the statistics user range feature of some

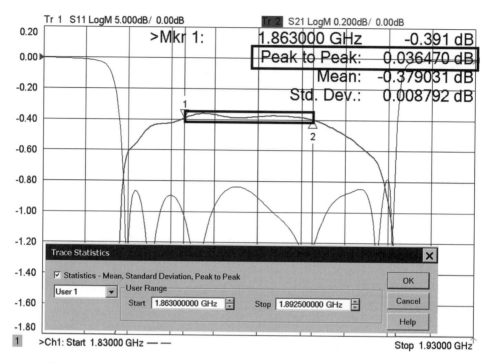

Figure 5.35   Using trace statistics to report the peak-to-peak ripple in the passband.

VNAs. The peak-to-peak value of the trace statistics directly shows the ripple in the passband, as demonstrated in Figure 5.35. Other VNAs allow using a ripple limit line, which floats at the mean value of the trace and shows whether the ripple is in or out of specification.

In some systems, the frequency flatness response can be compensated for by using equalization, and so slope in the loss of a filter (or cable) can be removed, but amplitude deviation cannot. In this case, the flatness cannot be simply characterized as the peak-to-peak value over a region, but is defined as deviation from a linear response. In such a case, the slope of the amplitude response should be removed. There are many methods for fitting a straight line to an amplitude response, and perhaps the most common is a least-squares fit. In some modern VNAs, least-squares approximations are provided as a post-processing function (as an imported function of the equation editor, for example), and both the flatness (deviation) and the slope (sometimes called tilt) of the response can be directly displayed. Figure 5.36 shows the flatness (around a 0 dB reference) and slope (called the tilt parameter) of a filter in the passband, as well as the original S21 and the best-fit line. The tilt parameter is a single value that is equal to the slope of the best-fit line, and in the case of the figure, is the difference between marker 1 and marker 2 on the best-fit trace.

In some cases, the flatness computation can be restricted to a small portion of the underlying trace, allowing a single trace or channel to provide both in-band (flatness) and out-of-band (isolation) responses. These functions are commonly needed for amplifiers used in cable TV systems where the slope or tilt is set to compensate for cable loss, and the amplitude ripple after the slope compensation must be constrained to a small value to ensure high quality of service.

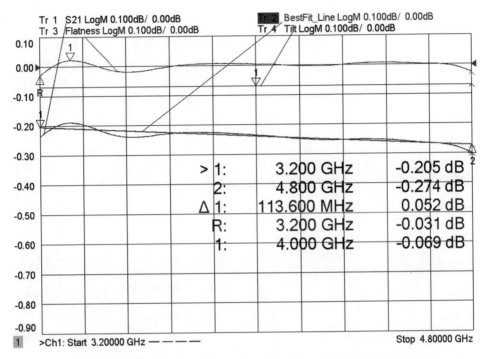

Figure 5.36   The flatness and slope of a filter, along with the S21 response.

5.2.5   High Speed vs Dynamic Range

In the transmission measurement of filters, there is a direct tradeoff between the speed of measurement and the dynamic range or noise floor of the measurement. When tuning filters for passband response, the dynamic range is not usually of concern, as the tuning is primarily performed on the passband response. But some high-performance filters have transmission zeros (or notches) in the S21 response, and on some occasions, these are required to be tuned. This typically occurs in coupled-resonator filters with an adjustable cross-coupling element. The strength of the cross-coupling sets the position and depth of the S21 transmission zero or null. In such a case, real-time measurements are required for both the passband and a portion of the stopband of the filter, to facilitate tuning.

The conflict occurs between measurement cycle-time and dynamic range due to the effect of the measurement IF bandwidth. Wider bandwidths provide for high-speed measurements, but have a higher noise floor, simply due to the wider bandwidth allowing more noise into the measuring receiver. Usually a transmission measurement has three regions of interest: passband, stopband high isolation (from the transmission zeros) and other stopband regions. The insertion loss in the passband is quite low so the signal-to-noise is very high and a wide IF bandwidth may be used. Over much of the stopband of the filter, only moderate dynamic range is needed, and only a moderately narrow IF bandwidth is required to obtain an adequate noise floor. However, over some of the regions of the transmission zero, very high dynamic range is needed, which requires very narrow IF bandwidths. Figure 5.37 shows the result of testing the transmission of a filter with three different IF band widths, 10 kHz, 1 kHz and 100 Hz.

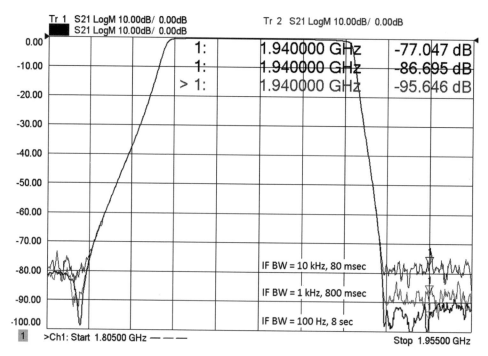

Figure 5.37 Transmission response with three different IF bandwidths and three different measurements speeds.

The figure clearly shows the improvement in noise floor using the narrower IF bandwidths, but what is not normally apparent in the plot is the change in sweep time that occurs. For the trace shown, and using a full two-port calibration, the measurement cycle time is 80 msec for 10 kHz, 800 msec for 1 kHz and 8 seconds for the 100 Hz IF bandwidth. The noise floor goes down 10 dB for each 10 times decrease in IF BW while the measurement cycle time goes up 10 times. For a linear sweep covering the full transmission range, using an IF bandwidth narrow enough to see the transmission zero will cause the sweep to slow down so much that it is completely unusable for filter tuning.

The VNA noise floor may also be reduced by increasing the source power. Modern VNAs have very linear receivers, but the receivers of many older VNAs would compress at maximum source power (up to 0.5 dB compression at max power or more) causing the passband insertion loss readings to be in error. Such compression, while not very consequential in the stopband, may cause many filters to fail their passband specifications, so such a high power level could not be used on a trace measuring both the passband and the transmission zero region.

5.2.5.1 Segmented Sweeps

Most VNAs provide a convenient solution to the problems described above, which can be succinctly stated as: in the passband region, lower power and wider IF bandwidths can be used, and high point density is needed to ensure that the band edges are properly measured, but in the stopband region, very narrow bandwidths and high source power levels are required to obtain a low enough noise floor. What is needed is a way to provide both, and such a method is found in the segmented sweep feature.

The segmented sweep feature originally provided a way to have different point densities over different regions of the measurement sweep. For a filter, the passband could have a high point density, and the stopband regions could have lower point density, with points placed right at the band edges to ensure proper evaluation of the limit tests. Starting with the HP 8753 (and at the insistence of a large cellular-base-station company), the segment options included the ability to assign different power levels and IF bandwidths to individual segments. Now a sweep could be defined that had high power in the stopband regions, to give good dynamic range, lower power but wide bandwidths in the passband regions, and a few points with very narrow bandwidths where the transmission zeros were to be tuned. An example of a segmented sweep is shown in Figure 5.38 for the same filter as the previous figure.

In this case, the same point density is maintained in the passband with 200 points and a 10 kHz IF bandwidth, but the lower and upper stopbands have only 40 points each with a higher power and narrower 1 kHz IF bandwidth, and the transmission zero region has 32 points so that the transmission zeros are clearly seen, also with high power but with a 100 Hz IF bandwidth. The overall measurement cycle time for this segmented sweep is 1.1 seconds, not quite the desired real-time but much better than the 8 second alternative above.

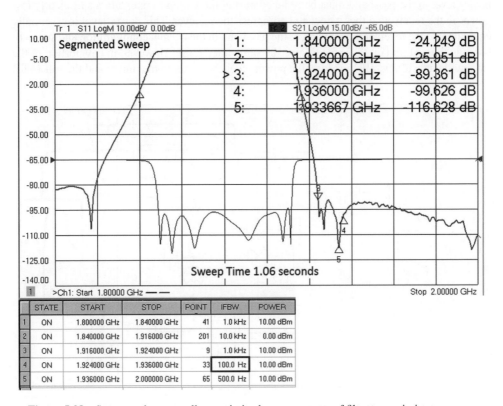

Figure 5.38 Segmented sweeps allow optimized measurements of filter transmission response.

A further optimization of the measurement setup can be utilized to achieve the desired cycle-time, as discussed in the next section.

5.2.6 Extremely High Dynamic Range Measurements

For filters with very high to extremely high dynamic range requirements, there are other modifications to the test system that can provide the needed speed and dynamic range. The modifications are only available on VNAs that provide an option that is commonly referred to as a configurable test set. In this option, the cables which interface to the test-port coupler are routed to the front panel so that the user has access to the coupled port and the input port (or coupler-thru) of the coupler, as well as the test port. With this access, it is possible to reroute the port 2 coupler so that it provides a lower noise path directly to the port 2 receiver, at the cost of lower source power available from port 2. This is commonly referred to as "reversing the coupler", and is illustrated on port 2 in Figure 5.39.

In this configuration the VNA receiver is connected on the thru-arm of the coupler to the test port, so it has substantially more sensitivity, typically about 14 dB more. This will improve the noise floor sufficiently to allow the IF bandwidth to be increased about 30 times, making the sweep rates much faster for the same dynamic range. In the example above, the sweep rates can be as fast as about 160 msec. The passband sweep rate cannot be increased much because the power level in the passband must be reduced during the sweep to maintain a sufficiently low power in the receiver so that it does not compress the receiver. This power level is essentially equivalent to the receiver power level of the passband in the normal configuration and so the IF bandwidth must remain the same to give the same trace noise. However, in any portion of the stopband with sufficiently high loss, the power level can be set to maximum and the full 14 dB

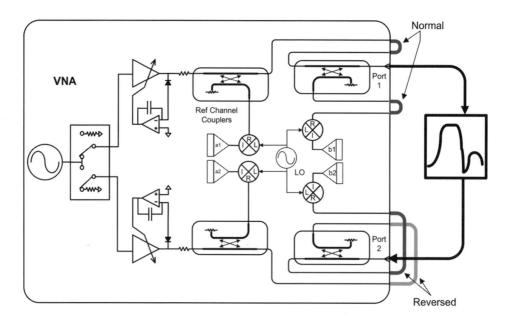

Figure 5.39 Block diagram of a VNA with configurable test set and reversed port 2 coupler.

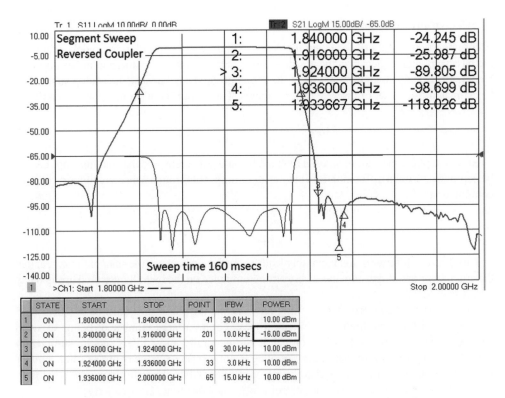

Figure 5.40   Increased dynamic range and speed using a reversed coupler.

improvement in noise will become apparent. Figure 5.40 shows the measurement result on the filter with a 160 msec sweep time, which fits the requirement for near real-time adjustment, and provides for the ability to see the transmission zeros down more than -115 dBc.

When reversing the coupler, it is critical to use segmented sweeps with adjustable power levels for each sweep, to avoid overdriving the test port receiver in the passband. Also, one must be aware that the dynamic range and noise floor on the reverse (S12) sweep will be degraded by the same 14 dB. Since filters are almost always linear and bilateral, $S21 = S12$ and there is no need to measure the reverse transmission. On the reverse sweep, the source power is dropped by 14 dB, hitting the DUT port 2, but the lower source power is offset by 14 dB more sensitivity on the port 2 receiver so the signal-to-noise of the S22 measurement is not changed by reversing the port 2 coupler. The higher noise on the S12 measurement for high-loss region has almost no effect on the corrected value of the other parameters; the contribution of S12 to the error correction math is always as a product of $S21 \cdot S12$. Since both are very small values, the product is small indeed. Such is not the case when testing high-gain amplifiers where the product can be greater than 1 if the noise of the reverse path limits the S12 measurement; this is discussed at length in Chapter 6.

Other test parameters are important besides the insertion loss and isolation measurements of the filter. In many cases the phase response and other derived measurements form a critical part of the filter specification. Some examples of these are given in the next section.

5.2.6.1 Group Delay Measurements

A common figure of merit for many filters is group delay, in particular group delay ripple in the passband, and on occasion the absolute group delay. Group delay measurements on filters by default are only an approximation of the actual group delay because the analog value must be estimated from a set of discrete measurements. The definition of group delay is classically

$$\tau_{GD} \stackrel{\Delta}{=} -\frac{d\phi_{Rad}}{d\omega} = \frac{-1}{360}\frac{d\phi_{Deg}}{df} \tag{5.13}$$

However, phase measurements of a filter are by nature discrete, and the analytic slope cannot be computed through differentiation, but must be computed through a discrete finite-difference computation:

$$\tau_{GD\_meas} = -\frac{\Delta\phi_{Rad}}{\Delta\omega} = \frac{-1}{360}\frac{\Delta\phi_{Deg}}{\Delta f} \tag{5.14}$$

This discrete differentiation inevitably leads to confusion because the span of the Δf, called the aperture or delay aperture, has a very strong influence on the overall measured group delay response.

The first issue of concern for measuring filters is that the phase shift must be less than 180° per point, and for some high-order filters, it may require a large number of points to avoid aliasing issues as described in Section 5.1.5. But generally, most filters require sufficiently close frequency spacing so that under-sampling is not an issue. On the contrary, often a major issue in making group delay measurements on a filter is that the data points are so closely spaced that the group delay becomes inordinately noisy due to a very small Δf.

In most legacy VNAs, the group delay was computed by simply taking the change in phase at each point divided by the frequency step. This causes two issues: first, it produces a slight skew in the data as the actual delay on each point is offset by one-half of the spacing, if only two points are used. Second, because there are N − 1 frequency segments, there are N − 1 group delay computed points for an N-point sweep. In most legacy analyzers, this is handled by making one of the delay points repeated, usually the first point. One can avoid the skew issue by always using an odd number of points for the delay or smoothing aperture.

If the frequency step was very small, even very slight trace noise on the phase trace would cause very large trace noise in the group delay trace. Figure 5.41 shows the group delay response of a filter with two different frequency spans in the upper window, and for various numbers of points in the lower window, where the reference position is offset by one division for each setting to clarify the resulting traces. For each measurement, a trace statistics function is shown for the middle 60 MHz of the group delay response, and the relative value of trace noise can be seen in the Std. Dev. result.

A wide span trace has less noise than a narrow one, roughly in proportion to the span. Traces with fewer points show less delay noise than greater point traces roughly in proportion of the number of points. In all cases, the trace noise on the phase trace is the same, but the divisor in Eq. (5.14) changes. All of these results are attributes of the group delay aperture and the fact that it changes with span and number points.

While the trace noise in the phase responses is essentially a constant level, the changing number of points changes the aperture with a corresponding increase in the trace noise of the group delay trace. One way to avoid this issue is to use fewer points on the group delay trace,

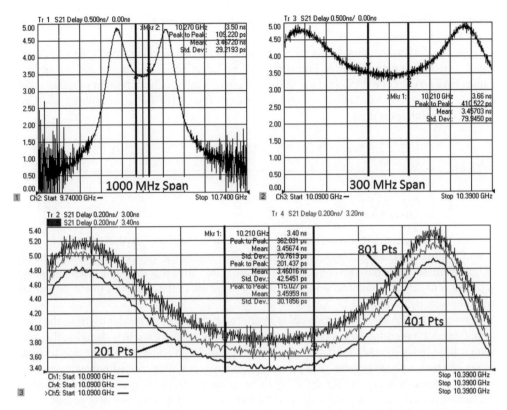

Figure 5.41 Group delay on a filter with various numbers of points and various spans.

but in many instances a large number of points is required to ensure that there are no issues in the amplitude response, and it would be convenient to have the group delay computed on the same trace. Most legacy VNAs incorporate a smoothing function that applies a moving average smoothing on a trace, and this can be used to set a wider effective aperture on a group delay trace. The smoothing function operates on a moving average window with

$$Y_{n\_Smooth} = \frac{[Y(n-m) + Y(n-m+1) + \cdots + Y(n) + \cdots Y(n+m-1) + Y(n+m)]}{2m+1}$$

(5.15)

where $2m + 1$ is the smoothing aperture, in points. Smoothing is often given in percent of span, and this is converted to smoothing points by

$$m = \mathrm{int}\left[\frac{N \cdot (smoothing\ percent)}{2}\right]$$

(5.16)

where N is the number of points.

When this is applied to most traces, each point forms an average of the surrounding points, and is similar to the video bandwidth function of a spectrum analyzer. In *most* cases, smoothing is an invalid way to reduce noise, as it can also remove important structures in the response. In the case of an amplitude response, smoothing can be used to eliminate mismatch ripple due to a poor calibration, but it can also hide the fact that a response has some excess ripple due to the DUT response.

In group delay traces, however, a quirk of mathematics makes using smoothing *identically equal* to reducing the number of points around the target point, and thus any intermediate values have no effect on the delay; a simple example of applying the definition of group delay from Eq. (5.14) to the definition of smoothing from Eq. (5.16) for a five-point smoothing (m = 2) illustrates this point:

$$D(n)_{Smo} = \frac{\left[\frac{\varphi_{n-1} - \varphi_{n-2}}{\Delta f} + \frac{\varphi_n - \varphi_{n-1}}{\Delta f} + \frac{\varphi_{n+1} - \varphi_n}{\Delta f} + \frac{\varphi_{n+2} - \varphi_{n+1}}{\Delta f} + \frac{\varphi_{n+3} - \varphi_{n+2}}{\Delta f}\right]}{360 \cdot 5}$$

$$D(n)_{Smo} = \frac{\varphi_{n+3} - \varphi_{n-2}}{360 \cdot 5\Delta f} \tag{5.17}$$

From this it is clear that none of the intermediate points contribute anything to the computation of the smoothed delay. Unlike smoothing in other traces, the values of the measurements of intermediate points have no consequence for the result, such that the value after smoothing is identical to the value that one would see if only the phase of the endpoints of the smoothing aperture were measured.

In more modern VNAs, the group delay aperture can be sent independently from the trace smoothing, so it is no longer necessary to turn smoothing on and off when changing the trace format from delay to other formats such as log magnitude. In some VNAs, the aperture can be set either as a point aperture, or a percentage of span, or a fixed delta-frequency aperture. In these cases, it is necessary for the delay to be computed from interpolated phase points if the fixed delta frequency or percent of span does not lie exactly on measurement points. Using a fixed delta frequency is a very convenient way to specify aperture because the resulting group delay trace, and noise on the trace, does not change with changes to frequency span or the number of points. Figure 5.42 shows the resulting response when a fixed frequency group delay aperture is applied the measurements of Figure 5.41. Here the aperture of 3 MHz was used, which matches approximately the default aperture for the 201 point trace above. Looking at the Std. Dev. numbers, one sees that they are nearly identical for all traces. Thus specifying the aperture in terms of delta frequency yields the most consistent results.

In most filters, the absolute value of the group delay is not important; only the group delay ripple matters to the communications channel. Group delay ripple can cause distortion in modulated measurements. However, in some classes of filters, the absolute group delay is critically important, and these are called, not surprisingly, "delay filters". These types of filters are often found in feed-forward amplifiers to provide a fixed delay to a signal applied to an error amplifier that is eventually combined at the output and used to cancel the distortion of the main power amplifier. These filters must be tuned for a precise delay, as well as other characteristics.

5.2.6.2 Long Delay (SAW) Filters

Some filters, particularly SAW filters, have very long delays. The delay of a filter can be approximated as one over the bandwidth, times the number of resonators. SAW filters use a transducer that can have many effective resonators, high Q, and narrow bandwidth thus

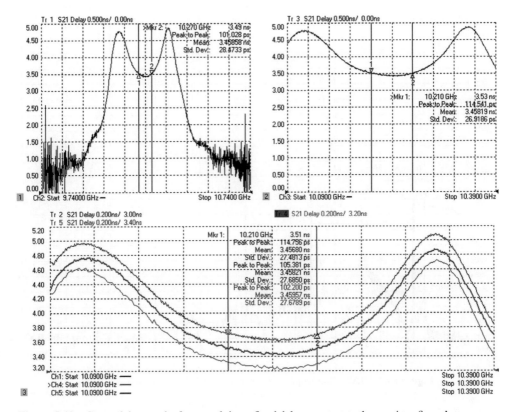

Figure 5.42   Group delay results from applying a fixed delay aperture to the previous figure's response.

producing the long delays. SAW filters are often tested in an on-wafer form, before being packaged, and the isolation response may be limited by the RF leakage that occurs directly from probe to probe. In the packaging process, the physical ends of the SAW material are treated to prevent the acoustic waves from reflecting, but before they are packaged, or if they are poorly packaged, a re-reflecting acoustic wave, called triple travel, will degrade the isolation and increase the ripple in the passband. In some instances, such as on-wafer test, it is desired to see the response of the filter in the absence of the triple travel signal. This effect can be eliminated by using time domain techniques [6].

Since the group delay of SAW filters is so long, the response from the leakage can be easily distinguished from the main transmission response. Similarly, the triple travel response will occur at a time sufficiently long that it is also easily distinguished from the main response. Thus, the response of the main lobe in the time domain may be gated to exclude the leakage and the triple travel, and yield a response of the filter that depends only upon the design of the transducers. Figure 5.43 shows the response of a SAW filter.

The upper window is the gated and ungated frequency response of the filter. It is clear that the gated response has better isolation, indicating that the transducer is well designed to provide the isolation. The ungated response shows the effect on the isolation of the RF leakage and triple travel. The lower window shows the ungated time domain response along with the

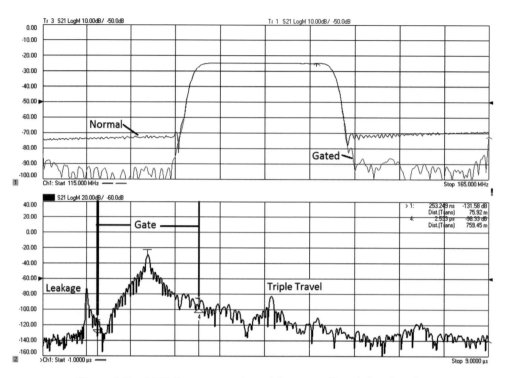

Figure 5.43   SAW filter response (upper) frequency, (lower) time domain.

region where the gate will be applied. The RF leakage and triple travel are easily identified in the time domain response.

5.2.6.3   Deviation from Linear Phase

An important transmission response related to the group delay is the deviation from linear phase. This is a more direct statement of the phase error or phase flatness, and is sometimes specified instead of group delay ripple. In fact, group delay ripple is probably a better measure of the phase flatness, as a very sharp change in phase over a small change in frequency might still pass a peak-to-peak phase ripple specification, but would cause a large peak in the group delay ripple measurements. Nonetheless, many systems specify the phase deviation as a figure of merit, and for phase deviation measurements, one must normally remove the linear slope portion of the phase response to show just the non-linear deviation.

Most VNAs provide a convenient feature, sometimes called "Marker−>Delay", which computes the electrical delay of the DUT within ±10% of the marker position, and removes that amount of linear delay from the phase response. This is usually accomplished through a scaling function called "electrical delay"; it provides a mathematical normalization of the phase response to the phase slope associated with an ideal transmission line that has the entered

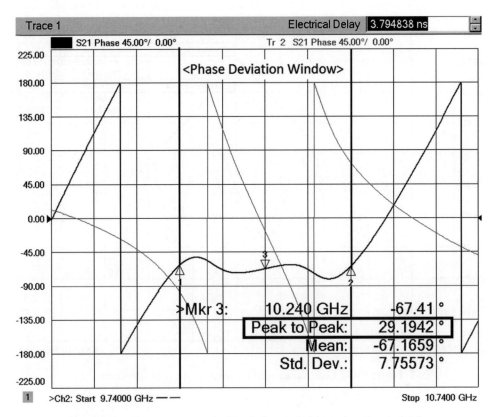

Figure 5.44 Phase response of a filter before and after setting electrical delay.

delay. The phase response of the filter from Figure 5.41 is shown in Figure 5.44, before and after applying Marker−>Delay, with the electrical delay function value shown.

The Marker−>Delay feature is convenient for providing a quick offset, but it is clear from the trace that some adjustment to the electrical delay will provide a flatter response with less peak-to-peak phase deviation, due to the downward tilt of the phase trace in the phase deviation window. Typically, this fine adjustment is performed by manually adjusting the electrical delay while monitoring the peak-to-peak phase ripple using trace statistics. The trace statistics have been limited to the narrower user range as shown by the phase deviation window, just as in Figure 5.35.

In some modern VNAs, equation editor functions are available that compute a variety of phase deviations and offsets directly. The simplest form is similar to the flatness feature of amplitude response, where a least squares fit is computed. This is very easy to compute numerically, and provides a quite good answer for the best fit line through a phase ripple trace, but not the optimum answer. It is possible to compute mathematically the best fit line minimizing the maximum peak-to-peak error (min-max), and some VNAs provide this as an additional equation editor function. The results of both the least squares fit and the min-max fit are shown in Figure 5.45.

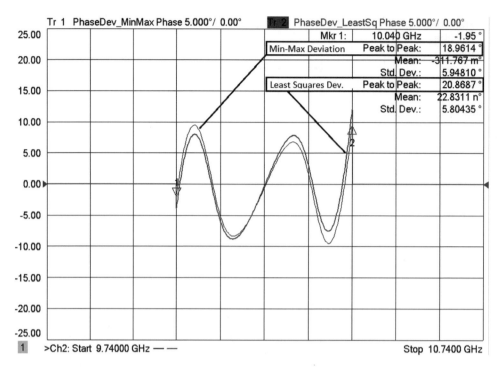

Figure 5.45   Least squares and min-max fit of a phase deviation.

The min-max fit is very computing-intensive, and will create very slow updates if the number of points is large. The least squares fit provides better computational speed, at the tradeoff of slightly less than optimum results. In this case, min-max provides two degrees lower deviation. Each one provides about ten degrees improvement on the basic marker to delay function. For both of these functions, only the portion of the phase trace inside a user specified window is displayed.

5.2.7   Calibration Considerations

In measuring filters that have both high isolation and low loss, the stimulus settings to use during calibration may be different from those that are used during measurement. One of the common misconceptions with regard to calibration is that it is presumed to be a requirement to exactly match the same stimulus conditions during calibration as during measurement. This misconception most likely came about because of an unfortunate choice of annunciators used in some of the earliest network analyzers. After calibration, it was common to put a "C" or "Cor" annunciator on the screen to indicate that the measurements had error correction applied. If *any* stimulus settings were changed, the annunciator would change to "C?", leading many users to presume that this indicated a questionable calibration. To avoid this unpleasant label, many users would choose to select settings for calibration that gave poorer results, after being applied, than would be the case if the settings were changed after calibration.

An excellent example of this is apparent in the segmented sweep settings for Figure 5.40. It is not possible to calibrate at the maximum power used in the isolation segments, because during the thru portion of the calibration, the VNA receiver would be in hard compression. An unwitting user would select a lower power to calibrate the isolation portion, just low enough to avoid compression (about -6 dBm) and then leave that power for the measurements. However, this would result in at least 10 times the noise in the isolation case, perhaps many dB of noise degradation. The other alternative, calibrating at a lower level then changing the power after calibration, would result in a few hundreths of a dB error at most in the isolation measurement, but would produce the regrettable "C?".

Since the days of the legacy VNAs, the calibration annunciators have become more informative; it is common to see C^* to indicate that the error correction array has been interpolated, and $C\Delta$ to indicate that some setting has been changed, but the underlying calibration is still valid. The "C?" annunciator is no longer used.

Another common error in calibration for filters is to set the power very high (usually in a non-segmented sweep) to ensure that isolation will have low noise without changing the power level after calibration. This can cause one of two distinct, and differing, errors in the transmission passband.

If the calibration uses mechanical standards, the thru portion might put the VNA receiver into compression, causing the receiver response to read lower than it should be for the calibration trace. If the filter to be tested has some substantial loss, then the receiver will *not* be in compression during the measurement and will read a lower insertion loss value than the filter actually has, giving an optimistic view of the filter loss. While this will allow filters to appear to pass specification more easily, it is not valid.

If the calibration uses electronic calibration modules, the loss of the calibration module may take the VNA receiver out of compression during the calibration portion, but a very low loss filter will put the receiver into compression during the measurement. This will lead to a reading during the measurement that presents a higher insertion loss than it should, causing some filters to fail their limits when their performance is actually satisfactory.

Thus, in many cases, it is best to use a moderate power for calibration and change the power afterwards, to provide the optimum tradeoff between dynamic accuracy and trace noise in the measurements. Some of the most modern VNAs have receiver linearity that is so good that there is no need to be concerned at all about varying the power after calibration, so the best power for a low-noise calibration should be chosen.

5.3 Multiport Devices

Many linear devices have more than two ports, but most discussion of measurements focus on the two-port case for convenience. In the past, the measurement of multiport linear devices proved very difficult when only two-port VNAs were available. The termination effect on the ports not connected to the VNA could have substantial effect on the measurements, and the process of changing cables to access all the combination of ports was tedious and error prone.

In the very late 1990s the first VNAs with more than two ports were introduced. These systems were four-port systems that added an additional two-port test set to the standard VNA. Soon after, the integrated four-port VNA became a standard offering, especially in the RF frequency range where differential devices were being developed. Now four port measurements systems are available up to THz frequencies, and extension test sets are available that provide full calibration capability up to more than 32 ports.

So, while the problem of multiport measurement systems has largely been solved, there are some important considerations when making measurements on multiport devices that will be discussed below.

5.3.1   Differential Cables and Lines

Differential devices have become very common in the RF and even microwave frequency ranges. Details about active devices will be discussed in later chapters, but a very common passive component is a differential transmission line or differential cable pair used to connect between differential devices. Often, these are used only for test purposes, but their careful characterization is necessary to provide accurate measurements of devices when these test lines or cables are used.

Differential lines are by their nature coupled, but the definition of a mixed-mode measurement system presumes in its definition that a four-port measurement system has essentially uncoupled ports. A simple mental test to decide if a pair of ports is uncoupled is to consider what happens if power is applied to just one of the ports that comprise the pair. If power is transmitted to the other port, then they are *not* uncoupled. A differential probe, for example, will likely have some leakage from one signal line to the other, and so is not completely uncoupled. A four-port VNA will likely have extremely high isolation between ports, and so is completely uncoupled. When ports are uncoupled, the two modes of transmission from the port pair, common mode and differential mode, are easily defined, and this is the basis for most differential measurements. The details of differential measurements are discussed in Chapter 8.

5.3.2   Couplers

Directional couplers are three- or four-port devices that are used for monitoring and signal separation. In the past, measurements of directional couplers were typically performed using two ports at a time, while loading the other ports with good terminations. This works quite well, but some special cases occur, and using a three- or four-port VNA can provide substantially more convenience. One case is when the test port of the coupler cannot be well terminated. The isolation of a directional coupler depends upon a good match at the test port, when a two-port VNA is used. However, if a three- or four-port VNA is used, multiport error correction can be applied that provides a full correction for the mismatch applied at all ports. This allows a good measurement of isolation even if a high-quality load is not available for that port. Some cases that this applies to are testing PCB or integrated circuit couplers that must use fixtures or probes for testing.

Another example where using a three-port VNA for coupler testing is useful is testing and tuning high performance couplers. Because a coupler's directivity depends upon all three of the path measurements – main-arm loss, coupling, and isolation – all three must be measured and accounted for when measuring directivity. Often, the overall isolation measured comes from a combination of the intrinsic isolation of the coupling structure and mismatch at the coupler output or test port. Normally mismatch acts to degrade the isolation but in some cases it can work to cancel some intrinsic leakage and thus slightly mismatching the coupler output can improve the overall isolation. For high-performance couplers, where the directivity approaches 40 dB, some tuning is typically required. In such a case, making all three measurements on a three- or four-VNA allows direct computation of directivity so that it can be adjusted in real time. Figure 5.46 shows an example measurement of a coupler.

In the figure, the directivity is computed in the equation editor using the formula Eq. (1.90). In the case where the test port is at port 1, with forward coupling to port 3, and the main thru arm at port 2, then

$$Directivity = \frac{Isolation}{Coupling \cdot Loss} = \frac{S_{31}}{S_{21}S_{23}} \tag{5.18}$$

An example where a four-port VNA is useful is in the case of a four-port coupler that uses an external fixed load instead of an integrated load. Often the load is attached in a semi-permanent way (typically painted over) and the coupler is treated as a three-port (see Figure 1.29). However, during the manufacturing process, or if the attached load is damaged (perhaps due to high power) then the coupler's intrinsic characteristics can be determined by a four-port measurement without using a load. Equation (5.18) still applies, but the load for

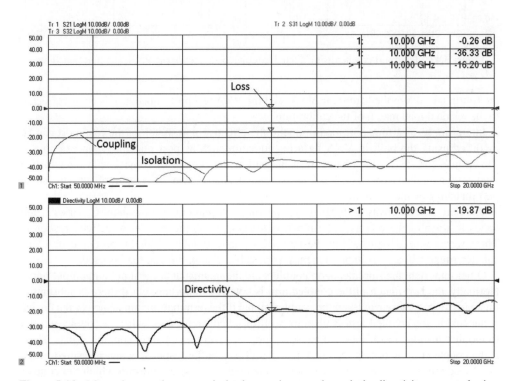

Figure 5.46 Measuring couplers: upper is the three main terms, lower is the directivity computed using an equation editor function.

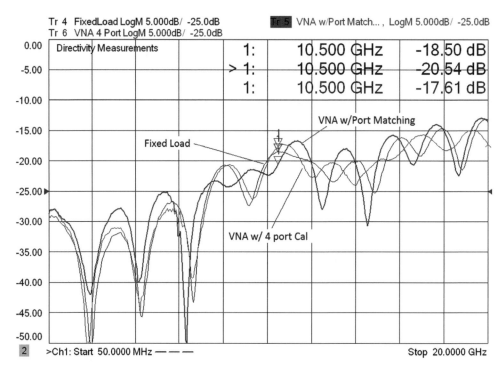

Figure 5.47   Four-port coupler using a fixed external load, a four-port VNA, and port matching.

the fourth port is supplied from the VNA, and the VNA test port impedance is error corrected to the residual load match of the calibration.

Interestingly, as a four-port measurement, the effective port match of the load port can be modified using the built-in VNA fixturing function such as port matching. An example is shown in Figure 5.47 which has a measurement of the coupler directivity with the attached load, and again with a four-port VNA measurement. Also shown is the result of applying some port matching elements to improve the directivity.

The port matching was performed by adjusting a virtual circuit of a shunt C, series L and series R, to show the effects of termination impedance on the coupler directivity response. These were adjusted to provide the best directivity up to 10 GHz. With this port matching, one can determine that a 3 dB improvement was possible if an optimized load is used.

Other important measurements on couplers are the mismatch at the input and output ports. Normally, the port match of the coupled arm is not very important, but if there is substantial mismatch, it can cause ripples in the coupled signal if the termination impedance of the device connected to the coupled arm is also not well matched. In the case where a coupler is used to monitor a signal level, this will cause a ripple error in the apparent coupled signal. Thus, if the coupled arm is poorly matched, it will still have a flat response when measured into a good load, but the flat response will not be maintained if the circuit into which the coupler is eventually terminated does not provide a good match. In contrast, if the coupled arm has a good match, the match of the receiving circuit connected to the coupled arm will not have a strong effect on the ripple seen in the coupled signal.

Ripple in a wideband coupler is to be expected as they are often designed to have some equal-ripple response if the coupling bandwidth is wider than one octave. Thus, a common specification on a coupler is peak-to-peak ripple in the coupled signal.

5.3.3 Hybrids, Splitters and Dividers

Hybrid is a common name for a circuit which provides a nearly lossless split between two signal paths, and sometimes also called 3 dB hybrids, splitters, dividers or baluns. They can be considered a special kind of coupler, where the coupling factor and loss are equal and around -3 dB. Loss in a hybrid is often specified relative to the ideal loss, so it is not uncommon to see a specification for a hybrid of -1 dB maximum loss, and this is to be interpreted as being less than 4 dB loss from the input to the output.

Hybrids come in three distinct types: $0°$, $90°$, and $180°$. Many hybrids are four-port devices which have two inputs in addition to having two outputs. One input provides an equal power, $0°$ phase split to the two output ports, and the other input provides each output with equal power and a $180°$ phase split. The $90°$ hybrids have similar versions with two inputs, one of which provides $0°$ and $90°$ to the two outputs, the other providing $0°$ and $-90°$. The $90°$ hybrids are often used to provide signals for I/Q mixers or image-reject mixers. Example illustrations of both types of hybrids are shown in Figure 5.48.

Common measurements on hybrids include input and output match, isolation between the two output ports and, most importantly, loss, and magnitude or phase imbalance of the two output ports. For the imbalance measurements, the equation editor function of VNAs can be used to directly compute the balance of the splitter. Using the equation

$$Bal = \frac{S_{21}}{S_{31}} \tag{5.19}$$

one can plot on separate traces both the magnitude balance and the phase balance. For all hybrids, the ideal balance is 0 dB (equal power). For splitters, the ideal phase is $0°$. For $180°$ hybrids, it is obviously $180°$, and of course $90°$ for $90°$ hybrids.

For four-port hybrids, the balanced can be computed for both the inverting and non-inverting ports, or for the $\pm90°$ ports. Two examples of measurements for hybrids are shown in the following figures.

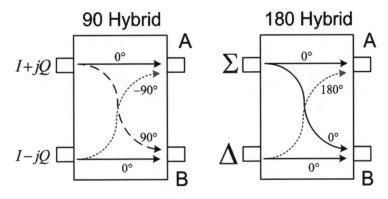

Figure 5.48 A four-port $90°$ hybrid, and a four-port $180°$ hybrid.

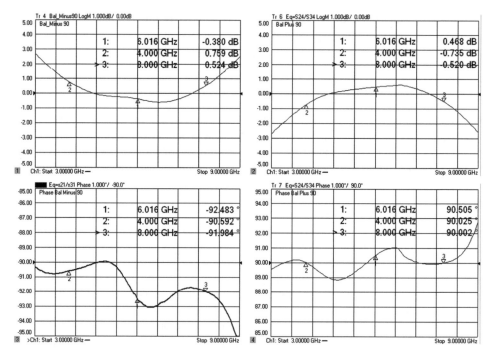

Figure 5.49   Response of a 90° four-port hybrid.

5.3.3.1   Ninety-Degree Hybrids

Figure 5.49 shows the response of a four-port −90° hybrid. The upper window shows the individual magnitude balance, and the lower windows show the phase balance. The ±90° responses are generated by driving different ports on the input of the hybrid (port 1 for −90°, port 4 for +90°), so different equations are used based on the source port, to show the relative phase. In this case, as is most common, the hybrid has frequency limitations on the order of one octave, ranging from 4 to 8 GHz.

5.3.3.2   Balanced Hybrids

Often, balanced hybrids are used to drive differential devices. In such a case their measurement is typically reported as mixed mode S-parameters. An ideal balance-to-unbalanced hybrid will have two key parameters: common mode gain from the summing or sigma port (Scs21) and differential mode gain from the differential or delta port (Sds21). Even though it is a four-port device, when used as a balun it is presumed to be used to drive either a differential or common mode signal, so it is modeled as one of two different three-port devices. Hybrids are used for this purpose as they allow matching of both the common mode and the differential mode simultaneously. Transformer type baluns do provide for a differential drive, but the common mode impedance at the balanced port is usually either infinite (for a four-port simple transformer) or zero (for a center tapped transformer). For more details on differential and mixed mode behavior, see Chapter 8.

5.3.3.3 Splitters and Dividers

Splitters and power dividers come in lossy or lossless versions. Lossy versions split the signal while maintaining good match on all ports by use of resistors. The two main lossy splitters are three-resistor splitters (sometimes called power dividers), used when the power is simply divided to go to two different loads, and two-resistor splitters, used when the ratio of the two outputs paths are used, as in the reference arm of a network analyzer.

Lossless splitters use many different structures to divide the power between two paths. One type of lossless splitter is a 3 dB coupler, which in fact has a fourth port that is the internal load of the coupler.

Figure 5.50 represents the layout of a different kind of splitter; this particular splitter uses a Wilkinson structure [7], and has a "hidden" fourth port (a resistor of value $2Z_0$) that provides for a matched condition and only absorbs power when the two output lines are not equally terminated, or as a power combiner, when lines are not driven in an balanced way.

Figure 5.51 shows the frequency response of this splitter. The line length sets the center frequency of the splitter. While the isolation and match are degraded at low frequencies, the loss is still relatively flat. An equation editor function incorporates Eq. (5.19) for each trace in the lower window, with the format set to logmag on one trace and phase on the other.

If the device is driven in the forward direction to isolated test ports, the balance is directly discerned. The definition of balance in this case is

$$Bal = \frac{S_{21}}{S_{31}} \tag{5.20}$$

However, in the case where the splitter has an unequal drive into the split ports, some signal will be absorbed inside the internal load of the splitter. Thus, this style of hybrid is not ideally lossless, and in fact it is to be expected that non-matched impedances on the split ports will result in reflections into the splitter that will be absorbed by internal load. It can be shown that for any three-port network, it is not possible to have the network be lossless and matched at each port. This type of splitter/combiner demonstrates that, with certain drives or loads, it does absorb power in its internal load and is thus not lossless. But in normal operation it is

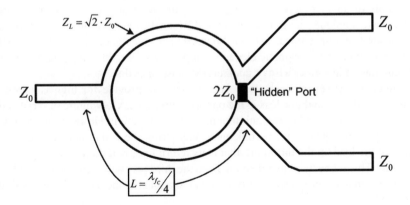

Figure 5.50   Typical form of a Wilkinson power splitter.

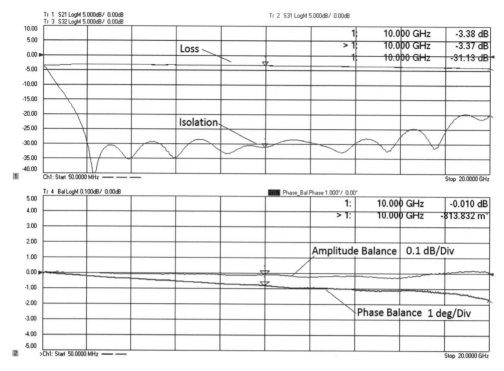

Figure 5.51   Response of a splitter.

nearly lossless. Also, the amplitude and phase balance is very good, with less than 0.1 dB of amplitude imbalance and less than 1° of phase imbalance.

5.3.4   Circulators and Isolators

Circulators have much in common with directional couplers, in that the transmission response from one port to another depends strongly upon the load or return loss on a third port. Circulators have a unique role in microwave and RF as they are perhaps the only component which is at the same time linear and *not* bilateral. The circulation function is a result of a non-uniform response of a magnetic element that provides for a differing phase delay when a signal is circulating in a clockwise or counterclockwise direction.

Most circulators cover approximately one octave, and provide for high isolation in one direction of signal flow, and low loss in the opposite direction. A key figure of merit is the loss in the low-loss direction and the isolation in the reverse direction. However, using a two-port network analyzer presents similar problems to measuring the isolation of a coupler, in that the quality of the load on the third port, that is its return loss, will directly affect the isolation measurement. This is because any reflection from the third port will continue to circulate in the low-loss direction and cause a leakage-type signal to appear at the isolated port, as illustrated in Figure 5.52.

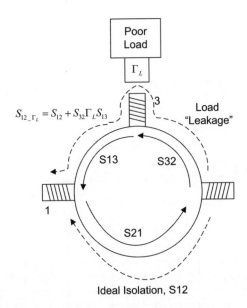

Poor
Load

Γ_L

$S_{12\_\Gamma_L} = S_{12} + S_{32}\Gamma_L S_{13}$

3

Load
"Leakage"

S13   S32

1

S21

Ideal Isolation, S12

Figure 5.52   Isolator behavior in the presence of a non-ideal load.

The measurements of an example isolator are shown in Figure 5.53. The upper window shows the isolation measurements, and the lower window shows the insertion loss measurements. For comparison, the isolation measurement of S21 is also shown in the presence of a non-ideal load applied to port 3, in the left upper window. In this case the non-ideal load presents approximately a −32 dB return loss. This is a narrowband high-isolation circulator, and, with the full three-port calibration, the true isolation of the S21 path is shown to be well over 40 dB at the center frequency.

For a high isolation circulator, a poor load can cause a substantial degradation of the isolation measurement. If the isolation approaches the return loss of the load, the worst-case signal can be as much as 6 dB higher than the actual isolation, as shown in the trace marked "Isolation w/-32 dB RL load". As a reference, when combining a large signal with a small error signal, if error signal is 19 dB below the other, it will cause a 1 dB error in the response measured for the larger signal. These errors become important when looking at the isolation response of circulators.

Finally, it is apparent that this isolator has a very small mode (on the order of 0.1 dB) around 3.448 GHz (marker 2). There is evidence of this mode also in the isolation paths.

5.4   Resonators

5.4.1   Resonator Responses on a Smith Chart

One key measurement made with Smith charts is the measurement of resonators to find the Q and center frequency. For one-port resonators, the procedure to find the Q and center frequency is relatively simple. Consider the circuit of Figure 5.54. The key figures of merit for

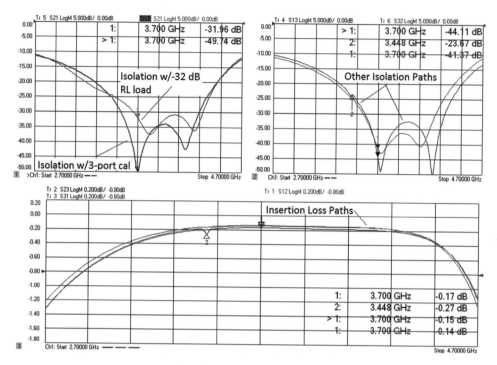

Figure 5.53 Measurements of an isolator.

the resonator are f_0, the center frequency, and Q_0, the unloaded Q. The center frequency and Q can be found from

$$f_0 = \frac{1}{2\pi\sqrt{LC}}, \quad Q_0 = 2\pi f_0 R_0 C \tag{5.21}$$

The common measurement of Q is to look for the 3 dB loss points of a transmission (S21) response, relative to the center frequency, where the loaded Q is defined as

$$Q_L = \frac{f_0}{f_2 - f_1} \tag{5.22}$$

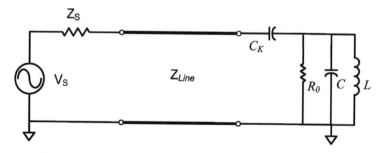

Figure 5.54 Schematic of a one-port resonator with coupling capacitance.

where f_2 and f_1 are the frequencies of the lower and upper 3 dB down points. Loaded Q means that the circuit is loaded by the external resistance of the VNA source, typically 50 ohms. The usual figure of merit for a resonator is unloaded Q, or Q_0.

In a one-port measurement of Q, the measurement is S11, so the normal definition doesn't apply. When a high Q resonator is measured with a one-port S11 measurement as in Figure 5.54, there may not be any point below 3 dB so the concept of looking for a magnitude response fails to find a Q value. When viewed on a log magnitude return loss plot, the reflection from a high Q circuit is hard to see, as the return loss is very nearly 1 for all frequencies. Therefore, it is common to add a coupling structure, often a very small capacitance, C_K which transforms the impedance of the lossy elements to match the impedance of the test system. Figure 5.55 shows the return loss of a relatively low Q resonator with a direct connection and with a coupling capacitance added to match the circuit to Z_0. If the resonator was of a higher Q, the direct connection trace would show almost no change in return loss and would appear as a flat line of S11 nearly equal to 0 dB return loss.

When plotted on a Smith chart with a direct connection, the resonator forms a perfect circle and crosses the real axis at f_0. But in real world measurements, there is almost always some external transmission line that shifts the response of the resonator as shown in Figure 5.56. Also shown is the same resonator with the same value of coupling capacitance as used in Figure 5.55. From these Smith chart plots, the Q factor can be directly computed [8].

The center frequency is found by looking for the point where the trace crosses over itself at the outside of the Smith chart. Drawing a line from this point through the center of the Smith chart marks the position on the trace trajectory that represents the resonant frequency, f_0. From

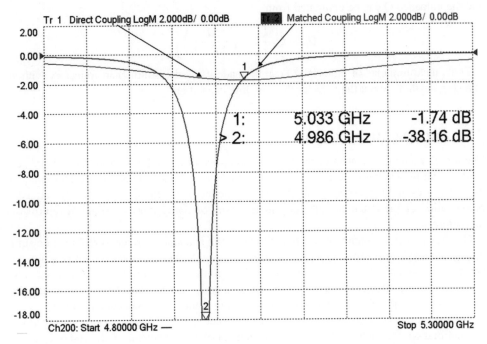

Figure 5.55   Return loss plot of a resonator with direct coupling and with matched coupling.

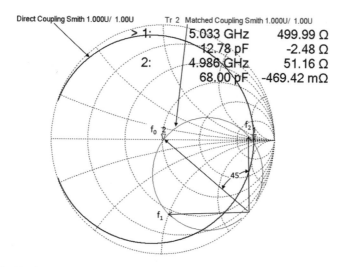

Figure 5.56 Smith chart plot of a directly connected resonator and one matched to Z_0 using a coupling capacitance; the Q is determined from f_1 and f_2, taken as $45°$ from the diagonal chord to f_0.

this line, a position is marked at $\pm45°$ angles and a line drawn from the crossing point, at these angles, until it crosses the Smith chart trace at frequencies f_1 and f_2. One can use these frequencies to measure the loaded Q of the circuit as in Eq. (5.22). The diameter, d, of the circle formed on the Smith chart is a measure of difference between loaded and unloaded Q, and the unloaded Q can be computed from

$$Q_0 = Q_L \left[1 + \left(\frac{d}{2d - 1} \right) \right] \tag{5.23}$$

Thus, measurements of unloaded Q can be made using the Smith chart trajectory without knowing anything about the value of the coupling capacitance.

5.5 Antenna Measurements

The topic of antenna measurements could encompass an entire book, from the area of near field and far field patterns, to phased array measurements to antenna efficiency. However, for many RF uses, the key aspect in measuring an antenna is determining the correct frequency of operation of the antenna, and the principal way of doing this is through a return loss measurement.

One key aspect of antenna measurement is that the antenna to be measured must be sufficiently removed from any conducting surfaces that might reflect energy back into the antenna and cause the apparent impedance to change. The ground conductor of the test cable can also act to change the effective radiation of the antenna and thus care must be used to ensure that an effective ground plane is used which emulates the mounting of the antenna in use.

The measurement of an antenna return loss is rather straightforward, requiring only a one-port calibration. Figure 5.57 shows the measurement of a whip antenna at the end of a coaxial

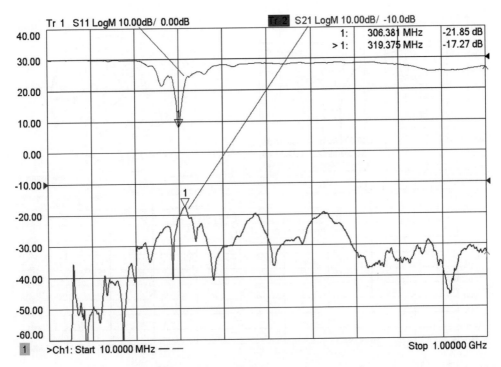

Figure 5.57 Measurement of an antenna return loss.

cable. This one is suitable for use at about 300 MHz. Also shown is a transmission measurement where port 2 is just an small pickup coil loop some distance away. It is interesting to note that the S21 peak is just off frequency from the antenna's apparent center frequency.

One source of uncertainty on the antenna frequency is caused by the directivity of the directional coupler. While normally a directivity error is assumed to cause the return loss to be larger than actual, in fact it can add to or subtract from the value of the return loss. In the case of an antenna measurement, the return loss causes a characteristic null at the tuned frequency of the antenna, but this only happens because that is the point at which the antenna's impedance matches that of the VNAs reference impedance. If there is a directivity error, or if the antenna's effective impedance is different from Z_0, the effective impedance of the VNA will be either lower or higher than the antenna as the frequency sweeps, and the null from the antenna will subsequently move lower or higher in impedance, as the impedance of an antenna is not fixed, but varies as the frequency varies. Thus, the apparent frequency of an antenna depends entirely on the directivity, or effective impedance of the VNA. Since the effective impedance of a VNA is set by the calibration, errors in calibration will lead to errors in the apparent tuned frequency of an antenna, as illustrated in Figure 5.58. Also, if the characteristic impedance of the antenna is not 50 ohms, it will cause an apparent shift in the S11 null of the antenna.

In the figure, the lower plot shows the real and imaginary parts of the antenna impedance. If one chooses the reference impedance to equal the point where the imaginary part crosses zero,

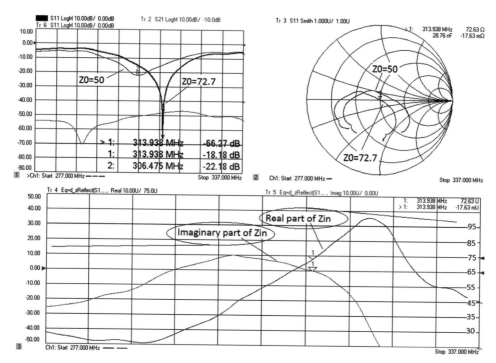

Figure 5.58 Change in apparent tuned frequency due to directivity errors or change of reference impedance.

then the real part has a value of 72.7 ohms. The upper plots show the antenna response for the both $Z_0 = 50$ ohms and $Z_0 = 72.7$ ohms. Interestingly the shift in the S11 null corresponds better to the peak in the S21 response seen in the upper left window.

5.6   Conclusions

While many aspects of testing linear devices have been discussed here, no one chapter can possibly cover every aspect of every linear device. However, attending to some key guidelines as outlined in many of these sections should allow extending the specific techniques described to almost any other situation.

A practice important enough to be re-emphasized here is the practice of pre-measuring a device before starting any calibration or serious measurements. Almost all modern VNAs have a built-in factory calibration that provides approximately correct results for most measurements. In pre-measuring a DUT, countless hours of wasted calibrations and measurements will be saved. During the pre-measurement, cable connections can be tested for stability, the insertion loss measurements can be evaluated for trace noise and appropriate averaging or IF bandwidths can be set, delay measurement aperture effects can be evaluated, and even unnecessary measurements on damaged or incorrect parts can be avoided.

The basic principle of measuring linear devices is to understand the system stability and noise, understand the best calibration to be applied, and understand the interactions between the measurement system and the DUT. This chapter provides the foundation for such an understanding.

References

1. User's Guide: Agilent Technologies 8753ET and 8753ES Network Analyzers, June 2002, available at http://cp .literature.agilent.com/litweb/pdf/08753-90472.pdf.
2. Dunsmore, J.P. (2007) Transmission Response Measurement System and Method of Using Time Gating, Agilent Technologies assignee, Patent No. 7170297, Jan. 30, 2007.
3. ANSI/SCTE 05 2008, Test Method for "F" Connector Return Loss In-Line Pair, available at http://www .scte.org/documents/pdf/Standards/ANSI_SCTE%2005%202008.pdf.
4. Rowell, J., Dunsmore, J. and Brabetz, L. Cable Impedance and Structural Return Loss Measurement Methodologies, Hewlett-Packard white-paper, available at http://na.tm.agilent.com/8720/applicat/srlpaper.pdf.
5. ANSI/SCTE 66 2008: Test Method For Coaxial Cable Impedance, available at http://www.scte.org/documents/ pdf/standards/ANSI_SCTE%2066%202008.pdf.
6. Agilent Time Domain Analysis Using a Network Analyzer Application Note 1287-12, available at http://cp .literature.agilent.com/litweb/pdf/5989-5723EN.pdf.
7. Wilkinson, E.J. (1960) An N-way hybrid power divider. *IRE Transactions on Microwave Theory and Techniques*, **8**(1), 116–118.
8. Kajfez, D. (1994) *Q Factor*, Vector Forum, Oxford, MS.

6

Measuring Amplifiers

When considering active device measurements, most engineers naturally think of testing amplifiers. These devices form the heart of any communications, radar or satellite transponder system, and are required in almost every other RF or microwave system. Amplifiers are often considered in the system design as a unilateral gain block used to boost signals, and for the inexperienced system designer, that is often as far as the amplifier characteristics are considered. Unfortunately, RF and microwave amplifiers have complicated and subtle behaviors that challenge such a simple view.

Amplifiers can be roughly split into two categories: low-noise amplifiers used in receiver applications, and power amplifiers used in transmitter applications. While amplifiers for each application might share many attributes and require many of the same measurements, the attributes of the tests and the relative importance of various factors of the test system vary greatly depending upon the particular requirements for each amplifier and application. A power amplifier used to drive the probe of a nuclear magnetic resonance (NMR) imager has very different requirement from an ultra-low-noise cryogenically cooled amplifier used in a space-based radio telescope. However, both have many similar issues including such things as stability, compression, power consumption and distortion. This chapter will explore most of the attributes of amplifier measurements important to an RF or microwave designer or test engineer.

6.1 Amplifiers as Linear Devices

The most basic attributes of an amplifier are often considered as purely linear, thus the fundamental description of most amplifiers is the S-parameter matrix of the amplifier. However, amplifiers differ from passive linear circuits in that the power scattered from an amplifier will always be greater than the power incident on the amplifier. The basic definition to call a circuit an amplifier is that the RF power out of the device is greater than the RF power incident on the device, or

$$\sum |b_n|^2 > \sum |a_n|^2 \qquad (6.1)$$

Handbook of Microwave Component Measurements: With Advanced VNA Techniques, First Edition. Joel P. Dunsmore.
© 2012 John Wiley & Sons, Ltd. Published 2012 by John Wiley & Sons, Ltd.

When applied to the forward signal of a two-port device, this can be interpreted as

$$|S_{21}|^2 > 1 - |S_{11}|^2 \tag{6.2}$$

For amplifiers in the linear mode of operation, the incident and scattered waves are defined exactly the same as for passive devices, described by Eq. (1.17) and repeated here

$$
\begin{aligned}
b_1 &= S_{11}a_1 + S_{12}a_2 \\
b_2 &= S_{21}a_1 + S_{22}a_2
\end{aligned}
\tag{6.3}
$$

The principal measurement performed on an amplifier is the small signal gain measurement, followed by the input match, output match and isolation measurements. These are traditionally made in the same manner as for linear passive devices, stimulating the DUT one port at a time, measuring the resulting waves, and computing the S-parameter error-corrected results. But this method relies on the DUT being a linear device, so one might ask how to define the linear mode of operation.

6.1.1 Pretesting an Amplifier

For most simple devices, the linear mode for amplifiers as applied to Eq. (6.3) occurs at approximately 20 dB below the 1 dB compression point: there is approximately 0.1 dB compression at 10 dB below the 1 dB compression-point, and 0.01 dB compression at 20 dB below. At such a level, other practical limitations such as drift – of both the amplifier and the test system – render any distinction between linear and non-linear meaningless as far as computation of the S-parameters is concerned.

Figure 6.1 shows the relative gain of two different amplifiers for a power sweep at a fixed frequency, with compression markers shown at 1 dB (marker 1), 0.1 dB (marker 2) and 0.01 dB (marker 3). One amplifier is a normal class A amplifier operating at a lower power, which follows the rule that compression drops 10 times (in dB) for every 10 dB back-off from the 1 dB compression point. From this, it is clear that 20 dB back-off is a good estimate of the linear range in an amplifier.

The other amplifier is a high-linearity amplifier with about a 10 dBm higher compression point, designed so that the bias changes at higher powers to increase the gain, thus creating a more linear transfer curve. As is common in these types of compensated amplifiers, there is some expansion before compression occurs. Even here, the clearly linear region is about 20 dB below the 1 dB compression point. The x-axis of the plot is different for each amplifier; both are scaled at 4 dB/division, and the stop power is set just above the 1 dB compression point. By placing both on the same grid, it is easy to see the difference in compression curvature between the two designs of amplifier.

Before measuring the S-parameters of an amplifier in earnest, it is a good practice to try to ascertain some of characteristics such as its linear operating point first. This may be simply done by starting at a low power and measuring raw (uncorrected) S21. If the amplifier is to be tested over some narrow frequency range, it is wise to extend this range to ensure that there are no frequencies at which the amplifier is unstable under the source and load conditions of the VNA being used. Unstable operation will show itself as a large gain peak at some frequency, or possibly even a large spike in the normal gain plot if the device happens to be oscillating.

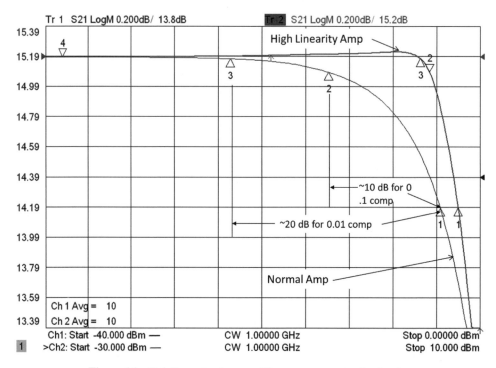

Figure 6.1   Relative gain of an amplifier versus compression level.

These investigations can be done with wide bandwidths, no calibration and large number of points to quickly detect problems.

Once it is confirmed that the device is not actively oscillating when being measured, then the linear operating region may be determined. For this pretest, the VNA source power is set to a low level where it is expected that the amplifier must be linear, and the S21 trace is placed into memory and the data over memory traces are displayed. The VNA source power is then increased, while a marker is used to track the minimum point on the trace. It is also useful to have a second trace of output power (typically using the b2 receiver) with a marker tracking the maximum output power level, monitoring to ensure that the power into the VNA receiver does not drive it out of its operating range. A third trace of input power may be added, and provides a convenient trace of drive power that can be displayed using the trace markers. Continue raising the source power until a clear offset in the S21 trace occurs. The most common figure of merit is 1 dB compression, but in some cases it will not be possible to achieve 1 dB compression from the level of source drive available from the VNA. Figure 6.2 shows, in the upper plot, a trace of the wideband measurement of the S21 of an amplifier, at low power (for linear response, marked "S21_Lin") and at high power (for compressed response, marked "S21_Comp"). In the lower plot is shown the results of looking at the input and output power, as well as the compression. The compression trace is the high power gain divided by the memory trace, which was saved at a linear power. All these measurements are before a correction and so rely on the VNA factory cal.

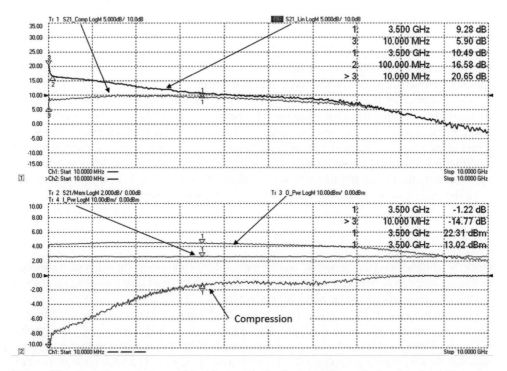

Figure 6.2 Amplifier pretest: a wideband sweep looks for instability, and changing the drive power looks for compression.

One object of particular interest in this figure is the gain peaking in the linear trace (upper plot) at the very low frequency range, around marker 3. This shows that the gain peaks 4 dB in only 90 MHz, likely indicating some feedback mechanism which is limited in low frequency response, probably by an AC capacitor. This wideband plot shows that the low frequency gain may be an issue if the low frequency matching is not properly considered.

6.1.2 Optimizing VNA Settings for Calibration

If investigations into the compression point of the amplifier are anticipated, now is the time to adjust source and receiver attenuators in the VNA to ensure sufficient power to drive the amplifier fully into compression (more than 1 dB compressed at all the operating frequencies) and to ensure that the VNA receiver is not compressed by monitoring the VNA output receiver's measured maximum value and ensuring that the receiver attenuation is set sufficiently to keep the power into the receiver below the top 5–10 dB of the VNA maximum operating power. In some VNAs, the receiver has automatic over-range detection and will post a message if the VNA receiver is overloaded. For high-power amplifier testing, where the overload might exceed the damage level of the VNA, the overload detection may be set to shut off the source power if overload conditions are detected. In Figure 6.2, the maximum power from the amplifier is about +23 dBm, and with an expected gain of about 10 (from the raw S21), the maximum source power is about $+13$ dBm, meaning that the 0 dB source attenuation state must be used. The maximum operating power for the VNA receiver is about $+13$ dBm, and so to ensure that the test power is at least 5 dB below this value, 15 dB of receiver attenuation is required. For the amplifier shown, the source power for linear operation is about -10 dBm, approximately 20 dB below the compression point power.

While monitoring the S21 trace at the linear drive power, the IF bandwidth should be adjusted to ensure that the trace noise over the desired range of the amplifier is sufficiently small. Because the source power control of the VNA occurs before the reference coupler, the power in the reference channel might be at the low end of the receiver's normal range when the amplifier is in the linear operating region. At higher source powers, the trace noise reduces at approximately the rate of 10 times for each 10 dB increase in power. So near the compression point the trace noise will be about 100 times smaller than at the linear operating region.

6.1.3   Calibration for Amplifier Measurements

Once the amplifier has been pretested, and the VNA settings optimized, the calibration methods and acquisition can be performed to produce a calibrated instrument state providing the highest quality of measurements for the amplifier under test.

For most moderate gain amplifiers (gain below 20 dB), the calibration does not require any special care. Before calibration begins, it is wise to look at the S21 response of a thru, or the thru state of the Ecal if it is to be used, and determine if the IF BW setting is such that the trace noise on the thru state is acceptable for the measurement results. The trace noise of the thru state will be embedded in the response calibration, and measurement results of the amplifier will never be more smooth than this trace. For this reason, it is common to reduce the IF BW even more than needed for good amplifier measurements to obtain good calibration results (one should always use stepped-sweep mode if changing the IF BW, as discussed in Chapter 5). Perhaps a better alternative is to add averaging during the calibration. With averaging applied, the trace noise is reduced and also some other, subtler, dynamic effects may be reduced as well. Trace deviation (drift of the S21 trace) can occur for many reasons such as non-coherent spurs in the source or receiver, external noise, environmental drift or dynamic effects in the Ecal. Using trace averaging to reduce these effects during calibration may be more effective than reducing IF BW as the effects may come and go at a rate sufficiently slow that the sweep-to-sweep averaging will have a good effect where the one-shot IF BW measurement will not.

6.1.3.1   Source Power Calibrations

If power measurements of the amplifier are desired, it is best to do the power calibration before the S-parameter calibration. Almost all VNAs support some form of source power calibration. The details of source power calibration are described in Section 3.6. For older, legacy VNAs, the source power calibration is an independent function, which creates a source output offset table that would be applied to the factory source power setting to provide the correct power at the power calibration plane. A power meter is used to measure the power for each frequency, and the source power is normally iterated until the target power is reached within some tolerance. This source power setting is valid for power changes around the calibration region, but if the power is changed too much (more than perhaps 10 dB), source linearity errors can cause significant errors in the drive power. For this reason, it is usually wise to perform a reference receiver calibration immediately after the source power calibration, as the reference receiver can be used to continuously monitor, and even level, the drive power during measurements. This will be discussed in detail in Section 6.1.4.1.

6.1.3.2 Receiver Power Calibrations

The reference receiver calibration can be simply a response calibration that is acquired on the reference receiver, with the source power calibration applied. In older VNAs, the response calibration presumed 0 dB for the reference point, so it was required that the source power calibration be performed at 0 dBm. Most modern VNAs have explicit receiver power calibrations that allow any power for the reference power and automatically use the last source power calibration value. For the reference receiver cal, it is best to leave the power meter in place to ensure the same match condition for the receiver calibration as the source calibration.

The VNA output receiver may be calibrated for power at this time, by providing a thru connection between port 1 and port 2, and acquiring a receiver response cal for the output receiver, typically the b2 receiver. The details of receiver power calibration are discussed in Section 3.7. A simple response calibration method does not remove mismatch effects from the receiver response calibration, so that if the VNA measurement system does not have good source or load match, the response calibration can have significant error. For example, consider a system with 15 dB source and load match at some frequencies (not untypical for a VNA with test port cables), the receiver response calibration would be in error by approximately 0.27 dB. If the DUT also had a 15 dB return loss, the receiver power measurement error could be as much as 0.52 dB!

6.1.3.3 Advanced Techniques and Enhanced Power Calibration

Sections 3.6 and 3.7 discuss some advanced power calibration techniques that greatly improve the source power calibration and the receiver response calibration. The basic idea is to make return loss measurements during the power meter acquisition, so that mismatch effects can be removed from the source power settings. The reference receiver or a1 power calibration uses the power meter mismatch and knowledge of the source match to remove the mismatch effects from the reference receiver calibration. The step of calibrating the output or b2 receiver is not performed explicitly. Instead, the b2 receiver response calibration is computed by combining the a1 calibration with the S21 tracking from the full two-port calibration, yielding a match-corrected b2 receiver response calibration. Given the VNA scenario described above, with 15 dB source and load match, a typical receiver response error would be less than 0.03 dB using full match correction.

These advanced techniques can be used on legacy VNAs by using the details of Section 3.7 to apply the mathematical corrections to error terms acquired during the power calibration and S-parameter calibration using remote programming commands, and downloading the corrected receiver response arrays to the VNA calibration memory.

The chief difficulty is in situations where the power meter cannot be connected directly to the VNA, such as in the case of an on-wafer or in-fixture measurement.

Fortunately, some of the most modern VNAs now support a guided power calibration as part of the normal calibration engine. As part of these advanced calibrations, even the situations where the power meter cannot be directly connected to the port are nicely handled by using an additional calibration step. If the power meter cannot be connected to the measurement port, such as an on-wafer probe, or in the case where only a coaxial power meter is available for calibrating a waveguide system, the user is prompted to add (or remove) an adapter to allow the power meter to be connected. In the case of an on-wafer measurement, the coax connection from the VNA to the probe might be removed and the power meter connected directly to the VNA. In the waveguide case, a waveguide-to-coax adapter might be added to the test port and the power meter connected to the adapter. After the source power calibration, the user is prompted to perform a one-port calibration at the power meter measurement plane using the coax standards appropriate for that power meter. From the information of this one-port cal, and a later two-port cal at the measurement plane, the adapter's full characteristics are determined and removed from both the source power calibration and the receiver calibration.

It should be noted that it does not matter at all if an adapter is added, or an adapter is removed, the mathematics works exactly the same. In the case of an added adapter, the de-embedding network computed has loss, as would be expected. In the case of removing the adapter, the de-embedding network has gain. In fact, one adapter may be removed and another added, for example, in an on-wafer probing situation, an RF switch may be added to the cable path from the VNA to the wafer prober, with the switched port set to apply the power meter and corresponding one-port calibration. The adapter from the probe to the switch is effectively removed and any cable from the switch to the power meter is added, but the de-embedding network computed for transferring the power cal becomes the difference (in a full S-parameter matrix manner, including mismatch) between the two paths.

With these recent improvements in power calibration techniques, the enhanced power calibration should be considered the new gold-standard in power measurements. In fact, one can demonstrate that except for the case of a DUT having both a nearly ideal output impedance and an output power near 0 dBm, the VNA measured power using these calibration methods is more accurate than a power meter, in measuring output power. And if the DUT has a high harmonic content, the error in power meter measurements versus VNA measurements may even be greater.

6.1.3.4 S-Parameter Calibrations for Amplifiers

After the power calibration, or as part of a guided power calibration, the S-parameter calibration is performed. It is important in legacy VNAs to perform the power calibration first to avoid a "C?" or "CΔ" indicator as the source power calibration level is one of the stimulus attributes that signal the indicator change. In fact, for some VNAs, if the S-parameter calibration is performed first, then the source power cal and then the receiver calibration, the stimulus difference between the S-parameter cal and the receiver cal will cause one of them to be marked invalid in the cal arrays.

If the amplifier is only to be measured in its linear region, there is no real need to perform a source power cal as most VNAs have built-in factory calibrations on the source that give a reasonably flat and correct power. Some older VNA systems, such as the HP 8510 or Wiltron 360, used external sources so that the power at the test port was substantially different from the source power settings. Care is needed in such a case to ensure that the proper power is set at the test port.

S-parameter calibration for amplifier test is relatively straightforward. Any of the techniques such as Ecal, SOLT, TRL or unknown thru may be used. In the case of Ecal, care should be taken that the drive levels for the Ecal don't exceed its recommended levels, which might occur in the case of a high power amplifier, where the drive power is set quite high. This is even a concern for the SOLT calibration where high power might cause heating in the load element; more details on calibration issues for high power cases are discussed in Section 6.4.

One method to reduce noise while calibrating is to increase the source power during calibration by approximately the gain of the amplifier. It was the common wisdom of users of older VNAs that the source power should not be changed after calibration, or a questionable cal ("C?" indicator) would appear. Common, but wrong. The "C?" indicator showed that a change in power had been made, but in most cases, the measurements that resulted would be more accurate than ones made at lower power levels.

To understand the tradeoffs, consider the two competing effects that affect measurement accuracy with changing power levels. First is dynamic accuracy, which is another way of stating the ability of a receiver to measure correctly a change in power level. If the calibration is performed at a higher power level, and the source is then set to a lower level, the power in the reference receiver has changed and so it will have a dynamic accuracy error. However, in the case of an amplifier measurement, the b2 or output receiver will have the same power as calibration, and so it will have no dynamic accuracy error. On the other hand, if the calibration is performed at the lower level, the b2 receiver will have a change in power during the measurement due to the amplifier gain and see a dynamic accuracy error. Further, most VNAs have more attenuation in the reference receivers than the test receivers, so the test receivers typically show more dynamic accuracy errors than do the reference receivers. Thus, the first tradeoff is no tradeoff at all: the measurements will always be more accurate on an amplifier if the source power is set near the expected output power of the amplifier, provided that it does not exceed the power handling capabilities of the calibration standards.

The second tradeoff is also no tradeoff: higher source power during calibration will yield lower noise during calibration; period.

Unfortunately, most VNAs still report at least a "CΔ" indication, which concerns many users if they see it after calibration. On older VNAs this can be overcome by starting the calibration with the low source power settings, which will be saved as the stimulus value, then raising the power level or adding averaging and resuming the calibration. One caution with calibrating at a higher power: after the power is lowered for measurement, if the calibration is reapplied (turned off and back on again) it can sometimes reset to the original calibration power. To avoid overdrive in high-power situations, it is often wise to turn off the RF source power prior to making any changes, and only turn it on again when one has confirmed that the proper settings have been made.

Finally on some VNAs, there is a remote command that allows one to simply copy the calibration files to a new calset that has the current stimulus settings applied. This allows the new calset to be used with whatever stimulus changes the user desires (source power, IF BW, averaging, even interpolation, fixturing or de-embedding) as though there were no change. This is sometimes called "flattening the calibration" as the effects of fixturing and de-embedding are notionally stacked on top of the original calibration, and the fixture math "pushes" them down into the calibration array. Creating a new calibration array with these attributes embedded has, in a sense, flattened the calibration and fixturing state. In fact, at least one VNA vendor even uses the word "Flatten" in the remote command. This has the added benefit of embedding the new stimulus in the calset, and avoids the unfortunate situation of turning on a calibration and having the high power used for calibration destroy one's amplifier.

Calibrations should be done at the densest point spacing expected for any future measurements. If measuring a wideband amplifier, and one anticipates some need to zoom in to inspect a small region, the calibration point spacing should be sufficient to allow good measurements in any region. Interpolation of the calibration can be used if the point spacing is sufficiently dense to allow an accurate acquisition of the VNA raw response. As a rule of thumb, the point density should be greater 200 points per GHz per meter of test port cable with a minimum of 200 points per GHz span. For example, a 6 GHz span with normal (less than 1 m long) test port cables should have 1200 points in order to use interpolation. At this point spacing, there is about a 12° phase shift between points, so that circular interpolation generally works well and the error from interpolation is on the same order as the uncertainty from the calibration standards. One caution about interpolation is that some VNAs change characteristics over different bands, and the change at the band edge in the raw response is abrupt. Many VNAs have factory calibrations to remove most of this effect, but it can still be seen when using interpolation, as small discrete errors.

6.1.4   Amplifier Measurements

Once the calibration is complete, the basic S-parameter measurements may be directly made. For simple testing, this is all that is required. More complicated testing might include power measurements, compression measurements, voltage and current measurements to compute efficiency, as well as other changes in test conditions. Further analysis may be performed on a combination of these measurements, which is discussed more fully in Section 6.1.5. Some particular details of measurements are discussed below.

6.1.4.1   S-Parameters, Gain and Return Loss, Input Power and Output Power

The S-parameters are the most common measurement of amplifiers with S21 or gain being the principal defining feature of an amplifier. For the most part, all four S-parameters are important for further analysis, and on most VNAs, all four are measured and saved as part of the full correction. It's convenient to create multiple windows to group various amplifier measurements. For example, S11 and S22 are often displayed on the Smith chart, while S21 and S12 are displayed principally in a LogMag format. It may be convenient to show the gain and isolation on separate grids as the gain is often viewed at small scale (e.g., 1 dB/div) to look for ripple, but the isolation is typically viewed at 10 dB/div or more.

When measuring power, it is good practice to show both the measured input power and output power. The source power setting should not be relied upon for giving the input power as it is not typically corrected for mismatch effects. The output power should ideally be the input power times the gain, and so should have the same ripple as the gain, but error in the input power setting will cause additional ripple in the output power. Figure 6.3 shows a typical measurement configuration with the S11 and S22 plotted on a Smith chart, isolation on a

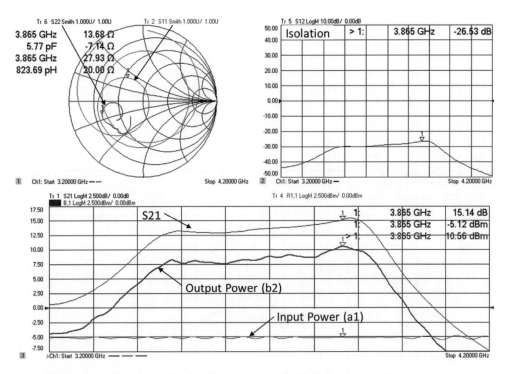

Figure 6.3 Typical plot showing S-parameters, gain, isolation, input and output powers.

large log-scale plot and gain, input power and output power on a separate log-mag plot with a smaller scale.

This measurement illustrates the results using a simple source and receiver power calibration; the measurement is performed at nominally −5 dBm input power. The input power and output power show substantially more ripple than the S21 trace. The S21 trace, of course, is fully corrected for mismatch errors, but the input and output power are not corrected for mismatch.

Enhanced power calibration provides for input and output match correction of the power measurements, but the results from VNAs without such capabilities can be manipulated to provide the same corrections provided that the source and load match terms are available. In modern VNAs, there often exists the ability to display one or more error term results in the VNA display. If so, the built-in equation editor function can be used to apply match correction to the result using the formulations in Eqs. (3.65) and (3.70).

For measurements of input and output power, the equations cited above can be applied in the equation editor, or in offline analysis. For more modern VNAs, the match-corrected power measurements are available as a part of a guided power calibration, and are displayed along with the corrected S-parameters. An example of such a guided power calibration is applied in Figure 6.4 to the amplifier from Figure 6.3 to give an improved measurement of input power (IPwr) and output power (OPwr). In the upper window, three traces are shown. One trace shows the gain computed by taking the output power divided by the input power (labeled "OPwr/IPwr") is very nearly the same as the match corrected gain of S21. The slight difference is due to the fact that the "loop" term of S21·S12·ELF·ESF is not fully corrected for, although

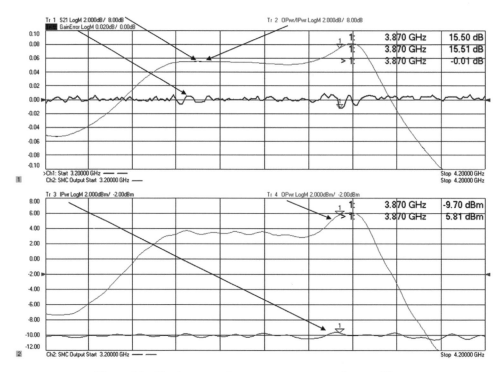

Figure 6.4 Match corrected power measurements of an amplifier.

the use of one-port corrected Γ_1 and Γ_2 compensates for most of the effect of this loop term; with S12 being small, the error is small indeed. This error is shown as "GainError" in the plot, and is less than 0.01 dB. At the scale of 2 dB/division, the comparison of S21, and gain computed from OPwr/IPwr, are indistinguishable.

The output power (OPwr in the lower window) trace still shows some excess ripple when compared to the S21 trace, and this is directly due to the input power ripple, seen in the lower plot of Figure 6.4. This ripple in the input power occurs even though the source power calibration was performed at 0dBm, with virtually no error. However, when the amplifier is connected to port 1 during the measurement, even at the same source-power setting, the input power shows ripple because a different match is applied to port 1: the effective input match of the amplifier terminated in port 2 of the VNA, or Γ_1.

From Eq. (3.65), we see that the input power measurement properly accounts for this error, and the input power displays the actual incident power, but that power is not exactly -10 dBm. Instead it has ripple associated with the amplifier input match and the source power match (see Section 2.2.2). The source ALC does not respond to the change in match – either from the power meter used in calibration or to the DUT input match – in a way that completely compensates for the mismatch, so a ripple in the actual incident power occurs. This ripple in the input power causes an equal ripple in the output power which is measured at the port 2 receiver. Both of these power measurements are exactly correct, as the incident power is really changing and the output power is really changing, so the gain computed as the ratio of the two is almost exactly the same as the fully corrected S21. However, it is desirable that the incident power exactly match the source power setting, and for this an additional measurement adjustment must be made in the case of devices with even moderate reflections at the input port.

The cause of the ripple is that while the reference receiver *does* properly detect the change in incident power due to DUT mismatch, the source ALC loop does not. This can be corrected for by using the a1 receiver, with its match correction, as the reference for the source ALC loop rather than the internal source detector. A function, sometimes called receiver leveling or Rx-leveling, provides such a capability. This can be performed either in hardware, by adding a detector to the a1 receiver path, or in software by iterating the source power until the desired level is achieved. In either case, the detected signal is only the raw reading, and some correction must be applied to ensure that the a1 leveling provides the proper output level. In software Rx-leveling, match correction to the a1 receiver reading can be easily added so that the source power is adjusted to provide a true incident power. And because the a1 receiver is used as the source level detector, the power source match and the ratio source match are identical, meaning that the incident power level is exactly corrected.

Figure 6.5 shows an example where the Rx-leveling method is applied to the amplifier from the previous figure. The input power is now nearly exactly −10 dBm (within the tolerance setting of the software Rx-leveling function) and the output power has almost exactly the same shape as the gain or S21, offset by 10 dB. Of course, the output power will be different from the gain since the source power is not 0 dBm. The gain computed by taking output power divided by input power (marked "OPwr/IPwr" in the figure) matches the S21 gain almost exactly.

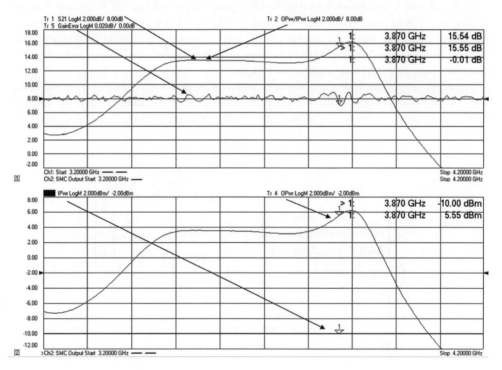

Figure 6.5 Match corrected powers with software Rx-leveling.

This final figure demonstrates the highest level and performance of calibrated amplifier measurements with input power, output power and gain all fully corrected for test system mismatch, and source power set exactly to the desired incident power. Note that the gain error, while very small (less than 0.01 dB) remains the same as in the previous figure. This gain error represents the fact that while the output power is corrected for load match of the VNA, and the incident power is corrected to be provide the same power as a perfect 50 ohm source, there is a re-reflected term that is not fully accounted for, but it is very small.

6.1.4.2   DC Measurements

A final set of measurements on amplifiers is often required to evaluate the efficiency of the amplifier. These are measurements of the DC voltage and DC current into the bias pins of the amplifier under test. While it is common to just measure the DC at a single frequency or power using a DC power analyzer or source measurement unit (SMU), the actual DC power consumption of an amplifier can change over its frequency range or power range, if it is in a non-linear mode of operation. Almost by definition, the DC operating point cannot change if the amplifier is operating in a linear region, as the definition of linear implies that the signal level is so small as to not affect the operating point of the amplifier.

However, many amplifier test scenarios are with the amplifier in or near compression and so in such cases the DC operating point must be measured at each RF frequency point and power point.

Some modern VNAs provide integrated DC measurements as built-in functions. The DC readings of voltage and current can then be displayed simultaneously with the RF input and output power measurements. Figure 6.6 shows a typical connection scheme for measuring the DC power consumption of an amplifier. In this case, the VNA has two built-in ADCs that can measure two analog input signals (AI1 and AI2).

Some voltage scaling resistors are needed if the DC meters have limitations on maximum voltage. The current sense resistor should be scaled to give a reading sufficiently large to provide a difference voltage that is easily detected by the DC meters. The AI2 signal is the

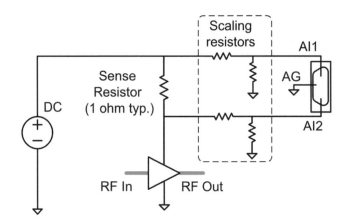

Figure 6.6   Typical configuration for measuring DC power consumption.

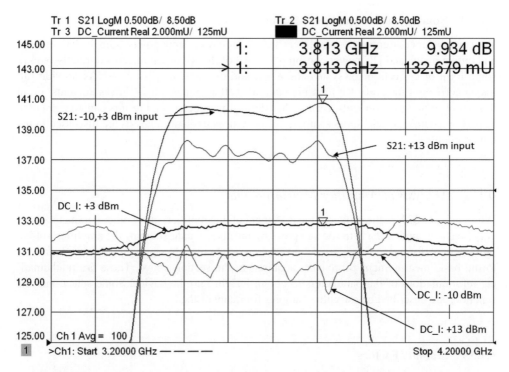

Figure 6.7 Measurement output power and DC current, for three different input power levels.

voltage drive to the amplifier. The DC current is computed using the equation editor function by taking the difference in AI1 and AI2, and dividing by the current sense resistor. If the current requirement of the amplifier is approximately known, the offset from the current sense resistor can be programmed into the DC power supply. If not, then the AI2 voltage can be measured and DC power supply adjusted to provide the correct DC supply to the DUT.

An example measurement is shown in Figure 6.7 displaying the gain (S21) and DC current of an amplifier for three different RF input power levels.

In this case, the current consumption changes with output power so that the efficiency also changes. It is interesting to note that while the S21 trace doesn't perceptibly change between a drive level of −10 dBm and +3 dBm, the DC does change indicating that the power of +3 dBm is sufficient to increase the DC operating point. At higher drive power, near the 1 dB compression drive, the DC current actually drops below the level of a small signal bias point.

These DC measurements may also need some calibration, to account for offset or gain errors in the DC meter, or for offsets from the measurement setup. For example, if a DC resistive divider is needed to limit the voltage to the DC meter, then the current consumption of output divider needs to be subtracted from the current reading (the current draw of the input divider does not affect the voltage across the current sense resistor). Further, if there is DC resistance, from the voltage sense at the output side current sense resistor, to the amplifier bias input, then the voltage drop associated with the current passing through that resistance also needs to be compensated for. This compensation would normally need to be iterative, if a particular output voltage is required at the DUT bias pin.

6.1.4.3 Conclusions on Amplifier Measurements

The principal measurements of amplifiers are output power, input power, gain, isolation, input and output match and DC power consumption. All of the RF measurements normally require some form of correction for systematic error, and the most modern of VNAs provide full match correction for every RF measurement, as well as integrated DC measurements. From this set of basic measurements, many other derived parameters and results can be computed as part of a data analysis. Some examples of post measurements analysis are discussed in the next section.

6.1.5   Analysis of Amplifier Measurements

While gain, or S21, is the principal measurement of an amplifier, many other parameters have very practical application in the design process, and also in understanding the amplifier's effect on system performance. Many useful results can be obtained by applying additional analysis to the basic measurement parameters acquired in the section above. These are mathematical reformulations of the parameters to obtain insight into the device. Some select analysis functions are described below with examples for many of them.

6.1.5.1 Stability Factors

For engineers having only a casual familiarity with RF circuits, gain and S21 are used interchangeably. This imprecise usage works because many systems utilize matched gain-blocks which have S11 and S22 very near zero. In fact, many hours of design are absorbed by the tasks of producing both the desired gain and power from an amplifier while at the same time providing a matched condition. However, many more hours of troubleshooting occurs when the true nature of the amplifier is not understood and both in-band and out-of-band matching conditions provide issues with the overall design.

When placed in a circuit, the behavior of an amplifier can be rewritten as in terms of various gain attributes such as maximum available gain (G_{MA}) when conjugately matched as

$$G_{MA} = \frac{|S_{21}|}{|S_{12}|} \left(K \pm \sqrt{K^2 - 1} \right)$$

where the sign is negative if

(6.4)

$$1 - |S_{22}|^2 + |S_{11}|^2 - |S_{11}S_{22} - S_{21}S_{12}|^2 > 0$$

where K is the stability factor, Eq. (6.5). [1]. For many designs, especially at high microwave and mm-wave frequencies, the goal is to achieve this maximum gain, and since the design is often defined for matched source and load conditions, the design task is principally creating a matching network, or transformer, which changes the (normally) 50 ohm reference impedance into the conjugate match of S11 and S22. However, this matching condition is only true if the amplifier is unconditionally stable, that is, if K > 1. Thus, measuring the stability of an amplifier is one of the key linear measurements.

In many modern VNAs, the stability can be computed and displayed on the screen directly using either a built-in function or using equation editor math on the S-parameters. The stability function most commonly used is the K factor, and is defined as

$$K = \frac{1 - |S_{11}|^2 - |S_{22}|^2 + |S_{11}S_{22} - S_{21}S_{12}|^2}{2|S_{21}S_{12}|} \tag{6.5}$$

An amplifier is unconditionally stable if $K > 1$, *and*

$$|S_{12} \cdot S_{21}| < 1 - |S_{11}|^2 \text{ and}$$
$$|S_{12} \cdot S_{21}| < 1 - |S_{22}| \tag{6.6}$$

The latter two conditions are met if $|S_{11}S_{22} - S_{21}S_{12}| < 1$. This last condition, often written as

$$|\Delta| = |S_{11}S_{22} - S_{21}S_{12}| < 1 \tag{6.7}$$

is sometimes forgotten, but is an important criteria and becomes very significant in out-of-band amplifier cases where S11 and S22 can become quite close to one, due to tuning conditions.

Both conditions must apply; if so, the amplifier is called unconditionally stable, and no combination of source and load impedances will cause the amplifier to oscillate.

If the amplifier does not meet the stability factor conditions, it is often erroneously called "unstable". More correctly, such an amplifier should be called "conditionally stable"; that is, the amplifier will not oscillate under some conditions of source and load impedances.

A conditionally stable amplifier is measured in Figure 6.8. The upper left window shows the S11 and S22 on a Smith chart. It is clear that S11 is poorly matched. The upper right window shows the gain and isolation, S21 and S12. The lower left window shows both the K factor (dark trace) and Δ term (called delta, light trace), with limit tests for values. The K factor fails the stability criteria, so this amplifier is only conditionally stable. In the lower-right plot, the G_{MS} is computed as $|S_{21}|/|S_{12}|$ (called maximum stable gain, G_{MS}) and is displayed along with G_{MA} based on Eq. (6.5); but here G_{MA} is set equal to G_{MS} as $K < 1$.

A common approach is to measure the four S-parameters and save the S2P file for further investigation using some offline simulation tool. However, using the built-in features of many modern VNAs, a great deal of investigation of other circuit topologies and matching scenarios can be applied in real time.

From the Smith chart trace of S11 in Figure 6.8, it is clear that this might be a parameter that affects K to a great extent, since it is so large. One aspect of a conditionally stable design is that it generally cannot be made stable by reactive matching alone; typically some resistive matching is needed.

In the case of the circuit from Figure 6.8, a small amount of series and shunt resistance was added to the input port using the fixture simulator function called "port matching" (see Figure 5.10). Port matching gives several circuit topology choices plus the option to import an S2P file to create a matching network between the test port and the DUT. In this case, adding about 10 ohms of series resistance and about 330 ohms of shunt resistance improved the K factor dramatically. The resistor values were chosen to keep the K factor above 1.5 everywhere. The result of this fixture simulation is shown in Figure 6.9

This does have the effect of lowering the maximum gain (shown along with the computed GMS value which is identical to the previous figure), and the input and output matches are still not very good.

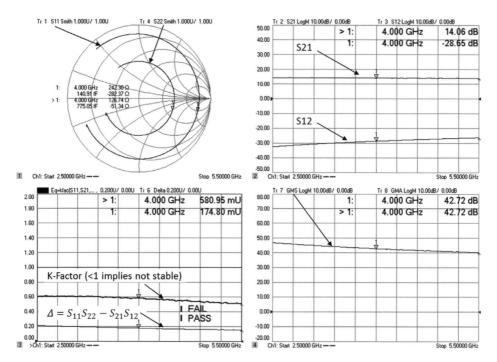

Figure 6.8 S-parameters, K factor and max stable gain.

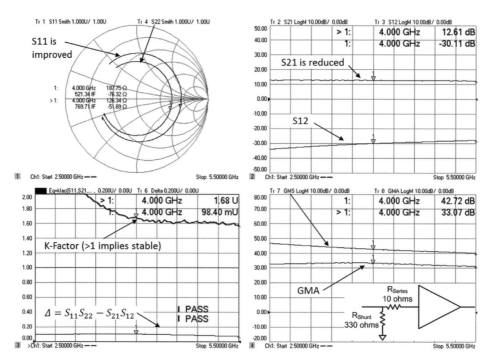

Figure 6.9 Resistance is added to the input network to improve the stability.

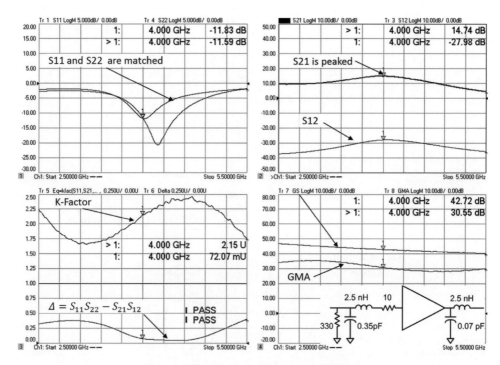

Figure 6.10   Circuit response after matching.

The reactive elements in the fixture simulator can be utilized to improve the reactive matching. From the G_{MA} value, which is much larger than the S21, there is sufficient room to make matching choices that would allow an increase in the gain. Here, the matching is changed to recover the gain lost from the stabilization of adding the resistive elements. After some trial and error, final values were obtained and the overall response is plotted in Figure 6.10. The matching that was employed has the effect of peaking the gain at the marker value above the nominal 50 ohm gain obtained in Figure 6.8. The stability factors are well away from any limits, and the input and output match are tuned to a reasonable value.

While this is not an amplifier design text, it's useful to note that with modern VNAs, the normal workflow of measurements and modeling can be combined into real-time analysis. This also facilitates other investigations such as K factor vs bias.

6.1.5.2   Stability Circles

In many cases, it is not desirable to use resistive matching to improve the stability. Resistive matching can degrade the noise figure at the input of an amplifier and degrade the maximum output power at the output of an amplifier. To avoid these degradations, designers often choose to use a conditionally stable device, but must find the proper impedance transformations to ensure that they are stable in operation. Since most amplifiers are used in a matched environment, the stability criterion is to add matching elements so that Z_0 is a stable impedance for both input and output ports.

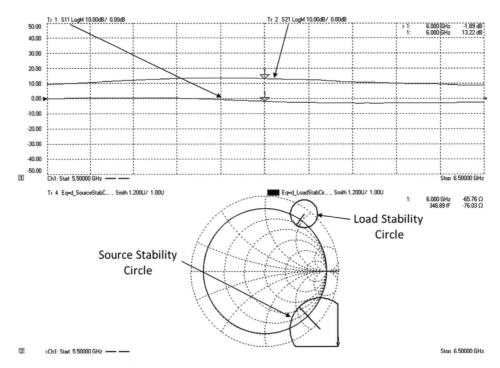

Figure 6.11   Stability circles at the center frequency.

Stability circles are often used to assist in determining the proper matching. These circles lie on the locus of impedance points where the amplifier circuit is just on the brink of instability. On one side of the circle the amplifier is stable; on the other side it may oscillate if either port of the amplifier is terminated in unstable areas. These circles can be drawn for both the input and output ports. Circuits that are designed to be oscillators have a stability circle that covers nearly the entire Smith chart, with Z_0 inside the unstable region. Circuits designed to be amplifiers strive to have the stable region as large as possible and, of course, encompassing the Z_0 point, or center, of the Smith chart.

In the past, creating stability circles was quite tedious, and simulation tools were used to create sets of stability circles for frequencies across the simulation bandwidth. In modern VNAs, one can once again turn to the equation editor function to create stability circles directly on the response traces of the VNA [2], as shown in Figure 6.11, with the upper window showing the S-parameters (S11 and S21) and the lower window showing the stability at the center frequency.

For the stability circles, the line to the center indicates an unstable region. This amplifier is conditionally stable, and a high reflection with the correct phase on either the input or output port could cause it to oscillate.

6.1.5.3   Mu Factors

An alternative stability factor has been derived which is finding favor with many designers based on a geometric analysis of stability circles. The mu factor is essentially a measure of the distance from an unstable region to the center of the Smith chart. If the distance is greater than 1, the amplifier must be stable, as the nearest unstable region is outside the Smith chart.

There are two mu stability factors, originally called μ and μ', but more recently called mu1 and mu2, or μ_1 and μ_2. These are the input and output mu factors, and they are related to the maximum distance for stability of the input or the output load. The factors are related, and if either one is greater than 1, the device is unconditionally stable. If they are less than 1 but greater than zero, the device is conditionally stable including the matched Z_0 point. If μ_1 and μ_2 are negative, then the Z_0 point is not stable. The magnitude of mu indicates the distance, on a Smith chart, from the origin to the closest unstable impedance. From this definition, it is clear that a magnitude greater than 1 means that the device is stable for any points inside the unit circle of the Smith chart.

The formulas for or μ_1 and μ_2 are

$$\mu_1 \equiv \frac{1 - |S_{22}|^2}{|S_{11} - S_{22}^*\Delta| + |S_{21}S_{12}|}$$

$$\mu_2 \equiv \frac{1 - |S_{11}|^2}{|S_{22} - S_{22}^*\Delta| + |S_{21}S_{12}|}$$

(6.8)

where Δ is the same as defined in (6.7). These formulas could be created directly in the equation editor of a VNA, but some modern VNAs provide this as a built-in function. Figure 6.12 (lower) shows the result of plotting mu1 and mu2 for the amplifier of Figure 6.8, along with the S-parameters (upper left) and K factor (upper right).

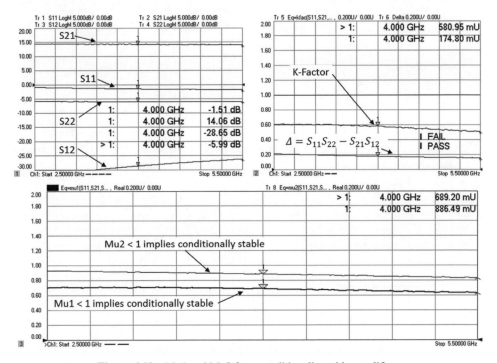

Figure 6.12 Mu1 and Mu2 for a conditionally stable amplifier.

Just as with the K factor, the stability of an amplifier with respect to the mu factors can be evaluated using the port matching functions of the VNA. Using the same techniques as above, the conditionally stable amplifier may be evaluated with the port matching applied, as shown in Figure 6.13. Here the values of the loss elements are adjusted until mu1 is just greater than 1, and one can note that mu2 also becomes greater than 1, and the same reactive matching as in Figure 6.9 is used to peak the gain.

6.1.5.4 Gain Factors

Besides S-parameter gain, which is defined as the gain from a matched source into a matched load, there are many other gain factors that provide useful insight into how a part may be used. Maximum stable gain and maximum gain are discussed above as part of stability, but some other important factors are available gain and transducer gain.

The transducer gain gives the ratio of power available from the source to power delivered to the load, and so both the source and load match values of the system must be known

Available gain, defined in Chapter 1 as the gain that an amplifier can provide to a conjugately matched load from a source or generator of a given impedance, is

$$G_A = \frac{\left(1 - |\Gamma_S|^2\right) |S_{21}|^2}{|1 - \Gamma_S S_{11}|^2 \left(1 - |\Gamma_2|^2\right)} \tag{6.9}$$

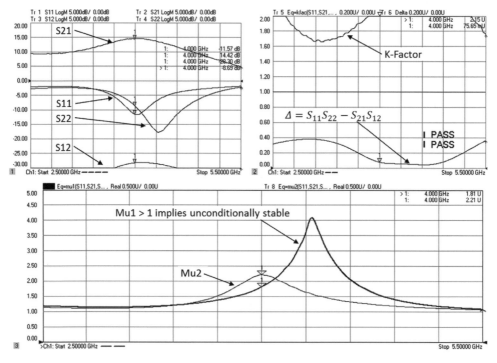

Figure 6.13 Mu1 and Mu2 for an amplifier after port matching to make it unconditionally stable.

Since this analysis depends upon the particular source impedance of the previous stage, this additional information must be provided and can be obtained by measuring the output match of the previous stage and saving it in a different trace, or, on some of the more modern VNAs, loading the output match into a trace using and S1P or S2P file. This response can come from a measurement, or from a simulation of the matching at the input or the output. Often, the preceding stage may be from a filter or other element whose match varies with frequency, such that the available gain also varies with frequency even if the amplifier under test has a flat frequency response.

Figure 6.14 shows the computed available gain (labeled "GA") compared with the S21 gain, and the source impedance (labeled "G_S" for Gamma_Source), utilizing the equation editor function of the VNA.

In this case an intermediate result of the Γ_2 of the amplifier is computed and shown as well, called out as the trace labeled "G_2", for Gamma_2.

Note that where there are peaks in the S11 of the amplifier, there are also peaks in the available gain, and everywhere the available gain is greater than S21.

The transducer gain gives the ratio of power available from the source to power delivered to the load, and so both the source and load match values of the system must be known. Just as in available gain, both the source and load impedance of the system into which the amplifier will be placed can be loaded into the VNA trace data, and equation editor functions can be used to directly display the transducer gain.

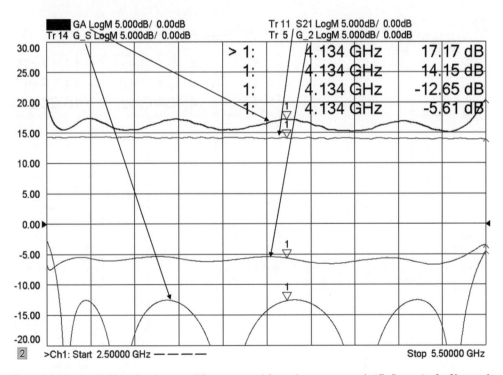

Figure 6.14   Available gain of an amplifier computed from the output match (G_S trace) of a filter and measurements of the DUT S-parameters.

Transducer gain is computed as

$$G_T = \frac{\left(1 - |\Gamma_S|^2\right)|S_{21}|^2\left(1 - |\Gamma_L|^2\right)}{|1 - \Gamma_S S_{11}|^2 \cdot |1 - \Gamma_2 \Gamma_L|^2}$$ (6.10)

An example of transducer gain (G_T) computation for an amplifier placed in a system with an identical filter on either end is shown in Figure 6.15. For reference, both the S21 of the amplifier, and the available gain (GA) of the amplifier are shown.

Available gain and transducer gain were developed before the widespread use of simulation technology and they created a way to evaluate the effect of source and load matching of system elements on an amplifier's gain response. However, these design techniques have been almost entirely replaced by simulation where the complete S-parameter matrix of system components can be embedded in a single measurement result.

Consider the amplifier from above, with a known source and load impedance from a pair of filters before and after the amplifier. If the filter characteristics are known, then the full effect of the S-parameter matrix of the filters at the input and output of the amplifier can be embedded in the response using the T-matrix math provided in the port matching function of the VNA – see Eq. (3.5) – as shown in Figure 6.16 (labeled S21_Simulated), along with an equation showing simply the product of S21 of each of the elements (labeled S21_Product), which would be the supposed result if one ignored the mismatch effects. Using the port matching is a good way to simulate the overall gain effect, and can essentially replace other forms of analysis of source

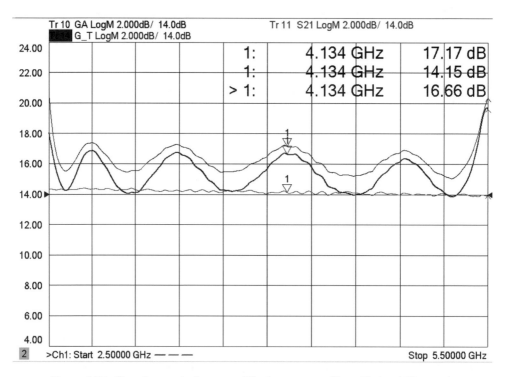

Figure 6.15 Transducer gain for an amplifier between two filters (dark, middle trace).

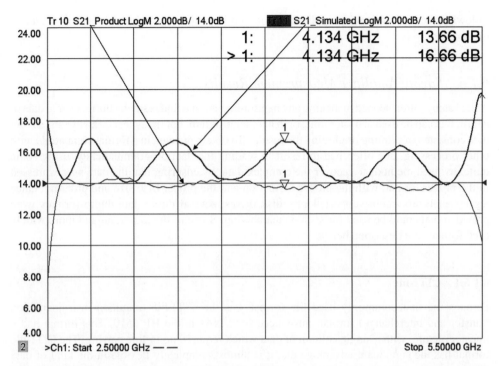

Tr 10 S21_Product LogM 2.000dB/ 14.0dB   S21_Simulated LogM 2.000dB/ 14.0dB

1:        4.134 GHz   13.66 dB
> 1:   4.134 GHz   16.66 dB

>Ch1: Start 2.50000 GHz — —   Stop 5.50000 GHz

Figure 6.16   The overall gain from embedding the filter response using port matching, compared to the product of S21 of the amplifier, input and output filters.

and load influences on amplifier gain; since the filter loss is small, the simulated concatenation of the filters and the amplifier gain (S21_Simulated) is nearly identical to the transducer gain in Figure 6.15.

This gain, including the full effects of port matching, does not give the same answer as taking the product of S21 of each of the components whenever the match of the filters and amplifier becomes large, especially if the amplifier is conditionally stable. This difference is a source of many errors in system design when mismatch effects are ignored.

6.1.5.5   Analysis Conclusions

Sophisticated real-time analysis of DUT performance can be created and displayed on modern VNAs using a variety of built-in functions. Fixturing applications allows the application of port matching or embedding of the response of other components on the measurement results of the amplifier under test.

The equation editor function is another method of providing virtually unlimited types of analysis on the underlying data, including RF or DC or any other measured parameter. The resulting analysis traces can update in real time, allowing for tuning of the response of the amplifier on more than just the S-parameter results. An example would be tuning for improved stability or improved maximum gain. Other important analysis examples are presented in the following sections on specific applications of amplifiers including high-power, high-gain and low-noise.

6.1.6   Saving Amplifier Measurement Results

A final step is often needed in making any measurements, and that is saving the measured data in a form that can be used at a later time, or in another program. For many legacy analyzers, saving data consisted of printing or plotting the results. To obtain the data in a format compatible with other processing, one often had to resort to reading the data programmatically, using some quite involved methods to recover the complex data, by ensuring that the data was expressed in the proper format, with the proper correction and the proper byte size and ordering. Once legacy analyzers could support floppy disk drives, several direct data dump formats were provided that could be used for such data archiving. Some of the important and more widely used formats are described below.

6.1.6.1   CITI-file

Perhaps the first commonly used file format was the CITI-file (common instrumentation transfer and interchange) format, introduced for VNAs in the HP 8510. This format had a linear structure, with a single column of data which required no in-line parsing. The CITI-file format has the significant advantage that it is almost completely flexible in the kind of data that it can represent. The data is formed into packages, based on the stimulus. The structure of each package is a header with information about the instrument model and firmware revision, date and time, the names of the data traces, the independent variable (or stimulus) type and the format of the data. The body of the file has variable lists with a "begin" and "end" keyword. Multiple channel files are all sorted into such packages according to channel number. If a channel contains two different stimulus domains, all the traces from each domain are combined into a package, and a separate package is created for each stimulus type. A sample CITI-file is shown below in Table 6.1, for a two-channel measurement, where the stimulus is different between channel 1 and channel 2, and channel 2 has one trace with time domain. This creates three packages. The table has sorted the file into three columns, for clarity and comparison, but in reality all three packages follow one after another in a single column in the actual file.

The example shown is an output from an Agilent PNA-X analyzer. As with some modern analyzers, this provides the user with several options on how the CITI-file is formatted, and what data is saved. The choices include whether to save all the displayed traces or just a single trace. An "Auto" choice saves all the corrected data for the selected trace; for example, if an S11 trace is selected which has a two-port calibration applied, the auto setting will also save S21, S12 and S22.

The save function also allows the format of the saved data to be different from the format of the displayed trace, by forcing all traces to either LogMag/phase, LinMag/phase or real/imag. An auto format choice saves all the data in the format of the currently selected trace.

While CITI-file is a very flexible format, and was the format used in the first VNAs, several newer formats have come about that have found common use.

Table 6.1   CITI-file output format; the file output is broken into three columns only for comparison

CITIFILE A.01.01	CITIFILE A.01.01	CITIFILE A.01.01
!Agilent	!Agilent	!Agilent
Technologies	Technologies	Technologies
!Agilent N5242A:	!Agilent N5242A:	!Agilent N5242A:
A.09.42.08	A.09.42.08	A.09.42.08
!Format:	!Format:	!Format:
LogMag/Angle	LogMag/Angle	LogMag/Angle
!Date: Sunday,	!Date: Sunday,	!Date: Sunday,
November 13, 2011	November 13, 2011	November 13, 2011
05:26:09	05:26:09	05:26:09
NAME CH1_DATA	NAME CH2_DATA	NAME CH2_2_DATA
VAR Freq MAG 5	VAR Freq MAG 5	VAR Time MAG 5
DATA S[2,1] DBANGLE	DATA S[2,1] DBANGLE	DATA S[1,1] DBANGLE
DATA S[1,1] DBANGLE	VAR_LIST_BEGIN	VAR_LIST_BEGIN
VAR_LIST_BEGIN	2500000000	-4e-009
1000000000	2750000000	-2e-009
1250000000	3000000000	0
1500000000	3250000000	2e-009
1750000000	3500000000	4e-009
2000000000	VAR_LIST_END	VAR_LIST_END
VAR_LIST_END	BEGIN	BEGIN
BEGIN	-0.38341591,	-35.606487,
-0.18913588,	91.595734	-87.11322
107.97729	-0.19906346,	-26.621368,
-0.2556681,	-152.15067	-126.83741
-134.61461	-0.60013449,	-35.606487,
-0.29677463,	-34.987034	-87.11322
-17.619053	-0.52427602,	-26.621368,
-0.25021815,	81.872543	-126.83741
99.132004	-0.43853623,	-35.606487,
-0.38517338,	-161.16911	-87.11322
-143.36929	END	END
END		
BEGIN		
-23.06007,		
-128.98009		
-35.266006,		
-0.99580592		
-28.660841,		
151.64325		
-41.02335,		
-23.198109		
-32.739304,		
-28.003317		
END		

6.1.6.2   S2P or Touchstone® Files

Originated by EEsof before it was acquired by Hewlett-Packard, the Touchstone format – or S2P format as it is most commonly called – was created as a compact form to encapsulate S-parameter data of a two-port device. There are also higher port count versions, and as a class these are often referred to as SnP files, where n denotes the number of points. The structure of the S2P file is shown in Table 6.2. The Touchstone format has been officially accepted by several organizations including the IBIS (input output buffer information specification) open forum group.

The version 1 Touchstone format was never formally adopted, partly because it has one major deficit in that the reference impedance for the S-parameter file must have only one value. This does not present any theoretical limitation, because if a network were measured with a different reference impedance on each port, it is a simple transformation to change one of them so that they are both the same. However, this shortcoming likely prevented formal adoption of the version 1 standard, which is found on many websites marked as "draft", and has been at the draft state for 10 years. In 2009, the IBIS Open Forum officially adopted version 2 of the Touchstone format, which includes modifications to support different port impedances on each port, and includes a format specification for mixed mode S-parameters. At this time, very few instrument or EDA companies have implemented the new format.

For higher port count, the SnP file is used, and the format of the file differs from the above in that the data appears as a first line of five-to-a-line data (frequency plus the first four data points) and the remainder of the data is presented four-to-a-line, all in the normal matrix order of Sxy, with x being the row number and y being the column number.

One curious attribute of the S2P file is that the standardized parameter order is a bit odd. It does not match the normal row/column description of the S-parameter matrix. Rather, it matches the order of parameters that is the default for many legacy VNAs: S11, S21, S12, S22. This has the unfortunate effect that the S2P file parser must be different from the SnP file parser.

Table 6.2   S2P data format

```
!Agilent Technologies,N5242A,US47210094,A.09.42.08
!Agilent N5242A: A.09.42.08
!Date: Sunday, November 13, 2011 06:01:33
!Correction: S11(Full 2 Port(1,2))
!S21(Full 2 Port(1,2))
!S12(Full 2 Port(1,2))
!S22(Full 2 Port(1,2))
!S2P File: Measurements: S11, S21, S12, S22:
# MHz S dB R 50
1800 -25.33 -132.64 -14.87 -46.52 15.25 12.14 -37.27 -64.08
1850 -26.51 -74.60 -14.98 -7.89 15.35 12.41 -39.20 -43.02
1900 -31.96 15.85 -15.06 31.22 15.43 12.45 -33.63 -77.49
1950 -24.41 -107.31 -15.06 70.42 15.44 12.64 -30.44 -44.29
2000 -22.84 -27.07 -15.03 109.00 15.42 12.83 -30.57 -29.91
```

One remaining deficit of the SnP file is that only S-parameters can be represented. If a VNA has other traces representing power, or time domain, or any other parameter, the SnP file cannot be used to save the data.

6.1.6.3   CSV Files and Exporting Data to Excel

One new format to appear in some modern VNAs is the comma separated values (CSV) format. This is a very generic format and in some ways matches the CITI-file format in that it sorts the data into similarly grouped packages. An example of a CSV file is shown in Table 6.3.

Table 6.3   CSV file example

```
!CSV A.01.01
!Agilent Technologies,N5242A,US47210094,A.09.42.08
!Agilent N5242A: A.09.42.08
!Date: Sunday, November 13, 2011 06:33:10
!Source: Standard

BEGIN CH1_DATA
Freq(Hz),S21 Log Mag(dB),S11 Log Mag(dB)
1800000000,-14.877501,-25.325617
1850000000,-14.982286,-26.499651
1900000000,-15.063152,-31.963058
1950000000,-15.069975,-24.410412
2000000000,-15.031799,-22.855028
END

BEGIN CH2_DATA
Freq(Hz),S21 Log Mag(dB)
2500000000,-0.37367642
2750000000,-0.19583039
3000000000,-0.59756804
3250000000,-0.52162892
3500000000,-0.43824977
END

BEGIN CH2_2_DATA
Time(s),S11 Log Mag(dB)
-4e-009,-35.611397
-2e-009,-26.620441
0,-35.611397
2e-009,-26.620441
4e-009,-35.611397
END
```

The format is completely flexible, just like CITI-file, but it has the advantage that for a single channel, if the stimulus data is consistent, all the data for each parameter is on a single line for the each stimulus value. The file can store any trace, including power traces, or the results of any equation editor function. Of course, a channel may contain traces with more than one stimulus – for example, channel 2 in the above table contains a frequency and time domain trace. In such a case, a package for each stimulus is created. The CSV data is convenient because it may be directly read into an Excel spreadsheet. The CSV file output function can give several choices such as only outputting a single trace, or all the displayed traces. The data may be formatted automatically, similarly to CITI-file, or the format can be set to a specific format such as LogMag/phase, LinMag/phase or real/imag. For the most flexibility, one can also choose the currently displayed format for each trace.

With these more flexible and comprehensive data save formats, it becomes much less necessary to read data from a modern VNA using the programmatic interface. Instead, a simple "store data as CSV" manual front-panel function can provide a file that has all the desired data formatted in the desired way. And, of course, the same command can be sent programmatically. This removes the need to create data buffers, select traces and read data programmatically over the bus, saving substantial programming time for a test engineer.

6.2   Gain Compression Measurements

The previous section describes a wide variety of linear measurements that are commonly made on amplifiers, but amplifiers are often used in a non-linear portion of their operating range, and there are several important non-linear descriptions that are found in common use. Perhaps the most fundamental measure of non-linear performance is gain compression.

6.2.1   Compression Definitions

Just as the name implies, gain compression is a measure of the degradation of an amplifier's gain as a result of increasing input power. While amplifier non-linear gain is almost always referred to as compression, it is not uncommon to find some amplifiers that have a slight expansion, or increase, in their gain before the compression comes into effect.

Figure 6.1 showed an example of trying to identify the compression point of two amplifiers, before optimizing the VNA settings to ensure that driving the amplifier into compression will not overdrive the VNA. One amplifier shows only compression, while the other shows some slight expansion before the onset of compression occurs.

For the most basic compression measurement, the same technique can be used to find the onset of compression. After calibration, and starting from a power level sufficiently small to ensure linear amplifier operation, the S21 response is put into memory and the ratio of S21/memory is displayed. The power level is stepped up while monitoring this normalized response until the onset of compression is clearly found. This onset can be easily tracked by adding a marker and setting the marker search function to continuously search for the minimum value of the trace. In Chapter 5, the response of the VNA to increasing the source power by 20 dB was shown to reduce the trace noise by 10 times. Thus, if one wants to optimize trace noise and measurement speed, one should increase the averaging or reduce the IF BW for the initial linear response by a factor of 100. Since compression comes on slowly, the first power step can

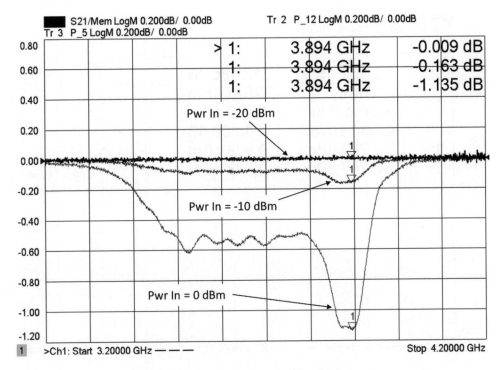

Figure 6.17   Detecting the onset of compression.

be 10 dB higher than the initial power, and the IF BW can be increased or averaging reduced by a factor of 10. If the onset of compression is not apparent, the power may be increased in increments until the marker shows some compression on the order 0.5 to 1 dB. An example of this progression of measurements is shown Figure 6.17. The trace is normalized at −20 dBm and the trace noise is evident in the measurement. The progression of power at −10 dBm and 0 dBm input is shown, and the compression versus frequency is clearly seen. Since this is a filtered amplifier, the out-of-band area shows no compression. The peaking in the compression is likely related to peaking in the gain. The fact that the compression changes dramatically where the gain is peaked could also indicate that the value of S22 is affected by compression as well, and the gain peaking is reduced the most (thus more compression) at higher powers for frequencies near marker 1 as compared with other frequencies.

6.2.1.1   Compression from Linear Gain

Once the onset of compression is found, the marker's frequency value can be entered into the fixed (CW) frequency for a power sweep, and a power sweep over the linear to non-linear power range can be acquired. Such a compression sweep is shown in Figure 6.18. Also shown in the plot is a trace of input power and a trace of output power. The compression of the amplifier's gain is clearly shown in the S21 trace. For a power sweep, there are three key powers: the linear power or power at which the reference gain is measured, the maximum

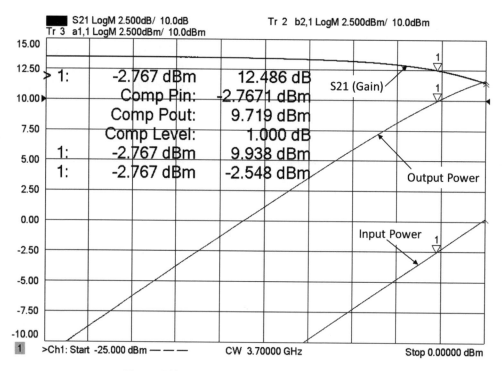

Figure 6.18   CW power sweep to find compression.

power or stop power for the power sweep, and of course the power at which the gain is compressed by 1 dB.

A marker can be used to manually determine the 1 dB compression point (the most common definition of compression) or, in many modern VNAs, the compression point can be found automatically using a gain-compression marker function. In this method, the definition of compression is the point on the S21 trace that is 1 dB below the S21 value at the linear power measurement, usually defined as the first point of the power sweep. The compression marker will search for the 1 dB down point and then report the gain at that point, as well as the source power setting (the x-axis stimulus value), and the output power computed from the source power and the gain at the marker. This is in keeping with the definition of a trace marker in that it only operates on the data from the measured trace itself, rather than from other data, such as input power or output power traces. This is a quick and quite good measurement, but it has some small drawbacks.

The main issue is that the x-axis values or marker stimulus values represent the source setting, but not necessarily the input power. However, if the input power and output power traces are also shown, as in Figure 6.18, then the markers can be coupled together, and the compression marker search will place the markers on the input and output power at exactly the right point to show the exact input and output power at compression. In this case, the output power at compression is about 0.2 dB higher than the marker search value reported on the S21 trace. This is because, from the input power trace, one can see that the actual measured input power is about 0.2 dB above the stimulus setting for that point (-2.5 dB vs -2.7 dB).

Alternatively, receiver leveling can be turned on so that the source setting exactly equals the input power reading.

While the definition of compression above is the most common, there are several other definitions which are also found in various industries, as listed below.

6.2.1.2   Compression from Max Gain

Some amplifiers experience an expansion or increase in gain just before the onset of compression, often due to a subtle re-biasing of the amplifier that slightly increases the gain. In fact, some amplifiers are designed particularly for this effect as a means to extend the linear operating range. In these amplifiers, the gain versus drive power peaks and then compression quickly ensues (refer back to Figure 6.1). For these types of devices, the gain compression is sometimes specified as compression from maximum gain, rather than compression from linear (or low power) gain. This definition is somewhat more conservative than the prior one, in that it will report a lower output power for 1 dB compression than will compression from linear gain. Note that for an amplifier following a normal compression curve, where maximum gain is at the linear power, the definitions are identical.

6.2.1.3   Compression from Back-off or X-Y Compression

Another compression measurement that harks back to the first days of measuring non-linear behavior is compression from back-off, or the so-called X-Y compression. There are different implementations of the same method, and essentially define compression as a defined change in gain (or the y-axis of the S21 plot) over a specified change in input power (or x-axis). Compression from back-off looks at the S21 versus power curve, and looks for a point where the gain drops by a specified value (usually 1 dB) over a specified change in drive power (usually 10 dB). In the past, compression was measured by moving a 10 dB pad from the input to the output of an amplifier, while monitoring the output power. The input power was increased for each iteration until a power was found where moving the pad changed the output power by 1 dB, giving the same result as the back-off method. The X-Y method, which is functionally equivalent, looks for a specified change in output power (usually 9 dB) over a specified change in input power (usually 10 dB). These compression methods are illustrated in Figure 6.19.

In some ways, this is the best method of finding compression because it is insensitive to noise at the linear power range and it incorporates a concept of looking for a change in gain over a nominal change in input power. Compare this with the max gain or compression from linear methods: Max gain requires the maximum gain be determined, which might require a very dense point spacing in the power sweep to ensure that the maximum power is correctly measured; compression from linear means that trace noise at the very low power of the linear measurements will directly affect the compression computation at high power. Further, for some amplifiers, the gain drops very slowly with increasing power, so that the power sweep range must be very large to ensure that the linear power is achieved. However, the X-Y or back-off compression is always found within the X dB (e.g., 10 dB) range of the compression point. For many modulated signals, which maintain some average power level and have some specified peak-to-average ratio, the compression from back-off provides a more

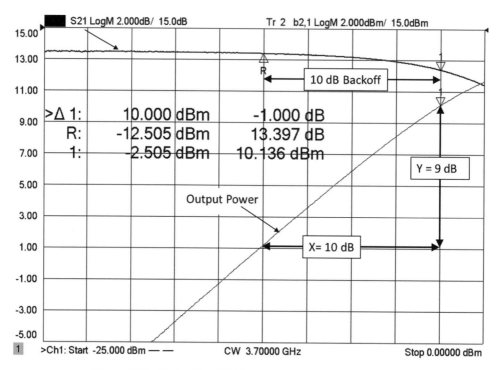

Figure 6.19 Back-off and X-Y methods of finding compression.

real-world use case for a compression measurement. If the modulated signal provides an average power, it is probably more reasonable to use that power as reference for the measurement of compression at the peak than it does to use a very low linear power, which some modulation formats will never impart. Notice that X-Y or back-off always gives a larger compression value than compression from linear; perhaps this is one reason why it is favored by many amplifier manufacturers.

6.2.1.4 Compression from Saturation

While somewhat of a misnomer, compression from saturation is a method that is applied to amplifiers that are normally used at or near their saturation point. For some amplifiers, such as traveling wave tube (TWT) amplifiers, a very clear saturation point occurs in the input-power/output-power curves. The amplifier is operated backed-off, just below this saturation point, with a specified back-off level. Sometimes this offset level is quite low, such as 0.03 dB below saturation (or max power) typically found in TWT amplifiers. This level is very close to the maximum output power level, but one key operating point metric is the input power for maximum output power, sometimes called the normal operating point (NOP). When the saturation curve is very flat, even the slightest noise in the trace can cause large swings in the input power measurement at saturation, so backing off just a slight amount, such as 0.03 dB, provides a much more stable number for the NOP.

Sometimes the back-off is more substantial such as 8 dB often found in solid state power amplifiers (SSPA) that are used in some satellites, replacing TWT amplifiers. The method of finding this back-off level from saturation is identical to the TWT method, with the only change being the value of the back-off. Note that while the same term is used, the value of back-off here is unrelated to the term used in a similar sense in Section 6.2.1.3.

6.2.2 AM-to-PM or Phase Compression

When viewing the effects of gain compression on a complex modulated signal, a common figure of merit is error from the intended magnitude and phase. The total error in this signal is the vector difference between the desired value and the actual value, and this includes both amplitude errors and phase errors; the size of this error is called the error vector magnitude (EVM). In some cases, the change in phase at the compression point can cause a larger vector error than the change in magnitude. To quantify these effects, a common measure of compression's effect on phase is expressed as AM-to-PM, although two distinct definitions have emerged.

One definition is the change in phase from a reference value at the defined amplitude compression point. Thus, displaying the phase versus drive along with the magnitude response will provide a direct measurement of AM-to-PM. If the marker compression search is used, the phase marker can be coupled to the compression marker and the phase at 1 dB compression can be simply read from the marker display, as shown in the first marker of Figure 6.20.

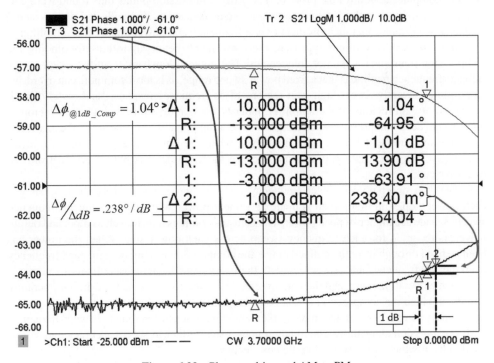

Figure 6.20 Phase vs drive and AM-to-PM.

Here the X-Y, or 10 dB back-off, method is used to find the 1 dB compression point. It is clear from the phase trace that trace noise at the lowest power will add significant error to the estimate of phase change at the 1 dB compression point. This is because the reference would have almost 1° of trace noise. Using the 10 dB back-off as a reference eliminates much of the noise issue.

An alternative definition of AM-to-PM is the phase slope in degrees/dB, at the 1 dB compression point. This can also be directly displayed, although in a roundabout way. An S21 phase versus drive trace is used with one marker set to 0.5 dB below the input power of the compression point, and another at 0.5 dB above the power. One of the markers can be set to be a reference marker, and the other will read directly the change in phase for a 1 dB change in drive power. This measure of AM-to-PM is shown in the bottom marker set of Figure 6.20.

6.2.3 Swept Frequency Gain and Phase Compression

The compression definitions used so far are single-frequency measurements where the power is swept and the compression point is determined. However, in many cases the compression of an amplifier, particularly a narrowband tuned amplifier, can change over the frequency band of the amplifier. In this case it is desirable to measure the gain compression across the entire frequency band.

Traditionally, the swept frequency gain compression measurement was performed by repeating the CW power-sweep method over each frequency in the frequency range, using an external computer to control the process. The 1 dB compression points thus found were collected and displayed as the 1 dB compression versus frequency. The collected data essential creates a two-dimensional measurement over a frequency and power range of the amplifier.

While this is a very direct approach, there are much better algorithms for finding the compression point over a span of frequencies. The biggest drawback from the swept-power stepped-frequency method is that a fixed power sweep range, as is most common, can overdrive an amplifier severely across high gain regions of its frequency response, and under-drive it across lower gain regions. If the amplifier under test is overdriven in its high gain region, it may go into deep compression and change the operating point or temperature of the amplifier; if so, the measurement for the next frequency, when the power sweep is reset to a lower power, gives an inaccurate gain reading because the amplifier has not recovered from the overdrive condition.

Rather than sweep power at each frequency, a better approach is to sweep the frequency at each power step. The first sweep is made at a linear power, and each subsequent frequency sweep is made at a higher stepped power. The gain and compression, as well as input and output power, is recorded for each frequency sweep across each power step. When the power has been stepped through its entire defined range, the compression is computed for each frequency from the 2-D frequency and power recorded. In this way, the linear measurements are all made at the same operating point, and the compression levels are likewise made at similar operating points, provided the compression is relatively constant across frequency.

Recently, some VNA vendors have implemented variations of automated swept-frequency 1 dB compression algorithms. One method goes a step farther in speed, accuracy and safety, the details of which are described in the next section.

6.2.4   Gain Compression Application, Smart Sweep and Safe-Sweep Mode

For a swept-frequency gain-compression measurement, the data acquisition can be performed in two ways: sweep power and then step frequency, or sweep frequency and then step power. The former method is most commonly applied, but has several serious drawbacks, the main one being that the DUT is exposed to maximum power at the end of one power sweep just before the measurement changes to the new frequency and minimum power. This often causes a change in DUT behavior which makes subsequent linear power acquisitions incorrect. On the other hand, sweeping frequency starting at the minimum or linear power, and then stepping up the power for a new frequency sweep avoids any issue with overdrive.

For the very fastest measurements of compression, it is not necessary to measure every power level between the linear power level and the compression level. An initial measurement of the linear power is first acquired. Then a first guess at the power compression level can be made. For example, choose a power halfway between linear and maximum power, and measure the compression. A second guess can be made at a higher power. From the measurement of these two powers, a third guess can be computed for the level at which 1 dB compression will occur. A new frequency sweep is performed with the new power settings, and another compression acquisition is obtained. This continues until the desired compression level is achieved within a specified tolerance. One VNA vendor refers to this as Smart-Sweep in gain compression, because the sweep attributes are adaptive on a point-by-point basis to the DUT behavior.

At this point, the input power table to which the source is adjusted is exactly the input power for 1 dB compression, sometimes called CompIn or CompIn21 to identify the path of the compression measurement. The gain at this compression level is called the CompGain21, and the output power is called the CompOut21. These represent the fundamental compression measurements. An additional parameter is often useful to view, the DeltaGain, which is the measured value of compression. Ideally, this should be exactly -1 dB, but if the amplifier is not driven into compression or the tolerance is large, it can vary from that value.

An example of a swept frequency 1 dB compression measurement is shown in Figure 6.21.

From the DeltaGain21 trace, it is clear that the amplifier is not in compression at the band edges, due to the filtering function at the input of the amplifier not allowing sufficient drive power to compress the amplifier. If the format for DeltaGain21 is changed to phase, the result is the phase deviation at the 1 dB compression, which is one definition of AM-to-PM. Thus the frequency-dependent AM-to-PM is easily displayed as well. The CompOut21 shows an interesting result in that the compressed output power is not constant, but is higher where the gain is peaked. A marker on the CompOut21 trace reads the value of output power at 1 dB compression at any frequency along the trace.

6.2.4.1   Safe Modes of Measuring Compression

While the iterative method of finding the 1 dB compression point can be very fast, it does have a drawback in that it can also overdrive a DUT if the DUT has a lot of gain variation, and the power settings for the initial gain readings are high enough to overdrive the amplifier. Also, if the amplifier does not follow a normal compression curve, it is possible that the predicted

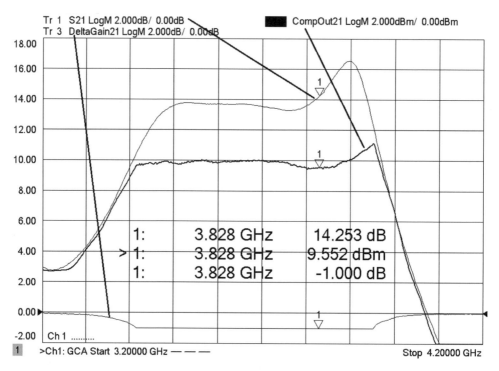

Figure 6.21   Swept frequency 1 dB compression measurements.

power for the next iteration will overdrive the amplifier. Particularly for very high power amplifiers, an overdrive condition must be avoided both to protect the DUT and to protect the test equipment connected to the DUT.

In such a case the iterative method may be modified to provide a safe mode of operation. Such a safe mode would need to have some defined limits to the power setting and the output power; in particular, there should be limits set so that the input power is not increased in the next iteration if the output power has exceeded a predetermined limit, even if the amplifier is not compressed. Also, the step size of the power change should be limited to avoid jumping from an underdrive to an overdrive condition in one step.

One method is to specify a maximum power step size, say 1 dB. Then during the iterations, the input power could not increase by more than the specified step, ensuring that no overdrive could occur by more than this step. However, if the linear power is far from the compression point, a large number of iterations will need to occur before the compression level is found. A smarter method is to define two step sizes: a course step (perhaps 5 dB) and a fine step (perhaps 1 dB). In addition, a compression threshold should be defined such that when an amplifier exceeds some safe compression threshold (perhaps 0.5 dB), the step size automatically switches from course step to fine step. Such a scheme ensures that the amplifier will never exceed a specified output power (thus protecting the external equipment) and never be overdriven by more than the fine step size (thus protecting the amplifier).

6.2.4.2 Full 2-D Gain and Compression Characterization

For some cases it is desirable to have a full description of the amplifier's performance, over a predefined set of power points and at every frequency point, as a two-dimensional array of gain versus power and frequency. The adaptive sweep methods above do not give a deterministic number of power points as the power steps depend upon the amplifier's response. Thus, it would be convenient to modify the gain compression acquisition to make a full two-dimensional array of frequency and power. The safe mode would support this if the fine step and course step are set to be the same size, but the safe mode stops increasing the source power after the compression point has been passed, and subsequent iterations simply zero in on the compression point, so the data is not regularly formed in the data set.

Instead, a separate mode of operation should be used to sweep the frequency and step the power for each one of a defined power sweep range, even if it overdrives the amplifier. In this way a regular array of Pin, Pout and gain data can be created. Some VNAs already provide 2-D sweeps, and these also allow the choice of sweeping power and stepping frequency, or sweeping frequency and stepping power. Except in a rare circumstance, the frequency should always be swept as the power is stepped from low to high, to avoid overdrive-related issues with measurements of the linear power.

If a data-set with this two-dimensional array of gain versus input power and frequency is exported, a surface plot may be simply created using a variety of plotting tools such as MATLAB, to create a three-dimensional surface of compression versus power and frequency, as shown in Figure 6.22.

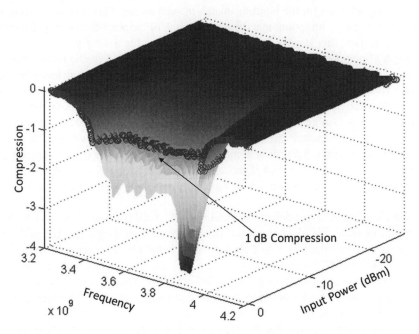

Figure 6.22 A 3-D surface of compression versus frequency and input power.

Also shown in the figure is a selection of points that represent the points closest to the 1 dB compression value.

6.2.4.3   Calibration in Compression Measurements

One important point to note is that for gain compression measurements, the error correction math of Eq. (3.3) does not strictly apply. The concern about this math function is that the input reflection errors are not properly accounted for when an amplifier is in compression. Take the extreme case of an amplifier in saturation. The output power is limited at a fixed level by the saturation. If there is mismatch ripple at the input which causes the incident source power to be greater than that of a matched source, there will be no corresponding increase in the output power. However, the error-correction function of Eq. (3.3) supposes that an increase would occur (S21 being linear in the assumption) and so the corrected gain and output power (which is sometimes computed as input power times gain) are reduced by the input mismatch error. In fact, the input mismatch error causes no change at all in the output power (the amplifier being saturated), and thus the gain will show ripple which it does not really have. Therefore one should not use normal error correction for amplifiers tested in their non-linear mode of operation.

Instead, an alternative computation of gain and output power can be performed using the definition of gain as the match-corrected output power over the match-corrected input power. In such a definition, there is no assumption of linear behavior for S21 and so that any error caused by mismatch at the input will not be reflected in output power. The gain of an amplifier will be properly adjusted for the input mismatch in such a case. This type of correction does have one problematic assumption: that the output match of the amplifier does not change with drive power. In fact, this is not likely to be the case, but for a simple compression setup, this assumption cannot be avoided. More advanced non-linear VNA methods can address such situations, as discussed in Section 6.8 on X-parameters. In the case where an amplifier's output impedance varies greatly with drive level, the best approach to getting quality measurements may be to add some small attenuator at the end of the port 2 cable, to improve the raw match.

Another issue that is more prevalent when an amplifier is just starting to go into compression is the error in the input power due to input mismatch. Because of this effect, the DUT will have a higher than expected power at some frequencies and a lower than expected power at other frequencies. This is true even if the source power of the VNA is calibrated, because the calibration was performed into a well matched power meter but the DUT may present a poor match.

An example of the S21 at linear power and near the 1 dB compressed power is shown in Figure 6.23. For each case, the normal full two-port correction is applied, but in the case of the compressed power, the S21 gain (labeled "S21 Comp") shows excessive ripple. Also in the figure is a trace of S21 with receiver leveling applied, near compression. In this case, the input power is controlled to provide a flat, leveled power even in the case of mismatch, and so the compressed gain shows a smooth response as one might expect. A reference trace of S21 for linear drive powers is shown as well.

From this figure, one can see that the apparent ripple in the "S21_Comp" trace is not truly representative of the gain at that power, due to the non-linear change in S21 responding to the mismatch-induced ripple of the input power.

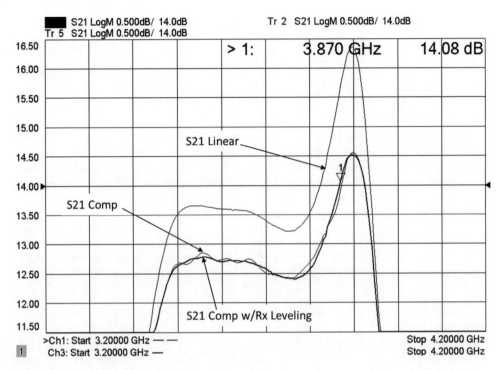

Figure 6.23 S21 gain of an amplifier in compression, normal and with match corrected Rx leveling applied.

6.2.4.4 DC Power Analysis

For many amplifiers, the efficiency of the amplifier in terms of creating RF power from the DC power consumed is a critical performance parameter, with the most common form being the power added efficiency, (PAE), which is defined as

$$PAE = \frac{Output\_Pwr - Input\_Pwr}{DC\_Pwr} = \frac{(S21 - 1)}{DC\_Pwr} Input\_Pwr \qquad (6.11)$$

where the power is expressed in watts or milliwatts. Since the DC power of a linear amplifier is essentially constant (one definition of linear being that the RF signal level is so low that the DC operating point of the amplifier is not changed by it), the PAE will increase with increasing input power until the amplifier becomes non-linear and gain compression occurs. For many amplifiers, the measurement of PAE in the non-linear region is a key design parameter, and this measurement depends greatly upon other aspects of the design such as the termination impedance of the amplifier at the fundamental, second and third harmonics. Often, PAE is measured as part of a load-pull measurement to find the optimum load for power efficiency. Once this load is determined, a matching network is designed and the overall matched amplifier response must then be re-measured, into a matched impedance load (normally 50 ohms).

As described in Section 6.1.4.2, on measurements, some VNAs provide a means to measure the voltage and current of an amplifier synchronously with the frequency or power sweep. In such a case, the equation editor function can be used to implement Eq. (6.11) directly, or there

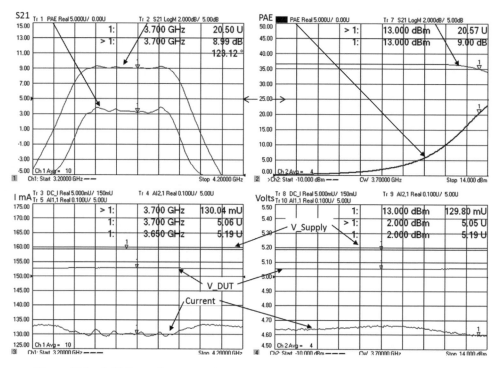

Figure 6.24 (left) Swept frequency PAE; (right) swept power PAE, lower are DC readings.

are even some built-in functions that have the definition of PAE already specified. An example of a swept-frequency PAE measurement, as well as gain, is shown in the left upper trace of Figure 6.24, along with voltage and current in the left lower.

A similar measurement is shown in the right windows of Figure 6.24 for a swept power PAE measurement. For this device – as is common in most amplifiers – the efficiency improves as the device approaches compression.

When a full two-dimensional gain-compression measurement is performed, it is convenient to also acquire the DC voltage and current for each frequency and power point. Some VNAs provide this as part of the built-in gain compression data acquisition, and the saved data can be processed to create power added efficiency surfaces that show the PAE as a 3-D surface function of input power and input frequency, an example of which is shown in Figure 6.25.

6.3 Measuring High-Gain Amplifiers

Most amplifiers are low to moderate gain, in the region of 10 to 30 dB of gain. For these amplifiers, no special techniques are needed in setup, calibration or measurement as the normal setup and source ALC range will allow a reasonable calibration at the maximum of the source power, and then a reasonable measurement at a lower source power. At these gains, the input and output match are quite reasonable to measure as well as the reverse isolation. However, as the gain of the amplifier becomes high, the input power must be greatly reduced

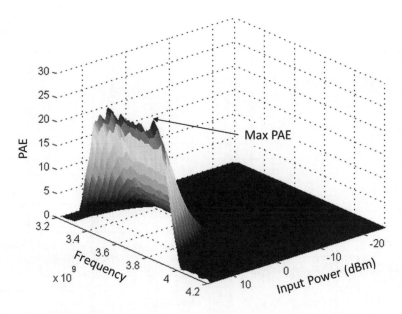

Figure 6.25   PAE versus power and frequency on a 3-D surface.

and the raw measurements of the other parameters can become very noisy. This in turn affects the quality of the error correction, resulting in poor measurement results for all S-parameters.

Consider an amplifier with greater than 60 dB gain and 10 dBm compression level, as might be commonly found in a communications system LNA. The input drive level for linear operation of this amplifier is found by subtracting 20 dB from the compression point, and 60 dB due to the gain, to yield −70 dBm input drive level. From a preset condition, if the power level is simply changed to −70 dBm very poor error-corrected measurements will result. An example of such a measurement is shown in Figure 6.26, with both raw measurements (uncorrected) and corrected measurements. The default setup for most VNAs is to couple the port powers, so setting the port 1 power to −70 dBm also sets the power of port 2 to −70 dBm. Notice that the raw S21 measurement is much less noisy than the corrected measurement and the same is true of the S11 measurement. The S12 measurement is essentially just noise, in both raw and corrected. The most interesting result is the S22 measurement: while the raw S11 looks OK (with some noise), the S22 is completely full of noise and is in fact showing a measurement greater than 0 dB. To understand how this can be, consider also the b2,2 trace, which is a measure of the b2 power received during the S22 sweep. Since the port 2 test port power is −70 dBm, one would expect the b2 power to be below −70 dBm; instead it is around −58 dBm. The reason for this is that the high gain of the amplifier produces high noise at its output, even with no input signal, and so this high noise swamps the measurement of S22, and both the raw and corrected results show just noise. Methods for removing these effects and improving these terrible measurements are described in the following sections.

The question remains: why are the S11 and S21 corrected measurements so bad, since their raw measurements are relatively noise free? The root cause of this is that for the low drive power, the S11 measurement has some noise, the S22 measurement has terrible noise, and the

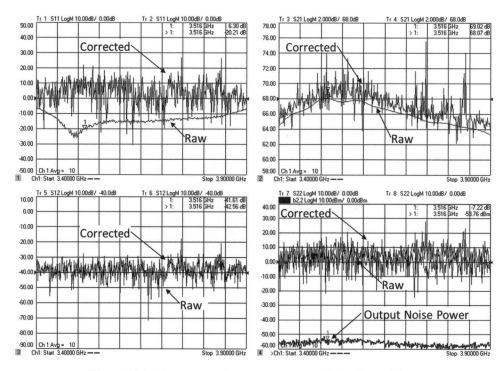

Figure 6.26 Error-corrected measurements on a high-gain amplifier.

S12 measurement is substantially all noise. This is because the low input drive power from port 2 does not allow enough signal to capture the behavior of the amplifier. Consider S12: The S12 of the amplifier is actually lower than -110 dB, but with the drive level of port 2 being just -70 dBm, and the noise floor of b1 receiver (used for S12 measurements) being -110 dB, the apparent S12 is only -40 dB ($-110 - (-70)$) due to the noise floor limitations. Thus the "loop-gain" term for error correction from Eq. (3.3) becomes very large as the product $S21 \cdot S12$ is nearly 30 ($\sim$30 dB), much greater than one (0 dB) that is required for a stable amplifier. In such a case the error correction essentially adds the noise of S12 onto the S11 and S21 traces.

In addition to the noise from S12, noise in the raw S11 and S22 will also be translated into noise on the S21 trace through the error correction mathematics.

6.3.1.1 Setup for High-Gain Amplifiers

Avoiding these noise issues is quite simple if the setup of the measurement is slightly modified. For any amplifiers, but most especially for high-gain amplifiers, the test port powers for each port should be uncoupled. The power for port 1 of course must be set to the linear power level, but the power for port 2 does not need to be set to the same level. In fact, during normal operation, the power level of port 2 should be set to the linear input power, plus the gain, less about 10 dB. This ensures that the power is sufficiently large to avoid noise issues, but is always small enough to ensure linear operation of the amplifier, as it is much below the normal output power of the DUT. Raising the port 2 power has the benefit of reducing the noise in both the S12 and S22 traces. In this case, the noise is reduced by the equivalent of 50 dB, or about 300 times reduction in trace noise.

In some cases, the S11 trace remains noisy even after raising port 2 power; the noise on the S11 trace is imparted on the S21 corrected results yielding a noisy S21 trace. An additional modification of the test setup can help reduce this noisy effect. If the VNA being used has a configurable test set, that is, if the connections between the VNA source and receiver to the directional coupler are exposed as loops externally, then the port 1 test coupler can be "reversed" in much the same manner as described in Section 5.2.6, shown here in Figure 6.27. With the source routed to pass through the coupled port, a higher reference channel power will occur for the same test port power. In addition, the b1 reflection receiver is connected to the through arm of the test-port coupler lowering the loss to the b1 receiver. In this way the signal-to-noise ratio of the S11 measurements will improve by about 14 dB, resulting in around a 25 times improvement in S11 trace noise. This same change also improves the signal-to-noise of the S12 trace, yielding further benefits for noise reduction in the overall response.

6.3.2 Calibration Considerations

When testing high-gain amplifiers, the source power must be reduced, but this leads to low signal levels during calibration and excess noise in the calibration error terms which are imparted to the corrected trace appearing as a stationary noise-like error or ripple.

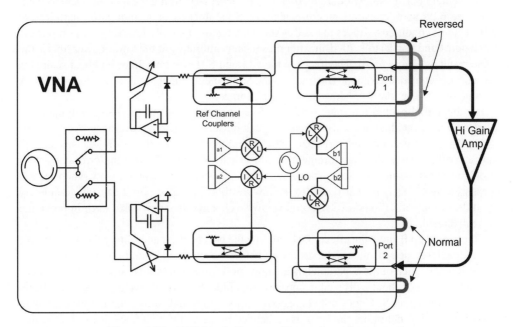

Figure 6.27 VNA block diagram with port 1 coupler reversed.

For modern VNAs, the receiver linearity is so good that there is no degradation raising the source power for calibration and then lowering it for measurement. In setting the measurement for optimum results, the source attenuator should be set so that the linear power level should be at the very bottom of the ALC range for that attenuator setting. Presumably, there is no reason to measure the amplifier below the linear power. Then, for calibration, the source power level is raised to nearly the top of the ALC range. For very best results, the top 5 dB should be avoided in some older VNAs due to some compression of the reference receiver, but if the linear power is very low, even the top 5 dB of the ALC range should be used as the compression may be much lower than the trace noise.

While the idea of calibrating at a high power and measuring at a low power is disturbing to those who were brought up on the idea that changing power after calibration invalidated the calibration, it is very easy to demonstrate that in almost every case, the error from raising the power for calibration is orders of magnitudes less than the error due to noise in the calibration.

Further, if the gain of the amplifier is greater than the change in source power, additional averaging or IF BW reduction should be enabled during calibration at the rate of 10 times change for each 10 dB of DUT gain above the power offset. For example, a 60 dB gain amplifier should have the power set to maximum during calibration; during measurement the power will be set approximately 40 dB lower. To ensure the noise during calibration does not limit the quality of the measurements, and additional 20 dB worth of noise reduction should be used, meaning 100 times IF BW reduction or 100 times averaging. Of course, this supposes that during the amplifier pretest (when the power levels and IF BW should be determined for the particular amplifier used) the IF BW was set so that the trace noise during amplifier measurement is just at the desired level. Finally, depending upon the setup, the noise in the reference channel may limit the measurement results when the power is reduced, so some extra averaging or IF bandwidth reduction may be necessary during the measurement as well.

Figure 6.28 shows several measurements of S-parameters of a high-gain amplifier for different test conditions. Traces marked A show the result of simply lowering the port powers (both port 1 and port 2) to −70 dBm, after preset, and calibrating at the lower power (this is the same result as Figure 6.26). The traces in the S11 and S21 windows have identical scale/div, but each offset by one division. The traces in the S12 window (upper left), have identical scales with no offset. The lower window shows the S21 trace.

For this initial calibration state, while the S21 trace has about the correct amplitude, it is extremely noisy, with about 5 dB of peak-to-peak noise, even after 10 averages.

The S12 trace is completely noise at −40 dB, and the S11 trace shows only noise centered above 0 dB! Clearly, this measurement is almost entirely invalid, with the error-correction itself causing substantial noise, as shown in Figure 6.26, when compared to raw (uncorrected) measurements. This is primarily due to the low test port power of port 2 causing completely invalid readings of S12 and S22. Several stepwise changes are shown that each improve the corrected result.

Traces marked B show the effect of uncoupling the port powers, and setting the port 2 power to a higher level, at 0 dBm (which is still well below the expected saturation of the amplifier). In the B traces, each of the S-parameters are now more valid, but there is still substantial noise on the S21 trace of about 2 dB peak-to-peak (here, unlike trace A, no averaging is used). Here, the residual noise in S21 does not change sweep-to-sweep, indicating that the measurement itself is not noisy, but rather the noise is embedded in the calibration trace due to low source power used during calibration.

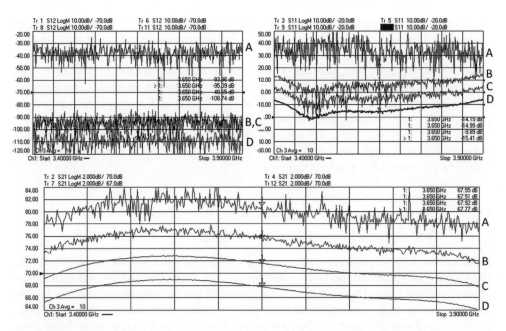

Figure 6.28 S21 noise on a high gain amplifier with various settings; S11 and S21 reference offset 1 and 2 divisions, respectively, for clarity.

Trace C shows the result of performing the calibration at a higher power while measuring at a lower power. Now the S21 trace is clean (the trace noise on the thru-tracking term during calibration has been reduced), but this has no effect on the S11 trace, nor on the S12 trace, and they are unchanged relative to the B traces.

Finally, trace D shows the result of all the above settings, plus reversing the port 1 coupler. The port 1 attenuator and power setting is changed to increase nominal power by 13 dB to provide the same −70 dBm incident power to the DUT. With this configuration the S12 trace drops by approximately 13 dB, indicating improved dynamic range in the reverse direction, and the S11 trace has much lower trace noise, indicating improved S11 sensitivity in the reversed coupler case, for the same −70 dBm signal applied to the DUT. There is even an improvement in the S21 trace noise as the effect of a noisy S11 trace is eliminated. Thus the configuration of Figure 6.27 shows the clear benefit of proper setup and settings for calibration and measurement of high-gain amplifiers.

6.4   Measuring High-Power Amplifiers

High-power amplifiers are widely used in radar and communications systems. For purposes of this book, amplifiers are considered high power if they cannot be measured using the normal configurations of the VNA, and require either external-booster amplifiers, external couplers and attenuators, or both.

High-power amplifiers can be segmented into some classes that dictate the changes required to the setup for the VNA. Amplifiers with high drive requirements will require a booster amplifier in the source path. Moderate power amplifiers, below 1 watt (+30 dBm) output, can generally be tested directly at the test port of many VNAs, sometimes with some small amount of simple padding at the port.

Medium-power amplifiers, between 1 watt and 20 watts (+30 dBm to +43 dBm) can be tested with many VNAs directly using the built in test-port couplers, but require some reconfiguration behind the test-port coupler to add isolators or pads to reduce the signal level to components behind the test-port coupler.

High-power amplifiers, above 20 watts, generally need to have external couplers and external high-power isolators and attenuators connected to the VNA source and receiver, essentially bypassing the internal VNA test set entirely.

6.4.1   Configurations for Generating High Drive Power

6.4.1.1   Moderate Drive Levels (Less than +30 dBm)

There are two basic configurations for creating high-power drive signals from a VNA. The first simple configuration is available with some modern VNAs that provide rear-panel loops which allow direct access between the VNA source and the reference channel coupler. Adding high drive power for this configuration is as simple as adding a booster amplifier in the loop. Drive levels up to approximately +30 dBm are possible with this simple scheme. Some other modifications might be required such as adding an attenuator in the reference path to reduce the signal to the a1 receiver, Figure 6.29. At sufficiently high drive powers, an additional attenuator may be needed in front of the b1 receiver; often a VNA has built-in switched receiver attenuators for this purpose.

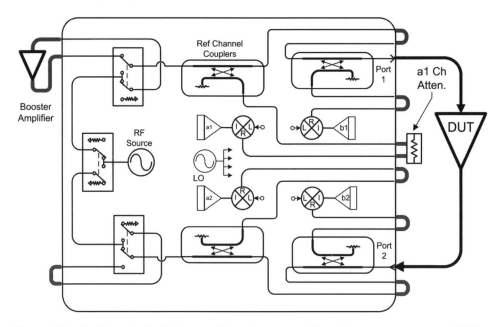

Figure 6.29   Configuration for high power drive using rear panel loops for test port powers to +30 dBm.

Typically, there is some setting in the VNA software, called source offsets, which allows one to provide an offset to the source power so that the drive levels are approximately correct even without calibration. This is very convenient to avoid overdrive issues that can occur when the source power settings don't match the test port power.

6.4.1.2   High Drive Levels (Greater than +30 dBm)

The second configuration, Figure 6.30, is required if the VNA source path components cannot handle the required drive power. In this version the booster amplifier comes from the source output through a loop between the reference coupler and the test coupler. The output of the booster amplifier is routed through a high power coupler that provides a signal to the reference receiver (a1) of the VNA. Some additional attenuation may be needed to avoid overdriving the receiver.

The through path of this high-power reference coupler is then either routed through the test-port coupler (if it can handle the drive power), or routed through a second high-power coupler, as shown in Figure 6.30. The coupled arm of this reflection coupler is routed to the port 1 reflection receiver, b1. The combination of coupling factor and some external attenuators should be added to provide a sufficiently small signal to the VNA receivers for the maximum drive signal from the amplifier. As a safety precaution, one should never set up a configuration that would allow the booster amplifier to overdrive the VNA test receivers. These receivers usually have a guard band of damage level that is 10–15 dB above their maximum operating range. In general, use the smallest power booster amplifier that will support the test needs.

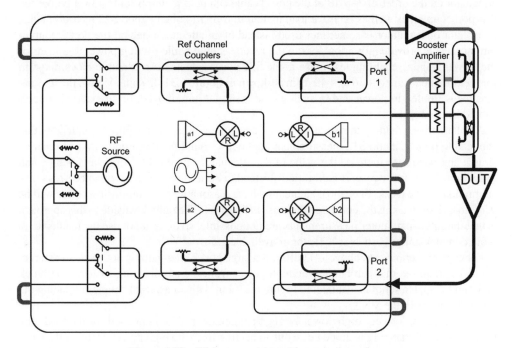

Figure 6.30   High power drive with external couplers.

In each of the configurations discussed, the full S-parameter functionality of the VNA is preserved. Of course, if the DUT has gain, then additional steps are needed to allow the VNA to receive high power, as discussed in the next section.

6.4.2   Configurations for Receiving High Power

Once a high-power test system has been designed to deliver sufficiently high power to the DUT, the configuration for receiving even higher power from the amplified signal must be considered.

For moderate power devices, less that +30 dBm, the amplifier can be connected directly to port 2, as long as the source and receiver attenuators are set sufficiently high to prevent overload on the internal components such as the source/load switch and the b2 test receiver. The internal attenuators are typically rated for about +30 dBm. Even for powers up to perhaps +36 dBm, the simplest solution is to add a high-power attenuator to the end of the port 2 cable. In such a case, the S22 measurements are generally valid even up to 10 dB loss in front of the test-port coupler. For even higher powers, larger attenuation may be used, but in such a case the reverse match, S22, becomes even more noisy and less reliable.

In cases where full two-port calibrations are required, and for powers up to about +43 to +46 dBm, the internal coupler of the VNA may still be used (check manufacturers' specifications for maximum power handling). In this case, external attenuators or isolations should be used to drop the drive power sufficiently to avoid overloading the VNA components such as the port 2 source/load switch. Typically, these components can handle power up to +30 dBm damage level, but operationally they are limited to perhaps +20 dBm. Again, adding a high power attenuator on the order of 3–6 dB at the port 2 cable can help to dump some power before the test-port coupler, as well as improve load match. In trying to obtain good S22 measurements, an isolator or circulator is sometimes used instead of an attenuator behind the port 2 coupler thru arm. This provides for a low loss to the reverse signal while routing power to a load for the forward signal, enabling less noise in the S22 measurements. An example block diagram for measurements up to +46 to +49 dBm is shown in Figure 6.31. In some VNAs, the internal bias tee is limited in power and in such a case it should be removed, or an external coupler should be used.

An alternative to adding the isolator is to use an attenuator behind the port 2 coupler, but that can lead to poor S22 and S12 measurement due to low signal from test port 2. An alternative simple approach is to increase the value of the attenuator on port 2 to some large value, and abandon using a full two-port calibration. Instead, a simple response or enhanced-response calibration can be used. Since most attenuators have reasonably good return-loss, there will be very little error in using the enhanced response calibration. The only downside to this approach is that the full S-parameter measurements are not possible, so one cannot use the results to do more complex analysis such as K factor or available gain.

For even higher powers, above +43 dBm, if a large port 2 attenuator is not used, an external coupler is required that can handle the high power from the amplifier under test. The block diagram for such a setup is identical to Figure 6.31, but with an external coupler replacing the internal coupler of the VNA.

In some cases, for very high power levels, the attenuator or isolator used for the load may have an issue of changing its impedance due to heating from the high power. If this is the case, then it may be necessary to reconfigure the test system to use three high-power couplers as shown in Figure 6.32. With such a configuration, the load impedance may be monitored using

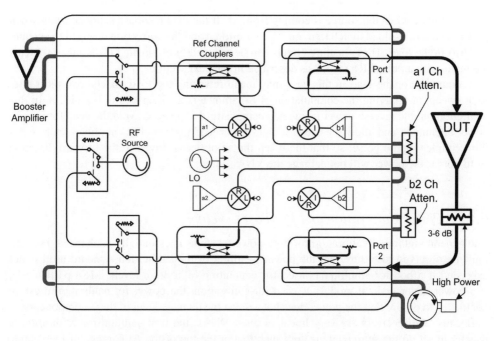

Figure 6.31 Measurement setup for +46 dBm maximum power.

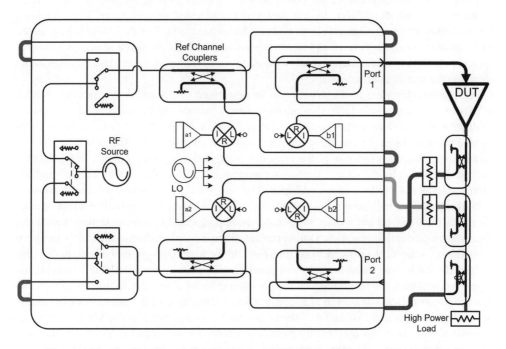

Figure 6.32 Configuration for high power test where the load changes with power level.

the a2/b2 trace when the source is coming from a1. If the load response is not constant when port 1 is driven, the load match term must be monitored and the correction changed to account for drift in the load. These heating effects can be reduced or eliminated if the measurements are made with a pulsed RF signal, as described in the next section.

Accounting for the load in this case may mean rewriting the load match error term after each sweep, or applying the computation of S-parameters as described in Eq. (1.21). In this configuration, the reverse power is injected using a third coupler to avoid the need for a high-power attenuator, and allow the use of a high-power load instead. The coupling factor of this third coupler should be set so that the when the amplifier is driven in the forward direction from port 1, the power will not damage the VNA source.

6.4.3   Power Calibration and Pre/Post Leveling

The general method of calibration for high-power systems is essentially unchanged, but care must be observed to ensure that the power handling of the calibration standard used is not exceeded if a booster amplifier is used for generating high drive levels. Most particularly, if a power meter or Ecal module is used for calibration, the power for calibration must be controlled to ensure that the power does not exceed the damage level of these components.

Because the receivers are very linear in most VNAs, the best solution for calibration is usually to set up the source attenuators such that at the top of the ALC range of the source, the output power is just enough to drive the maximum required output power of the booster amplifier. Then for calibration, the power may be reduced to near the bottom of the ALC range. In many modern VNAs, this can be up to 40 dB lower, or even more. Thus, for drive powers below about 40 dBm, this is no problem for either the Ecal or the power sensor, as the ALC allows the cal power to be as low as 0 dBm. Some VNAs provide a means to bypass or open the ALC loop altogether, and allow power control ranges of more than 70 dB without changing the source attenuator.

The power calibration of the system can be performed independently or, in some VNAs, as part of a guided power calibration method. Great care should be taken during the power calibration portion as the source power is iterated to find the proper drive level, and it is possible that the maximum source power may occur during the iteration process. The difficulty in power calibration is that if a booster amplifier is used, the actual source power out will be higher than the nominal power by the gain of the booster amplifier. Consider the case of a drive amplifier that can produce +35 dBm, with 25 dB gain. The source can be set so that maximum power is +35 dBm and minimum power is −5 dBm. A reasonable power such as 0 dBm might be used for the calibration, and the source power calibration performed to achieve 0 dBm, before the S-parameter calibration proceeds. During the initial setting and reading of the source power, it is common for the VNA software to set the source to the nominal requested power, in this case 0 dBm. But the booster amplifier gain will generate +25 dBm or more, perhaps destroying the power meter. Thus, it is recommended to add some external attenuator to the power meter so that the power meter is safe even at maximum drive. The value of the attenuator can usually be compensated for as a loss factor in the power meter configuration of the VNA.

Some VNAs avoid this problem by providing a source power offset entry, which accounts for external gain (or loss) so that the nominal source power setting includes the effects of the booster amplifier. This is usually a fixed offset, so there can still be difficulties if the gain of the booster amplifier has large peaking or is not flat. With this offset, the initial setting of the source is lowered by the gain of the amplifier, and this will avoid overdrive issues. To be safe, the maximum gain of the amplifier should be used for the offset value.

During the S-parameter portion of the calibration, the averaging factor should be increased, or the IF BW should be reduced, so that noise in the calibration does not degrade the measurement results. As a rule of thumb, the averaging should be increased, or IF BW reduced, by a factor of 10 for each 10 dB of power level reduction between calibration and measurement, and for each 10 dB of gain of the DUT, to ensure that the noise contribution of the calibration is on the same order of magnitude as during the measurement.

One final aspect of high power measurements is that for many amplifiers, the performance is specified at a particular output power, rather than some input power. Gain-at-rated-power is one example, where the gain of an amplifier is specified to be at or above some level for a given output power of the amplifier. For these measurements, it is critical that the measurements be made at exactly the output power of the specifications. To achieve this, the receiver leveling function described in Section 6.1.4.1 is modified so that the b2 receiver is used as the power detector, and the source power is iterated so that the output power is maintained at a constant level.

6.5 Making Pulsed-RF Measurements

For high-power amplifiers or amplifiers operating near compression, the RF dissipation of the DUT can cause heating effects that will change the measurement results. This is particularly true for on-wafer measurements where it is not possible to adequately heat-sink the die. In these cases, making pulsed measurements with low duty cycles can avoid the problem of device self-heating. Other devices are designed to operate only in a pulsed mode, and so pulsed S-parameter measurements are required.

In older systems, creating and synchronizing pulsed measurements was rather involved. The RF signal from the source must be sent to an external pulse modulator, which in turn must be driven by an external pulse generator, and the pulsed RF signal should be routed to both the reference and test channels to ensure that drift in the pulse modulator is accounted for; thus external couplers were required. Older, legacy VNAs had relatively narrow IF BWs, so that only wide pulses could be used, as the pulse on-time must be long enough for the VNA to capture a data point in the IF, and the triggering of the pulse generator, modulator and measurement receiver must be all synchronized, which might require special interface circuits. However, these methods are largely replaced now by modern techniques.

6.5.1.1 Wideband vs Narrowband Measurements

Wideband pulse measurements refer to the idea of using a wideband IF, with a very fast response time, to measure the RF signal during the pulse on-time. For example, if a 10 μsec RF pulse is used, the IF BW would need to be wider than approximately one over the pulse width, or wider than 100 kHz, to capture the energy of the pulse. Typically, an IF BW of 1.5 times is the minimum used to ensure that the entire IF measurement is made even if there are some timing errors and pulse delays. A 15 MHz bandwidth, the widest currently available on modern VNAs, can provide for measurements on pulses as narrow as 100 nsec. Figure 6.33

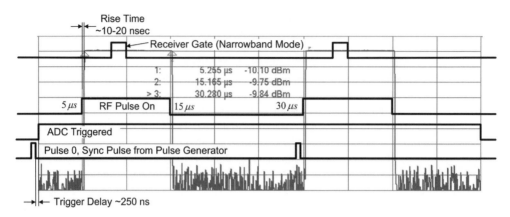

Figure 6.33   Timing diagram for wideband pulsed measurements.

illustrates the timing diagram for pulsed measurements, overlaid on an example measurement. Here the pulse is set to 10 μsec on-time with 25 μsec pulse repetition time. The RF pulse modulator signal is delayed by 5 μsec after the start-of-sweep to see the full pulse rise time. It is common to have some delay due to hardware path differences between the pulse generator's sync pulse (called Pulse 0) and the start of ADC data taking. Slightly offsetting the delay of Pulse 0 can compensate for these delays. Also, the RF power signal will be delayed by the rise time and response delay of the pulse modulators in the VNA (or external pulse modulators if the VNA does not have an internal pulse modulator).

For narrower pulses, or for systems with narrower IF bandwidths, an alternative method called the narrowband approach can be applied to achieve measurements on very narrow pulses, down to 10 nsec pulse width [3]. In the narrowband mode, the receiver of the VNA is time-gated to measure just a narrow portion of the pulse signal, as indicated by the upper line of the timing diagram. The narrowband approach relies on the fact that a repetitive RF pulse will have a spectrum of discrete frequencies related to the pulse repetition frequency, as shown in Figure 6.34. If a narrow band IF filter is centered on just one of these discrete frequencies, with filter zeros centered on each of the other frequencies, then the overall RF measurement can be made using very narrow IF filters, by accumulating a signal in the IF. The difficulty in this method is that custom IF filter bandwidths are needed for each different pulse period. In a sense, the narrowband mode averages the results of several pulses in the IF filter, and displays the result of this average. The only way this average gives a good response is if all but the central spectral element are removed from the average, through the use of the custom filters. Figure 6.34 illustrates the spectrum of a pulse and time domain measurement of the pulse profile, for a 3.3 μsec pulse with a 50 μsec pulse repetition time (20 kHz pulse frequency). The 20 kHz pulse repetition frequency implies that the spectrum should have 20 kHz spectral components; marker 1 is set at the spectral peak, and marker 2 is at the next spectral line, exactly 20 kHz away. A narrowband IF filter would need to pass the center of this pulse, and have a zero at 20 kHz intervals. In practice, even narrower IF filters can be used to lower the noise floor, as long as the IF filter has transmission zeros at each of the other spectral lines. Since the pulse width is 3.3 μsec, one would expect nulls in the pulse spectrum

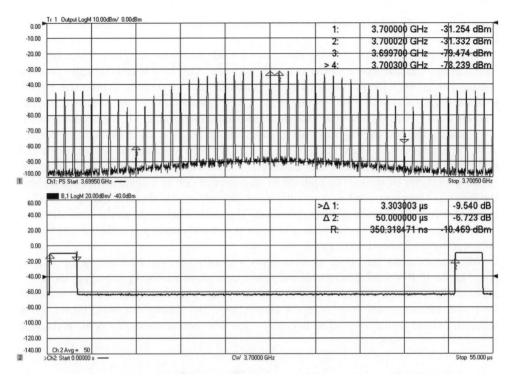

Figure 6.34 Narrowband pulse measurement spectrum and time measurement.

every $1/(3.3 \times 10^{-6})$ or 300 kHz. Markers 3 and 4 are positioned ± 300 kHz from the center of the spectrum and indeed have substantial nulling.

With narrowband pulse methods, the average power displayed is less than the true power by the ratio of the receiver gate time to the pulse repetition time. For power measurements in narrowband mode, the received power should be modified by this ratio to reflect the true power in the pulse, using a magnitude offset function or an equation editor function.

6.5.1.2 Point-in-Pulse Measurements

The most basic pulsed RF measurement is measuring the S-parameters and power of an amplifier in the center of an RF pulse, over a range of frequencies. This is essentially just a standard measurement, but with a pulsed RF stimulus. This is sometimes referred to as point-in-pulse measurement and refers to the fact that for each point on the trace for the frequency sweep, one RF pulse is measured, typically in the center of that pulse, as illustrated in the timing diagram in Figure 6.33. Thus, the normal frequency response of the amplifier is measured, and the RF pulse becomes just another stimulus setting.

Most modern VNAs now have a high-speed digital IF, some with sample rates up to 100 MHz. Some high-performance units have built-in pulse modulators and pulse generators, all internally synchronized, making pulsed RF measurements particularly simple. In fact, some application programs require that the user only enter the pulse width and pulse period (or pulse repetition frequency) and every other setting is automatically adjusted to create the proper pulsed measurement.

For many pulsed systems, the pulse modulator is inside the source ALC loop. Since the ALC loop typically has a response slower than the pulse, the ALC loop is automatically disabled and put into an open loop or sampled mode. Since the internal detector is not used, the RF level can have substantial errors in this mode. Once again the use of the reference channel receiver as a leveling detector can provide a substantial improvement in the accuracy of the drive power. The Rx-leveling function in pulsed mode uses the same method for leveling as in the normal mode, where background sweeps are used to iterate the source power and find the proper level; then a data-taking sweep is performed with the corrected source power. Figure 6.35 shows an example of a pulsed RF measurement on an amplifier, with and without Rx-leveling. The upper window shows the input and output pulsed power (along with S21), in the open-loop ALC mode, with no correction. The lower window shows the same stimulus, but with receiver leveling on the input. The input power and output power flatness are clearly improved by the Rx-leveling function.

Noise reduction for the power measurements is usually achieved by reducing the IF BW; however, the IF BW must remain wide for pulsed measurements. Sweep averaging for power measurements averages the power (rather than the voltage) because the phase is not coherent for power measurements on a sweep-by-sweep basis; this means that noise power is not reduced, and adds to the trace value. In contrast, for ratio measurements, the phase is retained

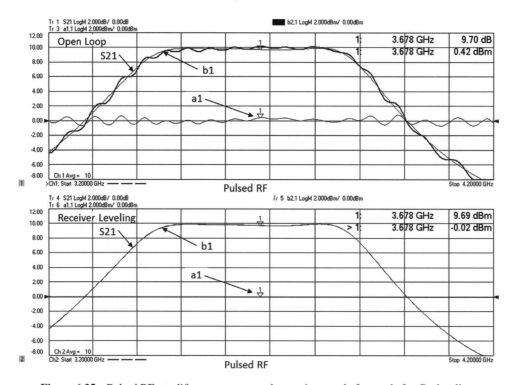

Figure 6.35 Pulsed RF amplifier measurement shows a1 power before and after Rx leveling.

on a sweep-by-sweep basis so the noise is averaged away in S21 for any averaging mode. However, some VNAs provide for point-mode averaging which takes many samples of the same point before moving to the next frequency. In the pulsed case, this means that several pulses are acquired and the results averaged. For power measurements, the phase is coherent from pulse to pulse and so vector mode averaging can be used to reduce the noise in pulsed power measurement. Thus point averaging provides a way to get low-noise power and gain measurements of a pulsed signal, where IF BW reduction is not possible.

In some VNAs, it is possible to use the current source settings of the Rx-leveling to update the existing source power calibration, so that the open loop mode can be used without the background sweeps of the Rx-leveling function. This is particularly important when pulse-profile measurements are made on amplifiers, as described in the next section.

6.5.2   Pulse Profile Measurements

For many amplifiers used in pulsed applications, the measurement of the amplifier's response to a pulsed RF input as a function of time is a key concern. These measurements are generally referred to as pulse-profile measurements, and show the gain, phase and power response of an amplifier to a pulsed stimulus versus time relative to the pulse. The measurement method differs significantly from the point-in-pulse measurement in that the entire measurement occurs over a single pulse. The use of a wideband digitizer sets the effective resolution of the pulse profile measurement to essentially the inverse of the widest bandwidth of the digitizer, that is, in wideband mode, the bandwidth of the filter sets the minimum step-size for time for the pulse profile resolution.

It is possible to set up pulse profile measurements using external pulse generators and modulators, but some modern VNAs have complete built-in pulse-profile functions that make the measurements completely simple. An example of a wideband pulse profile measurement is shown in Figure 6.36. In this case the noise floor and trace noise of the measurement is limited by the wide IF BW that must be used to capture the pulse. From the timing diagram of Figure 6.33, the pulse profile is measured by moving the receiver gate a small increment. In wideband mode this increment is the IF BW data taking time.

Trace-noise on ratio measurements such as S-parameters can be improved by sweep-to-sweep averaging, but the noise floor for power measurements cannot be lowered in this way. For the figure shown, 100 averages are used. In this case, a 3 MHz IF BW filter is used, so the resolution is about 330 nsec. The acquisition time for a 10 μsec pulse profile is 20 μsec including pre- and post-pulse measurements of 25% of the time axis. So 100 averages take only about 20 milliseconds of measurement time, and reduce the trace noise on the S21 trace substantially. The total acquisition time for the measurement in Figure 6.36. is about 30 msec due to retrace overhead.

Pulse-profile measurements are also possible in the narrowband mode. Narrowband measurements require many pulse acquisitions for each measurement result. In a pulse-profile measurement, the effective timing of the measurement is changed by incrementing the pulse signal to the receiver trigger (sometimes called receiver gating) so that each successive data point is acquired on a different relative position of the pulse. In narrowband mode the resolution can be set by the user; narrower gates means more pulses will need to be measured to average out noise due to a narrower gate causing lower signal level. The gate width (pulse

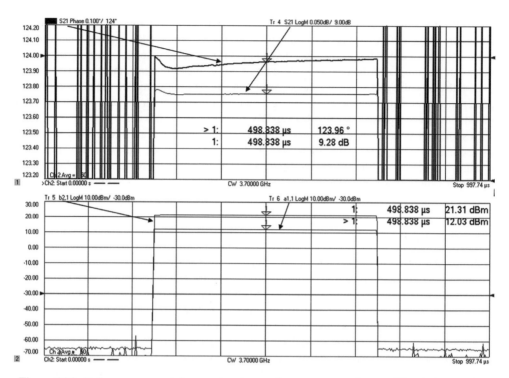

Figure 6.36   Pulse measurement shows gain, phase and output power of an amplifier during a pulse.

resolution) in narrowband mode is set by the pulse generator's minimum pulse width, or the minimum gate response time of the receiver gating circuit.

In this way pulse measurements are possible on pulses that are much narrower than the native IF BW of the VNA would allow. For VNAs with built-in pulse generators, this resolution can be a small as 10 nsec, or external pulse generators may be used where the limit is now the minimum gating time of the VNA receiver, perhaps as small as 5 nsec.

Older, legacy VNAs had narrower IF BWs, on the order of 10s of kHz maximum, so the minimum pulse time was limited to 100 μsec or so. Modern VNAs have high-speed digital IFs with greater than 10 MHz of bandwidth, so much narrower pulses can be measured in wideband mode, reducing the need for narrowband measurements in many cases.

An example of the narrowband mode pulse profile is shown in the upper windows of Figure 6.37. Note that this is really a composite response of many pulses, and represents an average of the pulse response over the pulses needed to acquire the IF BW of the narrowband mode. For comparison, a wideband pulse profile measurement is shown in the lower window.

In the example above, the pulse on-time is 1 μsec and the pulse repetition rate is 10 μsec, and a 500 Hz IF BW filter is used to remove the unwanted pulse spectrum. The upper-left shows the delta time from one point to the next in a highlighted box; here the step time is 11 nsec.

The lower windows show the same pulse measurement in the wideband mode, which has continuous sampling across the pulse. In this case, the VNA has a maximum IF BW of

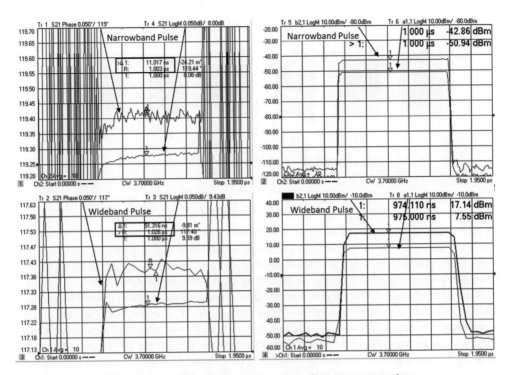

Figure 6.37   A narrowband mode pulse profile on a narrow pulse.

more than 15 MHz, which yields approximately a 50 nsec pulse resolution. In contrast, the narrowband-mode (upper windows) has a resolution that is limited only by the pulse generator resolution, in this case 10 nsec, so many more measurement points are available during the pulse period. Note that the pulsed power measurements (right windows) are offset in the narrowband mode (upper plot) by the duty cycle of the receiver gating to pulse period. In this case, this ratio is 11/10 000 for about a 59 dB offset. The narrowband method does allow a finer resolution to be seen on the S21 trace of the pulse.

6.5.3   Pulse-to-Pulse Measurements

Pulse-to-pulse measurements are used to measure the effect of device performance over time on a pulse train. In particular, on a high-power application such as a radar system, an amplifier may have more gain in response to a first RF pulse than to subsequent RF pulses due to heating or other effects. In a pulse-to-pulse measurement, the trigging is set to gate the receiver to measure only one point, on one pulse, for each measurement pulse trigger. The next pulse triggers a measurement on the same position in the next occurring pulse. If plotted versus time, this pulse-to-pulse measurement shows the droop or gain change from one pulse to another for the same point on the pulse. The idea is graphically represented in Figure 6.38. These measurements are usually performed in a single-sweep mode, and between sweeps the DUT is allowed to return to an ambient level.

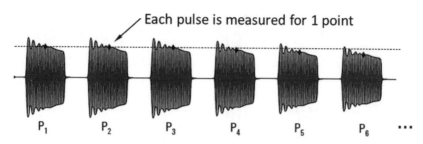

Figure 6.38   Pulse-to-pulse measurements.

6.5.4   DC Measurements for Pulsed RF Stimulus

A final aspect of pulsed amplifier measurements is measuring the DC voltage and current on an amplifier during the pulse, to compute the power added efficiency of a pulsed amplifier. This is particularly important when testing on-wafer where poor heat-sinking can cause significant amplifier performance variation due to heating in a CW measurement.

There are two basic DC configurations: the first configuration supplies constant DC voltage to the DUT, and the second pulses the DC as well as the RF. For some designs, the current dissipation doesn't occur until the RF signal is turned on. For these devices, constant DC sources may be used. Other devices require that both the RF and the DC be pulsed to avoid heating. For these devices, special DC switching circuits must be used as few commercial DC supplies provide for high-speed pulsed DC outputs. If the DC values can be measured during the pulse on-time, then the power added efficiency of a pulsed amplifier can be determined.

6.5.4.1   Pulsed-DC and Pulsed-RF Measurements

Pulsed DC stimulus can be combined with pulsed RF stimulus, provided care is taken to properly sequence the DC stimulus with the RF stimulus and measurement. A multichannel pulse generator is usually needed, to provide a wider DC pulse than RF pulse. The DC pulse usually must turn on before the RF pulse is applied. This sequencing can be created by making the DC pulse wider than the RF pulse, then delaying the RF pulse to center it in the DC pulse widow.

In either pulsed or fixed DC stimulus, the pulsed DC measurements are quite similar. As described in Section 6.1.4.2, some VNAs contain DC inputs that can also be gated along with the receiver inputs. Often the bandwidth of these DC inputs is fixed, so that the measurement time of the DC input is limited by the bandwidth. For a typical bandwidth of 25 kHz, the DC resolution time will be about 40 µsec. Since these ADCs are not synchronized with the internal DSP, it is difficult to use them for pulse-profile measurements of RF and DC, and their use is typically limited to swept-frequency point-in-pulse measurements.

Some VNAs, however, allow direct IF inputs into the high-speed digitizer used for the digital IF; in this case the DC pulse profile can be measured in a similar manner to the RF pulse, and higher speed measurements are possible. To allow this measurement, the digital IF center frequency is changed to 0 Hz, allowing direct measurement of the DC signal. Since the digital IF typically is centered at only one frequency at a time, the DC measurement must be made on one channel while the RF measurement is made on a different channel. This also

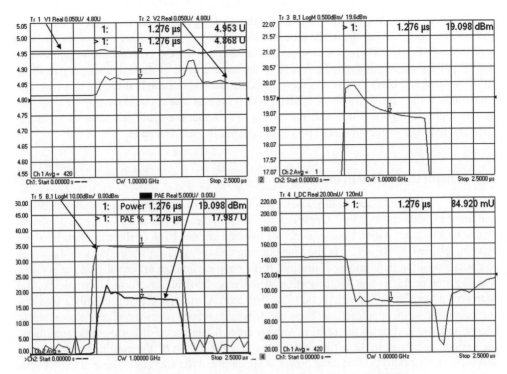

Figure 6.39 Pulse profile showing DC measurements and PAE.

presumes that the pulse response of the amplifier is consistent on a sweep-to-sweep basis, as at least two sweeps are required for acquiring both DC and RF power.

Figure 6.39 shows an example of a DC measurement on a 1 μsec pulse, using the two-channel approach with the high-speed digitizers measuring both the RF signal (from the VNA test ports) on one channel, and the DC signal (from the direct DC input) on another.

For this measurement, the DC input (V1 above) was connected to the DC source, and using a 1 ohm current sense resistance, the voltage on either side of the resistor routed was the IF input of the VNA, marked V1 and V2 in the upper left window. A 400:1 voltage divider was used to ensure the DC input did not exceed the maximum DC input of the digitizers; the series value of this divider was 10 kΩ, ensuring that very little current was absorbed by the DC measurement. Remote commands were sent to the VNA to set the IF frequency to zero, allowing DC readings with the IF inputs. The DC IF BW is set to be the same as the RF bandwidth, allowing noise reduction on the DC readings as well. The DC offset and scaling, as discussed in Section 6.1.4.2 is applied using the equation editor. When using the digital IF to measure DC in this way, the DC is represented as a complex number and the proper reading must be computed from $\sqrt{re^2 + im^2}$; this computation is performed by the equation editor in traces 1 (V1) and 2 (V2) in the upper-left window. The current is computed as (V1 − V2)/R, and is shown as I_DC in the lower-right window. The upper-right window shows a close-up scale of the amplifier output power and the lower-left window shows both the output power (at a large scale) and the power added efficiency, computed from the other traces, again using an equation editor function.

One aspect of interest on this amplifier is the peaking of the output power at the initial portion of the pulse. After the initial turn-on time, the output power decreases somewhat, also seen in the PAE trace. There is also some odd behavior in the DC current consumption when the RF pulse goes off; this may be related to the current draw of the amplifier (as part of an RF-dependent self bias) or it is possibly related to elements of the bias tee. In this case, the bandwidth of the bias tee is about 2 MHz, which implies that it could have some response to very high speed pulses. It is also interesting to note that the measured value of V1 is not constant, indicating some non-zero source impedance in the DC source.

6.6 Distortion Measurements

For basic amplifier characterization, distortion falls into two categories: harmonic distortion and intermodulation distortion. Amplifiers operating in the non-linear region create distortion products, which are a direct consequence of the gain compression of the amplifier. Looking at a single-tone CW RF signal in the time domain, compression causes the peak of the signal to flatten, resulting in harmonic distortion. If the compression occurs equally on the positive and negative peaks so that the distortion of the time domain waveform is symmetric, the harmonic distortion will be strictly odd (1st, 3rd, 5th . . .) order, and that is usually a sign that the bias point of the amplifier has been optimized. However, many amplifiers also generate second-order distortion in the form of second-order harmonics. For narrowband amplifiers, the harmonics created in the active device are often filtered out by the matching networks, so that there is essentially no harmonic power at the output of the amplifier. These amplifiers can still cause distortion of modulated signals, and two-tone or multitone signals can be used to ascertain the distortion quality of these devices.

6.6.1 Harmonic Measurements on Amplifiers

Traditionally, harmonic measurements have been performed by using a signal-source to stimulate an amplifier, measuring the resulting spectrum on a spectrum analyzer. For a direct display of harmonic distortion, the spectrum analyzer must be swept from the fundamental frequency up through the highest harmonic frequency; the other alternative is to make narrower bandwidth (or even zero-span) measurements at precisely the fundamental and harmonic frequencies, but this requires multiple measurements not normally supported on a spectrum analyzer. Some VNAs also provide a spectrum plot of frequency response. For either instrument, the broadband spectrum can take significant time to measure but the measurement results are very simply interpreted, as shown in Figure 6.40. The markers are used to measure the fundamental and harmonic powers. If a reference marker is placed on the fundamental, the harmonic markers can display their values directly in dB relative to the fundamental.

Most modern VNAs can be used to measure harmonics directly and quite simply, utilizing the frequency offset mode (FOM) capability built into many analyzers. In the FOM mode, the source and receiver frequencies can be directly and simply offset. One channel can be set to measure the fundamental frequency and another channel can be set to measure any of the harmonic frequencies. The dBc value of harmonics can be computed using the built-in equation editor, so that the entire measurement is captured in a single display. An example of this is shown in Figure 6.41. The fundamental trace is measured on one channel, and each

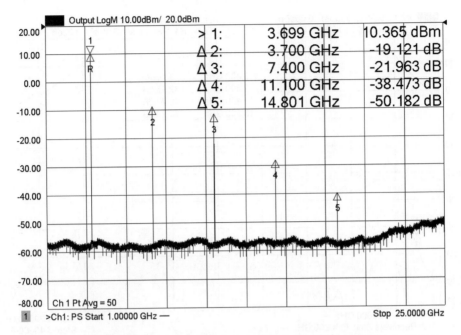

Figure 6.40   Spectrum plot of an amplifier's harmonic response.

harmonic is measured on a separate channel. In this way, a harmonic measurement across hundreds of frequency points can be completed in a fraction of a second.

For harmonic measurements, one must be sure that the source and receiver of the VNA does not produce harmonics in excess of the DUT harmonics. Most VNAs have rather poor harmonics, so some filtering is required between the source and the DUT. However, some high-performance VNAs include banded switch-filtering to produce exceptionally clean harmonic sources for distortion tests. One can evaluate the source and receiver harmonics by making two tests on the VNA with just a thru connection. The first test looks for source harmonics, so a large attenuator is added to the port 2 receiver to ensure that it is operating in a linear mode. The IF BW can be reduced to a very low value (even to 1 Hz) and the source harmonics measured in FOM mode for second-, third- and fourth-order harmonics. The upper window of Figure 6.42 shows a VNA measurement of harmonics (in dBc) with a 30 dB pad in front of the port 2 receiver (either an external attenuator, or an internal receiver-attenuator, if available, may be used). In this case, the source harmonics over the receiver bandwidth are lower than −62 dBc, measured at +10 dBm with a 10 Hz IF BW.

The lower window shows the same measurement, but with the attenuator removed. If the displayed harmonic level is above the source level, then one can assume that this is caused by the receiver harmonics. If the source harmonics are too great to evaluate the receiver harmonics, a filter may be used to evaluate a narrow portion of the receiver by removing any source harmonics. In this case, the receiver harmonics are as high as −37 dBc at +10 dBm input.

Calibration for harmonics can be derived directly from the receiver power calibration. The usual process is to turn off FOM mode and perform a power calibration over the source and

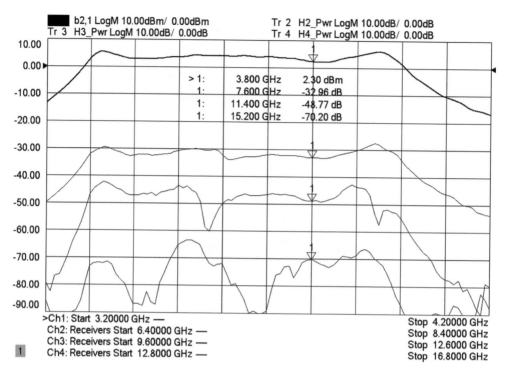

Figure 6.41   Harmonic measurements on a VNA.

receiver frequencies. When the FOM mode is turned back on, the calibration arrays are applied to the appropriate receiver frequencies automatically, performing interpolation on the arrays if required.

In some cases, the VNA receiver harmonics dominate the measurement and it is not possible to see the device harmonics because of this. This often happens with amplifiers that have built-in matching networks that filter out the harmonic response. The normal approach to improving receiver harmonics is to increase the attenuation to the receiver, and decrease the IF BW to make up for the loss of sensitivity. However, this method does not work for pulsed RF measurements, if the pulse width is narrower than the IF BW acquisition time. In such a case, the minimum IF BW is set by the inverse of the pulse width. This limits the maximum VNA receiver attenuation as the noise floor is set by the minimum IF BW. Unlike with a source measurement, one cannot simply filter the output of the amplifier to remove the fundamental because the harmonic level in dBc requires a measurement of both the fundamental power and the harmonic power.

In these situations, the solution may be to use a harmonic enhancement technique which utilizes a special circuit that provides some substantial but reasonable loss to the fundamental, while providing lower loss to the harmonics of the amplifier. An example of such a circuit can be constructed from power splitters, an attenuator and a high-pass filter, as shown in Figure 6.43, along with the frequency response of the circuit.

This circuit should be placed between the coupled arm of the port 2 coupler and the port 2 receiver (b2) so that any mismatch from the circuit is reduced by twice the test-port coupling factor. This technique is particularly useful in amplifiers with harmonic matching networks

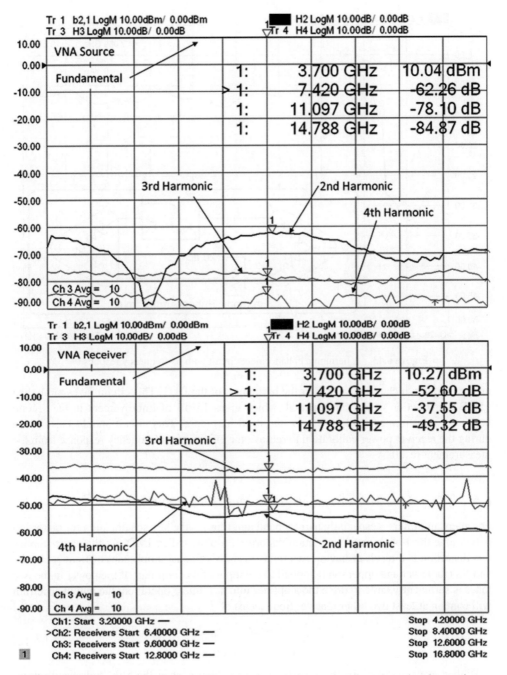

Figure 6.42 Evaluating VNA harmonics: (upper) source harmonics, (lower) receiver harmonics.

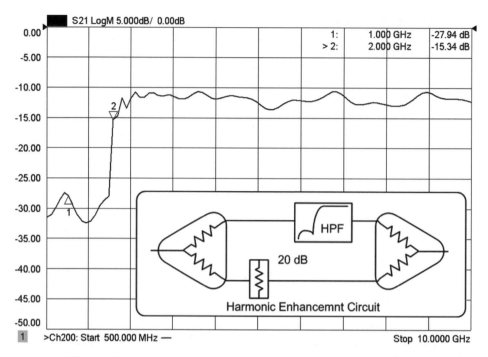

Figure 6.43   A harmonic enhancement circuit with its frequency response.

on the output, which effectively filter the harmonics of the DUT. This circuit gives more than 15 dB of reduction of the fundamental, which gives 15 dB of improvement in the second harmonic measurement and 30 dB for the third harmonic. This circuit should be in place during the receiver power calibration to remove the effects of its frequency response from the measurement results.

6.6.2   Two-Tone Measurements, IMD and TOI Definition

For many amplifiers – especially narrow band amplifiers – the key specification for distortion is two-tone third-order intermodulation distortion, usually called IMD. As the name implies this distortion is a result of having two tones (usually of equal amplitude) of frequencies F_L and F_U (for lower and upper tone) applied to the input of the amplifier. If the power in the two tones is sufficiently large to drive the amplifier into non-linear operation, the output spectrum will contain at least two other tones at frequencies of

$$F_{3U} = 2F_U - F_L$$
$$F_{3L} = 2F_L - F_U \tag{6.12}$$

Higher levels of drive will create higher-order products such as the fifth- and seventh-order products. If we define the center and delta frequencies as

$$F_C = (F_U + F_L)/2$$
$$F_\Delta = F_U - F_L \tag{6.13}$$

Then the Nth-order IMD product (for N odd) is defined as

$$F_{NU} = F_C + \frac{(N-2)}{2} \cdot F_\Delta$$

$$F_{NU} = F_C - \frac{(N-2)}{2} \cdot F_\Delta$$

(6.14)

There is some confusion over the definition and description of terms for two-tone intermodulation distortion; for the purposes here the following definitions will be used:

TOI:	third-order intermodulation product, the power in the third-order tone (also used to refer to this class of distortion measurements)
IMD:	intermodulation distortion, a general reference to this class of distortion measurements
main tone:	one of the two drive tones, measured at the output PwrMain
upper tone:	higher frequency main tone; aka hi tone, PwrMainHi
lower tone:	lower frequency main tone; aka lo tone, PwrMainLo
IM:	IMD product in dBc relative to the nearest product
IM3Hi:	upper third-order IM product in dBc
IM3Lo:	lower third-order IM product in dBc
Pwr3Hi:	upper third-order IM product in dBm
Pwr3Lo:	lower third-order IM product in dBm
IP3:	third-order intercept point
OIP3:	output referred IP3, usually used with power amplifiers
IIP3:	input referred IP3, usually used with LNA and receiver amplifiers
PwrMainHiIn:	upper frequency input power tone
PwrMainLoIn:	lower frequency input power tone

From Eq. (6.12) the IM3 products are created by modulation of one tone's second harmonic with the other tone's fundamental. This could be considered essentially a mixing product, and the power of the IMD product is related directly to the power of the two components of the mixing product. The power of the second harmonic changes 2 dB for every 1 dB of its fundamental, and the power of the other product changes dB for dB, so the powers of the third-order IMD terms change 3 dB for every 1 dB change in the pair of tone powers, and the dBc levels of the IMD products change 2 dB for every 1 dB of change in the fundamental tones.

Whenever dBc values are used for IM tones, it is always relative to the nearest main tone, if the two main tones are not exactly equal. Often the frequency response of an amplifier will cause the two main tones to be somewhat unequal. By using the definition of the nearest tone as the reference for dBc values, the value of the IM tone, in dBc, will be the same as if the two main tones were exactly equal. To understand this, consider the IMD products in Figure 6.44. This condition holds even if the tones are widely separated in power, for normal amplifiers.

Here the IM values are compared in two cases: in one, the IM tones are computed when the main power tones are equal, taking the lower IM value from the difference between the Pwr3Lo tone and the PowerMainLo tone. In the second case, the two tones are intentionally set different, with the upper tone raised by 5 dB and the lower tone reduced by 5 dB; this case is marked "Unequal". As indicated in the figure, the IM computed for this case is nearly

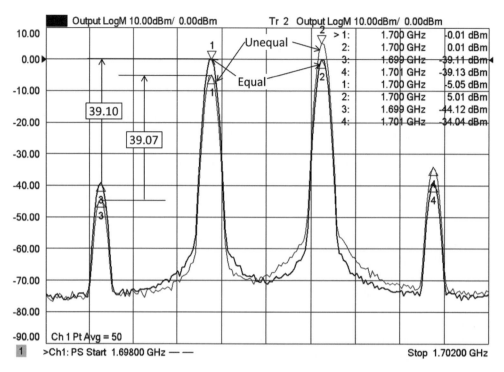

Figure 6.44 IMD measurements with offsets in the main tone power.

identical to that for the "Equal" tone case, even though the output tone powers are substantially different. The average tone power (averaged in the dB sense) is the same. Raising the upper main tone raises the upper IM tone by two times the dB number (in this case 10 dB) because it is related to the nearest tone by a second-order term, see Eq. (6.12); and raises the lower IM tone by one times, or 5 dB. In compensation, lowering the lower main tone by 5 dB lowers the lower IM tone by 10 dB, and lowers the upper IM tone by 5 dB. As a result, the two main tones are 10 dB apart as are the two IM tones, and the dBc value relative to the nearest main tone is exactly the same as before the main tone powers were changed. In this way, one can accurately compute the IM products, and the OIP3 value, even if the measurements of the two main tones are not equal, by computing the value for the average of the two powers. Of course, if only one main tone is changed in power, the average power is changed and the dBc value will be changed as well, to the value that represents the new average power.

6.6.2.1 Intercept Points: OIP3

The output referred intercept point is computed by projecting this three-to-one power response of the IM tones along a straight line projection with respect to an increased main tone power until they cross (or intercept) the projection of the main tone power, as illustrated in Figure 6.45. Note that this intercept point presumes a 3:1 slope of the third order IM tones versus main tone power, so that only one IM reading is needed to compute the IP3 point. The lower the

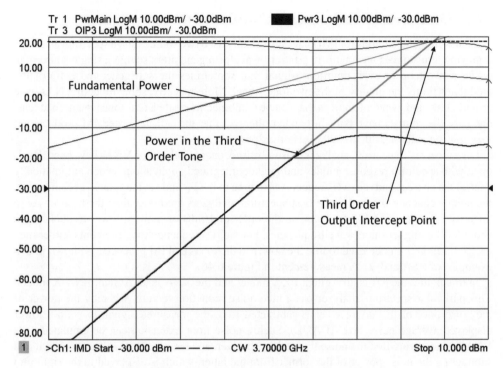

Figure 6.45   Projecting IM tones to obtain the IP3 point.

IM products are, the farther they will have to increase to theoretically intercept the main tone power, and thus the higher the IP3 power will be; one can therefore conclude that higher IP3 points are indicative of more linear amplifiers.

However, the IP3 point is simply a figure of merit; no actual intercept is measured. Its computation or value does not depend upon the actual linearity behavior of the amplifier's IM products. It is computed as

$$IP3 = PwrMain + \frac{|IM3|}{2} \tag{6.15}$$

For example, if at a power of 0 dBm output, the IM product is −38 dBc, then the IP3 will be +17 dBm. In fact, the IP3 value can and does change with drive level. Also shown in the figure is the computed OIP3 for each point of the power sweep. As expected, at lower powers the projected value for OIP3 and the computed value from the trace are the same. For simple amplifiers, the IP3 often decreases as the power increases. But some amplifiers are designed to enhance their IP3 value at some higher drive powers, and the IM tones may have some regions where they decrease before increasing again. In these areas the IP3 values can dramatically increase.

The IM products and intercept points for higher-order products follow a similar set of definitions, for example the fifth-order products simply replace the 3 with a 5 in the definition of the products, and the intercept point is computed projecting the IM produce with a slope of 5:1 rather than 3:1.

6.6.3 Measurement Techniques for Two-Tone TOI

The measurement methods for IMD have historically used two sources and a spectrum analyzer. Some modern sources have built-in arbitrary waveform generators so that a single source can create a two-tone or even multitone signal, but sometimes the self-generated IMD of the modulator limits the quality of the two tone signal. More recently, some modern VNAs have provided two independent sources that can be combined to create a two-tone signal; in at least one case the sources have built-in banded filters so that the self-generated IM products are very low, typically less than −90 dBc at max power.

For a receiver, a spectrum analyzer is usually used, but recently some VNAs have been developed with spectrum response modes and include integrated applications which automatically control the sources and receiver frequencies to tune to the F_C and set the span to accommodate the delta frequency. Typically, the best spectrum analyzers are better than the best VNAs in linearity for measuring IM products, on the order of 10 dB or more at lower frequencies to about 5 dB at higher microwave frequencies. For single-frequency measurements, either may be used. Markers can be used to find the power in the tones and IM products, as shown in two examples of a SA and VNA measurement in Figure 6.46.

In this figure, the SA is an Agilent PXA model and the VNA is an Agilent PNA-X model. This particular spectrum analyzer has a nice noise reduction feature, whereby the power of the noise floor of the SA receiver is subtracted from the measured signal, to yield a lower displayed average noise level (DANL). Such a noise floor extension can be emulated on a VNA by measuring first the noise floor of the receiver, and then the IMD of the amplifier, and subtracting the noise power of the former from the latter, which is displayed in the right plot of the figure above. This provides about 5 dB improvement in the displayed noise level for the VNA, and improves the measurement accuracy of power for the IM tone. Here the SA and VNA measurements agree within a few tenths of a dB. The signals shown were created with the VNA sources; using precision stand-alone signal sources will usually give lower phase noise. Also, since a similar synthesizer is used in LO for the VNA, the phase noise shown in the VNA plot is about 3–5 dB higher than in the SA plot. Spectrum analyzers typically have much better close-in phase noise than a VNA.

For fixed single-frequency measurements, there is no advantage to a VNA when using the IM spectrum mode of operation, and the VNA sweeps are usually slower than the optimized performance of the SA. Both SA and VNA in spectrum mode must sweep the frequency spectrum to search for the maximum signal level, even if the exact frequencies of the IM products are known. With external programming the SA may be put in a zero-span mode and tuned to exactly the frequency of each tone. But many VNAs have a built-in mode to sweep receiver frequencies separately from source frequency, thus the VNA can measure at only the tone and IM frequencies. As mentioned above, this is the frequency offset mode which can be used to support a swept IMD measurement.

6.6.4 Swept IMD

In swept IMD measurements, either the center of the tone frequencies is swept, holding the power level and the tone separation (or delta frequency) constant; or the tone frequencies are held constant and the tone powers are swept; or the center frequency and tone powers are held constant and the delta frequency is swept. For any of these cases, the receiver is set to measure

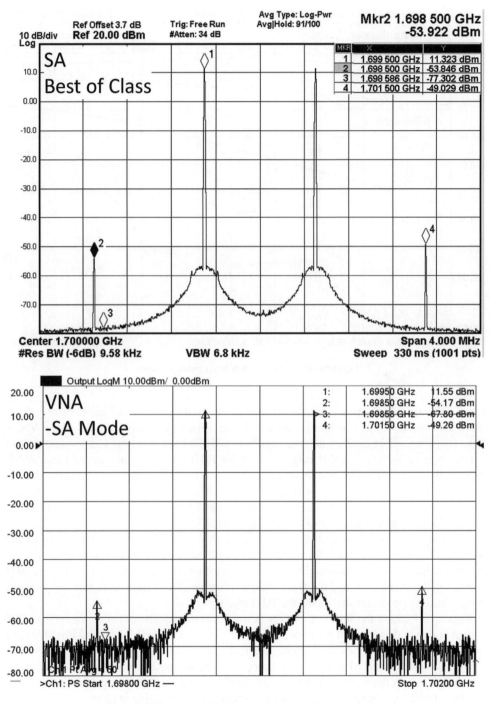

Figure 6.46   SA and VNA measurement of IM spectrums.

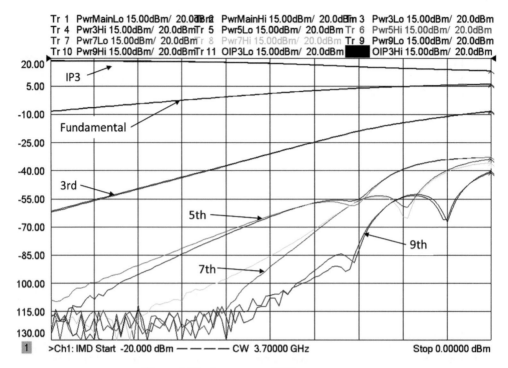

Figure 6.47   Swept power IMD measurements.

the main tones and IM products at each source setting. The swept variable is displayed on the x-axis and the IMD results are shown as traces versus the swept variable. Many VNAs have frequency-offset mode, where the source frequencies can be swept independently from the receiver frequencies. If this mode is used for IMD measurements, at least four VNA measurement channels will be required, one for each tone. However, some modern VNAs have built-in applications that provide a complete solution for the swept IMD measurement, meaning that a single channel can be used to measure and display all the tones and IM products, along with some precalculated functions for IM and IP3 values.

In this way, plots such as the IMD versus power plot shown in Figure 6.45 become very easy to create. For many amplifiers, the swept power IMD measurements are the most interesting because some of the various linearity enhancement techniques can yield vastly different IMD versus input power traces, and the plot of higher-order terms can sometimes give the designer insight into the processes that create the IMD results.

Figure 6.47 shows the result of a swept power IMD measurement. On the rather crowded screen is shown the power of the upper and lower main tones (PwrMainHi and PwrMainLo), the power in the IM tones up to ninth order (Pw3Hi, Pwr3Lo, Pw5Hi, Pwr5Lo, Pw7Hi, Pwr7Lo, Pw9Hi, Pwr9Lo) as well as the output referred IP3 point for the high and low tones. The inflections in the higher-order tones are likely caused by the mixing of different-order products (which generate higher-order tones) adding and subtracting in phase.

For the same amplifier, Figure 6.48 shows a plot of IMD measurements versus sweeping the center frequency of the IM tones, holding the tone spacing and power constant. The shape

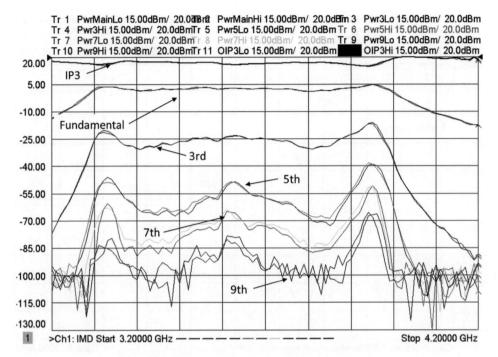

Figure 6.48 Swept center frequency IMD measurements.

of the input filter is clearly shown; for a constant input power, the IM tones are greatly reduced in the out-of-band area. It is interesting to note that the higher-order tones show significant variation in signal across the band, and peak at the filter band edges.

Finally, again for the same amplifier, Figure 6.49 shows the result of sweeping the delta frequency or tone separation frequency while keeping the center frequency and power constant. For this amplifier, there is some change in response versus delta-frequency in the fifth-order IM product. This might be due to some effect in the amplifier such as response of the bias network to the tone separation frequency that re-modulates the output in a way that causes some products to have variation with tone spacing, and may be the reason that the upper and lower tones generate different responses.

The unique advantage of using a VNA in IMD measurements is that both the sources and receivers are internally controlled and derived from the same reference. As such, very narrow IF BWs (the equivalent of a resolution bandwidth in an SA) can be used, as only one frequency point for each tone power is measured to compute the overall result. Also, the resolution bandwidth can be set wider for the main tones and narrower for the IM tones to optimize the measurement speed.

6.6.5 Optimizing Results

The measurement of IMD products may be limited by the linearity of the receiver, as well as by the IMD content and signal quality of the source. Optimization of the measurement setup can greatly improve the measurement results.

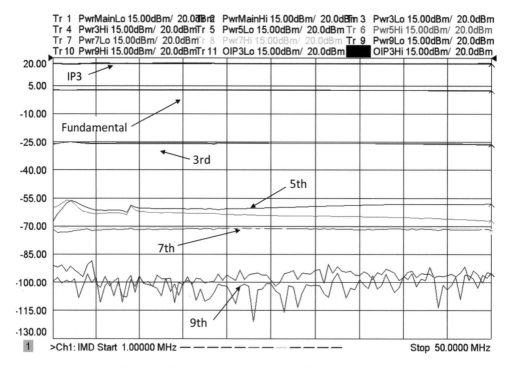

Figure 6.49   Swept delta frequency IMD measurements.

6.6.5.1   Source Optimization

Measurement optimization includes choosing the tone spacing, when such a choice is possible. Wider tone-spacing helps eliminate masking of the IM signals due to the phase-noise of the source. If the tone-spacing is sufficiently close, the phase noise of the source at the IM product spacing will swamp the IM product signal, unless very narrow IF BWs are used. If possible, selecting a wider tone-spacing will reduce or eliminate the effect of phase noise on the measurement result. In Figure 6.46, one can see the phase noise just dropping down into the noise floor at about 150 kHz off the tone frequency, for the VNA spectrum. Closer tone spacing would cause the IM product to be lost in the phase noise, especially in the VNA based measurement.

Another area of optimization consists of ensuring that the sources' self-generated IM products are below the level of the DUT products.

A possible cause of source-generated IM is due to harmonics (particularly second harmonic) of one-source mixing with the output of another. If two sources are combined to create the two-tone signal, adding a filter that suppresses the second harmonic can substantially reduce the self-generated IMD. Some VNAs and signal sources include many narrowband switched filters that provide for very low harmonic levels. Sources without integrated filters often must be followed with external narrowband or low-pass filters to remove the harmonics and these filters will limit the frequencies that can be measured.

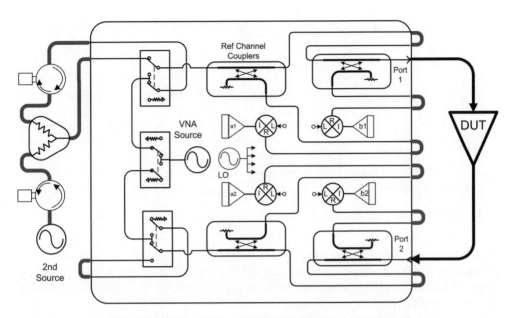

Figure 6.50 Block diagram of combined source with isolators.

Even with filtering, non-linear effects in the source's output amplifiers can cause source-generated IMD. Using a coupler instead of a simple combiner can provide dramatically increased isolation at the expense of power in one of the tones. Some modern VNAs use either a switched internal coupler, or use the test-port coupler from an unused port, to combine the two sources together. This does result in a loss of maximum power that can be applied with two equal tones to the DUT. If the power is too limited in this way, the internal combiner may be bypassed by an external equal-loss combiner. This will increase the available tone power by approximately the loss of the coupler less the loss of the combiner, or 16 dB minus 3 dB for around 10 dB improvement in most cases, as illustrated in Figure 6.50.

The main cause of source-generated IM products is direct cross-modulation between the two sources, and IM generation in the combining network. When a coupler is used, the isolation of the coupler prevents this source of IM generation, but when higher power is needed, and an external combiner is used, the cross-modulation of sources will likely occur. One way to improve the source direct cross-modulation (from one source to the other) is to add isolators between each source and the combiner, also illustrated in Figure 6.50. The combiner and isolator can be used in cases where higher power is needed, for example if a booster amplifier is applied to the either source. With very high powers, one must ensure that the combiner and couplers following the combiner themselves don't generate passive-intermodulation. In the diagram, an external source is used as the source for the second-tone, but some VNAs provide two or more internal sources that can be used as well.

Figure 6.51 shows the result of combining two sources with a low-loss combiner. In one case (left plot) the sources are directly connected, and in the other (right), the sources have an isolator placed between each source and the combiner network. The isolator doesn't cover

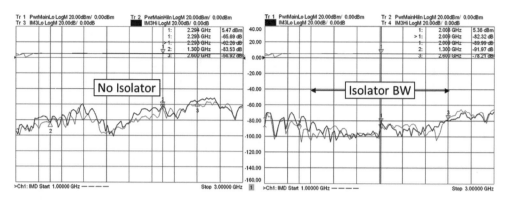

Figure 6.51   Source generated IMD due to direct cross-modulation with and without isolation.

the full band, so at the edges, the source-generated IM products are not improved. Here, over the bandwidth of the combiner, the worst-case source-generated IM products are improved by about 20 dB. In this measurement, the receiver is padded down by 25 dB and an IF BW of 1 Hz is used to obtain the best noise floor.

In some cases, including the case illustrated above, the source IM products are generated by some mechanism that is extremely sensitive to power, so that lowering the power by even 5 dB will cause the source IM products to drop to the noise floor. This can occur in circuits that have some limiting aspect to them; such limiting is sometimes placed internal to instrumentation sources to protect against reverse power from a DUT causing damage to the source. Thus, there is a tradeoff between maximum source power and IM generation.

Finally, one should avoid using tone spacings that match identically with internally generated clock signals of the measurement instrument, with the worst case example being 10 MHz spacing. Almost all instruments provide a 10 MHz reference oscillator and derive their own frequencies from this reference. Any leakage of this 10 MHz signal may show up in very low level IMD measurements if a 10 MHz tone spacing is used. Slightly changing the tone spacing, even by a few kilohertz, will eliminate this issue.

6.6.5.2   Receiver Optimization

For the most part, source effects can be removed to a negligible level with isolation and filtering between the sources, but receiver effects generally are more difficult to account for. The non-linear behavior of the receiver will generate IM products from the main tone powers hitting the receiver. In general, it is difficult to remove the main tone products as very narrow notch filters, or a pair of band-pass filters, are required. Thus, the linearity of the receiver often sets a limit to the level of IMD products that can be measured. The main method for reducing receiver-generated IM products is to add attenuation to the receiver channel.

Adding attenuation reduces the main tone values, and for every 5 dB of attenuation added, the IM products go down 15 dB, for a 10 dBc improvement. However, the noise floor comes up by 5 dB and so further IF BW reduction is needed to maintain the same noise floor.

The IMD properties of the receiver can be evaluated by changing the receiver attenuation value and looking for any change in the IMD product level, up to the point that the noise floor dominates. In a spectrum mode, the noise floor is set by the resolution bandwidth (RBW) and narrow RBWs will dramatically slow the sweep. However, in swept IMD mode, the narrow RBW is only applied to the IM tone, rather than sweeping the entire frequency range around the main tones and IM products, as the spectrum mode does, so the speed will be much faster.

For many systems, the source IM level can be improved sufficiently so that the receiver dominates. In such a case, a large source signal into the receiver, with low or no attenuation, will show the receiver-generated IM level directly. Adding attenuation should show the expected drop until the source generated level is reached.

Figure 6.52 shows the result of driving the receiver with a pair of ±5dBm tones, with 0 dB and 15 dB and 30 dB attenuation. The receiver-generated IMD is clearly visible. In this case, a 1 Hz IF BW filter was used to set a noise floor of approximately −120 dBm. At the highest receiver attenuation, the IM tones are limited by the source IM generation. At any particular frequency, the combination of source and receiver IM products can either add or subtract, depending upon phasing, so there will be ripple in the IM tone if contributions from both the source and receiver take a part in generating it.

It is typical for an SA to have a default minimum attenuation of 10 dB; this is to protect the input mixer. VNAs have a 13–16 dB input coupler, and so generally don't have any extra default receiver attenuation. For cases where the signal being measured is so small that the receiver will not generate any IMD products, the SA attenuator may be set to 0 dB using a manual override. On a VNA, the test-port coupler may be reversed (see Section 5.2.6) to

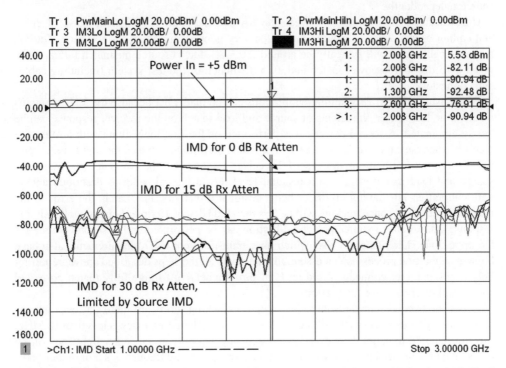

Figure 6.52 Receiver generated IMD in swept IMD mode, at +5 dBm input, with 0, 15 and 30 dB of receiver attenuation.

improve the sensitivity of the receiver. For microwave VNAs, whose couplers often have corner frequencies in the 500 MHz range, it may always be necessary to reverse the coupler for lower frequency IM measurements to remove the loss effect of the coupler roll-off. In some case, the coupler could roll off at very low frequencies (10 MHz) by as much as 35 dB. In these cases, the coupler may be reversed, but it might then be necessary to add some receiver attenuation to avoid distortion effects due to direct receiver access by the thru arm of the coupler.

6.6.6   Error Correction

Error correction of the IMD measurements is typically provided by a source power calibration and a receiver response calibration. For a two-tone stimulus, the source power calibration may need to be separated into two steps, one for the lower tone and one for the upper tone. If using sources and an SA, each source will need to be calibrated independently for cable and fixture loss, typically with a power sensor. The combiner should be in place to account for its loss. After one source is calibrated (over all frequencies needed for all tones) then a thru connection can be made to the SA and it may be calibrated. Finally, the second source may be turned on, and calibrated by the SA, leaving the first source on (and offset by the delta frequency for each cal point) to account for its mismatch effect on the output of the second source. If the match of the second source changes when it is turned on, the first source will need to be recalibrated as well. The SA is used as a frequency-discriminating receiver to calibrate each source independently.

Unfortunately, the mismatch between the SA and the sources is a main cause of error, and cannot be removed, so making the match very good is important for accuracy. The SA response calibration will typically be around ±0.3 dB due to source mismatch interactions, given a 15 dB source power match and a 15 dB match of the test system (including cables and connectors).

When using a VNA, very similar calibrations can be made, but because the VNA contains a built-in reflectometer, the effects of source and load match on the receiver response can be nearly eliminated. A source power cal is performed on the first tone, along with a reference receiver response calibration. Then a two-port cal is performed between the input and output. From these two calibrations the exact b2 receiver response can be computed removing the source and load match effects to a residual level. For a typical setup as described above, with calibration kits having 40 dB or better residual directivity and match, the error in the match-corrected response tracking will be less than 0.03 dB, or 10 times better than the simple response cal. Finally, a source power calibration can be performed on the second source for the upper tone. The modern VNAs which have built-in swept IMD measurements also have built-in calibration processes that essentially automate all the steps described above, requiring only a power meter connection and an Ecal thru connection to give fully match corrected source and receiver power corrections.

Experience shows that the source and receiver tracking terms are quite flat over narrow bandwidths, on the order of 0.01 dB per 10 MHz, so if the delta frequency is less than 10 MHz, it is not required to calibrate at every tone frequency, but only at the center frequency of the tones, and then interpolate the tracking term for each of the tone frequencies. For higher order of tones, this can dramatically shorten the calibration at virtually no degradation in quality of measurement.

6.7 Noise Figure Measurements

Noise figure, though seldom measured on power amplifiers, is a key measurement on amplifiers used for receivers and other low noise applications. Usually referred to as LNAs, these amplifiers typically have gain in the 10–20 dB range, although low-noise modules can have much higher gains as they are composed of multiple stages of LNA.

Because of the design constraints to create LNAs, the other parameters such as S11 or input match, stability, compression and IMD are sometimes sacrificed for better low-noise performance; these other parameters must still be measured, and occasionally there is a tuning step to trade off lower noise for better performance in one of the other areas.

The subject of noise figure measurements could support an entire textbook or more, but here the key concepts used in modern measurement methods are introduced, along with the key formulas for measurement and correction.

6.7.1 Definition of Noise Figure

The definition of noise figure is at the same time simple, but with some subtleties. In Chapter 1, noise figure is defined as (Eq. (1.75)

$$N_{Figure} = 10 \log_{10} \left(\frac{Signal_{Input}/Noise_{Input}}{Signal_{Output}/Noise_{Output}} \right) \tag{6.16}$$

from which a more useful definition can be derived as

$$N_{Figure} = 10 \log_{10} \left(\frac{DUTRNP}{G_A} \right) \tag{6.17}$$

DUTRNP is the available relative noise power from the DUT, where "relative" means relative to kT_0B, and is computed by dividing the available noise temperature from the DUT by T_0

$$DUTRNP = \frac{T_{A\_DUT}}{T_0} \tag{6.18}$$

Since available power does not depend upon the load impedance, and available gain does not depend upon load impedance, this definition of noise figure also does not depend upon load impedance.

Noise figure is always expressed in dB, but a related linear form of noise figure is called the noise factor, and is often shown as NF. This is a common source of confusion, and here the distinction will not be made between noise figure and noise factor; rather, the two terms are used interchangeably with the understanding that, unless explicitly indicated otherwise, the linear form noise factor is used in all computations and equations, even if it is called noise figure.

A related value is the incident relative-noise-power from the DUT (DUTRNPI). Here, incident means the noise power into a non-radiating non-reflecting load, or an ideal cold Z_0 load; this is the power measured by a Z_0 noise receiver. This is distinguished from the total power measured in the receiver (which includes the receivers own noise), defined as the SYSRNPI (or system relative-noise-power incident), and which represents the total raw noise power measured at the receiver output, including the noise from the DUT input, amplified by the DUT gain, and added to the input noise of the receiver, provided the receiver has a Z_0 impedance. Finally, for completeness, the total noise power measured at the receiver, available from the DUT, including the noise power generated in the receiver is called the available system relative noise power and is represented by SYSRNP. In newer instruments, where the receiver noise power can be calibrated out, the DUTRNP values are usually displayed. Older instruments did not have a means to automatically remove their own noise contribution, so they displayed only SYSRNP.

For most measurement purposes, the noise figure is *defined* from a 50 ohm source and into a 50 ohm load, so incident noise power makes the most intuitive sense to measure, if the system is Z_0. In most measurement cases, the source and load impedances are not 50 ohms. Using the available powers in calculations removed the load impedance mismatch effect, to the first order, but did not remove the source mismatch effect, which remains a substantial source of error. In the case that the system is not Z_0, the DUTRNPI can still be computed by applying the appropriate match correction factors, as described in the following discussion.

6.7.2   Noise Power Measurements

Noise power measurements are usually performed on a noise figure analyzer (NFA), spectrum analyzer or more recently, on a VNA with a noise measurement application. The NFA is really a specialized version of an SA, with more flexibility in the RF and IF gain path to optimize the noise measurement to the receiver's most linear region, including a built-in LNA at the input. SAs have an advantage in measuring noise in that the input is image-free, so the SA only measures the noise power at the selected frequency, and is insensitive to noise at other frequencies. Newer SAs have very similar capabilities to NFAs and can provide very similar noise figure measurements, provided they also have an LNA in front of the first converter. Calibration and measurement of noise power with either is essentially the same. The measured noise power in a receiver is a function of the receiver gain and bandwidth. For any given resistance, the noise voltage generated by that resistance is equal to

$$V_N = 2\sqrt{kTBR} \tag{6.19}$$

where k is Boltzman's constant (1.38×10^{-23}), T is the temperature in kelvin, B is the bandwidth over which the voltage is observed, and R is the value of the resistance. The available power from this resistor is the power that would be dissipated in a matched noiseless resistor (essentially an ideal noise receiver); half the noise voltage appears across each resistor, so power delivered to the load resistor is

$$P_N = \frac{\left(V_N/2\right)^2}{R} = kTB \tag{6.20}$$

Thus the available noise power from any passive source depends only on the temperature of the source, and not on its impedance.

A VNA can also be used as a noise receiver, under some special conditions. In general, the effective noise bandwidth of a VNA is twice as wide as the IF BW, since the VNA does not have an image protected mixer. That is, noise from one IF above and one IF below the VNA local oscillator frequency will be mixed in the final IF, and measured. If the IF frequency is high (e.g., 10 MHz) then the noise will come from the displayed frequency (LO + IF) and a frequency 20 MHz away (LO − IF), and leads to uncertainty in the measurement due to not knowing which frequency is the major source of noise. Recently, some VNAs have provided flexible IF structures, where the IF frequency can be arbitrarily set, even to 0 Hz. In such as case, the effective bandwidth is still twice the IF BW, but centered at the LO frequency (which is the same as the RF frequency for zero IF), thus having no uncertainty on which frequency the noise comes from. Often with zero-IF detection, there is an intentional notch in the IF response exactly at DC to avoid DC offset errors. One example of the IF response in a VNA used for measuring noise power is shown in Figure 6.53. For this measurement, the VNA source is turned on, fixed at the center frequency. The noise receiver sweeps across the source power and traces out the effective noise bandwidth. The markers show the 3 dB down points from the reference, with an apparent bandwidth of about 5.5 MHz, but with the notch in the center the effective noise bandwidth is about 4 MHz.

There can be additional sources of error in VNA measurements due to noise conversion on higher-order LO products such as the third or fifth harmonic of the LO. In such a case, a band-limited filter must be placed in front of the noise source or DUT to ensure that out-of-band noise is not converted into the IF. At least one manufacturer, Agilent, provides a special noise measuring receiver in a VNA as a hardware option, with a built-in switched-gain LNA and some built-in filtering to protect against images from LO harmonics.

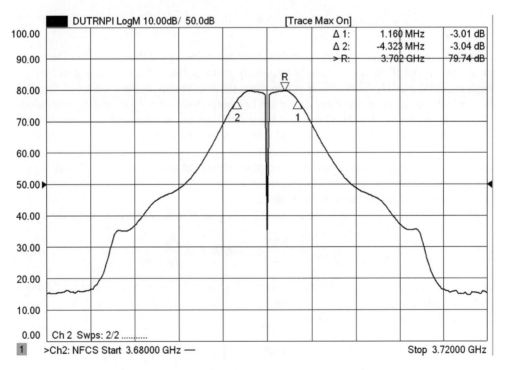

Figure 6.53   Noise bandwidth of a VNA with zero-IF receiver.

For all the noise-measuring instruments, the bandwidth sets the resolution of the measurement, that is, the ability to distinguish the noise figure at different frequencies. The most common bandwidth used is 4 MHz (historically, due to the original HP 8970 NF meter), with narrower bandwidths available for narrow channel measurements, and wider bandwidths available for better speed and lower jitter.

Since the NF receivers are measuring random noise, there will be a substantial variation in the measurement results from sample to sample. The variation in the noise goes down as the square root of the noise samples; the number of samples per unit time goes up as the bandwidth goes up. Wider bandwidths therefore give more samples in the same time, and thus have lower jitter for the same measurement time, sometimes called the integration time, as the noise power is integrated over the samples.

6.7.3 Computing Noise Figure from Noise Powers

6.7.3.1 Y-Factor Correction and Noise Receiver Calibration

A noise source is used in the calibration and measurement of a DUT from an SA or NFA by providing either a cold state (kT_0B noise) or a hot state, which produces a known amount of excess noise above the cold state. The total noise power detected depends upon the gain and bandwidth of the receiver. The noise source excess noise is defined as an excess noise ratio (ENR), and is related to the noise power (or noise temperature) above the kT_0B noise by

$$ENR_{dB} = 10\log_{10}(ENR) = 10\log_{10}\left(\frac{T_H - T_C}{T_0}\right) \tag{6.21}$$

This definition is slightly unusual in that if the hot noise equals the cold noise, the excess noise ratio is not 0 dB (as one might intuitively expect), but instead is the log of zero, or negative infinite dB. If the hot noise is two times the temperature of the cold noise, the ENR is 1 or 0 dB. So excess-noise ratio should not be interpreted as simply the noise power above kT_0B. The definition of (6.21) is unusual, but it simplifies the computation of the noise figure of a system measured with a hot and cold source. Normally, ENR is defined for T_C equal to 290 kelvin, but if the noise source is not at T_0, Eq. (6.21) accounts for the difference. (This presumes that the hot noise temperature, relative to the cold noise, changes with a change in the ambient temperatures; hot/cold noise resistors follow this rule, but some solid state noise sources do not, see the references for details.)

The noise figure of any system can be related to the pair of hot/cold measurements directly as

$$N_{F\_Sys} = \frac{ENR}{Y - 1}, \quad where\ Y = \frac{P_H}{P_C} \tag{6.22}$$

and P_H, P_C are the noise powers measured for the hot state and the cold state of the noise source, respectively. In the case where the noise source is not at 290 K temperature, a slight modification accounts for the difference as

$$N_{F\_Sys} = \frac{ENR - Y \cdot \left(\frac{T_C}{T_0} - 1\right)}{Y - 1} \tag{6.23}$$

where T_C is the temperature in kelvin of the noise source in the cold state, that is the ambient temperature, and T_0 is the reference noise temperature, defined as 290 kelvin.

To calibrate an NFA, SA or VNA, it is necessary to determine the contribution of the receiver noise to the overall system noise measured. The receiver noise figure is computed by applying the noise source directly to the input of the receiver and computing the system noise figure as shown above. Because the powers are used as a ratio, they can be expressed as either watts/Hz (noise power density) or simply noise temperature. The power conversion for the hot power to noise temperature is

$$T_H = \frac{P_H}{kB} = \frac{P_{H\_density}}{k} \tag{6.24}$$

Where k is Boltzman's constant ($kTB = 4 \times 10^{-21}$ at 290 K, B = 1 Hz), B is the receiver bandwidth, and noise power density is defined as noise power relative to a 1 Hz noise bandwidth. The cold temperature is simply the ambient temperature of the noise source. Most noise measurement systems give noise power in terms of noise power density in dBm/Hz, or noise temperature, and noise temperature is sometimes more convenient. Noise figure may also be expressed as a temperature (called excess noise temperature) and is defined as

$$T_{E\_Rcvr} = T_0 \cdot (F_{Rcvr} - 1) \tag{6.25}$$

Once the noise figure of the system is known, the gain-bandwidth of the system can be computed by knowing the ENR of the noise source and the hot temperature measurement, as

$$GB_{Rcvr} = \frac{T_{H\_Rcvr}}{(ENR \cdot T_0 + T_C) + T_{E\_Rcvr}} = \frac{T_{H\_Rcvr}}{(ENR + 1) \cdot T_0 + T_{E\_Rcvr}} \bigg|_{T_C = T_0} \tag{6.26}$$

With these factors, the noise receiver can be calibrated to read noise temperatures directly using the receiver correction

$$T_A = \frac{T_{M\_Rcvr}}{GB_{Rcvr}} - T_{E\_Rcvr} \tag{6.27}$$

where T_A is the actual available noise temperature, and T_{M_Rcvr} is the raw measured noise temperature on the receiver, and the measured noise power density can be computed as

$$P_{A\_density} = \frac{kT_{M\_Rcvr}}{GB_{Rcvr}} - kT_{E\_Rcvr} \tag{6.28}$$

With this correction, the second-stage noise contribution of the receiver is removed.

6.7.4 Computing DUT Noise Figure from Y-Factor Measurements

After the calibration, the noise source is applied to the input of the DUT, and the output of the DUT is connected to the noise receiver. The hot and cold states of the DUT driven from the noise-source are measured and recorded and the noise figure of the overall system is computed as

$$N_{F\_Sys}^{DUT} = \frac{ENR - Y \cdot \left(\frac{T_C}{T_0} - 1\right)}{Y - 1}, \quad where \ Y = \frac{P_{H\_DUT}}{P_{C\_DUT}} \tag{6.29}$$

This represents the combined noise figure of the DUT and the noise receiver, not just the DUT by itself. If the noise power is corrected as in Eq. (6.28), then this is the corrected noise figure. However, some systems measure the overall noise power (SYSRNP vs DUTRNP) instead. If the DUT has high gain, the noise contribution of the receiver is not significant, and the noise figure is nearly the same as the DUT noise figure. For lower gain devices, the Fris equation can be used to compute the noise figure of just the DUT as

$$N_{F\_DUT} = N_{F\_Sys}^{DUT} - \frac{(N_{F\_Rcvr} - 1)}{G_{DUT}} \qquad (6.30)$$

This can also be expressed in terms of excess noise temperature as

$$T_{E\_DUT} = T_{E\_Sys}^{DUT} - \frac{T_{E\_Rcvr}}{G_{DUT}} \qquad (6.31)$$

To find the noise figure of the DUT, all that is needed is to find the gain of the DUT, which can be computed from the two sets of noise power measurements as

$$G_{DUT} = \frac{P_{H\_DUT} - P_{C\_DUT}}{P_{H\_Rcvr} - P_{C\_Rcvr}} = \frac{T_{H\_DUT} - T_{C\_DUT}}{T_{H\_Rcvr} - T_{C\_Rcvr}} \qquad (6.32)$$

Much of this is automated when used with an NFA or SA with noise figure personality. Most VNAs do not rely on the Y-factor method, as they can measure gain independently. An example of the Y-factor technique is shown in Figure 6.54.

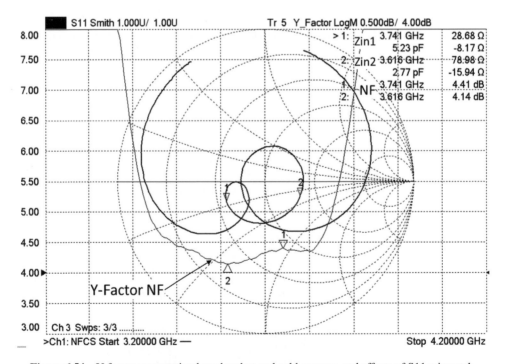

Figure 6.54 Y-factor computation based on hot and cold sources, and effects of S11 mismatch.

The Y-factor result here is a direct computation from two noise power measurements of hot and cold power using Eq. (6.29), after applying the correction of (6.28). A particularly interesting observation is that the noise figure has a ripple in the middle of the response and the peak and valley correspond at the frequencies exactly at the extremes of impedance as shown with Zin1 and Zin2 on the Smith chart trace. This indicates a likely error due to the noise source not being a perfect 50 ohm source. Thus, the measurement displays a lower noise figure for the Zin2 than it does for the Zin1, likely due to the noise parameters of this particular amplifier.

A few notes on the Y-factor technique:

First, the gain measurement is not the normal S21 gain, but the insertion gain of the DUT. This differs from the S21 gain due to the mismatch at the input and output of the DUT during the measurement.

The Y-factor measures the noise figure of the DUT in the impedance of the noise source, and (for the most part) it is the cold impedance that is important. The noise figure of a DUT depends upon the source impedance (see next section) and so if the noise source is not exactly 50 ohms, it is as if the noise figure had been changed or pulled by the noise source off its 50 ohm noise figure. The hot state of the noise source is designed to measure the gain, and it usually overwhelms the DUT noise figure, so pulling of the DUT during the hot state measurement is not so important.

For Y-factor measurements, it is usually recommended to use a low-ENR noise source. This has two benefits: the lower ENR sources are manufactured by adding high-quality attenuators to the high-ENR sources, thereby improving both hot and cold matches. Further, one source of error in Y-factor measurements is the linearity of the receiver, so a lower ENR causes less of a change in the power measured than a high ENR source. However, the DUT noise figure should not be significantly bigger than the ENR source or there will be too little change in the measured noise power for the on-state for the receiver to measure.

The Y-factor is very effective for very high gain devices, as only the change in noise is measured, not the absolute noise power, so drift in output cables or adding attenuation at the output has almost no effect on the measured noise figure (although it will have an effect on measured gain).

6.7.5 Cold-Source Methods

The principal advantage of the Y-factor method is that neither the gain of the receiver nor the gain of the DUT needs to be known; they are computed as part of the measurement, although the computed value is susceptible to error due to system mismatch. However, when using a VNA in a noise measurement system, the determination of the DUT gain is easily and precisely given by the S-parameter calibration and measurement. Since the principal reason for using a hot noise source is in the determination of the gain of the DUT, such a hot source is not needed when measuring noise figure using a VNA. Rather, the so-called cold source method is used which can result in a simpler and faster measurement scenario.

In the cold source method, the noise figure of the DUT is computed essentially from first principles. The definition of noise figure can written as Eq. (6.17), and if the receiver can be calibrated to measure the relative noise power from the DUT, DUTRNP, and the available gain can be computed from the S-parameters of the DUT and knowledge of the source impedance, then the noise figure computation can be completed without resorting to hot noise measurements. Since noise measurements are usually quite slow compared with S-parameter measurements, due to the averaging needed to reduce jitter, the cold source method is nearly twice as fast as the Y-factor method.

In the cold source method, the gain-bandwidth product of the receiver must be characterized. This can be characterized using a noise source and a pair of hot/cold measurements as described in Eq. (6.26). If the noise power measured by the receiver is converted to an equivalent temperature, then the effective noise temperature can be computed as

$$T_E = \frac{T_A}{G_A} - T_0 \tag{6.33}$$

where T_A is the available noise and G_A is the available gain. This is clearly recognized as the excess noise temperature at the input above T_0. The noise factor is similarly computed

$$NF = \frac{T_A}{T_0 \cdot G_A} \tag{6.34}$$

This can be converted to S-parameters and incident noise power as

$$NF = \frac{T_{Inc}}{T_0} \cdot \frac{|1 - \Gamma_S S_{11}|^2}{\left(1 - |\Gamma_S|^2\right) |S_{21}|^2} \tag{6.35}$$

and $DUTNPI = \frac{T_{inc}}{T_0}$, so $DUTRNPI = DUTRNP \cdot \left(1 - |\Gamma_2|^2\right)$, where Γ_2 is the output match of the device connected to the noise receiver.

If the VNA source impedance is matched (Z_0), the noise figure is simply

$$NF = \frac{DUTRNPI}{|S_{21}|^2}, \quad N_{Fig\_dB} = DUTRNPI_{dB} - S_{21\_dB} \tag{6.36}$$

Thus for a cold source measurement the noise figure, given a 50 ohm source impedance, is simply the excess incident noise power (in dB above kT_0B) minus the S21 gain in dB. For example, if a DUT has 22 dB of excess noise and 20 dB of gain, it has a 2 dB noise figure.

Figure 6.55 shows the cold source measurement of an amplifier, computed as in Eq. (6.35) as well as the Y-factor measurement, and the measurement presuming an ideal 50 ohm source, from Eq. (6.36). From the results, it appears that the Y-factor measurement has the most ripple (~0.25 dB), as it does not account for any correction for mismatch. The trace labeled "Matched-Source Approximation" takes a simple ratio of DUTRNPI and $|S_{21}|$ which is exactly correct if the source match is 50 ohms, though here it also shows some effect due to source-mismatch. In this case, the ripple is considerably less than the Y-factor (~0.1 dB). The smallest ripple, as well as the lowest consistent noise figure, is from the fully corrected cold-source method of Eq. (6.35). The ripple here is on the order of 0.05 dB, and the trace shows a response consistent with the inverse of the gain of the amplifier, which is to be expected. In this case, the cold source measurement is improved by adding an external 6 dB attenuator pad to the test-port cable, to improve the raw source match. This is a good practice to reduce the residual error from the non-ideal source impedance.

Equation (6.35) gives the noise figure for a source that is not 50 ohms, but it is *NOT* the noise figure commonly understood, N_{F_50}. It is instead the noise figure for the DUT in the

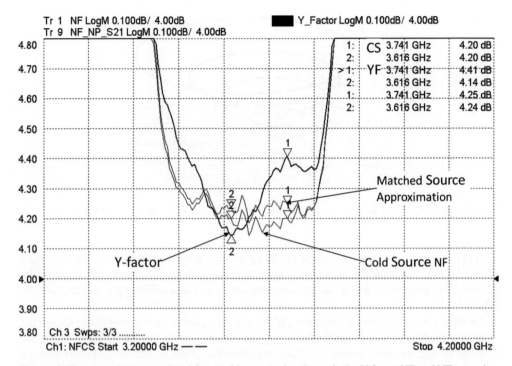

Figure 6.55 Noise figure computed from cold source; also shown is the Y-factor NF and NF assuming a matched source.

reference impedance of the source, just as the Y-factor gives the noise figure of the DUT at the impedance of the cold-state of the noise source. And, with the information so far discussed, it is not possible to predict or compute the exact 50 ohm noise figure from this value. However, methods based on the cold-source measurement can be used to find the noise parameters of a DUT, and from those, the exact 50 ohm noise figure may be accurately computed, as discussed in the next section.

6.7.6 Noise Parameters

The noise figure for amplifiers varies as the source impedance changes. For most amplifiers, the noise figure is specified at Z_0, usually 50 ohms. However, many amplifiers have a lower noise figure at some different impedance, and a key design task is to create a matching network that transforms 50 ohms to the optimum impedance for minimum noise figure.

Noise parameters provide the basis for understanding how the noise figure of an amplifier changes with the source impedance. The noise figure of a device at any reflection coefficient is described in Eq. (1.90) using four values as

$$N_F = N_{Fmin} + \frac{4R_n}{Z_0} \frac{|\Gamma_{opt} - \Gamma_S|^2}{|1 + \Gamma_{opt}|^2 (1 - |\Gamma_S|^2)} \tag{6.37}$$

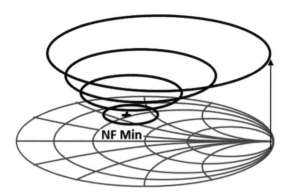

Figure 6.56   Noise parameters describe the noise figure as a function of source impedance.

This function can be describe intuitively as a parabolic shaped surface above a Smith chart, where the distance from the Smith chart represents the noise figure, the lowest point of the parabola lies above Γ_{Opt} and its height represents the minimum noise figure, as illustrated in Figure 6.56.

An amplifier with higher R_n will see the noise figure increase more quickly as the source reflection coefficient moves farther from Γ_{Opt}. In the illustration, the arrow shows the noise figure of an amplifier with source impedance near the edge of the Smith chart, which represents the condition of an amplifier with an open circuit at the input. The circles are typically set to represent 1 dB steps in the degradation of the minimum noise figure, and are typically plotted directly on the Smith chart for a single-frequency noise circle plot.

6.7.6.1   Noise Parameter Measurement Systems

Noise parameter measurements require, in addition to normal noise figure measurements, some method for changing the impedance at the source of the amplifier-under-test. Traditional noise parameter measurement systems were comprised of a set of equipment including a VNA, NFA, noise impedance tuner, noise source and switches to connect the various pieces of equipment for various measurements, as illustrated in Figure 6.57.

The source-impedance tuners are typically constructed by creating a well-controlled "slabline" transmission line with a slot in one of the ground planes that allows a capacitive probe to come close to the line. The height of the probe sets the magnitude of the reflection coefficient and the distance along the line sets the phase. These systems worked by pre-measuring the impedance of the tuner at many different positions to find the desired impedance at each frequency, generating hundreds or thousands of measurement points. Then, with the amplifier inserted, the noise figure for each impedance is measured at many pre-selected impedance points, from which the over-determined sets of measurements can be used to solve Eq. (6.37). Unfortunately, older systems operated on a point-by-point basis, going from one CW frequency to the next, and moving the impedance tuner to a variety of impedance states for each point. This process was very slow, and a typical system could take up to 20 minutes per frequency point including the time to calibrate the tuner.

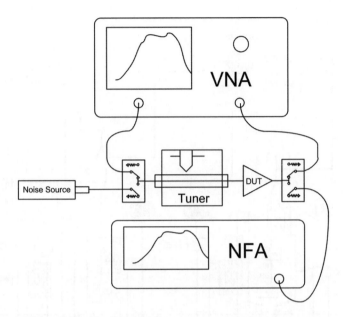

Figure 6.57   Traditional noise parameter measurement system.

With the introduction of noise figure measurements in the modern VNA, the process was greatly simplified by removing the need for external switching and noise sources. The measurement algorithms were optimized to select tuner position rather than impedance points, with sufficient spread of the positions so that they would not overlap over the frequency range of interest. In this way, the tuner could be positioned one time for an entire sweep of frequency for both S-parameters and noise figure. The tuner was then repositioned and data taken again. Since the tuner only a moved to a few positions (in one manufactures' case, 21), the precharacterization was much faster, as was the measurement, resulting in measurement times as short as a few seconds per point, or several hundreds of times faster than old techniques.

The noise parameters don't require a mechanically based tuner, and at least one manufacturer has integrated an electronic tuner (similar to an Ecal module) to provide a single connection measurement of noise-parameters, from which the exact 50 ohm noise figure is derived by using noise parameters to correct for the error caused by the non-ideal VNA source match. An example of such as system is shown in Figure 6.58, where a noise source is used only during the calibration portion, and the tuner is realized with an external Ecal module.

The system shown uses the electronic tuner to produce just a few states, on the order of seven distinct impedances, to characterize the noise parameters of the DUT. These states surround the 50 ohm impedance point, but do not extend to the edge of the Smith chart. In such a case, there can be some substantial error in determining the values for N_{Fmin} and Γ_{Opt}, if Γ_{Opt} is far from 50 ohms, but these errors are correlated in such a way that the computation of the corrected 50 ohm noise figure has very little error, due to the impedance states surrounding the 50 ohm point. Since the magnitude and phase of the source impedance affects the noise figure, this correction method is sometimes called a vector-corrected noise figure measurement, in contrast to the scalar (magnitude only) correction from Section 6.7.5. A measurement of the

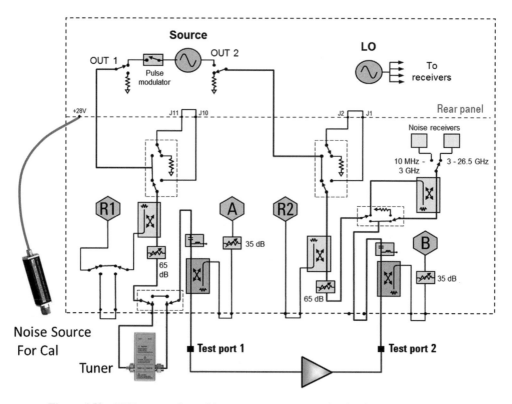

Figure 6.58   VNA system for making vector error corrected noise figure measurements.

50 ohm vector-corrected noise figure measurement is shown in Figure 6.59, along with a cold-source scalar-corrected measurement of the same amplifier.

The ripples in the measurements are an indication of the source pulling effect of non-ideal source impedance. It is much larger in this figure as no 6-dB attenuator was used on the port 1 for the scalar noise correction, so the full mismatch of the VNA source pulls the noise figure of the DUT. This mismatch is only properly corrected when using the noise parameter correction method, the details of which are discussed below. In this case, the source match, at the end of the VNA test port cable, is about 15 dB. The full vector-calibration improves the effective match to better than 40 dB.

6.7.7   Error Correction in Noise Figure Measurements

Error correction for noise figure measurements consists of a few parts. In Y-factor measurements, the calibration measures noise at the NFA receiver input, and the correction consists of removing the effect of the NFA noise from the overall noise figure. However, the receiver noise figure can be "pulled" by the output match of the DUT so that the correction has some residual error if the DUT output match is not near 50 ohms, or rather, if it is not near the impedance of the noise source used to characterize the NFA receiver. The Y-factor also is

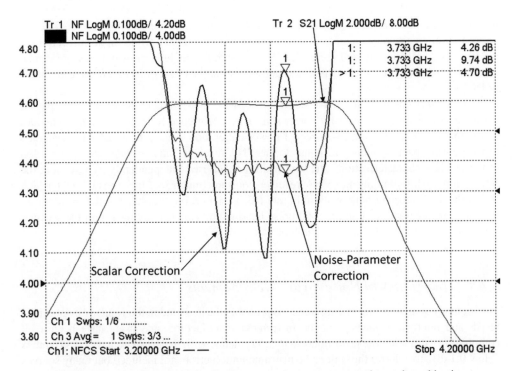

Figure 6.59 Vector-corrected noise figure measurement, compared to scalar cold noise.

susceptible to the errors due to the impedance of the noise-source driving the DUT being different from the reference impedance. The errors in impedance will also reflect in errors in the gain as measured by the Y-factor technique. Thus, systematic raw errors can cause significant overall errors in measurement; Y-factor measurements rely on having good, well matched source and load impedances. This is one reason why isolators are often used in precision Y-factor test systems.

For the scalar-corrected cold-source method, the gain and noise power are measured separately, and error correction is also applied separately. Normal S-parameter error correction is used for the gain computation, and so the errors from gain are the same as for S-parameter, which can be reduced to very small errors using good techniques. For the noise measurement, the receiver noise can be measured using the hot/cold method, but this does not remove the noise pulling effects of the receiver. However, if a tuner is used, the noise-parameters of the VNA receiver can be characterized, and the noise power measurements can be corrected exactly for the noise of the second stage as it is pulled by the output match of the DUT. For this correction, an additional S22 measurement is required to ascertain the DUT output impedance, and provide an exact determination of the second stage receiver noise. In this way, for low gain devices, the cold source method has less error than the Y-factor method. Of course, for high-gain devices, the second stage noise figure has a much smaller detrimental effect.

For vector-corrected cold source measurements, that is measurements that include a tuner on the input, the pulling effects of the test system source impedance are also corrected, in addition to the gain and noise receiver correction. Thus, all systematic errors in a noise figure measurement are corrected for in this method. This makes the vector-corrected noise-figure measurement very suitable in cases where the impedance at the input our output of the test system cannot be well controlled. Some examples include on-wafer measurements, systems with multiport switching matrixes before or after the DUT, or in-fixture measurements of DUTs.

Recently, new methods for VNA based noise-figure measurement systems have replaced the noise-source with the use of a power sensor as a reference for receiver gain-bandwidth determination. A guided match-corrected power calibration is used to transfer the accuracy of the power sensor frequency response to the VNA receiver. This provides the gain portion of the receiver characterization. The bandwidth is measured independently by sweeping the VNA source while holding the VNA receiver constant (much like the measurement which shows the noise receiver bandwidth in Figure 6.53), to trace out the shape of the IF response; the shape is integrated to find the total noise bandwidth. This is combined with the receiver gain to produce the total gain-bandwidth product.

6.7.8   Uncertainty of Noise Figure Measurements

The accuracy of noise figure measurements is probably the least understood and most optimistic in RF and microwave measurements. Instrument manufactures and users have been a bit complicit in estimating the error in noise measurements, with the desire to have uncertainties below 0.1 dB. To achieve this value, the instrumentation errors are included but many sources are of error are often not described except in supplemental documentation. To be fair, the instrumentation manufacturers cannot control some of these external effects, but they really must be included in any assessment of overall uncertainty. Instrumentation errors are primarily linearity and mismatch pulling of the receiver used to measure noise figure, and is the most often quoted error. However, in many cases, the instrumentation error is perhaps the smallest contributor to the overall error. Fortunately most manufacturers now supply a spreadsheet uncertainty calculator which does include all the important effects; since uncertainty depends upon the actual characteristics for the DUT, one must compute a new uncertainty for each new device based on noise figure, gain, input and output mismatch, noise source used and receiver used. In fact, one cannot really predict the uncertainty of a particular noise figure measurement unless all the noise-parameters of the DUT are measured, as the noise pulling effect of the input match cannot otherwise be properly accounted for.

A principal error in the uncertainty computation is the error associated with the ENR of the noise source which translates almost directly to noise figure error. This is sometimes ignored in a published specification of an instruments noise uncertainty, with the rationale that an arbitrarily good calibration can be obtained for a noise source, and the noise source is not part of an NFA instrument. But in practice, most users buy off-the-shelf noise sources without special calibrations. This error affects all methods of noise figure measurement, except VNA systems calibrated with power sensors. In that case, the power sensor calibration factor becomes the primary uncertainty.

The mismatch error at the output of the DUT, between the DUT and the receiver, primarily effects the gain measurement in the Y-factor method, but does not have a large effect on the noise figure. This is because the error due to mismatch affects both the gain and noise power in a similar way, and is effectively cancelled. The mismatch of at the output does affect the noise-figure of the NFA receiver, and can cause some pulling effects that may be significant for a low-gain, low noise-figure device.

In many practical systems, the largest error is causes by input mismatch. This error can only be fully understood if the noise-parameters of the DUT are also known.

6.7.9   Verifying Noise Figure Measurements

Unlike S-parameters, it is rather difficult to verify noise figure measurements, and no widely agreed upon method exists for such verification. Verification normally entails measuring a device with a known value of the attribute to be verified, and then one can determine the quality of measurement from the difference between the measured and expected value. Unfortunately, noise figure is somewhat complex, and the errors associated with the measurement come in two forms that cannot be easily distinguished.

A common verification standard for noise figure is to measure a known passive attenuator, whose noise figure will be 1/S21, and compare the measured result with the expected S21. This method does not really validate noise figure measurements, as it only tests noise power measurement at exactly the kT_0B level, and as such really only verifies the S21 measurement. If the noise receiver has some uncompensated added noise (for example, the noise figure of the NF receiver depends upon the match of the DUT), then the noise power will not display the proper kT_0B value, but something either higher or lower, and the error can be quite substantial depending upon the averaging used during the noise figure calibration and measurement. In this way measuring a passive device can give some indication of the sensitivity of the receiver noise power to S22 of the DUT. In order to see this effect clearly, a mismatched device must be used.

A measurement where the noise power of the DUT is below that of the measuring receiver, as is always the case with a passive device, is almost completely unrelated to the proper reading of noise power when testing an active device, where the DUT excess noise exceeds the noise generated in the NF receiver. If the DUT produces no excess noise, then the noise power reading is simply a re-measurement of the NFA self-generated noise power, and testing a passive noise figure is simply testing the stability of this noise power reading, and is almost unrelated to the accuracy if the noise power is from a real DUT is higher. For example, if the residual noise floor after calibration is 10 dB below the DUT noise power, the NF error is around 0.5 dB due to jitter; if the DUT noise is 15 dB above the residual system noise, the error is only 0.17 dB. Thus, unless one knows the actual total system noise figure, one cannot predict the accuracy of noise figure measurements, and the use of a passive device is of almost no value.

A better technique to validate noise figure measurements, especially cold source measurements, is to understand the individual contribution errors to the overall result. In a cold-source measurement there are two factors: error in the gain (S21) measurements, and error in the noise power measurement. The S21 measurement error can be validated using well known techniques to understand the S21 uncertainty. Noise power measurements require measuring a known noise source. In fact, one of the best verification methods is to measure the noise power of a different noise source from the one used for calibration; the error in its measurement is a combination of the calibration error and the error in the verification noise-source ENR uncertainty. In some VNAs or NFAs, there is a direct ENR measurement parameter so errors in the noise calibration can be readily discerned by this measurement.

The other key source of error in a NF measurement is input mismatch. (Because output mismatch error affects noise power and output power in the same way, the composite error in the NF measurement is negligible due to output impedance mismatch.) One way to characterize the quality of a NF measurement with respect to input mismatch is to add a small, low-loss, high-quality airline to the input of the DUT. Since the airline adds no mismatch and almost no loss, any change in the NF measurement after adding an airline can be associated with an error in the effective source match and the measurement of NF of the DUT.

Full vector noise correction (using noise parameters) can minimize this error. Figure 6.59 provides a clear illustration of the improvement that a full-vector noise-parameter correction provides versus the scalar correction. Clearly the full correction gives the best results.

6.7.10   Techniques for Improving Noise Figure Measurements

6.7.10.1   Improving Y-Factor Measurements

The overall error is dominated by the input mismatch error, and this can be minimized by using an attenuator or isolator at the input. Isolators are narrowband by nature, but attenuators are broadband and adding them at the end of the port 1 connector improves overall accuracy of the measurement system with few drawbacks. If an isolator or attenuator is used before the DUT, its loss must be accounted for, and since it is resistive, the temperature must be accounted for as well. For a Y-factor measurement, loss before the DUT can be accounted for by modifying the ENR from the noise source to account for the loss. The modification can be computed simply as new hot temperature from

$$T_{H\_Loss} = T_H \cdot L + (1 - L) \cdot T_L \tag{6.38}$$

Here the loss is expressed as a linear power loss, and T_L is the temperature of the lossy element, provided that the loss is resistive. If the loss is purely reactive, just the first term is used.

Because mismatch cannot be corrected in Y-factor measurements, the use of cables or adapters should be minimized. Loss between the DUT and noise source should be compensated in the same manner as in (6.38).

Loss after the DUT must be compensated in a similar manner, and can be accounted for in the computation of the effective temperature of the DUT, but typically, the loss after the DUT can be included in the calibration of the noise system when measuring the noise figure of the second stage. In fact, it adds directly to the second stage noise figure.

For Y-factor measurements, using a low ENR noise source improves the accuracy of the measurement in two ways: first it typically provides a better match at the input of the DUT, and so reduces the mismatch error. Second, it provides a smaller change from hot to cold state, so the receiver is measuring smaller differences in noise and is theoretically more linear. However, many modern receivers with digital IF receivers have very good linearity over wide ranges and this error is relatively small. Using a small ENR requires more averaging to reduce jitter in the noise measurement, and will give a less accurate measure of the DUT gain. Since Y-factor relies on measuring the slope of a line, the wider spacing between hot and cold, the less fixed measurement error affects the result. On the other hand, for very high gain devices, the hot noise should be low enough so that the total noise power from the noise source plus the gain does not put the DUT into compression.

6.7.10.2   Improving Cold Source Measurements

The cold-source method can provide very accurate noise figure measurements, but some careful attention is needed to avoid common problems.

Since gain is measured using the VNA S-parameter method, one must be sure that the settings for S-parameter measurements do not affect the accuracy of the noise figure. In particular, for high-gain or low-power amplifiers, one must be sure that the input power does not compress the amplifier under test. For most devices, this means using a power at least 20 dB below the 1 dB compression point. In setting the power, one must be sure that the calibration for S-parameter has low trace noise, so using techniques described in Section 6.3 is recommended. In particular, for high-gain amplifiers, one must be sure to set the port 2 power higher than the power 1 power, by approximately the gain of the amplifier.

For the noise specific calibration of the VNA receiver, one must be sure that the ENR of the noise source is sufficiently high to overcome the noise figure of the VNA and any loss between the DUT reference plane and the VNA. For most cases, using a noise source with *high* ENR will yield the best results. This is the exact opposite of the recommendation for Y-factor. In Y-factor, the difference in power between the hot and cold level can cause some error due to the linearity of the noise receiver. In the cold source method, the excess noise from the DUT will be measured in one region of the receiver linearity curve. For optimal results, the ENR from the noise source should match the DUT excess noise to reference the same point. And unlike Y-factor, the cold source method measures the output match of the noise source and corrects for its effect, so that there is no reason to use a low ENR source to improve its match.

For very high gain devices, with excess noise exceeding 60 dB, or very wide bandwidth devices, the noise from the DUT can overwhelm the noise receiver. In such a case, an attenuator on the output of the DUT can be used to reduce the overall gain. Cold source systems typically support de-embedding of loss from either the input or the output of the test system, provided the S-parameters of the loss are known.

If the test system has high loss at port 2, such as testing a high power DUT or testing at the end of a very long cable, the loss from the reference plane to the VNA will be so great that the ENR from the noise source used for calibration will be absorbed in the loss, and there will be no excess noise left at the end of the loss. In such cases, the noise source should be connected directly to the VNA for calibration and the external path-loss separately measured, and de-embedded after calibration.

Alternatively, new methods have been introduced which allow using a power meter to calibrate the VNA receiver for gain-bandwidth product. This new method is not sensitive to loss at port 2. As part of the advanced power calibration techniques, loss either before or after the reference port planes can be removed using a multistep calibration and de-embedding process which is built into the calibration method, making it suitable for on-wafer, in-fixture and waveguide calibrations.

The cold source method (with scalar correction) relies on the VNA source providing a good matched Z_0 (50 ohms) reference impedance to get good noise figure results. Adding an attenuator to the end of the port 1 test cable can substantially improve the match, and allow very good noise figure measurements. Unlike the Y-factor method, the calibration process naturally removes the effect of the attenuator loss and no special care is needed. A value of 6–10 dB is very reasonable; higher loss attenuators give better match but make the S11 more susceptible to cable drift and degrade the raw directivity.

For the full vector-correct noise figure measurements, having a good constellation of impedance states is important. Often the impedance tuner is set behind the test-port coupler, so if there is significant loss in the port 1 cable, the impedance states will collapse to the center of the chart and not provide good separation for extraction of the noise parameters. However, the tuner or Ecal may be placed directly in front of the DUT, or anywhere in the port 1 path. The only drawback to this is that the loss and linearity of the Ecal or tuner may degrade the maximum power or create distortion if other measurements such as gain compression or IMD are desired for the same setup. These issues are avoided by at least one VNA implementation as the manufacturer provides a bypass switch to remove the tuner from the path during non-noise-figure measurements.

6.8 X-Parameters, Load Pull Measurements and Active Loads

For amplifiers operating in the linear mode, the effect of the load impedance on the output power is completely described by the S-parameter definition from Eq. (1.17), in particular

$$b_2 = S_{21}a_1 + S_{22}a_2 \tag{6.39}$$

In the case where the amplifier is terminated by some load impedance, Γ_L, this can be rewritten as

$$b_2 = \frac{S_{21} \cdot a_1}{1 - S_{22}\Gamma_L} \tag{6.40}$$

The power delivered to the load is

$$P_{del} = |b_2|^2 - |a_1|^2$$
$$= |a_1|^2 \cdot |S_{21}|^2 \frac{\left(1 - |\Gamma_L|^2\right)}{|1 - S_{22}\Gamma_L|^2} \tag{6.41}$$

Thus the power delivered depends upon both the load impedance and the S22 of the amplifier.

However, at high powers, S21 and S22 can change with the drive power and with the load impedance, so that predicting the output power is not possible using normal techniques. In these cases, the most common method of evaluation is to drive an amplifier to the desired level while modifying the load to determine which load provides the maximum output power and often, also the PAE of the amplifier at the load. The load impedance is modified by using an impedance tuner at the output, or more recently by providing an active load. This measurement method is usually called "load pull".

6.8.1 Non-Linear Responses and X-Parameters

One aspect of the amplifier's non-linear response is that both the gain and the effective output impedance, and sometimes the input impedance, can change as a function of the input drive level, and it may change as a function of load impedance, even at a constant drive power. For most devices, the actual operating point is more complex than can be expressed using linear S-parameters, but rather require a new set of parameters – the so-called X-parameters[TM] – to fully define the non-linear behavior of a DUT [4, 5].

Without going in depth into the mathematical derivation for X-parameter models, a narrow-band (only considering the fundamental input and not harmonics) definition can be derived which is useful in identifying the reason that the operating point, and its linear representation S22 and S21, changes with load impedance. The most basic version of X-parameters forms a spectral map of input waves to output waves, but each X-parameter element has a dependency on input drive power. The output power of an amplifier is described as

$$b_2(A_1) = \underbrace{X_{21}^F(|A_1|)P}_{Gain\ Related} + \underbrace{X_{22}^S \cdot a_2 + X_{22}^T P^2 \cdot a_2^*}_{Match\ Related}$$

$$(6.42)$$

$$where\ X_{21}^F(|A_1|) = S_{21}(|A_1|) \cdot A_1, \quad P = 1 \cdot e^{\phi_{A1}}$$

Here, A_1 is a large signal whose value affects the operating point and X-parameters, and a_2 is the small-signal reflection from the load. This simple definition of X-parameters includes two terms, a gain term and an output match term. The gain term, X_{21}^F, is essentially the output power of the DUT for any input power, P, contains the phase of the incident signal. It may seem unusual that X_{21}^F has units of power rather than gain, but this is done for computational efficiency in its use; for practical purposes it can be considered as the related to power-dependent gain. The superscript F denotes the fact that this is the response due to large signal inputs – in this case only at the fundamental frequency – that is, the response as if there were no reflections at the input or output at the fundamental and any harmonics, for a particular input powers. This could as easily be defined as the gain, but convention has settled on the power as the fundamental response term. Here, A_1 is the large signal input, and the use of a capital letter implies such; the response of the DUT depends upon the magnitude of this stimulus. However, the second portion uses the lower case a_2 and implies that the small-signal response of the amplifier is substantially linear with respect to amplitude of a_2.

The second set of terms identified as match consist of X_{22}^S and X_{22}^T (but informally called S and T terms) and form the pair of terms that describe the change in B2 power as a function of the signal reflected from the load, and together they form the effective output match $S_{22}^{eff}(|A_1|, a_2)$, which can be called the true "hot S22" of the amplifier. Now, the effective value of S22 appears to change with the magnitude and phase of the load, but this is only because of the way the S and T portions add and subtract as the magnitude and phase of a_2 vary. The value of X_{22}^S and X_{22}^T are presumed constant in the X-parameter model for any particular value of $|A_1|$ and any magnitude or phase of a_2. This is provided a_2 is small enough to not affect the actual operating point of the amplifier, an assumption that depends upon the particular device measured. The X_{22}^S changes only slowly if at all with $|A_1|$; at low input powers it devolves to small-signal S22, and at high powers it is essentially what is measured as part of a traditional, but incorrect, hot S22 measurement. But such a measurement ignores the T term which adds in as a conjugate multiple of the a_2 signal. The T term is created through a high-order non-linear mechanism, similar to an intermodulation product, and its level changes as a second-order power of the input signal. Figure 6.60 shows the relative response of the S21 term, the S22 term and the T22 terms of an amplifier as a response to the input drive level. The S21 term shows compression at higher power, the S22 terms shows some small variation as a function of power, but the T22 term shows substantial variation following the 2:1 curve expected of a higher-order product. In fact, at some levels it becomes larger than S22, and thus has a larger effect on the output power than does the S22 term.

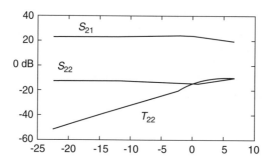

Figure 6.60   S21, S22 and T22 terms as a function of input power.

To understand how the conjugate operation of a_2 on the X_{22}^T affects the apparent total reflection from the DUT, consider the vector-diagram in Figure 6.61, where a_2 is the result of some non-matched load. The figure on the left shows the total signals coming from the DUT, as a result of the X_{21}^F (large light error), the $X_{22}^S \cdot a_2$ (smaller darker arrow) and $X_{22}^T \cdot a_2^*$ (also small darker arrow); the vector sum of these terms (constructed with the light dashed arrow) forms what could be called the true "hot S22". The total power, b_2 is the vector sum of all the X terms, illustrated by the dark black arrow on the figure. The figure on the right has identical X-parameters, but now the phase of the load reflection, a_2, is changed, as might happen through rotating the phase by adding a length of transmission line.

For this second figure, the relative value of $X_{22}^S \cdot a_2$ versus the fundamental output X_{21}^F, shifts as a_2 shifts in phase. This is expected and occurs just the same as in a linear device. But the $X_{22}^T \cdot a_2^*$ shifts in opposite phase as a_2 shifts, causing the relative amplitude of the effective DUT reflection to change as the phase of the load changes. In this way, the hot S22 (S22 under the condition of forward drive) does appear to change with the load, and one might infer that as such, the operating point of the amplifier changes with the load. But in fact, the S and T terms are constant; only their vector-sum changes due to the conjugate phase behavior of the T term and S term interaction, which causes a change in the apparent response of the hot S22.

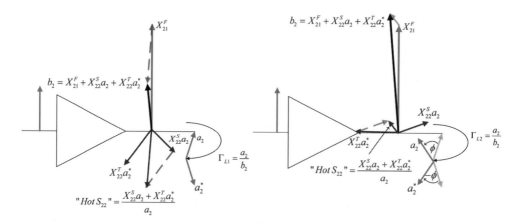

Figure 6.61   The vector effect of a2 and a2* on apparent hot-S22.

This also causes the b_2 value to change in a manner that is not consistent with a fixed value of S22.

For amplifiers that are pre-matched with impedances relatively close to 50 ohms, used in narrow band applications such as cellular phone handsets, the fundamental response may be sufficient to provide good agreement between modeled behavior and actual response, and predictions made for maximum power and efficiency based on X-parameter simulations may yield matching networks that give similar results between simulation-optimization and real application. However, for "bare" transistors, those that require substantial matching to transform them into a 50 ohm environment, the fundamental X-parameters measured at 50 ohms may not sufficiently characterize the DUT. Instead, load dependent X-parameters may be required, where the X-parameters become a function of both the amplitude of the input signal $|A_1|$, and the load reflection coefficient, Γ_L. In such a case the X-parameters provide a kind of perturbational or differential characterization around each load, with the S and T terms describing the behavior of the DUT at each load point characterized.

In addition to the load dependency, many amplifiers also have a dependency on the reflection of the harmonics at the input and the output of the DUT, and dependence on the DC operating point. Each harmonic term has its own S and T terms that characterize how any particular harmonic (say the lth) generates an output at any other harmonic or fundamental (say the kth), and from any port (perhaps the jth) to any other port (as in the ith). Thus, in general, X-parameters have four indexes, two of which describe which port and of which harmonic the input signal is incident to the DUT, as in a_{ik}, and two of which related to which port and which harmonic or fundamental is scattered from the DUT, as in b_{jl} so that in general, the X parameters are represented as

$$b_{ik} = X_{ik}^F (DC, |A_{11}|, \Gamma_L) P^k + \sum_{j,l} X_{ik,jl}^S (DC, |A_{11}|, \Gamma_L) P^{k-l} \cdot a_{jl}$$
$$+ \sum_{j,l} X_{ik,jl}^T (DC, |A_{11}|, \Gamma_L) P^{k+l} \cdot a_{jl}^* \tag{6.43}$$

This generalized equation describes a multidimensional dependency of frequency and amplitude of the input signal, DC operating point and output impedance point. The method for determining the X-parameters is beyond the scope of this book, but in general an X-parameter extraction is performed for each of the independent variables to determine the multidimensional response. During simulation, the X-parameter result of any arbitrary input encompassed by the data are determined by a multidimensional interpolation of the X-parameter data.

6.8.2 Load Pull, Source-Pull and Load Contours

Load pull measurements create a map of the output power of an amplifier as a function of the output load. If an amplifier were responding in a linear way, this map would form circles centered on the maximum output power, which would occur at a load impedance equal to the conjugate of S22 of the amplifier.

In practice an amplifier operating in a non-linear range does not produce maximum power at the same impedance as predicted by the linear measurement of S22, but rather the effective output impedance moves as the power is increased and, in the past, the impedance for maximum power could only be assessed using a load pull system. From the previous section, it is now understood that the apparent change in output impedance with increasing drive power is a consequence of the match portion of the X-parameters. As such, the need for direct load pull measurements may be reduced in the future as the understanding and quality of X-parameter characterizations improve. However in many cases, particularly for very high power or under the conditions of complex modulation, load pull characterization remains a good option for determining the effects of output impedance on the performance of an amplifier. Load pull systems are classified as mechanical, active or hybrid, depending upon how the load impedance state is created at the output of the DUT.

6.8.2.1 Mechanical Load Pull Systems

Mechanical systems, or true load pull, use an impedance tuner at the output of the DUT as an impedance transformer, transforming the VNA port 2 impedance into some other impedance on the Smith chart. Some systems use multiple tuners or multisection tuners to supply a desired match for the fundamental and harmonics. Harmonic load pull, used to determine the impedance match for harmonics to optimize amplifier performance, is a relatively new and active area of research. Recently some studies have shown that the impedance of the source at harmonic frequencies can also have a significant effect on amplifier performance, so source pulling has been more common as well. An example of a source-pull and load-pull system is illustrated in Figure 6.62.

Source pulling, as described in Section 6.7.6.1, is often used for noise parameter characterization. It is also sometimes used to investigate the stability of amplifiers, especially in an out-of-band region. For these investigations, the source tuner is set to a high reflection value and the output of an amplifier is monitored on an SA to look for oscillations or even a rise in the noise floor which indicates a tendency to oscillate.

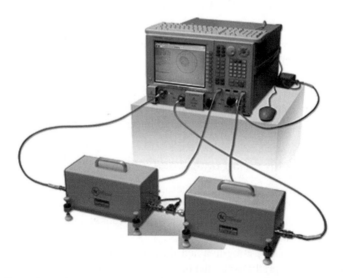

Figure 6.62 Illustration of a source and load pull system. Reproduced by permission of Maury Microwave Corp.

For investigation of maximum gain or maximum power transfer at the fundamental, it is not at all necessary to do actual source pulling as the effect of a source pull is simply to change the effective voltage applied to the input of the DUT. This could just as easily be performed by computing the output power from a 50 ohm source that would result in the same voltage, and adjusting that voltage. This is identical to doing an active source pull on the source, and would yield the same results as a tuner for CW measurements. However, the same cannot be said for harmonic source pull, which must be relative to the second harmonic generated in the DUT, and so the VNA must provide an actual effective reflection coefficient at that frequency, although the reflection can be created by another active source.

6.8.2.2   Active Load Pull

Active load pull or active tuning is an alternative to using mechanical tuners, and relies on utilizing a second source driving the output of the amplifier, the amplitude and phase of which is controlled to create a reflected a_2 trace that provides the desired load reflection coefficient, as illustrated in Figure 6.63.

In active load pull, all four waves from the DUT must be monitored. The process starts with applying an input signal to the DUT at the desired power level, and monitoring the output waves b_2 as well as the raw reflection for the VNA load, a_2. The effective reflection coefficient is computed from these waves, and a second source at the output is activated to provide an additional signal to add to or subtract from the raw reflection to create the desired reflection coefficient. The S-parameters are computed from the two stimulus signals applied using the formulation from Eq. (1.21). From the first measurement, the value of S22 is not known, and so the effect of applying the second source at port 2 on the b_2 power is not known. To solve for

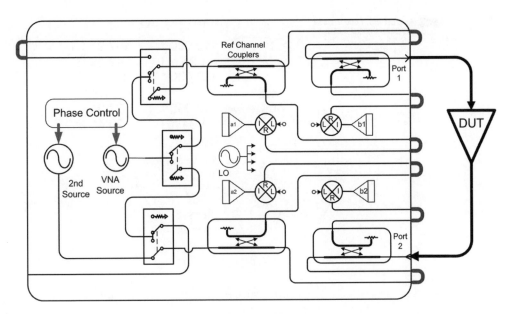

Figure 6.63   Block diagram of a VNA with active load pulling.

all four S-parameters, a change in the output signal must occur, so it is common to change the phase of a_2 to provide a second set of equations to solve. After solving for the S-parameters, a good estimate of S22 is established, and a new estimate can be made for the second source amplitude and phase. This process is repeated until the correct value of effective reflection coefficient is achieved within predefined limits.

Note that the S-parameters computed in this manner are truly "hot S-parameters" provided that the change in a_2 at the output is small enough to not affect the operating point of the amplifier. The S22 value gathered is the total effective reflection coefficient from the amplifier's output and represents the sum of the X_{22}^S and X_{22}^T for that particular value of load, again provided that the change in the a_2 wave is near that load value. In fact, this is a similar process to that used to determine the X-parameters of a DUT, which differ only in that multiple values of the a_2 wave are applied to allow solving the X-parameter defining Eq. (6.42).

One simple but informative example of an active load pull response is looking at the effective reflection coefficient, or true hot S22, while sweeping the load impedance around the Smith chart. An example is shown in Figure 6.64 where the one trace (marked Γ_{Load}) shows the value of the active load. Also indicated on the chart is the value of a lower power measurement of S22 of the amplifier (the trace being nearly constant), and a locus of points illustrating the value of the hot S22 as the phase of the load is rotated around the Smith chart. In this case, the magnitude of the load was set to match the magnitude of the linear S22, or about -12 dBc ($\Gamma = 0.25$). It is interesting to see the trajectory of the hot S22 as the load circulates. Because there are two vectors moving for hot S22 (S and T terms or the a2 and a2* signals), there will be two loops in the hot S22 for one loop in the load response. This shows the complex nature of the true Hot S22, but one must realize that the underlying S and T terms in fact do not change with load reflection.

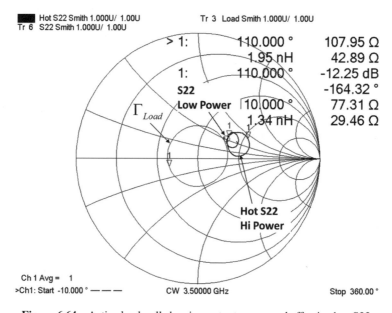

Figure 6.64 Active load pull showing output power and effective hot-S22.

Active load pull has some advantages and disadvantages relative to the passive or mechanical system. The key disadvantage is that the output power supplied from port 2 must be the same as the power from the DUT if the active load is to reach the edge of the Smith chart, or full reflection. The key advantage is that it is often much faster, perhaps up to 100 times faster, than a mechanically tuned system. An additional advantage is that the active load will not induce any oscillations in the device because oscillations occur when some load state applies an impedance that lies in an unstable region of the amplifier. The oscillations start because thermal noise is amplified by the near infinite gain, producing the oscillation signal. But in active load, there is no load reflection (other than the raw VNA reflection) applied to the amplifier except at the exact frequency of the CW source, and the noise enhancement due to unstable gain is limited by the maximum power of the port 2 source used to create the active load. Once that maximum is reached, the effective active load is limited and any oscillation is suppressed. That does not mean one cannot detect these regions of instability; in fact, these will be areas on the Smith chart where active load pull will fail to achieve the desired value of reflection coefficient. This gives the user a strong hint that the DUT will be unstable in that region of the Smith chart.

6.8.2.3   Hybrid Load Pull Systems

In some systems, a mechanical tuner for the fundamental is combined with active tuning for the harmonics, providing fine/fast tuning of the load. For very high power amplifiers, it may not be practical or possible to create an active signal that can drive the DUT to the desired state. In these cases, a mechanical tuner is used as a pre-matching network to create an approximate value of the desired load impedance. In fact, when X-parameters are measured at impedances other than 50 ohms, this is the method used to create the load impedance states. The second source is used to do a small load pull around the load impedance, by which the X-parameters are extracted.

In other applications, the mechanical tuner is used to create a load match at the fundamental, and the active load is used to provide the harmonic load impedance states. In such a system, the mechanical tuner can provide a high power reflection at the output, and the power level for the active load, which is relative to the power at the harmonics, can be much lower, making such a system quite practical. One attribute of many mechanical tuners is that while they have a high reflection at the fundamental frequency, they provide essentially a low-loss transmission line path at other frequencies, allowing the active source to be driven through the tuner at a different frequency, such as at the harmonics.

6.8.2.4   Power Contours

Load pull measurements in the most classic form are typically analyzed to show the impedance point of maximum power along with the locus of points representing lines or contours of constant power on the Smith chart. Power data from dozens or hundreds of impedance points are gathered to generate these contours, "connecting the dots" between impedance points that share a common power. Similar power contours can be created in simulation from an X-parameter model, with remarkably good correlation to actual measured results. A comparison

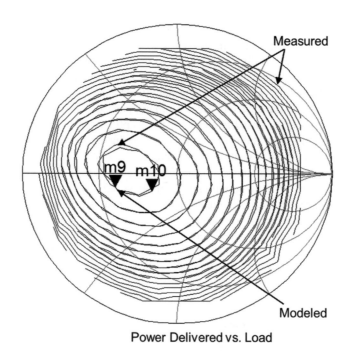

Power Delivered vs. Load

Figure 6.65 Comparing X-parameter simulated load pull with real load pull values.

between simulated load-pull based on an X-parameter model of an amplifier, and the actual load pull results, are shown in Figure 6.65.

6.8.2.5 Efficiency Contours

The load impedance applied to the DUT can also affect the PAE of an amplifier, especially one in a non-linear range of operation. If the power, voltage and current of an amplifier are measured, and PAE computed as a function of the load impedance, then load contours of PAE can be created as well as max power. While they tend to be similar (maximum efficiency will occur near maximum power), there can be some differences, and observing the power and PAE contours is a quick way of seeing the optimum load for a particular amplifier. These contours can also be created versus second or third harmonic load, or source load at second or third harmonic. An example of PAE contours is shown in Figure 6.66

6.9   Conclusions on Amplifier Measurements

An introduction to a wide range of amplifier characterization measurements is presented in this chapter, with comprehensive details focusing on gain and match-corrected power. A few key points should be emphasized: the time spent pretesting the amplifier-under-test and optimizing the VNA setup is very well spent in avoiding mistakes in biasing and setup, even in preventing damage to the DUT and VNA through overdrive conditions. For the most part, modern VNAs

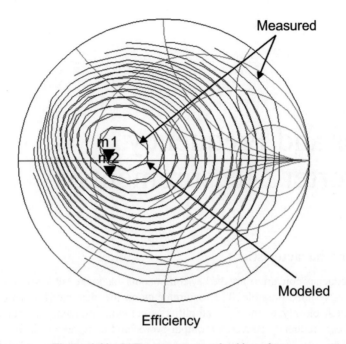

Figure 6.66 PAE contours versus load impedance.

can perform substantially all the basic characterization measurements required on amplifiers including distortion measurements, in the form of harmonics and IMD, noise and noise figure measurement, power and efficiency measurements, and even load pull measurements. A key advantage of using a VNA is that the high quality of calibration typical of S-parameter measurements can be extended to other measurements, resulting in both high speed and low error in the results. These methods, including some advanced de-embedding techniques, allow high quality measurements even with devices that have non-standard connections such as on-wafer and in-fixture devices.

References

1. Vendelin, G.D. (1982) *Design of Amplifiers and Oscillators by the S-parameter Method*, Wiley, New York. Print
2. Stability circles equation-editor function courtesy of Andy Owen, 2012 Agilent Technologies
3. Shoulders, R.E. and Betts, L.C. (2008) Pulse Signal Device Characterization Employing Adpative Nulling and IF Gating. Agilent Technologies, assignee. Patent 7340218. 4 Mar. 2008. Print.
4. Verspecht, J., Bossche, M.V., and Verbeyst, F. (1997) Characterizing components under large signal excitation: Defining sensible 'large signal S-parameters'. 4ninth IEEE ARFTG Conf. Dig., Denver, CO, Jun. 1997, pp. 109–117.
5. Root, D.E., Verspecht, J., Sharrit, D., *et al.* (2005) Broadband poly-harmonic distortion (PHD) behavioral models from fast automated simulations and large-signal vectorial network measurements. *IEEE Transactions on Microwave Theory and Techniques*, 53(11), 3656–3664.

7

Mixer and Frequency Converter Measurements

7.1 Mixer Characteristics

Mixers and frequency converters are the key components that make radar, wireless and satellite communication systems possible. The requirements on the characterizations of mixers and converters match closely to those of amplifiers, including frequency response and phase linearity of these frequency converters, output power and compression levels, noise figure, distortion and harmonics, and unique to frequency converters, higher-order mixing products, LO, RF and image rejection.

Just as amplifiers can be categorized by application, frequency conversion circuits can be separated into classes of devices grouping similar attributes and test requirements. At the broadest levels are mixers, which are simple three-port devices that have an RF, IF and LO port. These are most commonly created as a single-balanced or double-balanced device that uses a large external signal known as the local oscillator (LO) to drive a set of diodes into on/off conduction over the period of the LO waveform.

A single balanced mixer is shown in Figure 7.1. This simple mixer is balanced only at the LO port, and the LO signal causes the mixer diodes to conduct during the positive half cycle of the mixer. The balanced LO signal is created through the use of a transformer (at lower frequencies) or a balanced-to-unbalance (balun) circuit at high frequency, often consisting of coupled transmission lines. Current from the RF source flows in the IF load during this conduction, with the quarter-wave lines keeping the RF drive from shorting the LO signal at the diodes. Essentially the RF signal is commutated, sampled or chopped at the LO rate. Fourier analysis of the output waveform of the mixer will show frequency content at the sum and difference of the RF and LO signals and their harmonics. Since the LO is balanced, very little LO signal is present at the output. In the illustration, the lower frequency IF can be seen in the repetitive pattern of the IF output waveform.

A single balanced mixer is quite simple, and is often used at very high frequencies, but it has the disadvantage that the conduction is only 50% so it has lower conversion efficiency than other models. Its main advantage is that since the RF and IF ports are not balanced, no balun is required which can allow them to be quite broadband with simple construction.

Handbook of Microwave Component Measurements: With Advanced VNA Techniques, First Edition. Joel P. Dunsmore.
© 2012 John Wiley & Sons, Ltd. Published 2012 by John Wiley & Sons, Ltd.

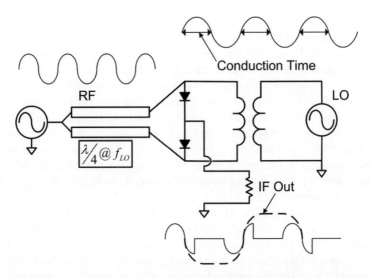

Figure 7.1   Input, LO and output wave forms from a single balanced mixer.

For most work in RF and microwave frequency ranges, conversion efficiency and isolation are more important than a simple design, and double-balanced mixers provide improvements in both areas. A typical double balanced mixer [1] is shown in Figure 7.2, with an illustration of the conducted signals at the output. The LO waveform alternately turns on pairs of diodes, and with each LO half-cycle, the RF signal's sign is changed as it is appears at the IF port.

Double balanced mixers have natural isolation of the transformer on each of the RF and LO ports, isolating these signals to the IF port. One variation on the double balanced mixer is

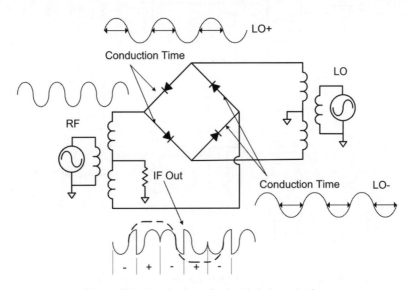

Figure 7.2   Conduction of a double balanced mixer.

splitting the RF and LO signals to two sets of diode rings, with inversion on the RF path, and combining the IF signal (which is now out of phase) in a third transformer. This is sometimes called a double-doubly-balanced mixer, dual-ring mixer, or a triple-balanced mixer. It has an advantage of splitting the RF signal between two mixers, thus reducing the RF level in each mixer and improving the linearity of the overall mixer by 3 dB. The tradeoff is more complexity and 3 dB higher LO power requirement for the same performance.

Other mixer forms include image reject, or single-sideband (SSB) mixers, that utilize two mixers, with 90° of phase shift on two of the mixer ports. The classic image reject mixer uses two mixers with a 90° phase shift on each of the RF and IF path, and an LO with an equal phase splitter to each LO input. These have the disadvantage of requiring 90° hybrids or phase shifters, which generally have limited frequency response, on each of the RF or IF path, as shown in Figure 7.3, upper.

Another configuration for image reject mixers has a 90° phase shift on each of the LO and IF outputs, so that the signal at the combined output suppresses the one sideband while enhancing the other. This version is often found in MMIC mixers, where the LO phase shift is created by multiplying the LO, or using a higher frequency LO, and then dividing it in such a way as to create 90° offsets; alternative forms use adjustable phase shifters to generate the offset on the LO port. The IF hybrid is often replaced by digitizers on each output (called the I and Q output), the 90° summation occurring numerically, also shown in Figure 7.3. The up-converter form of this mixer has two inputs, with assumed 90° phase shift, that are often used for complex modulation schemes where the output signal is centered directly on the LO. These are sometimes referred to as I/Q modulators as the LO signal is not translated in frequency, but rather any waveform on the I/Q input ports modulate the LO signal to generate the same baseband envelope waveform at the modulator output. In these mixers, the LO suppression is critical specification.

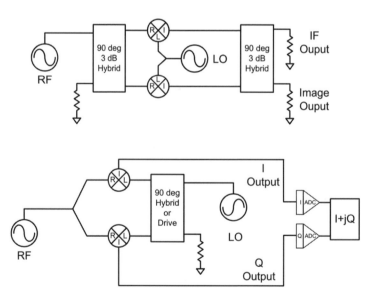

Figure 7.3   Image reject mixer topologies, (upper) standard topology, (lower) topology used in digital demodulators.

7.1.1   Small Signal Model of Mixers

Even though the very nature of mixers is strongly non-linear, for the most part the transfer of signal from the RF port to the IF port is linear in its behavior, and can be modeled much like S-parameters in an amplifier. Though the frequency is translated, it is linear in the sense that doubling the input voltage will double the output voltage, and if one applies a small modulated signal to the RF input, the same modulated signal will appear at the IF output, without distortion. The non-linear nature of the diodes produces the chopping of the RF signal which translates it to the IF frequencies, but does so in a linear way. Considering just the first Fourier component of the LO as a sine wave input signal, multiplied by a input cosine signal, the mathematical representation is

$$\cos(\omega_{In}t) \cdot \sin(\omega_{LO}t) = \frac{1}{2}\left(\sin\left[(\omega_{In} + \omega_{LO})\right]t - \sin\left[(\omega_{In} - \omega_{LO})\right]t\right) \qquad (7.1)$$

Thus the output will have frequency elements at the sum and difference of the two input signals. Of course, the LO signal has many harmonics (all odd if it is symmetrical signal) and so there will also be outputs at the sum and difference of each of the harmonics of the signals with the input signal as well, sometimes referred to as intermodulation spurs, or higher-order products. Thus either the sum or difference signal can be created. The sum signal always represents an up-conversion and by convention the input signal is referred to as the IF signal (the IF signal always being the lower frequency of either the input or output). If the difference signal is the desired output, the mixer can either be an up-converter (if the input is lower than the difference) or a down-converter (if the input is higher than the difference, or higher than the LO). This is shown graphically in Figure 7.4. The upper figure shows that this condition creates both and up and down converted signal; filtering on the output will determine if this is an up-converter or down-converter.

The lower plot shows the case where the mixer is only an up-converter, and the output can be either an image (LO − In) or normal (LO + In) mode mixer. An important aspect to note is that as the input signal moves up in frequency, the output signal can move up in frequency (call this the standard or normal mode) or down in frequency (call this the image mode or reverse

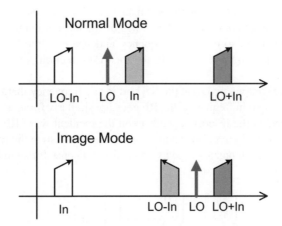

Figure 7.4   Graphical representation of signals at the input and output of a mixer.

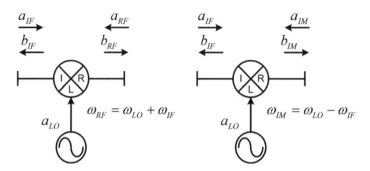

Figure 7.5   Schematic of a normal (a) and image (b) mixers showing incident and scattered waves.

mode). For the image case, the reversal carries over to a change in phase of the input signal: for a standard mixer, a positive change in phase at the input results in a positive change of phase at the output; for an image mixer, a positive change of phase at the input results in a negative change of phase of the output. This property is important to consider when determining how to cascade the effects of mixers, filters and other components, or to remove the effects of cables and connectors during a measurement.

A small-signal model based on incident and scattered waves can be developed for a mixer [2, 3] using the definitions shown in Figure 7.5.

For an ideal standard mixer, the input signal is translated to the output with no change in amplitude or phase, and no reflection from the port. Signal applied at the output of the mixer is translated similarly to the input frequency with no change in amplitude or phase and no reflection from the output port. Such an ideal mixer can be represented mathematically as

$$
\begin{bmatrix} b_{IF} \\ b_{RF} \end{bmatrix} = \begin{bmatrix} 0 & a_{LO}^* \\ a_{LO} & 0 \end{bmatrix} \cdot \begin{bmatrix} a_{IF} \\ a_{RF} \end{bmatrix}
\tag{7.2}
$$

for a standard mixer where $|a_{LO}| = 1$

Here the LO is assumed to interact with the mixer in such a way that a change in its power does not change the mixer conversion efficiency. For the normal operating point of mixers, this assumption holds true, so only the frequency and phase of the LO affect the transfer function. This represents an up-converter in the forward direction, and a down-converter in the reverse direction.

The scattered wave (output signal) at the RF port is at a frequency that is the sum of the IF and the LO; if either go up in frequency, the RF goes up, similarly for phase. Notice, however, that the scattered wave at the IF port depends upon the incident wave RF port, but if the LO goes up in frequency, the IF goes down; similarly for the phase, so in the reverse direction the IF response moves as the conjugate of the LO. The ideal response for an image mixer is subtly different, and described by

$$
\begin{bmatrix} b_{IF} \\ b_{IM}^* \end{bmatrix} = \begin{bmatrix} 0 & a_{LO} \\ a_{LO}^* & 0 \end{bmatrix} \cdot \begin{bmatrix} a_{IF} \\ a_{IM}^* \end{bmatrix}
\tag{7.3}
$$

for an image mixer where $|a_{LO}| = 1$

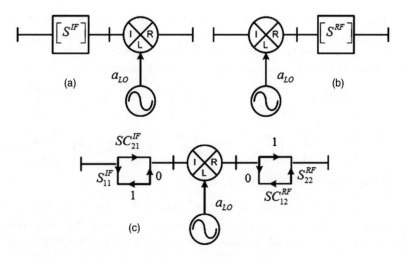

Figure 7.6 Schematic representations of mixers with non-ideal responses.

If the IF is taken as the lowest frequency, this is also an up-converter, but the image moves down in frequency as the IF moves up. And in the reverse direction, the IF moves down in frequency if the image moves up, so that conjugation of one of the LO terms and both of the image terms is required for proper treatment of the phase response.

These equations describes the frequency and phase response of an ideal standard mixer and an ideal image mixer, but real mixers will have reflections at the input and output, as well as transmission frequency response, and so must have a more complex description.

The additional response can be characterized in a few ways, as indicated in Figure 7.6. In (a) the non-ideal response is contained completely on the IF side; in (b) the non-ideal response is contained completely on the RF side; in (c) the non-ideal response is split between the IF side and the RF side, with the forward response assigned to the IF side, and the reverse response assigned to the RF side. All these representations are equally valid, and represent an "error box" which contains the non-ideal elements of the mixer behavior; and one may change from one form to another in a normal mixer. However, for an image mixer, special care must be taken when moving an error box from the input to the output.

From this figure one can define a set of scattering parameters that describe the behavior of these waves for a normal mixer under any load condition. The waves at the IF and RF frequencies for a standard mixer are represented by

$$\begin{bmatrix} b_{IF} \\ b_{RF} \end{bmatrix} = \begin{bmatrix} S_{11}^{IF} & a_{LO}^* \cdot SC_{12}^{IF} \\ a_{LO} SC_{21}^{IF} & S_{22}^{IF} \end{bmatrix} \cdot \begin{bmatrix} a_{IF} \\ a_{RF} \end{bmatrix} = \begin{bmatrix} S^{IF} \end{bmatrix} \cdot \begin{bmatrix} a_{IF} \\ a_{RF} \end{bmatrix}$$

$$\text{or} \quad \begin{bmatrix} b_{IF} \\ b_{RF} \end{bmatrix} = \begin{bmatrix} S_{11}^{RF} & a_{LO}^* \cdot SC_{12}^{RF} \\ a_{LO} SC_{21}^{RF} & S_{22}^{RF} \end{bmatrix} \cdot \begin{bmatrix} a_{IF} \\ a_{RF} \end{bmatrix} = \begin{bmatrix} S^{RF} \end{bmatrix} \cdot \begin{bmatrix} a_{IF} \\ a_{RF} \end{bmatrix} \tag{7.4}$$

for a standard mixer where $|a_{LO}| = 1$

Even though the S-matrix for $[S^{IF}]$ and $[S^{RF}]$ refers to different frequencies, the elements have identical values and one can say that $[S^{IF}] = [S^{RF}]$, remembering that they are indexed by different frequencies. Thus the real response of the mixer can be moved from one side to the other, allowing concatenation of mixer responses with other network elements. Applying this to an image mixer, however, is more complicated and subtle.

A similar scattering matrix can be defined for an image mixer as

$$
\begin{bmatrix} b_{IF} \\ b_{IM}^* \end{bmatrix} = \begin{bmatrix} S_{11}^{IF} & a_{LO} \cdot SC_{12}^{IF} \\ a_{LO}^* \cdot SC_{21}^{IF} & S_{22}^{IF} \end{bmatrix} \cdot \begin{bmatrix} a_{IF} \\ a_{IM}^* \end{bmatrix} = \begin{bmatrix} S^{IF} \end{bmatrix} \cdot \begin{bmatrix} a_{IF} \\ a_{IM}^* \end{bmatrix}
$$

$$
or \quad \begin{bmatrix} b_{IF} \\ b_{IM}^* \end{bmatrix} = \begin{bmatrix} S_{11}^{IM*} & a_{LO} \cdot SC_{12}^{IM*} \\ a_{LO}^* \cdot SC_{21}^{I*} & S_{22}^{IM*} \end{bmatrix} \cdot \begin{bmatrix} a_{IF} \\ a_{IM}^* \end{bmatrix} = \begin{bmatrix} S^{IM} \end{bmatrix}^* \cdot \begin{bmatrix} a_{IF} \\ a_{IM}^* \end{bmatrix} \quad (7.5)
$$

for an image mixer where $|a_{LO}| = 1$

Here again, even though the S-matrixes for $\begin{bmatrix} S^{IF} \end{bmatrix}$ and $\begin{bmatrix} S^{IM} \end{bmatrix}$ refer to different frequencies, the elements have similar values, but for an image mixer $\begin{bmatrix} S^{IF} \end{bmatrix} = \begin{bmatrix} S^{IM} \end{bmatrix}^*$, remembering that they are also indexed by different frequencies. SC is used for transmission terms to indicate conversion.

For the standard mixer, an overall equivalent circuit can be drawn that moves all of the response from the IF side to the RF side, including the effect of source match, as shown in Figure 7.7. For this equivalent circuit, the mixer is eliminated and the source changes frequency, but the values for the source match remain the same.

The same equivalent circuit can be drawn for an image mixer, but with a decidedly different result, as shown in Figure 7.8. Here, moving the IF scattering matrix to the output conjugates all the terms. What is remarkable, and until recently not well understood, is that moving the

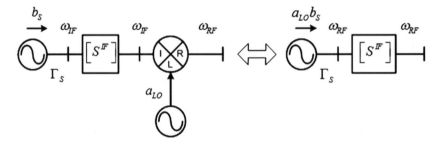

Figure 7.7   Actual circuit (a) and equivalent circuit at the RF (b) for a source and standard mixer.

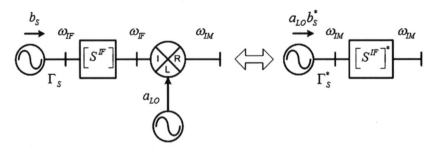

Figure 7.8   Actual circuit (a) and equivalent circuit at the RF (b) for a source and image mixer.

source to an equivalent version at the IM frequency requires that both the source wave, b_S, and the source match, Γ_S, must be conjugated. Thus, Eq. (7.5) also tells one how to cascade components in the IF or RF paths before or after the mixer. From this, one can create an overall response including IF effects and mismatch between elements.

This representation has important use in the computation of responses of reciprocal mixers, used in vector mixer calibration, for image mixer cases as discussed in Section 7.5.3.1.

7.1.2   Reciprocity in Mixers

Reciprocity in mixers has a special meaning when considering the definitions of Eqs. (7.4) and (7.5). For the amplitude response, the mixer is said to be reciprocal if $|SC_{21}| = |SC_{12}|$; this behavior can be verified using standard scalar measurements as described later in this chapter. Since the phase response from the input to the output of a mixer depends upon the LO, and a shift in the LO also shifts the output, it is difficult to explicitly state the phase response of a mixer, as it will change in time, if the input and output are not harmonically related. That is, the phase of the LO, even if known relative to the IF at some particular time, will rotate in time to some other relationship at some later time.

Thus, when referring to phase reciprocity, one generally means that the deviation in the phase response versus frequency of SC21 should match the deviation in the phase response versus frequency of SC12, or to be more succinct, the group delay of the SC21 should equal the group delay of SC12, for a mixer to be reciprocal. This definition is very useful, as one primary reason for caring about the reciprocal nature of a mixer is in the calibration process for measuring the group delay of a mixer or converter.

7.1.2.1   Notes on LO Phase Response

One interesting aspect of the LO phase response of a mixer can be demonstrated if one splits the source signal into two mixers, fed by the same LO, and compares the IF signals on a coherent receiver, such as the test and reference inputs of a VNA, as shown in the upper window of Figure 7.9. The relative IF signals will be stationary in time. If one changes the phase of the LO in the path of one mixer, one will see a resulting phase shift in that path's IF. Here, the LO and RF are swept through a 1 GHz span and the IF is fixed. Three cases are presented: the base mixer case, the case where a small delay is added to the RF path, and a case where a small delay is added to the LO path. Since RF and LO both sweep, this represents additional phase change as a function of frequency, and one might expect additional phase slope on the measured phase of the mixer. For the case of the RF delay, one does see additional slope, of about 36°. The additional delay was about 100 psec: the expected additional phase would therefore be

$$\Delta\phi = delay \cdot 360 \cdot \Delta f = 100ps \cdot 360 \cdot 1GHz = 36° \qquad (7.6)$$

However, when the same delay is added to the LO path, an opposite slope is observed, of about the same value. Thus the phase shift of the LO is translated directly to the IF signal, but as a conjugate of the phase change. For an image mixer, the opposite result is expected. This is exactly in line with the behavior described in (7.4). In this case, the measurement represents

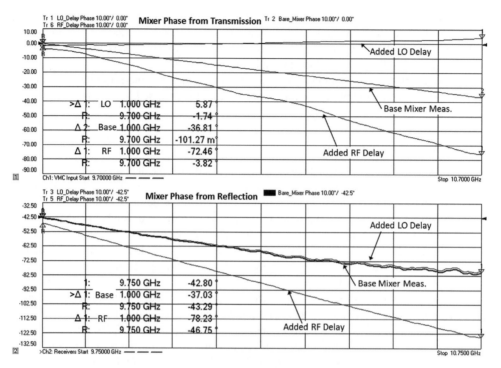

Figure 7.9   Understanding the effect of phase shift of the LO.

the RF to IF conversion, related to SC12 term. In the lower window, a characterization of the mixer phase is performed looking at the two-way reflection of a short placed at the end of the mixer, as one looks at the equivalent S11 of a mixer at the input, with a short at the output.

Ignoring other match terms, the reflection is essentially the product of SC21 and SC12. But the actual response includes the effects of the LO so that the proper description is

$$\text{For a standard mixer} S_{11}^{IF}\Big|_{P2\_Short} = a_{LO}SC_{21}^{IF}a_{LO}^*SC_{12}^{IF}\Big|_{S_{11}=S_{22}=0}$$

$$\text{For an image mixer } S_{11}^{IF}\Big|_{P2\_Short} = a_{LO}^*SC_{21}^{IF}a_{LO}SC_{12}^{IF}\Big|_{S_{11}=S_{22}=0} \tag{7.7}$$

Thus, if the mixer is reciprocal, the square root of this S11 measurement is the conversion loss of the mixer. From this, one can conclude that shifting the phase of the LO does not have any effect on the apparent phase of the S11 response of a mixer, as the up and down conversions create a similar positive and negative shift due to the conjugation function of the LO so that any phase shift is cancelled. The measurement in the lower window shows exactly this effect. Adding the delay to the RF input shows the same response as the upper window, but adding delay to the LO path shows virtually no change (the slight change is likely due to some small mismatch effect on the LO port). Thus, one cannot determine if the phase of an LO has been shifted by observing only the S11 input of the mixer, regardless of the terminating impedance. In one method of measuring mixers (see Section 7.5.3.1), the reflection of a termination on the output is used to determine the two-way conversion loss through the mixer. From Eq. (7.7), it is clear that the phase response of the LO port is not included in such a characterization.

7.1.3 Scalar and Vector Responses

The response of mixers is shown in previous sections to be complex-valued function with amplitude and phase response. For many applications, the only concern is the amplitude response, usually called the conversion loss. Other responses include the compression, the output power, output harmonics and spurious products. All of these responses are represented by scalar quantities and the characterizations are commonly called scalar mixer measurements. In the past, most scalar quantities were measured using simple systems of two sources (one for the LO and one for the input) and a spectrum analyzer (power meters are generally not used in mixer characterization as the other mixing products and LO feed through cause substantial error).

When mixers are used as part of a communications system, both the magnitude and phase response become important in some formats, so these systems require characterization of the mixer's phase or delay response in addition to amplitude. Since the responses are complex, they are commonly called the vector response of the mixer, and measurements are classified as vector mixer measurements. Until recently, entirely different systems were required for vector versus scalar, with many systems providing only the delay response, often through interpretation of modulated results. Newer techniques, described in the following sections, provide for measuring these complex values with a single system, with high accuracy in both magnitude and phase responses.

7.2 Mixers vs Frequency Converters

Frequency converter is a term used to describe a system composed of filters, amplifiers, isolators and mixers which are combined to create an overall frequency conversion system block. These can have one, two or even more stages of mixing, with amplification and filtering before and after each stage. These systems are developed to provide rejection of unwanted signals and images, to remove or isolate higher-order mixing products, and to provide the necessary gain and power required by the overall system design. Due to their specific attributes, the methods of measuring frequency converters are somewhat different from those used on "bare" mixers; the methods described here will be distinguished by the terms "mixer" and "frequency converter".

Mixers are understood to be lossy or low-gain (if active mixers), without input or output filtering, and have only a single conversion. Passive mixers are often reciprocal or nearly so, and can convert frequencies with nearly the same efficiency from the RF to the IF or from the IF to the RF. Passive mixers can have substantial higher-order products at the output, with some feed-through elements (such as LO) being as high as, or higher than, the desired output products. An example spectrum from a mixer is shown in Figure 7.10. where two measurements are made, with a shift to the input frequency to identify spurious products. Shifts in the output products indicate the order of the multiplication that generated the term. Explicitly called out in the picture are some of the harmonics, feed throughs and spurs. For example, the 2:1 spur, from $Spur_{2:1} = 2 \cdot f_{LO} - f_{RF}$ is shown at marker 5, and since it is near the IF, it might be in-band. The 3:1 spur is very large, only 13 dB below the main RF out; this indicates that the mixer has substantial conversion efficiency on the third harmonic of the LO. Also seen are many spurs that have second-order products of the RF, and one that has a third order: Marker 9 highlights the 5:3 spur, which changes power as it shifts down in frequency. Also shown at marker 4 is the primary sum product of RF and LO. One must realize that all of these spurs at the output can reflect off a non-ideal load and remix inside the converter.

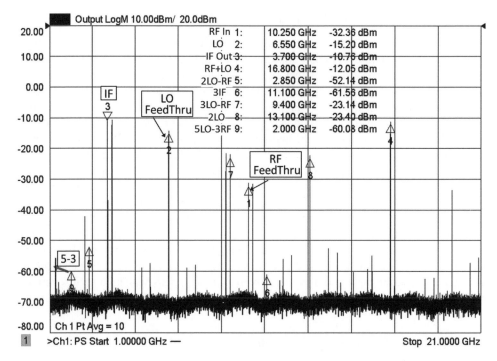

Figure 7.10 Typical output response of a mixer showing harmonics and spurious products.

Frequency converters, or simply converters, have filters that protect against unwanted image conversions, and typically have active stages that provide substantial positive gain and isolation between stages, and can have one or more stages of mixing. The multiple stages of conversion provide a way to create the same conversion as a simple mixer, but which allow spurious products to be eliminated by filtering between stages. The amplification stages provide substantial isolation in the reverse direction so that converters are essentially unilateral (one-direction) devices.

7.2.1 Frequency Converter Design

While this is not intended as a text on design, it is helpful to understand the principles of converter design to understand how the higher-order products of a mixer are created, and how they are eliminated using multiple conversion stages. Consider the diagram in Figure 7.11, which is roughly to scale. The up conversion of the input (IF) signal is shown above the upper line. In the lower line are shown the harmonics of the IF and LO. Below the upper line are shown the construction of the higher-order products: 2:1 and 3:2 and 4:3 products, and one can see that they must cross the desired output at some the indicated frequencies. In this construction, the crossing frequency occurs when the height of the higher-order product equals the height of the RF output.

This mixer will always have the spurious products, and in measurements of the conversion loss, when the higher-order product crosses the desired product, the conversion gain will show

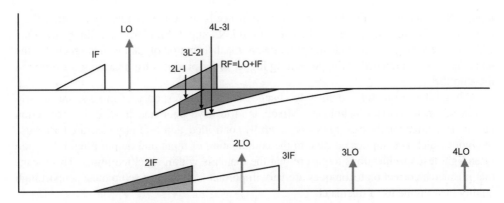

Figure 7.11   Diagram for mixer high-order products.

a discrete anomaly, or spur. This is *not* a measurement error; in fact, at that frequency, the mixer conversion loss really does change due to the higher-order product. These products can be eliminated with careful multistage conversions as described below.

7.2.2   Multiple Conversions and Spur Avoidance

If a multistage converter is constructed as shown in Figure 7.12, the major higher-order spurs can be avoided. The first stage consists or an up-converter with a high-side LO, so that the higher-order products lie outside the desired output. Depending upon frequency, multistage converters can have a wide variety of conversion configurations.

The second stage down conversion now occurs at higher frequencies, so that the higher-order products do not cross the desired output range. The first stage is an image mixer, and will cause a phase reversal; if the second stage is also an image mixer (high-side LO) a second phase inversion occurs and the overall mixer response will be normal. Filtering provided at each stage ensures that the overall response is not sensitive to out-of-band signals; amplification is required to make up for the loss of the multiple stages and filtering. Typically, an input amplifier is used to improve the noise figure of the mixer, but there is often a tradeoff between creating a low noise converter and creating distortion by overdriving the input mixer, so the

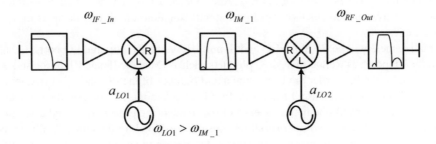

Figure 7.12   Multistage frequency converter.

input LNA interaction with the mixer distortion needs to be well understood. Often there is an input filter to restrict the frequencies at the LNA input. Similarly, amplifiers are often added between stages to provide the proper gain; the distribution of gain is made such that the noise figure is maintained while preventing higher-order products from causing distortion in the amplifiers.

Adding filters to mixers completely changes the overall response of mixers, and greatly affects the measurement techniques. Mixers tend to have poor matches, so the interaction between the filter and the mixer must be carefully controlled, and carefully measured. Because the input match is complicated (due to the combination of input and output filters) there are often tight test specifications to ensure that the match is within specified limits. This means that mismatch correction techniques are very important, and the test setup must support high quality measurements of mismatch.

On the other hand, bare-mixer measurements can have substantial higher-order products exiting all three ports (see Section 7.5 for details) and the interaction of these products with the test system can cause substantial errors. In some cases, a down-converter (which has, as a desired output, the difference term, RF − LO) can have a sum term (RF + LO) large enough to reflect off the VNA load, re-enter the mixer, reconvert to the input, and then remix with the LO to create a significant amount of IF signal. This will cause an apparent change in the conversion gain, just due to the output match of the test system at the sum frequency. In fact, a class of mixers called image-enhanced mixers is particularly designed to improve their conversion efficiency by making use of this image.

For normal mixers, there is no error correction method proposed, as yet, to remove this effect, so the best choice is to minimize the test system mismatch. Thus, with a bare-mixer, adding attenuators at the test ports may be required to obtain a sufficiently low mismatch to allow a good measurement. This issue only applies to higher-order products; the mismatch of the desired product with the test system can be characterized and removed, as discussed in Section 7.5.

7.3 Mixers as a 12-Port Device

In the most common understanding, mixers are treated as a simple three-port device, with an RF port having only RF frequencies, an LO port having LO frequencies, and an IF port producing only IF frequencies. This simple understanding is characterized by a conversion matrix that looks like Eqs. (7.4) and (7.5). And indeed, a properly designed converter does appear that way, but a mixer (without filtering and isolation amplifiers) behaves in a much more complex way.

In reality, there will be RF at the LO and IF ports, IF at the RF port and LO ports, LO at the IF and the RF ports, as well as undesired products such as IM at the IF and RF ports, and so on, so that even for the first-order products (sum and difference of the LO and input frequency), a mixer should be considered as a 12-port device [4]. Or rather, it is a three-port device with four "modes" at each port. Mathematically, each of the frequency elements is treated as a separate input at each of the ports, so the overall response is described by the equation illustrated in Figure 7.13. In fact, no one ever uses this mixer description to any practical effect, but it is an illustrative way to demonstrate all the possible mixing and remixing effects.

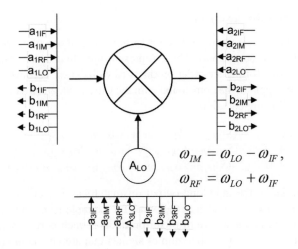

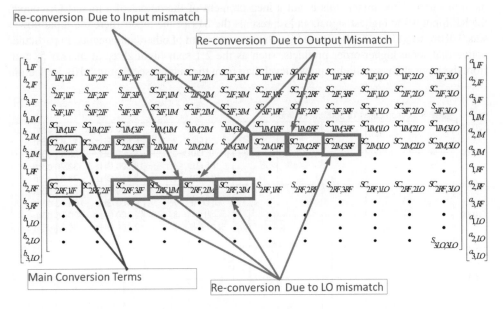

Figure 7.13 A mixer as a 12-port device to describe all first-order products.

7.3.1 Mixer Conversion Terms

7.3.1.1 IF to RF Conversion

The mixer described by the Figure 7.13 is an up-converter; the ports are identified by number and by the frequency present at that port number. If one assigns port 1 as the input, port 2 as the output and port 3 as the LO, then $SC_{2RF,1IF}$ is the first-order standard up-conversion gain (sum of LO and IF) and $SC_{2IM,1IF}$ is the first-order image (difference of LO and IF) up-conversion gain, as illustrated in the figure. Also illustrated are some examples of reconversion terms.

7.3.1.2   Reflection and Reconversion

The reconversion terms indicate how a reflection or mismatch at one of the ports will generate an RF or IM signal at the output port. These represent only first-order terms; there are sets of these conversion terms for higher-order products as well. These reconversion terms are an ever-present source of ripple in the conversion gain measurement, and a sample of these terms is illustrated in the following equation

$$
\begin{aligned}
\text{Errors} = {} & (\Gamma_{Source} \cdot (S_{1RF,1IF} \cdot S_{2RF,1RF} + S_{1IM,1IF} \cdot S_{2RF,1IM} + ..)\\
& + (\Gamma_{LO} \cdot (S_{3IF,1IF} \cdot S_{2RF,3IF} + S_{3IM,1IF} \cdot S_{2RF,3IM} + ..)\\
& + (\Gamma_{Load} \cdot (S_{2IF,1IF} \cdot S_{2RF,2IF} + S_{2IM,1IF} \cdot S_{2RF,2IM} + ..) + ...
\end{aligned}
\tag{7.8}
$$

The reconversion due to the input and output mismatch can be controlled while testing a bare-mixer by having very good match at the ports, for example from adding an attenuator at each port. To fully illustrate the nature of mixer products at each port, consider the spectrum shown in Figure 7.14. This shows a spectrum of signals that are emanating or scattered from the *input* port of the mixer; this is not a measurement of the output of a mixer! Of course, the RF input is the highest signal, and represents the S11 of the mixer (here the input power was 0 dBm, so S11 is −5 dB). But notice the rich spectrum of other components; in particular notice that some higher-order products, such as the 2:1 spur highlighted at marker 5, emit from the input at an even higher power level than from the output (see Figure 7.10). Reflection

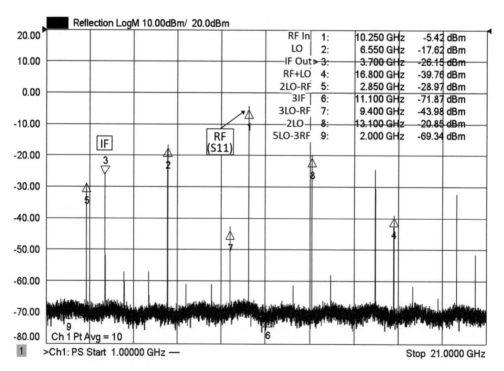

Figure 7.14   Mixer signals emitted or scattered (reflected) back from the input port.

of these higher-order products, and remixing in the MUT, is a source of errors in mixer measurements.

7.3.1.3 Image Enhancement

Some mixer designs, particularly narrowband applications, take advantage of the reconversion terms to employ a particular impedance matching at the image frequency to reflect it back to the mixer. This reflected signal is reconverted through the $SC_{2RF,2IM}$ term in Figure 7.13. This term is a measure of conversion gain of the image signal incident on the mixer output (port 2 in this case) to the RF signal exiting the mixer output port. In a normal ideal double-balanced mixer, the input signal is equally converted to RF and image (IM) signals. The conversion loss to the RF can be enhanced by reflecting the IM signal back into the mixer, where it can reconvert to RF (one may think of it as a double conversion, first to IF and then to RF). Though not widely used in practice, this is an example of reconversion that can have beneficial attributes. However, because phasing becomes very difficult, this benefit of improved conversion loss can turn detrimental as the bandwidth of the mixer is increased, and the phasing causes the output to cancel rather enhance, so that the conversion loss is degraded. As the bandwidth of the mixer is swept, the phasing between the mismatch and the mixer changes, and the enhancement/degradation appears as conversion loss ripple.

7.3.1.4 Conversion on the LO Port

The LO reconversion terms pose a greater problem, because LO drive levels needed for mixer measurements are often high levels, and many systems cannot afford the loss of an attenuator on the LO port. The use of isolators in the LO path does not necessarily improve the situation as the isolators typically are not well-matched at the RF, IM and IF frequencies, even if they provide a good match at the LO frequency. The input signal can leak to the LO port, and the output and image signals can also appear out the LO port; as they progress down the LO feed and reflect back into the mixer from the LO mismatch, they will be reconverted (in the case of input or IM) or leak (in the case of output) to the output and add-or-subtract to the main output signal. As frequency is swept, the phasing of this error signal will cause ripple in the measurement.

In converter designs, a filter is commonly used on the LO port and this will provide a high reflection at the other frequencies, but if the length between the filter and the mixer is small, the effect on conversion gain will be slowly varying and it prevents excessive ripple in the conversion gain. In many cases, an amplifier is placed very near the mixer to boost the LO signal; this has the added benefit or providing isolation to the other frequencies and so prevents these signals from reflecting with large phase shifts (due to longer delays) and thus minimizes ripple.

For a frequency converter, the 12-port mixer model is not needed. Filtering on the RF, IF and LO ports ensure that only signals at these frequencies exit the mixer, so only these signals interact with the test-system. That is the primary reason why measurements on frequency converters differ from measurements on bare-mixers: the wide range of other products coming from bare-mixers interact with the test system match to produce ripple; only by providing good, well-matched ports can high-quality measurements of bare-mixers be realized. With frequency converters, which produce only a single frequency at each port, any errors in match from the test system can be characterized and removed from the measurement results. These mismatch correction methods are discussed in the next sections.

7.4 Mixer Measurements: Frequency Response

7.4.1 Introduction

As with any other measurement, the quality of mixer measurements depends upon the quality of the measuring equipment. In the past, it was common to fully depend upon the instrument manufacturer's flatness performance of the source and the receiver (most often a spectrum analyzer), and it was up to the user to provide post-processing compensation for any cables or connectors used in the measurement.

Legacy VNAs were not used for mixer measurements except in the case of trying to measure relative phase between mixers. In the early 1990s, the first implementation of frequency offset mode was developed for VNAs that allowed the source to sweep over one set of frequencies while the receiver swept over a different set of frequencies. The source was phase-locked to the receiver at some offset frequency; for a mixer this amounted to the LO frequency.

This early implementation did allow conversion loss measurements, but measured only the output power, and relied upon a good source power calibration for accuracy. Only response calibrations were available. In the ensuing years the capability of VNAs to be applied to mixer measurements has dramatically improved, with high-speed measurements and high-quality calibrations, so that now VNAs are the preferred instrument for measuring mixers and frequency converters.

The discussion of frequency response measurement methods is divided into sections, first covering the measurement methods for amplitude response (7.4.2), phase response (7.4.3) which includes some group delay measurements, and group delay response based on modulation methods (7.4.4). In addition, some special consideration is given for swept LO measurements (7.4.5). This section only discusses the concepts of measurement; calibration for each of these measurements is discussed in the next section (7.5). While the calibration of the amplitude measurements is covered in the straightforward way, calibration for phase and delay are more complicated, where most phase measurements rely on a characterized mixer for calibration, with one exception. All the significant phase calibration methods are discussed in Section 7.5, and each calibration method may be applied to one or more measurement methods.

7.4.2 Amplitude Response

In general, there are two different ways in which mixers are used and these require two different stimulus setups to provide the characterizations of these mixers. The first is a swept-RF/swept-IF/fixed LO mode of operation, and this is the manner in which virtually all communication mixers and converters are used. Often, a frequency converter is used as a block down-converter to translate many channels or RF signals down to a common IF channel. Sometimes this is referred to as a "fixed IF" measurement, but that is a misnomer. For each fixed LO frequency, the RF-to-IF channel response is measured, and the measurement repeated for many LO frequencies.

In some cases, only the gain at the center of the channel response is measured, at each of the channels defined by the RF and LO steps. In this case, it is still not really a "swept LO" measurement, but more properly defined as a stepped LO, RF-to-IF measurement. The distinction is important primarily when considering the group delay response of a converter: the definition of delay refers to RF-to-IF transition, and so this implies that both the RF and IF must change (sweep in frequency) in order to have a defined group delay. One *cannot* simply sweep the RF and LO at a fixed IF and measure the phase change of the IF: the mixer definitions given in Section 7.1 show that while the LO phase will cause an phase shift in the IF, it is not related to the channel characteristics and so will distort the measurement of delay. For delay measurements of a converter, even over a range of LO frequencies, the LO must be stationary during the acquisition of the RF-to-IF transfer function.

In some limited applications, particularly in mixers used in radar systems, the RF and LO are swept together and in these cases, the relative phase and amplitude deviations between the RF and LO are important attributes. In this case, because the LO is not constant, a reference mixer must be used as will be discussed in Section 7.4.3.2. But in this case, only the relative phase can be measured, the delay result is not well defined as the IF frequency does not change at all, and therefore there is no "delta frequency" at the output from which to form a definition of delay.

7.4.2.1 Fixed LO Measurements

Amplitude response of a mixer is defined simply as the output power divided by the input power, or

$$SC_{21} = \frac{|b_{2\_OutputFreq}|}{|a_{1\_InputFreq}|}\Bigg|_{a_{2\_OutputFreq}=0} \tag{7.9}$$

In older, legacy analyzers, it was not possible to fully decouple the source from the receiver. Since the source would be phase-locked to the desired input frequency, by inserting the mixer-under-test (MUT) in the path between the source and receiver which was used as the source phase-lock reference, only the output power could be measured. In this case, it was not possible to measure the power of the input signal and so the value of input power was determined simply from the source power setting. In fact, the earliest implementation would require two different physical connections to first measure the source output power, over the input frequencies, on the reference channel with no mixer connected; this would be stored in memory. Next the MUT was added, and the frequency synthesizer was set to allow offset phase-locking so that the source remained at the input frequency while the receiver was tuned to the output frequency, and the output response was measured. The ratio of the measured response to the memory gives the conversion gain, provided that the source power doesn't drift, and the receiver response is calibrated. This method presumes that the local oscillator is provided separately; it did not require, however, that the LO be locked to the VNA reference. Any offset between the LO reference and the VNA would be absorbed by the receiver phase-lock process.

However, any change in source power would require a reconnection to re-measure the input power. A key limitation here is that the source was not truly independent of the receiver, but could only be offset by putting a frequency converter (the MUT) in the phase-lock path.

The HP8753 and HP8720 VNAs were the first commercially available VNAs to provide this capability.

Using modern techniques, the principal method for measuring the amplitude response of a mixer now is to connect the mixer directly from the VNA input to the VNA output. Many VNAs provide additional ports and additional sources; one of these can be used to drive the LO directly. A typical connection diagram is shown in Figure 7.15.

Two sweeps are made, one over the input frequency and one over the output frequency, and the ratio of these responses is computed to give the mixer conversion gain. This requires, of course, that the VNA receiver can be decoupled in frequency from the VNA source, and most modern VNAs provide this capability, either inherently or as an optional upgrade. During the first sweep, the reference channel measures the input signal, and the input match is measured as well. During the second sweep, the VNA receiver is offset and the power at the output is measured. If each sweep is performed on a separate channel, the equation editor can be used to display the conversion gain. Alternatively, the receiver can be switched from input to output frequencies on each point, providing a faster response between measurements and reducing the likelihood that the source power has drifted between measurements.

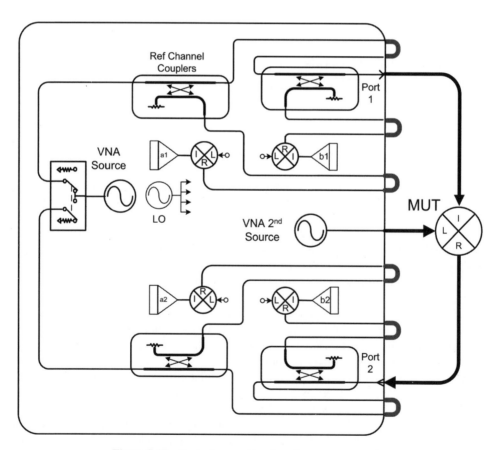

Figure 7.15   Typical connection for mixer measurements.

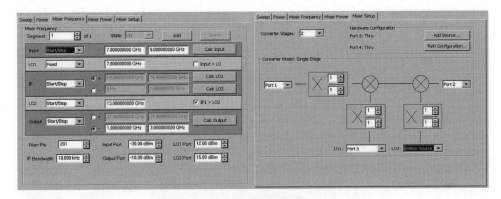

Figure 7.16   Mixer measurement GUI.

Some VNAs provide application-specific software to measure mixers automatically, including input and output sweeps, as well as automatically controlling the LO frequency and power, through a software graphical user interface (GUI). An example of one such mixer measurement GUI is shown in Figure 7.16. In this case, both single and dual conversion measurements are supported.

The response is still dependent upon the input and output match of the VNA, but with additional measurements of the input match at the input frequency and output match at the output frequency, correction for RF and IF mismatch can be accomplished as will be described in Section 7.5. Thus, the complete amplitude response of mixers, as defined by the conversion matrix from Eq. (7.4), may be obtained. An example of a mixer measurement, showing all the mixer "S-parameters" including forward and reverse conversion, input and output match, as well as input and output power in the forward direction is shown in Figure 7.17

This is an example of a single-mixer with input and output filters, and the conversion loss of the forward direction and reverse direction are very similar, indicating that this mixer is very nearly reciprocal in amplitude. Measurement of a mixer without any filtering may show it to be less reciprocal, due to remixing of higher-order products. Whenever a bare-mixer is measured, it is highly recommended to use a 3 to 6 dB pad on each of the test port cables to improve the port match. In some cases, even if the source and load match are well controlled for the frequencies used, out-of-band frequencies can see different reflections from the VNA. Adding pads at the test port helps to eliminate ripples caused by the reflection and remixing of these higher-order products.

Some examples of frequency converters will be demonstrated in the following sections on delay measurements (7.4.4) and high gain mixers (7.9.5). Other measurements besides frequency response, such a gain versus drive, TOI and noise figure are covered in Sections 7.6, 7.7 and 7.8.

7.4.3   Phase Response

With the increase of the use of complex modulation formats, the phase response of mixers and converters is becoming a key measurement parameter. In the past, measuring the phase (and from that computing the delay) of a mixer was very difficult. Several methods have

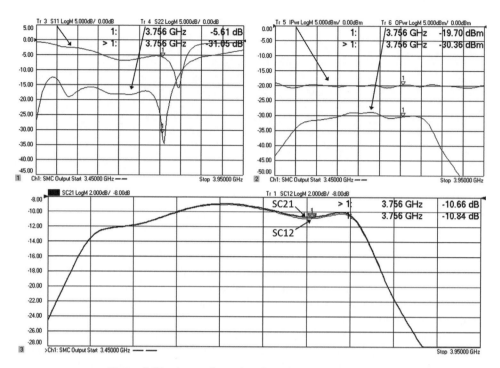

Figure 7.17   A complete mixer S-parameter measurements.

been proposed and until recently, each had substantial drawbacks. In the latest development in mixer phase measurements, a very new method removes many of the previous drawbacks and provides a simple but accurate calibration as well. The following sections describe the various mixer phase and delay measurement methods. All of the measurement methods have one or more associated calibration methods, which are described in Section 7.5.

7.4.3.1   Down/Up Conversion

The first practical means for measuring mixer delay employed the use of a reconverting mixer so that the input and output frequencies are the same. The overall response of a mixer pair was measured, and the MUT response was inferred by compensating for the effect of the reconverting mixer in some way (see Section 7.5.2). This measurement method is illustrated in Figure 7.18. Here the MUT contains a built-in band-pass filter, which dominates the frequency response and delay. The key benefit of this method is that it can be used on a standard VNA.

While very straightforward in concept, this method has some substantial difficulties:

- It requires a reconverting mixer that matches the frequency range of the DUT mixer but operates in the opposite conversion mode, that is, if one's MUT is a down-converter, the reconverting mixer must be an up-converter.
- It requires that the LO be shared between the up and down converting mixers; if either mixer has an embedded LO, or the MUT is a dual-stage mixer, the method is not practical.

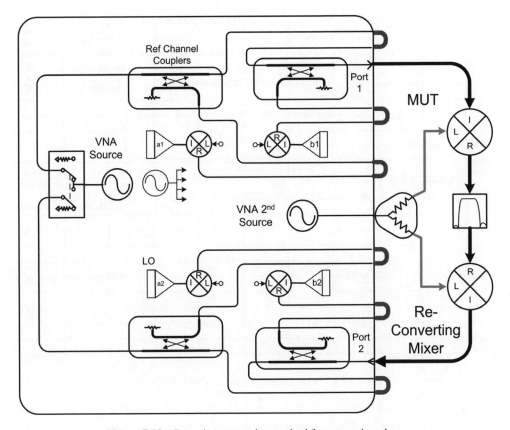

Figure 7.18   Down/up conversion method for measuring phase.

- For proper results, a band-pass filter must be used between the mixers to remove the image signal, or else the reconverting mixer will reconvert both the RF and the IM signal, resulting in an erroneous response. The effects of the filter response must be compensated for in the overall response, and mismatch between the filter and the mixers – and even between the two mixers in the passband – can lead to substantial errors in the overall result.
- Further, higher-order products that fall within the bandwidth of the image filter may reconvert in the second mixer and create an error in the overall response, which cannot be filtered out.
- The calibration and accuracy depends upon the characterization of the reconverting mixer; in most cases this mixer must be reciprocal to allow characterization.

Because of these difficulties, this method is seldom used except in cases where the VNA does not support mixer measurements, and so the input and output frequencies must be the same. It is most commonly used when measurements are made on converters with substantial phase response (large delay); in such a case the delay response of the reconverting mixer may be ignored. The measurement of phase deviation response of an example mixer – which will be utilized in the demonstration of other methods as well – is shown in Figure 7.19. The calibration method for this measurement uses the reciprocal mixer cal described in Section 7.5.3.1. After

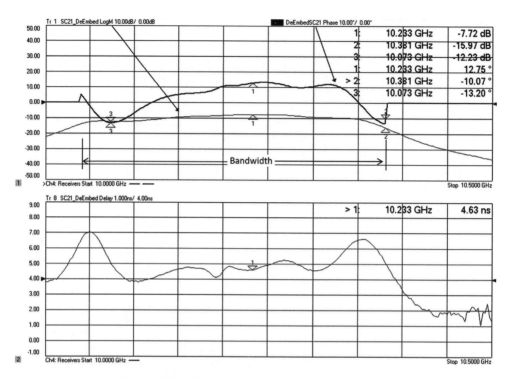

Figure 7.19 Phase response using down/up conversion.

a full two-port calibration, the equivalent S-parameters of the reconverting mixer, referenced to the MUT input frequency, are de-embedded from port 2 to create a calibrated response.

The upper window shows the magnitude response and phase response. In this figure, an equation-editor function is used to compute the deviation from linear phase over the bandwidth of the MUT, as illustrated in the upper window. The deviation from linear phase function allows one to look at the phase response just in the passband, without being distracted by the noise or excess response outside of the passband, and will be used here to compare several of the measurement methods.

The lower window shows the group delay response, with some odd-looking delay ripples in the mid-band. The built-in band-pass RF filter in this converter has a smooth delay response here. The extra ripples in this response are from interactions between the MUT mixer and the reconverting mixer.

The details for this measurement are as follows:

1. Precharacterize the reconverting mixer as described in Section 7.5.3.1.
2. Configure the down/up-converter path, using the MUT and the reconverting mixer and an image filter. It is usually best to have the VNA tuned to the highest frequency. If the MUT is a down-converter, connect the RF port of the MUT to port 1 and connect the output to the reconverting mixer, so the reconverting mixers goes from the MUT-IF to port 2. If the MUT is an up-converter, connect it to port 2, and connect the reconverting mixer to port 1 with its IF connected to the MUT IF input. It is a best practice when possible to use the

lowest frequency to connect the mixer pair, which often eliminates difficulties in removing the image signal.

3. Verify that the VNA can measure the down/up pair; check that the LO is split properly between both mixers. Note, this method will *not* work with a converter that has its own LO unless its reference can be locked the reference of the LO for the reconverting mixer. Check to ensure that neither mixer is compressing.

4. Once this setup is verified, disconnect the mixer pair and perform a full two-port calibration; this will allow the removal of some mismatch effects at the ports.

5. De-embed the reconverting mixer from its port using standard de-embedding math, referenced to the port frequency. De-embedding is discussed in detail in Chapter 9.

6. Reconnect and measure the mixer pair to obtain the results shown in Figure 7.19.

One note of caution: when choosing an image filter, one may choose a filter labeled with the appropriate bandwidth, but one which has a spurious passband at the image frequency (as the author did when first attempting to make the measurement of Figure 7.19). This is especially true for band-pass filters, which are notorious for having spurious passbands around the third harmonic of the filter. If this spurious passband occurs where the image signal (RF + LO) in the mixer occurs, erroneous measurements will certainly result.

With this method, the frequency response can be reasonably determined only for the low-power linear region of the mixer. Measurements of power-dependent parameters are difficult because of the series connection between mixers. This limitation can be eliminated using the other methods that follow.

7.4.3.2 Parallel Path Using a Reference Mixer (Vector Mixer Characterization – VMC)

Newer VNAs allow the source and the receiver to be independently tuned (without the external phase lock path required on older analyzers such as the HP 8753) and so it is possible to configure a setup with the MUT in the test channel, and a separate, similar mixer in the reference channel, as shown in Figure 7.20. The mixer in the reference channel is used to measure the source or input signal, but provides the same frequency signal to the reference receiver as the MUT does to the port 2 test mixer, so the phase relationship of the two represents the phase of the output signal to that of the input signal. The amplitude response can be discerned in a similar way. In the optimum configuration, a switch is provided in the reference path to allow the reference mixer to be bypassed so that normal S-parameter measurements, such as S11, can be performed.

Since the MUT is connected directly between ports, the effects of mismatch between port 1 of the VNA and the input of the MUT at the input frequency, and the mismatch between port 2 of the VNA and output of the MUT at the output frequency can be removed. This is accomplished in a method nearly identical to the full two-port calibration used in normal S-parameters, with the exception that the S12 term is set to zero, as the system cannot measure the reverse conversion loss of the MUT. Calibration of this system follows a normal two-port calibration method, but requires an extra step of placing a calibration-mixer in the thru path to establish the transfer phase of the reference receiver. Details on this are described in Section 7.5.3.1. Once the calibration is completed, the calibration-mixer is replaced with the MUT, and the conversion gain is directly measured. The calibration removes the mismatch

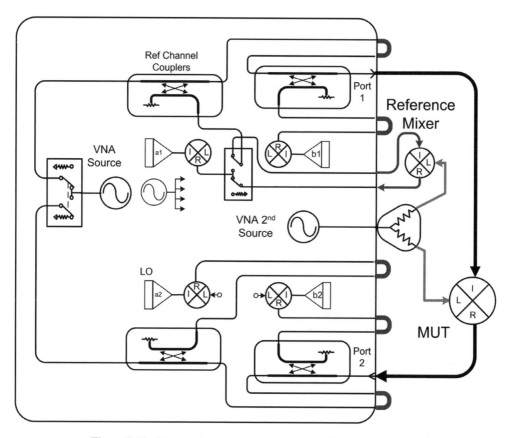

Figure 7.20   Vector mixer measurement system using a parallel path.

effects, but only for signals at the input and output frequencies. Other products that exit the mixer can still remix and create ripple on the SC21 trace if the VNA port match is not particularly good, and if the MUT is not filtered.

The parallel measurement system has several advantages over the down/up method described earlier:

- Match correction of the input and output signal can be performed to improve corrected performance.
- The reference mixer and calibration-mixers have the same conversion direction as the MUT. In many cases, users have several examples of each mixer, and so allowing one to be used as the reference when the other is used as test mixer is convenient.
- The measurement does not require that the MUT mixer have any filtering; the VNA naturally selects only the desired output frequency.
- One need not use the same LO signal to drive the reference and MUT mixer; in fact, they do not even need to be locked to the same frequency reference. In cases where the MUT has its own embedded LO, the VNA receiver can be programmed to track the output signal. Details on this are discussed in Section 7.9.4.

If the signal level to the reference mixer is small enough so that it is linear over all drive powers, then the gain and phase compression versus RF drive can be determined for the MUT. Also, if separate LO drives are used for the reference and the test mixer, then the mixer conversion loss versus LO drive can be determined as well.

An example of a mixer phase deviation measurement is shown in Figure 7.21 (again the phase deviation is only shown over the normal bandwidth of the mixer); the measurement is labeled SC21_VMC and this method of parallel measurement is sometimes called vector mixer/converter (VMC). Also shown is the response of the same mixer from Figure 7.19, using the down/up method and de-embedding (labeled SC21_De-Embed). The same second mixer used for the reconversion mixer in Figure 7.18 was used, in its down-conversion mode, as the calibration-mixer, and was characterized as described in Section 7.5.3.1. The up/down method with de-embedding shows a response with more ripple, which is very likely an error, compared to the VMC method. The VMC results for group-delay (labeled VC21 for Vector Converter 21) and phase-deviation match reasonably well with the S21 response of the filter that is used in the converter. The fact that the VMC trace is smooth and symmetric is a good indication that it has a better quality of calibration than the up/down method.

The VMC method has been used extensively in high-performance test of converters, and has the advantage that phase noise on the LO is cancelled between the reference and test channel; thus the VMC produces very low noise phase and delay responses. The VMC method provides very good measurements even for a lossy mixer, and one can see from the figure above that the group delay measurement is much smoother in the high-frequency, stopband

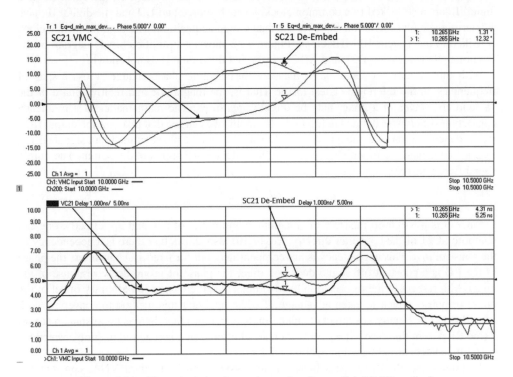

Figure 7.21   Phase deviation for a mixer using the parallel (VMC) method.

region. The main difficulty is that a calibration is required for every LO frequency used, so that if a converter has many channels, many calibrations must be performed, and for each LO frequency, the calibration-mixer must be separately characterized. The difficulties of calibration with multiple LO frequencies area essentially eliminated using the method described in the next section with calibration from Section 7.5.3.3.

7.4.3.3 Phase-Coherent Receivers

In 2010, new hardware and software innovations provided an advancement in VNA functionality that allowed the sources and receivers of some VNAs (e.g., Agilent PNA-X) to change frequency in a phase coherent manner; this allowed innovations in mixer measurements that vastly simplified the setup and calibration process.

The measurement system consists of a VNA with integrated sources and receivers, where both the source and receiver frequencies are determined by a pair of frequency synthesizers. One synthesizer provides the source stimulus signal and the other synthesizer provides the LO for the receiver. The difference of these two signals represents the IF frequency from the receiver, which is sampled by an integrated digital IF.

The synthesizers in this VNA are unique in that they use a custom fractional-N synthesizer chip-set with an integrated phase accumulator. When programmed to sweep frequency the phase accumulator will accumulate a certain amount of additional phase each clock cycle, thus providing a synthesized phase sweep coherent with the system clock. The DSP and digital IF are also locked to a common clock, so each source and LO and the digital IF have a deterministic phase relationship over the data sweep acquisition. The high frequency clock is locked to the system 10 MHz clock. Figure 7.22 shows a simplified block diagram of the fractional-N synthesizer, DSP and ADC control, but the digital IF is not shown.

With this system it is possible to measure the absolute phase change across a span of frequencies on the reference or test mixer. That is, the magnitude and relative phase of the "a" and "b" waves of a mixer may be directly measured. The mathematics of a frequency converting system was defined in Eq. (7.4). From this we can see that phase of the output (RF) signal depends upon both the phase of the LO signal, and the phase of the input signal (IF) as well as a mismatch term which depends on the reflected RF signal at the output. The output signal from a mixer can now be defined as

$$b_{RF} = a_{LO}S_{21}^{IF} \cdot a_{IF} + S_{22}^{RF} \cdot a_{RF} \qquad (7.10)$$

Since it is possible to measure the magnitude and phase of b_{RF} and a_{IF} directly, one can compute SC21 of the converter directly; Eq. (7.10) also shows that there is a dependency of the phase of a_{LO}. This dependency is not compensated, but for fixed-LO converters the phase of the LO is a constant offset and does not affect the phase or group delay response of the mixer with respect to the information bandwidth. Thus, there exists an arbitrary offset of the phase response, which is taken advantage of when applying this method to converters with embedded LOs. Finally, for direct computation one assumes that there is no mismatch at port 2 (the RF port in this case). If $a_{RF} \neq 0$, then an additional step is required to compensated for port 2 mismatch, as described in Section 7.5.3.3.

To understand the differences between this system and a standard VNA, consider the response shown in Figure 7.23 of a single receiver magnitude and phase measurement. The

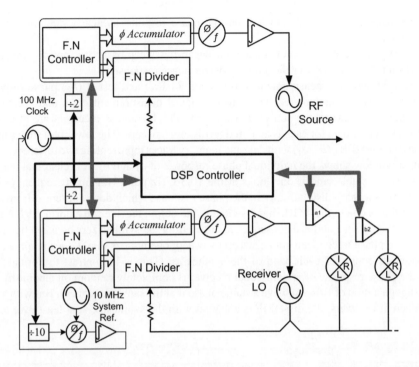

Figure 7.22 The synchronous sweeping is accomplished by a common reference and a digital phase accumulator in the synthesizers of the source and LO.

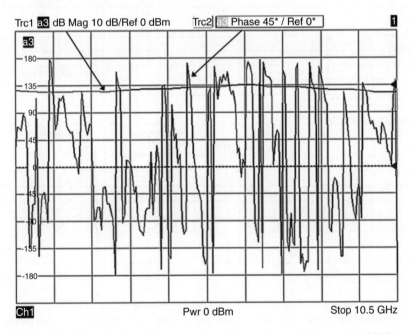

Figure 7.23 Amplitude and phase response of a single receiver in a normal VNA.

amplitude response is smooth and flat, but the phase response is nearly random across the frequency range.

Since the phase response is not stable on a sweep-by-sweep basis, it is not possible to find a correction method to recover the single-receiver phase response.

Figure 7.24 shows a measurement of reference R1 (a_1) and test B (b_2) phase responses, using a VNA with coherent receivers, in a standard measurement class and applying an offset frequency between the source and receiver. The R1 and B receivers are at different frequencies and the data for each is acquired on different sweeps. The lower plot shows change in phase response B/R1 (b_2/a_1) with data and memory traces representing two different sweeps. The upper window shows the individual phase response of a_1 and b_2. It is clear that the phase response is not stable, and if one looks at the 10.3 GHz point (for input frequency) in the sweep, there is a sharp discontinuity in the phase response which does not repeat from sweep to sweep; this is due to the 4 GHz band-switch in the output receiver. In a normal VNA mode, this is commonly seen and is a consequence of resetting the synthesizers between bands.

While the phase trace is not stable sweep to sweep, one notices that it does retain a sort-of continuous response for each band of the synthesizer. This is in contrast to normal VNAs where the phase response of an individual receiver is completely random on a point-by-point basis (though a ratio of B/R will have a stable phase if at the same frequency). But with a VNA with coherent receivers, the digital IF is coherent with the synthesizers, at least over a band,

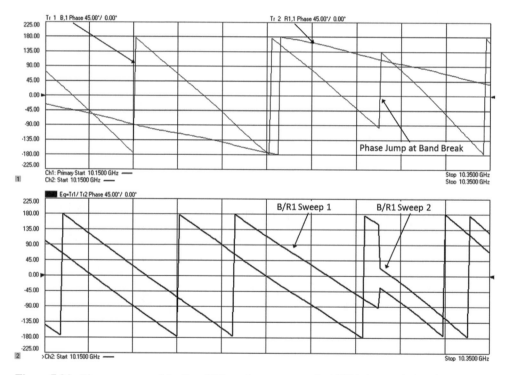

Figure 7.24 Phase response of the B and R1 receivers on a standard VNA (upper plot) and on the ratio of B/R on (lower plot) in frequency offset mode. The memory trace illustrates the change that occurs from sweep to sweep.

and the phase relationship of a single receiver is maintained inside a band on a single sweep; this attribute will be exploited for new measurement method.

One recognizes two important features of the traces in Figure 7.24: (1) the phase response at the start of the sweep changes with each sweep, but the point-to-point variation remains constant, except at the band break; and (2) the phase slope on each side of the band break remains constant even if there is a phase offset. The new method utilizes these two facts to remove the sweep-to-sweep variation of the phase response of the receivers. First, the phase traces are normalized to the phase of the first point of the sweep. From Eq. (7.10), the phase of the LO adds an arbitrary phase offset, so one can simply assign this phase offset to the first point of the sweep, removing the sweep-to-sweep variation. Second, the new method recognizes that the phase slope at each side of break forms the group delay of the response of the system near the band break, as shown in Figure 7.25. This can be used to extend the phase of the lower band break to ϕ_L, and from the upper side to ϕ_H, thus allowing a method of stitching together the phase offset at each band.

The offset is computed as

$$offset = \phi_H - \phi_L \quad where$$

$$\phi_H = \frac{\Delta f_H (\phi_N - \phi_{N+1})}{\Delta f} + \phi_N \quad and \quad \phi_L = \phi_{N-1} - \frac{\Delta f_L (\phi_{N-1} - \phi_{N=2})}{\Delta f} \quad (7.11)$$

Figure 7.26 shows the result of phase normalization and phase stitching. The input and output signals are labeled as IPwr and OPwr. One can see that the phase response is now continuous, and what is more, the phase response is stable on a sweep by sweep basis; the lower window shows the results of two consecutive sweeps, overlapped. This allows one to make uncalibrated phase measurements without a reference mixer, giving a vector version of SC21. But these responses also include the phase response of the source, the reference channel receiver and the test channel receiver and the phase response of the associated cables and connectors. Fortunately, these can be compensated with a few calibration methods, as described in Sections 7.5.3.1 and 7.5.3.3.

Computing from the markers on the input receiver (R1, called out as IPwr) and the output receiver (B, called out as OPwr), one can see that the SC21 phase is exactly the difference.

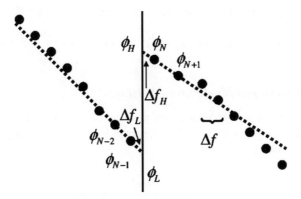

Figure 7.25 Phase stitching at synthesizer band breaks.

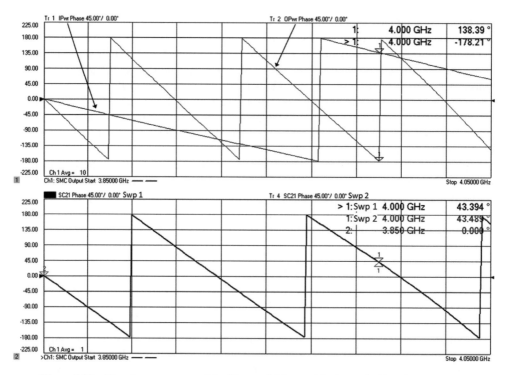

Figure 7.26 The phase response of the IPwr and OPwr, with phase stitching, is now stable.

Further, on a sweep-to-sweep basis, the phase drifts very little. The lower plot shows the results of two sweep-acquisitions, with only 0.005° difference. Since the reference and test measurements occur with two different sweeps, the phase noise of the source and LO are not accounted for and will add to trace noise. This can be effectively eliminated by adding more averaging; in this case 10 averages were used.

A calibrated measurement of the same mixer used in Figure 7.21 is shown for the coherent receiver test method in Figure 7.27. Here all three measurements (de-embedded, VMC and coherent phase) are shown and they compare very similarly, with VMC and coherent phase giving the very nearly the same answer.

The calibration details for these measurement methods are discussed in Section 7.5.

7.4.4 Group Delay and Modulation Methods

Group delay for mixers is computed in a similar manner as for linear devices, with the exception that the response for image mixers must be computed with the phase conjugated to show the delay as positive number, thus

$$\tau_{d\_std} = -\frac{d\phi_{Rad}}{d\omega} = -\frac{1}{360}\frac{d\phi_{deg}}{df}$$

$$\tau_{d\_image} = \frac{d\phi_{Rad}}{d\omega} = \frac{1}{360}\frac{d\phi_{deg}}{df}$$

(7.12)

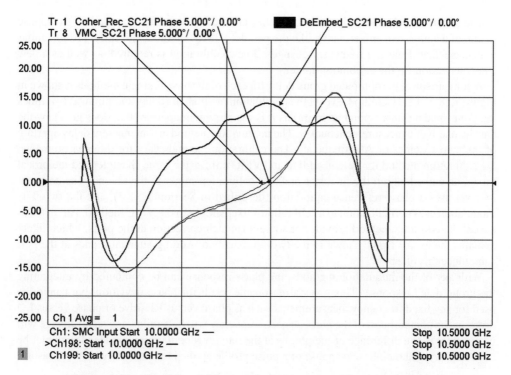

Figure 7.27 Comparison of three methods measurement for phase deviation.

Here, the group delay is computed from the phase response of the mixer. There is one additional technique that may be used to compute the group delay of mixers based on modulation techniques.

In one form, an AM signal is applied to the input of a mixer, and the AM envelope is detected at the input and the output, and compared. Any phase shift of the AM envelope can be associated with the delay of the MUT, and the delay determined in this way, computing delay as

$$\tau_d = \frac{\Delta\phi_{\text{deg}}}{360 \cdot f_{\text{mod}}} \tag{7.13}$$

In this measurement method, the higher the modulation frequency, the lower the error in measurement provided the phase detection is the same.

An alternative to this method uses a two-tone stimulus to create essentially a suppressed-carrier amplitude-modulated signal. However, rather than detecting the envelope, a receiver with essentially a dual channel detection path is utilized to measure the relative phase of two tones at the input, and compare them with the relative phase of the two tones at the output. In this way, if there is drift in the frequency of the LO for the MUT, the drift applies to both tones and the delay measurement is not affected to the first order. Thus the group delay can be computed by looking for an excess or difference in the change in phase of the measurement as

$$\tau_{d\_2-tone} = -\frac{\Delta\phi_{Input} - \Delta\phi_{Output}}{360 \cdot \Delta f_{2-tones}} \tag{7.14}$$

The calibration of modulation methods relies principally on comparing the MUT response to that of a calibration-mixer, much like the parallel (VMC) method, and suffers from similar drawbacks that if the LO frequency changes, a new calibration is required, as well as a new characterization for the calibration-mixer.

A comparison of the computed group delay for each of the measurement methods is shown in Figure 7.28. For each method, except the modulation method, the delay aperture was two times the point spacing (three-point aperture) as it is the minimum aperture that does not skew the data by one-half a measurement bucket. The modulation method has an inherent delay aperture of the tone spacings or AM frequency. The calibration methods vary for the measurements; the calibration method for de-embedding and VMC utilizes the same characterized reciprocal mixer, de-embedding in the first case and as a calibration-mixer in the second. The coherent receiver uses a phase reference calibration (described in Section 7.5.3.3), and the two-tone modulation uses normalization to a calibration-mixer similar to the VMC method. All methods except the two-tone method provide for some mismatch correction at the ports, which likely explains the excess ripple shown in the two-tone method. Adding attenuator pads at the port can reduce this effect.

With any of the modulation methods, the phase response can be computed by integrating the group delay response. This gives a reasonable result for phase deviation, but cannot be used for any fixed frequency measurements such as phase versus RF drive or phase versus LO drive, as the compression of the phase at each frequency will be similar, and so no change will be apparent in the difference of the phases at the output. Modulation methods also cannot be used in looking at the phase response of a pulse profile applied to a mixer for similar reasons.

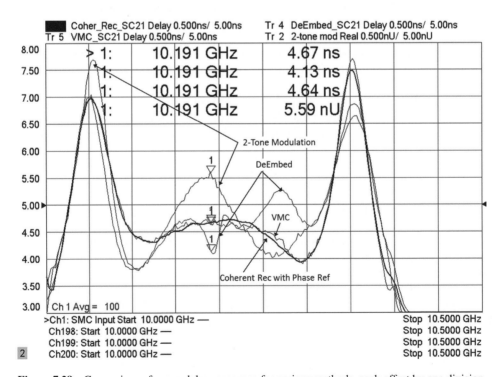

Figure 7.28 Comparison of group delay responses for various methods, each offset by one division.

7.4.5 Swept LO Measurements

In some cases, the mixer characteristics are defined over a range of RF or LO frequencies, with a common or fixed IF frequency. As described in the introduction of this section, in most cases the fixed IF measurement really refers to the final use case of a converter that has a fixed IF output channel, and the RF-to-IF transfer function is to be determined for each of a set of stepped LO frequencies representing the different channels of operation. In these cases, it is convenient to show the overall effect of RF or LO frequency on the nominal response (usually taken at the center frequency of the IF) and so a quasi-swept LO measurement is desired. For most measurement systems, including VNA based systems, the only change needed is to specify that the IF is a fixed frequency and the RF and LO as swept. One such implementation is shown in Figure 7.29. In older VNAs, it may not be possible to directly control an external LO; in such a case, the LO cannot be stepped synchronously with the RF, and instead a series of CW stepped measurements must be made using some external programming control. This is a common measurement on bare mixers, to allow a simple characterization of the response of each of the LO and RF ports.

A measurement of a wideband triple-balanced mixer with a different combinations of swept LO, swept RF and swept IF over a wide range of frequencies is shown in Figure 7.30. In the upper windows, the roll-off of the low frequency response of the mixer conversion gain is due to the LO or RF balun used in the mixer having a limited low frequency response. By measuring RF > LO (right upper) and RF < LO (left upper) one can infer that the RF corner is lower than the LO corner.

Independently fixing the RF, the LO or the IF while sweeping the other inputs is one way to determine the bandwidth limitation of the various ports of a mixer. For example, if the LO is held constant while sweeping the RF and IF one can determine the combination of RF and IF bandwidth characteristics. One can set the RF and IF range at a particular LO frequency and normalize the conversion gain response. After each sweep, the LO can be set higher or lower, keeping the IF sweep constant, thus changing the RF range by the change in the LO. If the entire trace response drops, that indicates a limitation of the LO bandwidth; if the conversion gain changes versus frequency, this indicates a limitation due to RF range. A similar method can be used to evaluate the IF range and the LO range. Figure 7.30, lower window, shows the effect of changing the IF frequency on the conversion gain. The low frequency corner moves

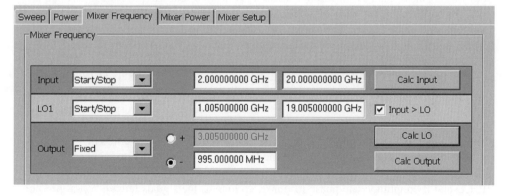

Figure 7.29 Defining a swept LO measurement using a GUI.

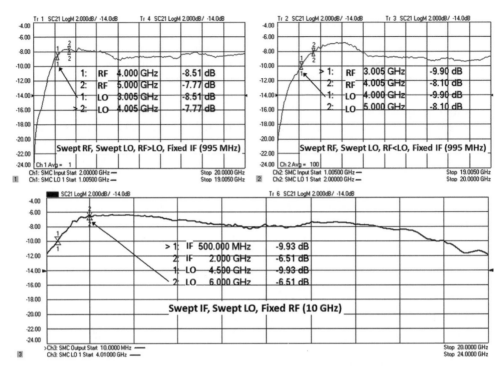

Figure 7.30   Fixed IF, swept RF/LO measurement of a mixer for two different IF frequencies.

with a change in LO and IF, with a constant RF range. Since the corner is at 2 GHz IF, but 6 GHz LO, one can presume it is limited by the IF balun.

7.4.5.1   Phase Measurements for Swept LO: Radar System Matching

In some cases, particularly for mixers used in radar or phased-array applications, it is desired to understand phase response of the IF path as the RF and LO are swept, including deviation from linear phase. That is, in these systems it is desired that the phase response apparent to the IF be identical for each of several converters. In operation, a comparison of the IF magnitude and phase is used for a variety of tasks including detection of objects and direction finding. It is through processing of these responses that tasks such as electronic beam forming are accomplished. The great difficulty in determining this response is that any phase change to either the RF or LO path will cause an apparent phase shift in the fixed IF. Sometimes, linear phase change (such as adding a line length to one of the paths) is of less concern than deviation of linear phase, but in some systems, the absolute phase-response measurement between several converters is required. The ultimate solution would allow a single converter phase response to be acquired so that it can be matched with other converters in a system. But this would require acquiring the phase of the LO in addition to the RF-to-IF phase response; in fact, some of the latest non-linear vector network analyzers (NVNA) systems do acquire RF, LO and IF amplitude and phase response, so long as the signals remain "on-a-grid" of a common phase reference base frequency (typically 10 MHz); that is, each of the RF, LO and IF frequency points must be related by a rational number.

Excluding NVNA systems, the best that one can do is a comparison of the phase response of one channel to other channels, or to a reference mixer, using the VMC techniques described above. In such a system, there are often multiple channels in a single converter and so a common channel can be placed in the reference path and the test path for each other channel goes in the MUT path, as illustrated in Figure 7.31.

This system can be calibrated using the calibration-mixer method, but there will still remain an ambiguous phase response due to the phase of the LO port, which manifests itself as a

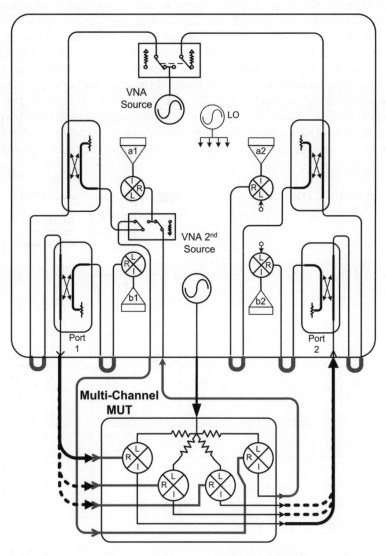

Figure 7.31   Measuring phase response of multichannel fixed IF converters.

fixed phase offset if the LO path is different between different systems. If the same system is calibrated in the same way but there is a path change in the LO drive to either the reference or the test mixer, then the MUT mixer results will show the path difference as a phase offset. In such a case, it is required to maintain exactly the same calibration-mixer from system to system, to compare various single-channel converter measurements among several test systems; that is, a "golden" calibration-mixer must be used as common reference between systems. The calibration-mixer characterization method does not respond to the phase of the LO drive (as described in Section 7.1.2.1), so if two calibration-mixers have differences in their LO path phase, these differences will incorrectly show up between converters measured on systems calibrated with different calibration-mixers.

Figure 7.32 shows a comparison of the phase response of two channels of a multichannel mixer used in a swept-LO, fixed-IF scheme. The two light traces are examples of measuring the same mixer two-times, with different LO paths (different LO cables), but using the same calibration-mixer. The overall error is about ±1° between systems calibrated with the same calibration-mixer. The dark trace (MXR1-2) shows the response of measuring a second, different, mixer which is intended to be matched to the first. In this case, the phase-matching between different mixers is on the order of less than 20°, and much closer at the center of the band.

The overall change in response is due to some differences in both the RF and LO path. The absolute phase offset of each measurement is determined by the particular characteristics of the calibration-mixer; changing the LO path on the calibration-mixer will result in an overall phase

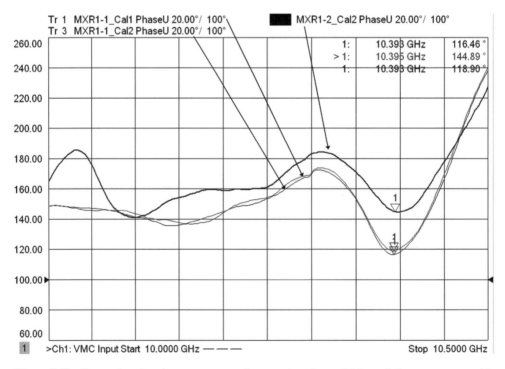

Figure 7.32 Comparing the phase responses of two paths of a multichannel down-converter with fixed IF.

shift of the individual traces, but the relative phase between channels will stay substantially the same. Small differences may occur due to differences in the mismatch of the various test mixers causing some mismatch-related phase shifts in the LO port. A similar effect occurs for RF and IF ports, but match correction removes these effects on the signal ports.

7.5 Calibration for Mixer Measurements

Each of the measurement methods for mixers can have one or more calibration methods applied. For amplitude response, the most direct and accurate measurement is using a power calibration on SMC measurements, as described below. Accuracy of the down/up method and the VMC method, for both magnitude and phase, depend upon the accurate characterization of a calibration or reconversion mixer.

For phase calibration, one must depend upon reference to a calibrated mixer for most measurement methods; the lone exception is the coherent-receiver measurement which can be calibrated independently for power and phase, relative to a known phase reference (discussed in detail in Section 7.5.3.3). All other methods extract a phase response from a presumably reciprocal mixer, through various methods, each of which is described in detail.

7.5.1 Calibrating for Power

The calibration for power measurements in a mixer utilizing the scalar mixer calibration (SMC) methods is performed independently from the calibrations for phase, and utilizes methods identical to those described in Section 3.7.1.4. Of course, the power calibration is needed for the reference receiver over the input frequencies and the test receiver over the output frequencies. The typical method for creating these power calibrations is to do a segmented enhanced power calibration, over both the input range and the output range. This generates a complete calibration map for both the forward direction and the reverse direction, and some implementations of the SMC calibration method support measuring the full four S-parameters of a MUT including the forward and reverse conversion, as was demonstrated in Figure 7.17.

The calibration proceeds in two steps:

1. A power meter is connected to port 1 of the VNA to measure the incident power from the source at both the RF and IF frequencies. At the same time, the reference receiver is calibrated to the same level as the power displayed on the power meter.
2. A full two-port S-parameter calibration is performed from port 1 to port 2 over the RF and IF frequencies. This acquires the input and output match correction terms, as well as the transmission loss between the two ports.

From these two measurements, a complete calibration of the VNA for measuring power at all ports is computed, as described below. Any connectors or adapters used between the power meter and VNA ports can be removed by an additional one-port calibration at the power meter reference. This allows one to use a coaxial power meter for the power calibration step, and then use any other calibration method, including on-wafer calibration, for the S-parameter step. The mathematical details are as follows:

The mathematical notation changes slightly for the mixer calibration as the responses of the inputs and outputs are at different frequencies, and so the tracking terms must be associated with the particular port's frequency of interest. The input frequency power correction is performed as

$$a_{1A\_MatchCor} = \frac{a_{1MUT\_In}}{RRF_{In} \cdot (1 - ESF_{In} \cdot \Gamma_{1MUT})}$$

where   (7.15)

$$RRF_{In} = \frac{a_{1M\_Cal\_In}}{P_{Meas}(1 - ESF_{In} \cdot \Gamma_{PwrMeter})}$$

Here, the ESF term is determined during a one-port or two-port calibration performed at the input frequencies.

The output power correction is similarly determined over the output port frequencies as

$$b_{2A\_MatchCor\_Out} = \frac{b_{2MUT\_Out}}{BTF_{Out}} \cdot (1 - ELF_{Out} \cdot \Gamma_{2MUT\_Out})$$

where   (7.16)

$$BTF_{Out} = ETF_{Out} \cdot RRF_{Out}$$

Here the BTF term is computed based on the transmission tracking from input to output, over the output frequency range, and the reference receiver tracking, also over the output frequency range, computed as in (7.15) but for output frequencies.

The correction for conversion gain is computed as

$$SC_{21\_Cor} = \frac{b_{2MUT\_Out}}{a_{1MUT\_In}} \cdot \frac{RRF_{In}}{BTF_{Out}} (1 - ESF_{In} \cdot \Gamma_{1MUT})(1 - ELF_{Out} \cdot \Gamma_{2MUT\_Out}) \quad (7.17)$$

This applies to standard mixers, but a small change is necessary when dealing with image mixers, to account for the effect of frequency and phase reversal. For image mixers, the correction is

$$SC_{21\_Cor} = \frac{b_{2MUT\_Out}}{a_{1MUT\_In}} \cdot \frac{RRF_{In}}{BTF_{Out}} (1 - ESF_{In} \cdot \Gamma_{1MUT})\left(1 - ELF_{Out}^* \cdot \Gamma_{2MUT\_Out}^*\right) \quad (7.18)$$

Thus the gain computation for a mixer and an amplifier are very similar, with the difference being only in a small portion of the mismatch correction term associated with the loop-term $SC_{21}SC_{12}ESF_{In}ELF_{Out}$. Unlike low-loss lines, mixers always have either an associated loss or an amplifier with isolation, which makes the loop term a very small effect.

7.5.1.1   Split-Port Cal for Amplitude Response

In some cases, it is not possible to perform the S-parameter thru step between the input and the output of an SMC calibration, usually because one of the ports is a connector (such as waveguide) that does not pass both the input and output frequencies. If the input connector is waveguide, then the BTF at the output cannot be computed. (On the other hand, if the output is a waveguide connector, and the input can pass the waveguide frequencies, then it is possible to add an adapter to make an unknown through calibration). One way to deal with incompatible connectors is to remove the waveguide adapter from the port requiring a waveguide connection to the mixer, and then performing the calibration entirely in coax, before replacing the adapter and de-embedding the adapter. In the past, this was the only method available for dealing with mixers that have a waveguide port. However, in the case of testing at mm frequencies, the waveguide test port is often a mm-wave extension head (sometimes called a mm-wave multiplier or simply a mm-wave head). In this case, the mm frequency is generated in the head and it is not possible to remove it to get to a coax port, nor is it possible to calibrate it for a mixer measurement with normal methods.

A new method, sometimes called "split-cal" allows a mixer calibration by splitting the power and receiver calibration into two separate power and reflection measurements. The basic concept is to perform a power meter calibration and a one-port calibration on each port. The port 1 power meter and reflection calibration allows the computation of the input power correction exactly as indicated by Eq. (7.15). The new technique then applies the same steps to the output port, at the output frequencies, as well to find the reference response reverse tracking term, RRR_{Out}, (the tracking response of the a_2 receiver) over the output frequency range as

$$RRR_{Out} = \frac{a_{2M\_Cal\_Out}}{P_{2\_Meas}\left(1 - ESR_{Out} \cdot \Gamma_{PwrMeter\_Port2}\right)} \tag{7.19}$$

From these measurements, the output b2 response tracking term is computed as

$$BTF_{Out\_SplitCal} = RTR_{Out} \cdot RRR_{Out} \tag{7.20}$$

The error correction math becomes

$$b_{2A\_MatchCor\_Out} = \frac{b_{2MUT\_Out}}{BTF_{Out\_SplitCal}} \cdot \left(1 - ESR_{Out} \cdot \Gamma_{2MUT\_Out}\right) \tag{7.21}$$

and the overall correction for SC21 gain becomes

$$SC_{21\_SplitCal} = \frac{b_{2MUT\_Out}}{a_{1MUT\_In}} \cdot \frac{RRF_{In}}{BTF_{Out\_SplitCal}}\left(1 - ESF_{In} \cdot \Gamma_{1MUT}\right)\left(1 - ESR_{Out} \cdot \Gamma_{2MUT\_Out}\right) \tag{7.22}$$

Here the source match replaces the load match, as it is not possible to measure the load match during the calibration process, because there is no way to supply a signal from another port. An alternative method would be to use the MUT measurement of a_{2_Out}/b_{2_Out} to determine the load match during the active measurement and correct for it in that way. However, in many cases, the difference between the source match and load match are very small, especially if there is any loss between port 2 DUT connection and the reference coupler.

7.5.2 Calibrating for Phase

In the late 1990s, a few new techniques for mixer phase measurements were developed. In one example, three mixers were used in three different measurements to extract the behavior of each mixer. In another, the response to reflections at a mixer output formed the basis of calibration. Very recently, new methods based on non-linear VNA techniques have been developed that promise to widely replace other methods due to the simplicity and flexibility of the calibrations. The details of each method and best practices of using the new methods are described below.

7.5.2.1   Three Mixer Method

One of the first methods proposed to characterize the phase response of a mixer using VNA measurements made use of three measurements on pairs of mixers. This method is used to find the phase of a calibration-mixer. Because it is a multistep approach, it is not very suitable for characterizations of mixers in a normal test sense. This method is illustrated in Figure 7.33. In this method, the MUT is mixer C. Mixer B is another mixer that can up-convert the signal from mixer C to the same input frequency as mixer C. Mixer A is a *reciprocal-mixer* that can be used both as a replacement for the MUT mixer (in the first line of the illustration) or as a replacement for mixer B (in the second line of the illustration).

Since the input and output are at the same frequency, full two-port calibrated S-parameter measurements can be used to determine the gains in the three cases, G_1, G_2 and G_3 respectively. The overall response for each gain (excluding mismatch effects) are

$$G_1 = SC_{21\_A} \cdot S_{21\_IF} \cdot SC_{21\_B}$$

$$G_2 = SC_{21\_C} \cdot S_{21\_IF} \cdot SC_{12\_A} \qquad (7.23)$$

$$G_3 = SC_{21\_C} \cdot S_{21\_IF} \cdot SC_{21\_B}$$

where S_{21_IF} is the loss of the intermediate frequency filter. If the image filter is not used, and the mixer creates nearly equal image signals, the overall gain can either be doubled (if the image adds in-phase) or can go to zero (if the image is out of phase and exactly equal to the standard conversion signal). This is an unacceptable error and must be reduced by using the image filter.

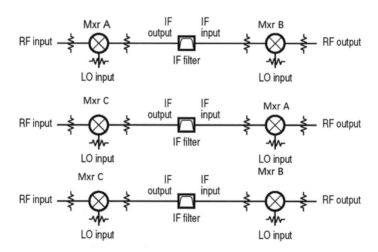

Figure 7.33   Three-mixer method for measuring mixer phase on a VNA.

From Eq. (7.23), the conversion gains of any of the mixers are computed as

$$SC_{21\_A} = \sqrt{\frac{G_1 G_2}{G_3 S_{21\_IF}}}$$

$$SC_{21\_B} = \sqrt{\frac{G_1 G_3}{G_2 S_{21\_IF}}} \qquad (7.24)$$

$$SC_{21\_C} = \sqrt{\frac{G_2 G_3}{G_1 S_{21\_IF}}}$$

Since all the gains are complex numbers, the phase response of the mixer can be computed as

$$\phi_{21\_C} = \frac{\phi_2 + \phi_3 - \phi_1 - \phi_{21\_IF}}{2} \qquad (7.25)$$

In the computation of the square root function for a phase response, care must be taken to choose the appropriate root (for details, see Section 9.3.1.1).

While this is a straightforward method of determining the phase of a mixer, it does require several steps and requires at least one mixer to be reciprocal. Further, in the method described, the gain excludes the effects of mismatch between the mixers and the IF filter. If the match of each were measured independently, it would be possible to include some of the mismatch effects, but typically a set of attenuator pads are used to lower the mismatch between the mixers. In such a case, the attenuator pad loss is lumped into the filter loss. And of course, one of the major difficulties is that three mixers must be used, and it might be uneconomical to create three mixers, especially at high frequencies.

An example of applying the three mixer technique is shown in Figure 7.34. Mixer A in this example is reciprocal, and will be used as the calibration-mixer in examples of the next section as well, and was extracted as the de-embedding mixer in Figure 7.19, as described in Section 7.5.3.1. It includes the IF filter, so that it is not necessary to add an additional IF filter or compensate for its S21. Rather than showing phase for the mixer pairs, the delay of each mixer pair is shown instead, with the understanding that it was derived from the phase response for each. The delay provides a nicely intuitive feeling for the relative response of each mixer.

In the figure, the upper left widow displays the magnitude result of a scalar measurement of the SC21 and SC12 of the "A" mixer. Also shown is the extracted up and down conversions, using the three mixer approach. It is clear that there is some substantial error (on the order of 2 dB or more) in this approach. The upper-right plot shows the group delay for each of the mixer-pair measurements. The "A" mixer is strongly filtered, but the "B" and "C" mixers are not so strongly filtered, and show a flat response when measured as a pair. The lower plot shows the group delay of SC21 extracted for mixer "A" with the three mixer method, as well as the deviation from linear phase (restricted to the passband region of the mixer). The maximum deviation here measures about ±18°. It is interesting to note that a mid-band ripple in the delay trace is apparent here, just as in the up/down method of Section 7.4.3.1. The three mixer approach is just as susceptible to high-order products affecting the result as the down/up approach.

From these results, any of the mixers' responses can be computed, and then any of the mixers, with their associated data, could be used as calibration-mixer. However, the ripple in the delay measurement suggests that there are some errors in the characterization. In the next two sections, improved methods for mixer characterizations are described, which don't show these error effects.

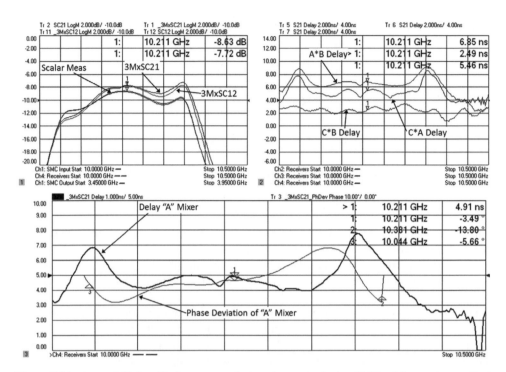

Figure 7.34 (upper left) Amplitude response of mixer A measured using SMC and three-mixer method; (upper right) delay of three mixer pairs; (lower) phase and delay of just mixer A.

7.5.3 Determining the Phase and Delay of a Reciprocal Calibration-Mixer

Several of the mixer measurement methods depend upon using a known mixer as a calibrated thru standard. The difficulty is in determining the qualities of the known mixer. One method utilizes reflections from the output of the mixer to infer the two-way response of the mixer, and then presumes the mixer is reciprocal to compute the one-way response. A similar approach is used in a second method that is related to the unknown thru calibration.

7.5.3.1 Reciprocal Calibration-Mixer Reflection Method

In the reflection method for characterizing a mixer, the presumption is that the conversion loss in the forward and reverse directions are the same, and not too great. In this method, a filter must be added to the output of the mixer which reflects or absorbs the undesired image, and passes the desired product with low loss. A succession of reflection standards are placed on the output of a mixer/filter pair, and the input match is measured for each reflection standard. Essentially, a one-port calibration is performed at the output of the mixer/filter combination, as illustrated in Figure 7.35.

In this case the IF− signal is the conversion that is desired; the IF+ signal (from the sum of RF and LO) will reflect off the IF filter rather than passing through, and so will not appear at the mixer output. For up-converting mixers, the filter should pass the IF+ signal and reject the IF− signal.

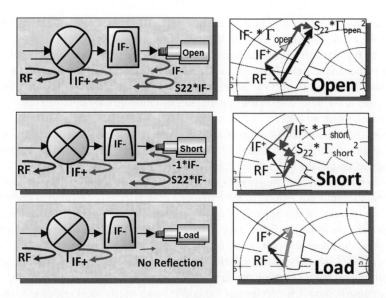

Figure 7.35 The reflection method of mixer characterization is essentially a one-port cal at the output.

From this one can see that, for each standard, there are up to four major contributors to the overall reflection response: reflection from the mixer S11 of the RF signal (marked as RF), reconverted reflection from the IF+ signal (which reflects off the stopband of the filter, marked IF+), reconverted reflection off the cal standard (marked $IF^- \cdot \Gamma_{Short}$, for example) and reconverted re-reflected signal off the cal standard and S22 of the cal mixer/filter combination (marked $S22^*\Gamma_{Short}^2$ for example). These are illustrated as vectors in the right half of the figure. If a one-port calibration is performed using each of the standards, the resulting error terms can be mapped to the mixer response. For each standard, the S11 and IF+ reflections are unchanged, and in the construction of a one-port cal, they represent the directivity error term; this is S11 of the mixer. The two-way transmission through the mixer is essentially the average of the open and short response, and so represents the reflection tracking term, or SC21·SC12. And the difference between the open and short is the source match term of the one-port cal, or S22 of the mixer.

In practice, this is measured with a non-ideal network analyzer, and so if a one-port calibration is performed before the mixer, and a second one-port calibration is performed after the mixer, adapter characterization math of equation 9.20 (see Chapter 9 for details) can be directly applied to yield

$$S_{11\_MUT} = \frac{(EDF_{MUT} - EDF)}{[ERF + ESF \cdot (EDF_{MUT} - EDF)]}$$

$$S_{21\_MUT} = S_{12\_MUT} = \frac{\sqrt{ERF \cdot ERF_{MUT}}}{[ERF + ESF \cdot (EDF_{MUT} - EDF)]} \qquad (7.26)$$

$$S_{22\_MUT} = ESF_{MUT} + \frac{ESF \cdot ERF_{MUT}}{[ERF + ESF (EDF_{MUT} - EDF)]}$$

Here again one notes the presence of a complex square root term, and the same cautions apply as in the example of Eq. (7.24). In the case of the second one-port calibration, after the calibration-mixer, the calibration kit terms must be used for the output frequencies and not the normal values of the input frequencies. If an Ecal is used, then the values of the Ecal must be taken from the output frequency range; in general this capability is not readily available to the user, so one must depend upon the manufacturer's embedded functions to perform a mixer characterization using an Ecal.

Some notes on this method: the mixer needed is essentially the same as the mixer A in the previous section, that is, a mixer that is reciprocal. In this case, the IF filter is embedded in the mixer response, and so the mixer/filter combination is always considered as a whole. Any re-conversion from the undesired image is captured in the overall response; thus the response may have substantial ripple, but as long as the true value of the ripple is captured in this characterization, the actual value of the ripple is of no importance. An example of a mixer characterized in this way is shown in Figure 7.36, and compared with the previous three mixer method.

In the figure, the upper-left plot shows the SC21 amplitude comparison; the SC21 extracted using the reciprocal mixer is almost identical to the SC21 response using an SMC calibration, with the three mixer method having the largest difference. In the upper-right plot, the deviation from linear phase is shown; the reciprocal method shows a nearly ideal response to the original filter shape. Similarly, in the lower plot, the delay of the reciprocal extraction method shows a very smooth and symmetric delay response as would be expected from the filter used, compared with the three mixer method.

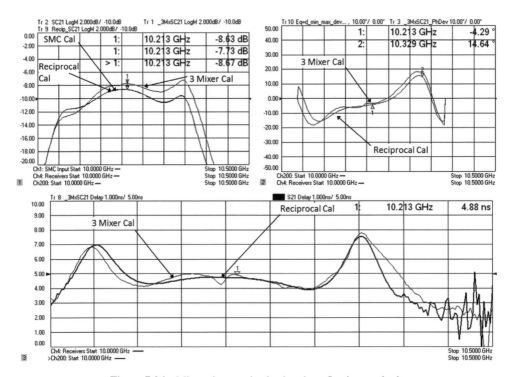

Figure 7.36 Mixer characterized using the reflection method.

Following the illustration of Figure 3.39 showing phase error as a function of amplitude error, one might postulate that if the amplitude response of a mixer is reciprocal within some limit, the phase response is similarly reciprocal within the limits proposed by Eq. (3.103). Thus the SMC measurement of amplitude of SC21 and SC12 gives an idea of the quality of the reciprocal calibration. In this case, the amplitude reciprocity error is less than 0.09 dB, implying that the phase reciprocity is less than $0.6°$.

Some notes on this characterization method: since the reflection off the mixer input is often the largest component of the signals measured, and the method requires this value to be consistent between standards, it is critical that there is no drift or instability in the measurements of S11 of the mixer. This characterization method works best when the calibration-mixer is connected directly to the VNA test port, with *no* intervening test cable. Also, loss in the calibration-mixer will increase this effect; for one-way conversion loss less than 10 dB, there is usually no problem. If the loss is between 10 and 15 dB, extra care is needed, with lower IF BWs, and more averaging necessary to obtain a good result. This method generally does not yield good results if the one-way loss is greater than 15 dB. This is evident in the noisy response of the delay in Figure 7.36 in the out-of-band region. In such a case, the method of the next section may be used. Since stability is very important, using an Ecal to generate the one-port calibration at both the input and the output removes the error associated with connector repeatability.

7.5.3.2 Reciprocal Calibration-Mixer Unknown Thru Method

An alternative method for characterizing a reciprocal calibration-mixer takes as its basis the unknown through calibration described in Chapter 3. For this method, a configuration is created that allows the signals from the MUT to be converted in such a way that the measurements of an unknown thru calibration can be performed. In one implementation, switched mixers in the forward and reverse reference path provide a reconversion of the incident signal so a phase measurement can be performed in each direction. In another implementation, a pair of mixers are added behind one of the test-port couplers, to reconvert the source and receiver signal at that port. In each of these instances, the LO must be split three ways to provide the same signal to each of the three mixers. The chief advantage of this method is that it performs better if the calibration-mixer has greater than 15 dB insertion loss. However, due to the complexity of the setup, neither implementation of this method has seen wide acceptance. Interested readers are referred to reference material at the end of this chapter for further information.

All three of the previous methods required reciprocal mixers, and the characterization must be performed all over again for each new LO frequency. However, recently, a new method of calibration for vector mixer measurements has been introduced, based on the coherent receiver test approach, that greatly simplifies the calibration process, and does not require any mixers at all, other than the MUT that is eventually tested; this method is introduced in the next section.

7.5.3.3 Phase Reference Method

With the coherent receiver measurement system described in Section 7.4.3.3, it is possible to measure the absolute phase change across a span of frequencies on an input or output signal of a mixer. Thus, the magnitude and relative phase of the "a" and "b" waves of a mixer may be directly measured, and the mixer conversion response can be computed directly from Eq. (7.10), and from this we can see that phase of the output (RF) signal depends upon both the phase of the LO signal, and the phase of the input signal (IF) as well as a mismatch term which depends on the reflected RF signal at the output. With the coherent receiver system, it is possible to measure the b_{RF} and a_{IF} directly, as well as recognize that a_{RF} is defined by any reflection off the test port 2 loading the mixer. Thus, one can compute SC21 of the converter directly. Equation (7.10) also shows a dependency upon the phase of the local oscillator. This dependency is not compensated for, but for fixed-LO converters the phase of the LO is a constant offset and does not affect the phase or group delay response of the mixer with respect to the information bandwidth. From this one can compute the SC21 (in this case, of an up-converter) directly as

$$SC_{21} = \frac{\left(b_{RF}/BTR^{RF}\right)}{a_{LO}\left(a_{IF}/RRF^{IF}\right)} \cdot \left(\frac{1}{1 - S_{22}^{RF}ELF^{RF}}\right) \cdot \left(\frac{1}{1 - S_{11}^{IF}ESF^{IF}}\right)\Bigg|_{|a_{LO}|=1} \tag{7.27}$$

Because the phase of the LO is not discernible, a particular frequency point (typically the center frequency) is selected to be a reference phase point and is set to a constant value, usually $0°$.

From Eq. (7.27) one notes that the reference receiver response tracking on the input and the test receiver response tracking on the output must be individually determined. From an understanding of standard S-parameter correction, one can recognize that

$$ETF^{RF} = \frac{BTF^{RF}}{RRF^{RF}} \quad and \quad ETF^{IF} = \frac{BTF^{IF}}{RRF^{IF}} \tag{7.28}$$

Thus if one can determine the phase response of either of the reference or the test channel over the RF and IF frequencies, then one can find the corresponding channel's phase through the S-parameter transmission tracking term.

In previous correction methods, the calibration separated the magnitude response of the a and b receivers by measuring their responses independently using a power meter as a reference. Then a calibration-mixer was added and the overall response of the system was measured. Independently, the source and load match of the system were measured, and finally a value for the phase of (BTF^{RF}/RRF^{IF}) was computed by solving (7.27) for a measurement of a known mixer above for that ratio. The amplitude was computed independently using the power meter calibrations. This method works well but requires knowledge of the calibration-mixer used, and determining this is a substantial problem that the phase reference method eliminates. With previous methods, the calibration was made with a particular calibration-mixer tuned to a particular LO frequency. Any other choice of LO frequency required a new calibration and new determination of the ratio of (BTF^{RF}/RRF^{IF}) for the calibration-mixer for each particular LO frequency. The phase reference method finds the individual phase response of each receiver without resorting to a known mixer.

This method is developed from an entirely different approach to calibration that is borrowed from the area of non-linear vector network analyzers. NVNA systems allow measuring the magnitude and phase of signals and their harmonics by using a harmonic comb generator as a phase reference into an additional channel, and comparing the phase of the harmonics to the phase of this known comb. The waveform of the signal can be accurately reconstructed from the fundamental and harmonics of the signal. Key to this reconstruction is knowing the

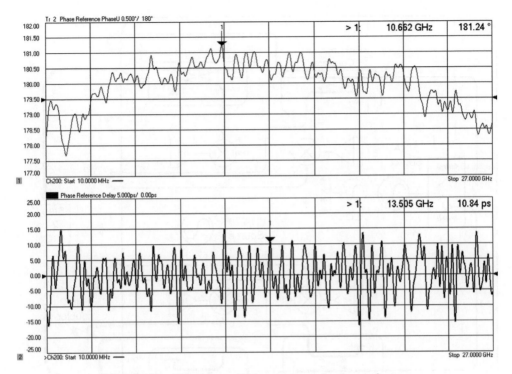

Figure 7.37   The phase and delay response of a phase reference.

relative phases of the harmonics of the comb generator, used as a phase reference. The phase of these harmonics can be accurately measured and traced to national standards with very small errors. Figure 7.37 shows the phase and group delay response from one such phase reference, an Agilent U9391 model, which has calibrated phase data. This provides the incident signal, $a_{Phase\_reference}$, to the receiver input on the VNA.

Since the measurement process has some noise, some smoothing may give a better result when using this phase reference in practice.

This phase reference provides the source of a signal, essentially a pulse, which delivers a known phase of a fundamental and all the harmonics. Since the coherent receiver system can measure the phase of a single channel response, these two elements are combined to create a single channel measurement receiver with calibrated phase. In the calibration process, the source for the phase reference is the 10 MHz reference oscillator in the VNA. The output of the phase reference is connected to the b_2 receiver of the VNA during calibration as illustrated in Figure 7.38. Here, one sees that the VNA is configured with a reversed port 2 coupler, to reduce the loss to the receiver. This is necessary for best performance as the individual comb-tooth signals are very small, on the order of −60 dBm. Thus, the coupler loss is removed from the measurement, which increases the signal to the receiver, but also adds an offset to the measurement. A full two-port calibration is performed as a second step, which effectively measures the offset loss of the coupler through-path. This offset will be compensated in an additional calibration step.

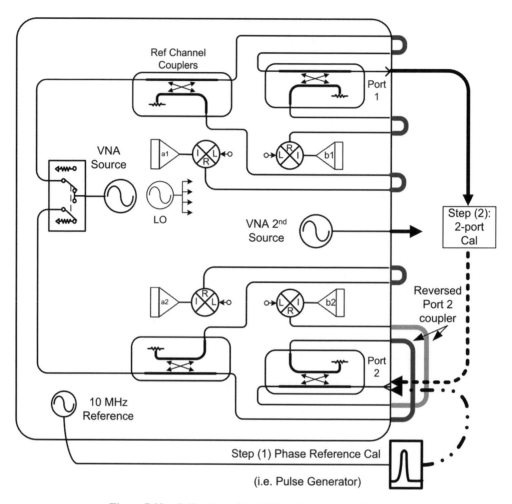

Figure 7.38 Calibration of the VNA using a phase reference.

The frequency stimulus is set to measure exactly every 10 MHz, which is the input reference signal into the phase reference. The b_2 receiver measures the phase reference (unlike the NVNA, no additional phase reference is needed). The result of this measurement is shown in Figure 7.39. In this example, the phase reference measurement is performed from 100 MHz to 26.5 GHz. If the MUT is low gain, the coupler may remain reversed. If the mixer is high gain, the coupler should be put into the normal configuration and a second two-port calibration performed. In the figure, the response is about 13 dB higher than the actual power (much more at low frequencies) as the VNA factory calibration is compensating for the expected coupler roll-off, which is not present when the coupler is reversed. The phase reference calibration of the receiver is then compensated by the difference in the S21 tracking terms, which captures the change in loss of the test-port coupler.

These results show approximately −50 to −60 dBm power response and phase response going from 0° to more than 250,000° over 26.5 GHz. It is more convenient to show the phase

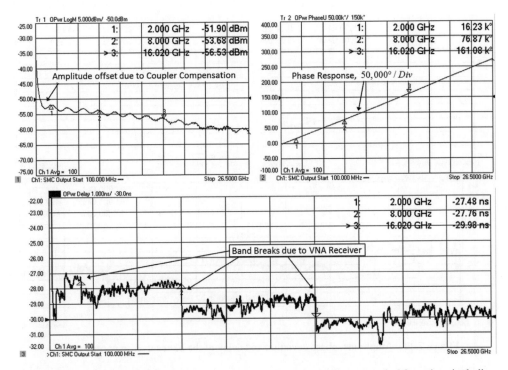

Figure 7.39 Measured amplitude and phase response of the phase reference at the b2 receiver, including reversed coupler effect.

response in terms of either group delay or deviation from linear phase. Remember that this is the raw amplitude and phase response, and represents the combination of the phase response of the phase reference as well as the phase response of the b_2 receiver including the directional coupler response in front of the b_2 receiver, and the phase response of the VNA mixer's LO (which of course also affects the measured phase in the VNA final IF).

The delay of the phase reference itself is very small and very flat across frequency, with a delay deviation less than 10 psec. Thus the phase response in Figure 7.39 is almost entirely due to the receiver response, plus some nominal delay due to cable lengths. Further, the fine-grain response of the delay is commensurate with the amplitude response variation measured on the receiver. The phase response of the phase reference, as well as its mismatch, can be removed, solving for the VNA receiver tracking response as

$$BTF = \frac{b_{2\_PhRefMeas} \cdot \left(1 - ELF \cdot S_{22\_PhRef}\right)}{a_{Phase\_reference}} \qquad (7.29)$$

where $a_{Phase\_reference}$ is the power and phase of the phase reference signal. The steps in the measured delay in Figure 7.39 are really from the VNA receiver, and not from the phase reference.

This receiver response is computed from the measurement above, and is shown in Figure 7.40. The amplitude error is quite small as there is already a factory-based amplitude correction on many VNAs.

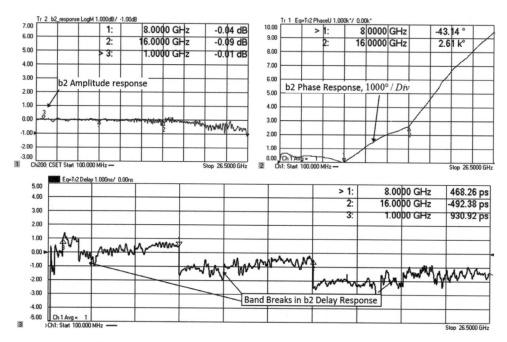

Figure 7.40   Amplitude, phase deviation and delay of the b2 receiver in normal mode.

In practice, the raw S22 of the phase reference is measured during its connection step. The next step is to perform an S-parameter calibration and extract the port error terms, which allow the S22 of the phase reference to be corrected, as well as providing the load match error term needed in Eq. (7.29). Also, the transmission tracking term, ETF, is determined during the S-parameter calibration. Since this is done at every 10 MHz, over the entire VNA frequency range, every RF and IF frequency span is included. Finally, from BTF and ETF, the RRF is computed. While the phase reference amplitude information is available, the power calibration described in 7.5.1 is usually more accurate and traceable than the phase reference amplitude, and should typically be used.

Thus, the response of each receiver, both magnitude and phase, is individually determined; the response of the overall system is easily computed for any mixer condition. Figure 7.41 shows three traces; the bottom two are the b_2^{RF} tracking and a_1^{IF} tracking computed for a mixer with a 6.5–26.5 GHz input and a 1.5–21.5 GHz output. Also shown is the computed ratio which is the SC21 tracking correction term (BTF^{RF}/RRF^{IF}).

These are formatted in delay for clarity, plotted versus mixer output frequency. Note the discrete jumps in the correction array. Investigating this further, one can plot the reference and test receiver tracking as a function of the frequency seen by each receiver. Figure 7.42 shows the tracking responses in this way, and we see that the discrete jumps are found at the same frequency for each receiver.

This implies a common cause to the jumps in delay between the reference and test receiver of the VNA, and is a particularly interesting result. The common factor is the LO used to drive both receivers. The VNA mixers must also act according to the response of Eq. (7.4)

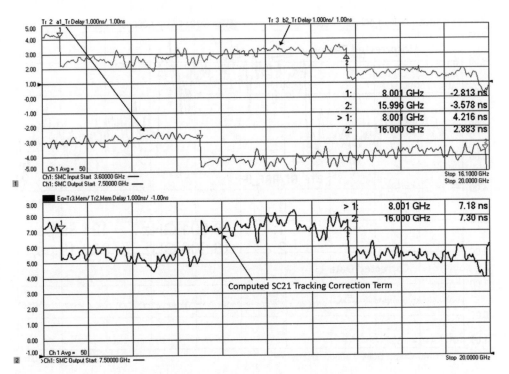

Figure 7.41 (upper) Individual input and output response of the VNA receiver for mixer frequencies; (lower) combined delay correction term for this setup.

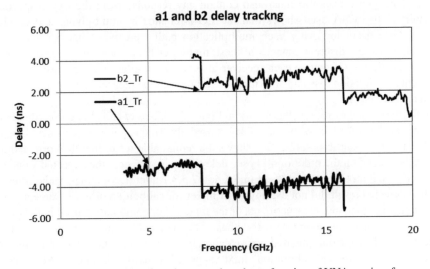

Figure 7.42 Receiver tracking for mixer test plotted as a function of VNA receiver frequency.

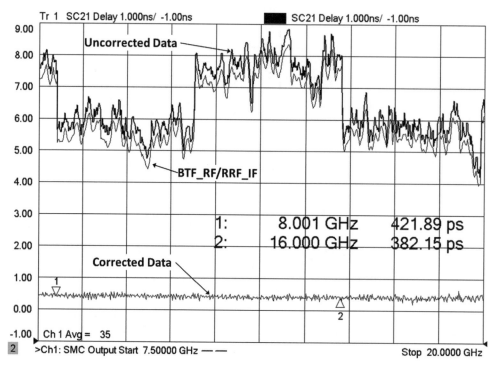

Figure 7.43   Raw and correct delay response of a mixer.

and thus their phase response must also contain any response from the LO. Inspection of the details of the VNA used shows that the LO is created by multiplying and dividing a base 2–4 GHz oscillator. After each multiplication path there is a filter, so it is entirely reasonable to expect discrete changes at these multiplier bands, and that is exactly what is shown in the figure. Of course each receiver also has a fine-grain response associated with its particular conversion loss and any ripple or response in the measurement path before each particular receiver.

Finally, the overall mixer result is shown in Figure 7.43, before and after correction. For this example, a wideband mixer with small delay is used; this frequency range cuts across some of the VNA receiver band-crossings. This shows the remarkable result that the corrected mixer response is flat and on the order of 400 psec delay. This is very near the expected value based on the reciprocal mixer approach used in Section 7.5.3.1 after accounting for the filter used. And this corrected result did not use any other mixer for calibration or measurement.

As another comparison, measurements of the mixer/filter combination used in the previous two calibration methods, along with the phase-reference calibration method, are shown in Figure 7.44. This is the only method whose calibration is independent of any other mixer.

Note that the "reciprocal-calibration" mixer approach is used in both the VMC correction method and for de-embedding the reconversion mixer in the up/down measurement method. The best response is certainly from the phase reference calibration, in that it is the smoothest

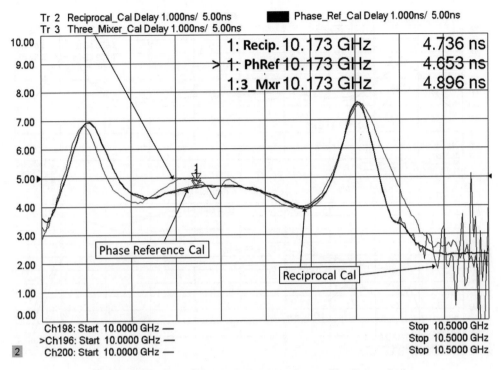

Figure 7.44   Comparing the three mixer phase calibration methods.

response and has the lowest out-of-band noise at high frequencies. In the passband, it compares almost identically to the reciprocal cal method, with a variation of less than 85 psec. In contract, the three-mixer method varies by about 500 psec.

And in a final comparison, the LO frequency of this mixer combination is changed, and the resulting delay and gain measurements are shown in Figure 7.45. This is the only method that allows changing the LO frequency of test without recalibrating the system or recharacterizing the calibration-mixer.

In this comparison, the input frequencies are kept constant, and different LO frequencies used which results in different output frequencies. In one case, the LO is shifted 100 MHz from 6.55 GHz to 6.45 GHz. In another case, the LO is moved from low side to high side, resulting in a reversal of the original output frequencies. With the small shift in LO frequency the delay is nearly the same (the delay of the RF filter is slowly varying across this range, with a slight slope, explaining why the double peak in delay is not exactly equal). When the LO is moved to the high side, at 13.45 GHz, the group delay response of the input filter is identical, but the response of the output filter is flipped, resulting in a big change in delay response. This is mirrored by an equally large change in the slope of the amplitude response (lower window). As one might expect, the amplitude response shows a slowly varying change on the left-hand side of the response, which is typically associated with lower peak group delay.

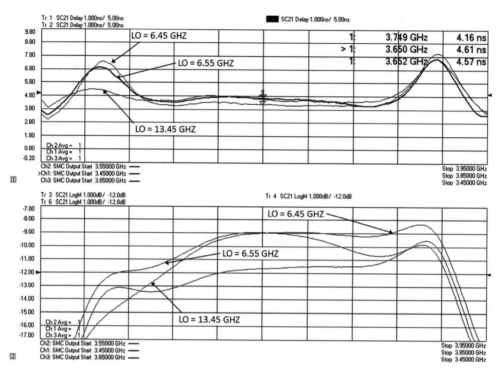

Figure 7.45 Delay and gain measurements with different LO frequencies.

7.6 Mixers Measurements vs Drive Power

Just as with other active devices, the behavior and characteristics of mixers depend upon the drive level applied to their ports. Mixers, however, have both a high level drive (the LO) and a low level drive (the signal input) and the mixer response is affected by both. In trying to optimize the operation of a circuit or system, determining the best operating point of a mixer requires characterizations versus both the LO drive and the RF drive of the mixer, as discussed in the following sections.

7.6.1 Mixer Measurements vs LO Drive

7.6.1.1 Fixed Frequency Response to LO Drive

Normally, a mixer is designated for operation at some particular LO level; mixer manufacturers often give some range of LO drive level over which the operation of the mixer, particularly conversion loss and distortion, is presumed to be relatively constant. Often the presumption is that higher LO drives will always give lower distortion or better conversion loss, but depending upon the details of the mixer design, this may not be the case. As such, it is often useful to perform tests of mixer characteristics over the LO power range.

With a traditional approach, these measurements are implemented using RF sources and spectrum analyzers, and are tedious to perform as the settings of frequency and power are either made manually, or a user must write their own program to control the individual instruments, gather the data, correct the data and present the results. In fact, many RF test engineers started their careers writing programs for mixer tests, including the author.

Today, many modern VNAs have frequency offset modes (FOM) or frequency-converter applications (FCA) that essentially automate all the required controls, provide calibration and directly present the significant results. This makes it very easy and practical for a user to evaluate and understand the behavior of their devices, using essentially the same setup that was created for amplitude characterization of the MUT, by simply changing the sweep modes and the stimulus settings.

Figure 7.46 displays a set of mixer measurements at a fixed frequency as a function of the LO drive level, for a simple mixer. The SC21 gain measurement is performed using the built-in FCA function. The S33 measurement shows the LO match of the mixer, also as a function of LO drive. In this case, the LO is obtained from the internal second source of the VNA. Since the LO response does not depend upon the RF signal applied, one could also measure LO match on a simpler, two-port VNA by applying a load to the input and output ports of the MUT, connecting the LO port to port 1 of a VNA, and sweeping the drive power. If the VNA does not have sufficient drive power to test the MUT over the entire LO drive range, the high-power setup described in Figure 6.29, can be used to increase the drive for this test.

What is interesting in this response is that from the SC21 response, one can see that there is a definite curvature of the LO drive for this particular mixer. Other mixer types may show

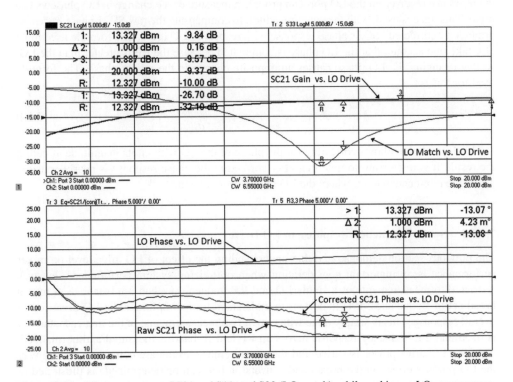

Figure 7.46   Mixer parameters SC21 and S11, and S33 (LO match), while making an LO power sweep.

such an optimum LO drive power for highest conversion gain, depending upon the internal limiting effects of the diode or switching structures in the mixer. The flatness and breadth of the SC21 response to LO drive gives one a good sense of this mixer's sensitivity to LO drive level, and with its broad and flat response, one can see that the assumption that LO drive does not, to the first order, affect the SC21 conversion gain is valid.

Also interesting to note is that the LO input match of the mixer depends strongly on LO drive, and the point at which the match is best is used as a reference point to determine the change in SC21 vs change in LO drive power, as indicated with marker R in the upper window. Marker 2 shows that a 1 dB change in LO drive from the point where the LO match is minimum results in a 0.16 dB change in SC21.

One can sometimes correspond the optimum drive level for minimum conversion loss with the LO match versus LO drive. The S33 plot clearly shows when the LO drive is sufficient to "turn on" the mixer diodes; at lower LO drive levels the diodes represent a high impedance and thus much of the LO signal is reflected. In this same region, the conversion loss is very poor. As the LO power is increased the mixer diodes turn on, and at some point provide a matched impedance. At even higher LO drive levels, the impedance of the mixer diodes becomes even smaller and so the impedance may no longer be matched.

The lower window shows the SC21 conversion phase vs LO drive level. Unfortunately this "raw" measurement cannot be completely trusted as the LO phase can also change with LO drive level. If an external source is used, the LO can experience step changes in phase if the external source steps through source-attenuator settings. Here, the internal source is used, and the reference receiver on the LO port can provide a measure of the change in LO phase vs LO drive. This trace is used, with the equation editor, to compensate the raw SC21 phase trace and produce a "corrected" SC21 phase response with respect to LO drive. The formula used must also take into account that the LO phase is conjugated because the RF is greater than the LO, so that changing the LO phase causes an opposite change in the output phase. The formula used for this is

$$SC_{21\_Corr} = SC_{21\_Raw} \cdot \frac{|a_{LO}|}{a_{LO}^*} \tag{7.30}$$

This formulation allows correcting for the phase of the LO without its magnitude affecting the result. It is interesting to note that after correction the phase response is nearly constant above the reference marker, where the LO match is optimized.

7.6.1.2   Swept Frequency Response to LO Drive

The previous section demonstrated that understanding the effect of LO drive level on mixer performance is an important consideration when optimizing system performance, but was restricted to measurements of a single LO and RF frequency. When a mixer is used over a broad set of frequencies, it is important to understand how to optimize the LO drive for an overall response. This is simply done with one of two measurements. In the first case, the LO is fixed and the frequency response across an RF and IF band is measured and normalized. Next the LO drive power is varied and the measurement results are compared. In this way the LO power's effect on the broadband conversion loss can be determined, as illustrated in Figure 7.47. The upper window shows the actual SC21 over frequency with different LO drive

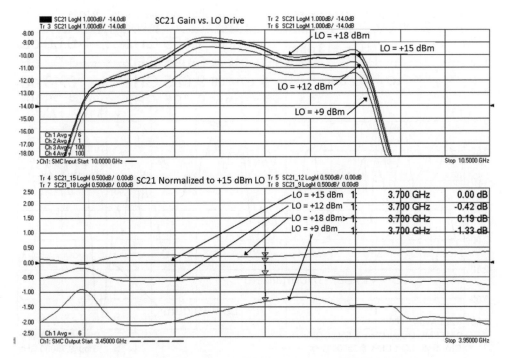

Figure 7.47   LO power effects on SC21 over a range of RF/IF frequencies.

levels. The lower window shows the same data, all normalized to the response of SC21 for a +15 dBm LO drive level. Clearly, this mixer has different responses to LO drives at different RF frequencies. One cause of this might be the change in match that the LO sees as a result of the input and output filters that are attached in this simple converter case.

Alternatively, a different manner of characterization is needed for a block converter, where the LO frequency changes as the channel being converted changes. In a similar manner to the lower window in Figure 7.47, the LO power is changed during a SC21 measurement, but in this case the LO and IF are swept, and the RF remains fixed, as shown in Figure 7.48. This measurement shows the channel conversion gain as a function of the LO frequency, for different LO powers, all normalized to the conversion gain for an LO power of +15 dBm.

Here again one can see that there is some tradeoff in the LO drive for maximum conversion as the IF frequency changes. In this case the LO drive of +9 dBm could be used at some frequencies, but has substantial degradation at certain frequencies across the band. For example, at the center frequency (LO = 8 GHz), there is less than 0.5 dB variation in SC21 gain for LO power changes from +9 to +18 dBm; however, at 9.085 GHz, there is almost 4 dB of variation over the same LO power change. Characterizations of this kind allow designers to fully understand the effects of optimization on overall system performance. It is clear that for this mixer, the single frequency characterization of Figure 7.46 is not sufficient to understand this mixer behavior over the wide range of LO frequencies used.

In this figure, one of the traces uses the output frequencies as its x-axis, and the others use the LO frequency. It is convenient to have the option to select the x-axis, and to have both

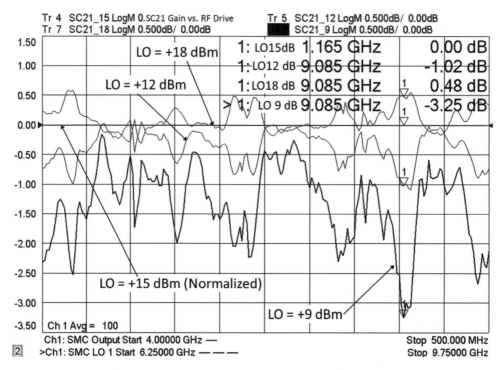

Figure 7.48 RF conversion gain vs frequency for various LO drive levels for swept LO measurements of a block converter.

scales available on the same plot. Also note that there is a reversal in the frequency axis for the RF output; this is caused by the fact that the LO is less than the RF, so that when the LO increases, the IF decreases in frequency.

7.6.2 Mixer Measurements vs RF Drive Level

The LO drive level used for a mixer application essentially sets the operating point for that mixer, just as setting the bias level of an amplifier sets the operating point of the amplifier. One aspect of mixer characterization is looking at the conversion gain of the mixer as the RF drive level is changed. Just as with LO drive characterizations, this can be done in fixed frequency or swept RF frequency.

7.6.2.1 Swept Frequency Measurements vs RF Drive

While the CW compression measurements give a very intuitive feel for the mixer compression curve, one may find that the compression of a mixer changes with RF frequency. As such, it is a good idea to start the investigation of mixer compression by measuring the change in conversion gain versus change in RF power, over a swept RF frequency, in a similar manner in which the amplifier swept frequency compression measurements were performed. For the same mixer used previously, at a fixed LO drive level of +12dBm several sweeps of conversion

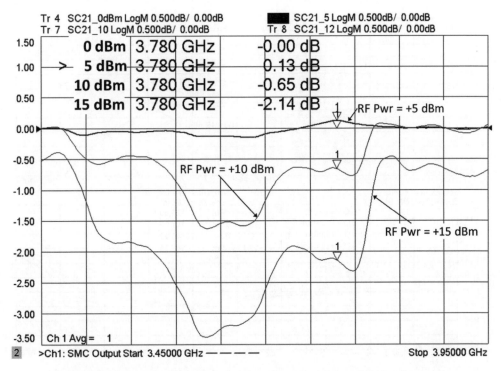

Figure 7.49 Normalized conversion gain vs frequency for various RF drive levels.

gain were performed for RF powers of 0 dBm, +5 dBm, +10 dBm and +12 dBm, as shown in Figure 7.49. Here the conversion gain is shown normalized to the 0 dBm RF drive level.

From this swept frequency response, one finds an interesting point at 3.78 GHz, where the higher RF power (+5 dBm) causes an improvement in the conversion gain. This frequency point can be further examined by performing a fixed frequency, swept RF power measurement, as shown in the next section.

7.6.2.2 Fixed Frequency Measurements vs RF Drive

The compression characteristics of mixers, that is, how a mixer conversion gain changes with signal (RF) drive level, is characterized in a manner quite similar to amplifiers, but with slightly different attributes.

Just as the LO drive can be varied at a fixed frequency, while monitoring the SC21 parameter, so too can the RF drive be varied to test the linearity of the mixer or converter. In Figure 7.50, the results are shown for testing conversion gain versus RF drive power, for several different LO powers. The figure shows both gain compression (in the upper window) and phase expansion (in the lower window) as a function of RF input drive level. As expected, the 1 dB compression point for a mixer, as well as the linear conversion loss, does depend upon the LO drive used.

As a very rough rule of thumb, the 1 dB output power compression is found at approximately 10–15 dB below the LO drive level. From the compression curves shown, the linear region of a mixer operates around 10 dB below RF compression level, or about 20–25 dB below the LO

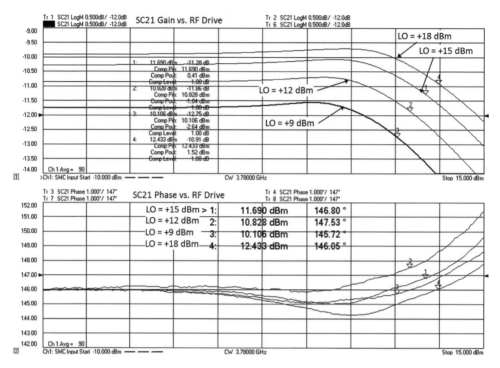

Figure 7.50 Conversion gain vs RF drive for several different fixed LO powers.

drive level. Note that because of conversion loss, the input power for compression is much higher. This is an important point to note when performing tests that depend upon the device behaving with linear gain, such as noise figure.

Another interesting detail on mixer-compression is evident here: for the LO drive levels, the conversion gain may first increase as RF drive is increased before it starts to go into compression, as demonstrated in this figure. This effect is more pronounced at lower LO powers. This effect is hinted at in Figure 7.50; at other RF frequencies, a normal compression curve is observed, but at 3.78 GHz, some expansion was noted in that figure. While some linearized amplifiers have a hint of this behavior, it is quite common to find this behavior in mixers. One likely reason is that for low LO drive levels, the mixer diodes are barely turned on and have high loss. Increasing the RF drive causes a change in the bias point of the diodes and effectively adds to the LO drive level to turn on the diodes harder, thus reducing their resistive loss. This change in diode operating point causes the conversion gain to increase at moderate RF drive levels, especially when the LO is under driven or "starved". Starving the LO is often a tradeoff that is accepted in a system design in order to improve the overall DC efficiency and reduce the complexity of a mixer, by eliminating LO driver amplifier stages.

Interestingly, starving the LO on a high-drive mixer may give either better or worse performance than using a lower LO drive mixer in the same applications. Higher level LO mixers have higher barrier diodes (or more diodes in series) and thus tend to be more linear when operated with the same RF input level as lower barrier diodes, provided that sufficient LO drive is delivered to allow the mixer to operate in its normal mode.

7.6.2.3 Automated Gain Compression Measurements on Mixers (GCX)

For many mixers, a key specification is the input power that produces 1 dB of gain compression, just as in amplifier testing. While the value can be obtained by repeating the fixed frequency or swept frequency measurements above, some manufacturers have created purpose-built applications that directly provide the swept frequency 1 dB compression (or any other chosen level of compression). These setups are almost identical to a combination of the gain compression application (GCA) for amplifiers, and the frequency converter application (FCA) for mixers, to form the gain compression application for mixers (GCX; X in a circle is a common symbol for mixers). For these applications, the calibration is identical to the calibration performed for an FCA measurements (in fact the same calibration can be used) but with the added stimulus parameter of start, stop and linear power. The measurement starts by sweeping the frequency response of a mixer at the linear power, to provide a reference gain. Next, the power is iterated from the start power (which may be higher than the linear power to shorten the measurement time) to the stop power. Just as in GCA (see Section 6.2.4) the sweep can be iterated in a smart way to find exactly the 1 dB compression point very quickly, or a two-dimensional sweep of frequency and power can be performed for every point, generating a 2-D data set that is evaluated for the 1 dB compression point.

Figure 7.51 shows the results of several automated gain compression tests, each one with the mixer driven at a different fixed LO drive level. Evaluations of this sort can be very useful in comparing mixers from various manufacturers or in optimizing system performance of a

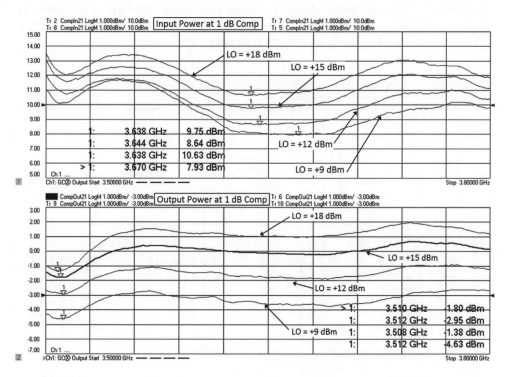

Figure 7.51 Automated GCA measurements for the same mixer with various LO drive levels.

particular mixer. The upper window shows the input power for 1 dB compression for a variety of LO powers; the lower window shows the output power for 1 dB compression for the same set of LO powers. For this mixer, changing the LO power from +9 dBm to +18 dBm improves the worst-case 1 dB compression point by about 3 dB.

As with amplifiers, gain compression is one measure of the linearity of a mixer, and non-linear behavior will lead to distortion in the output signal. A key measure of distortion is the two-tone IMD measurement, presented in the next section.

7.7   TOI and Mixers

Distortion measurements in mixers and frequency converters are similar to those of other active devices; however, the terms used are sometimes confused with other mixer measurements. One principal distortion measurement is the two-tone third-order intermodulation distortion measurement, sometimes called TOI. As with amplifiers, TOI sometimes refers to the third-order intercept point and so IMD is used instead. Unfortunately, the terms "intermodulation" or "intermods" or "mixer-intermods" are sometimes used to refer to higher-order products of the RF and LO. Usually the context makes the usage clear, but this book will refer to intermodulation as the mixing of two signals at the RF (or IF) input of a mixer, and the mixing products from RF and LO will be referred to as higher-order products. Both the RF and LO signal levels will affect the measured IMD, as discussed in the next sections. Figure 7.52 shows a typical setup for measuring mixer IMD using a VNA. It is also possible to do the same measurement using three independent sources and a spectrum analyzer, but many modern VNAs provided purpose-built applications, such as swept IMD (IMDX) that make measuring IMD measurements while sweeping frequency, or power, very easy.

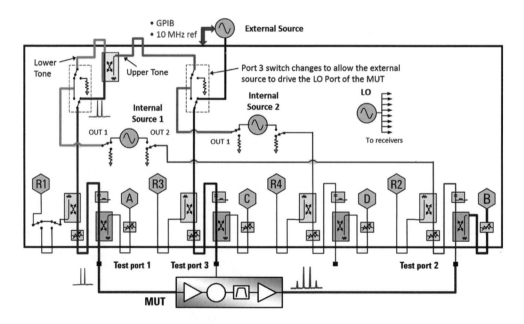

Figure 7.52   Setup for measuring mixer IMD.

For all IMD measurements, the same cautions apply as described in Section 6.6.5 on optimizing measurements. Principally this means ensuring that the sources don't create their own intermods into the mixer, and the power of the two tones hitting the measuring receiver are sufficiently attenuated so that the receiver's own intermods don't affect the result. In the block diagram in Figure 7.52, there are attenuators on the sources and the receivers to optimize each for a particular measurement. Using a coupler instead of a power splitter lowers the available two-tone power, but greatly increases the isolation between sources. A switch on the source going to port 3 allows flexibility in which source is used as an IMD source. With this configuration the VNA source, with its high sweep speed, can be used as a second tone for IMD measurements. By changing the switch position, the same source can drive the LO of the MUT for swept LO measurements. Finally, by routing the external source through the port 3 test port, the power and phase shift of the LO signal can be monitored on the R3 receiver.

Often, IMD is a critical measurement for receiving mixers used in low power and low noise applications. As such, a very common figure of merit is the input referred third-order intercept point, or IIP3. Since this is so common (unlike amplifiers that are typically specified with output IP3 or OIP3), the following examples all refer to the input intercept point. Output intercept point is easily computed by subtracting the conversion loss from the IIP3 value.

7.7.1 IMD vs LO Drive Power

Because it is the LO power applied to a mixer that sets, in some sense, the operating point of the mixer, one finds that the IMD response of the mixer depends strongly on the LO drive level. Figure 7.53 shows the third- and fifth-order IM product, in dBc, along with the measured output power and conversion loss, and the input referred third-order intercept point (IIP3). This point changes as both the IM product power and the gain of the mixer change with respect the LO drive power.

What is interesting in this figure is that the LO power for optimum third-order performance, in terms of IIP3, has two values, and the optimum power for the third-order products is not the same as for higher-order products; in fact, the IM5 is actually higher than the IM3 for the LO drive of +9 dBm. A spectrum plot of the mixer at these stimulus conditions (RF = −5 dBm, LO = +9 dBm) illustrates this point clearly, in Figure 7.54

For an LO drive level of 9 dBm, the IIP3 level peaks, and does not return to the same value until the LO drive is up to about +16 dBm. The IIP3 in the region of +12 to +15 dBm (the normal operating range of this mixer) is not as good as at a lower LO drive, for this particular RF drive level. However, one cannot say that a low LO drive of +9 dBm is good for all RF power levels; this must be investigated independently, as discussed in next section.

Finally, the same LO power is not necessarily best across the frequency span of the mixer, as shown later in Section 7.7.3.

7.7.2 IMD vs RF Power

The level of IMD created from an RF input follows the same rules in mixers as it does in amplifiers, provided that the input is sufficiently low. Normally changing the input power of the two tones each by 1 dB will increase the power in the third-order product by 3 dB, fifth-order by 5 dB, and so on. But this rule will break down in many mixers as the RF level gets high, especially if the mixer has a lower LO signal (starved LO) as the RF signal helps to

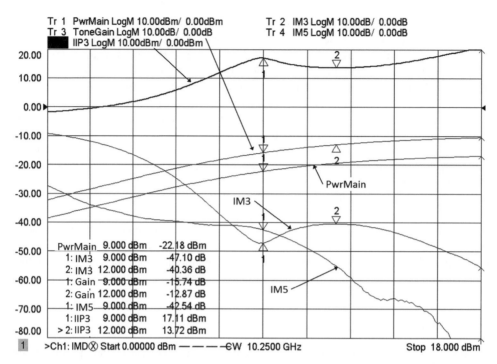

Figure 7.53   Third- and fifth-order IM product vs LO power, as well as output power, IIP3 and gain.

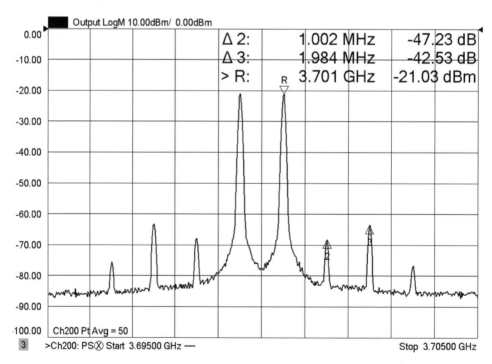

Figure 7.54   Spectrum plot of IM products at RF = −5 dBm, LO = −9 dBm.

turn on the mixer diodes. For this reason, it is often necessary to make swept RF power IMD measurements on a mixer to understand the actual IM level at the input power of interest.

Figure 7.55 shows the result of making a swept RF power measurement on a mixer while monitoring the third- and fifth-order products. Also shown is the IIP3 and the gain for this mixer, as well as the output power. Here again one notes that the IM powers follow the normal rules until the RF drive becomes significant relative to the LO power. Highlighted in the plot is the IM3 value for a gain compression of 0.11 dB: this follows a rule of thumb that the IM3 for 0.1 dB compression is approximately −40 dBc. Marker 1 is used on the tone-gain trace, reference to the R marker, to find the −0.1 dB compression point. Since the markers are coupled, the IM3 marker shows the relative IM3 level at that point.

Another remarkable attribute of Figure 7.55 is the behavior of the IIP3 as the RF power is increased. As described above, this mixer follows the behavior that the mixer distortion does not always increase with increasing RF power. In fact, the intercept point (one measure of relative distortion) is highest at an RF input level of −0.6 dBm. The value of intercept point is also strongly affected by the LO power.

Figure 7.56 shows just the main power, the third-order power and IIP3 for the same RF power sweep range, but this time for two different LO drive-powers. It is very common in mixer measurements to display many of the mixer characteristics such as gain, compression and distortion, as a function of several LO drive powers so that designers can make tradeoffs in the system performance between LO drive levels and other attributes. This is especially true

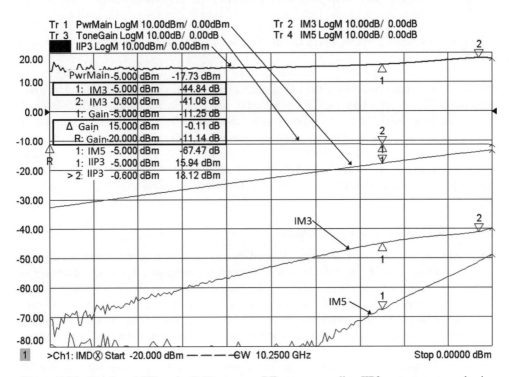

Figure 7.55   Third- and fifth-order IMD power vs RF power, as well as IIP3, output power and gain.

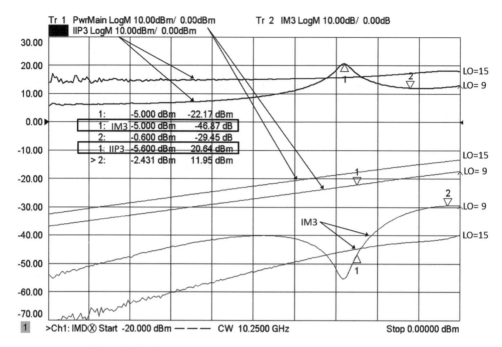

Figure 7.56   Swept RF power IM3 and IIP3 with different LO drives.

in low-power receiver circuits where LO drive power may be one of the largest contributors to battery drain.

In this figure one notes that the maximum input intercept point is not for the higher LO drive case, but for the case where the LO is almost starved. At +9 dBm LO, this mixer gives a region of RF powers where the IIP3 is high, but just outside this region the IIP3 is much worse than the higher LO drive condition. It is for this reason that system designers must have distortion characteristics as a function of several variables, to show the true behavior of a part. This particular mixer could be advertised as having greater than +20 dBm IIP3 at +9 dBm, which is true, as long as the RF drive power is very carefully controlled. Such behavior is not normally expected and would come as a surprise to system designers.

7.7.3   IMD vs Frequency Response

Since many mixers are inherently band-limited due to the nature of the baluns used in the mixers, or filters used in the construction of frequency converters, it is often necessary to characterize the performance of a mixer over its band of operation. While there may be little change at the center of the band of operation, there can be substantial performance difference at the band edges. Also, it is sometimes desired to use a mixer beyond its specified or nominal frequency range, and in such a case it is important to understand how the mixer's performance is degraded.

Figure 7.57 shows two examples of swept frequency IMD. In the upper window, the LO level is held constant and three different traces of mixer IIP3 are shown versus RF drive level.

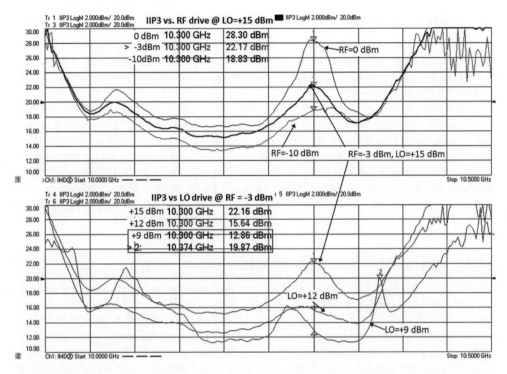

Figure 7.57 Upper: mixer IIP3 and gain vs frequency for three different RF power level; lower same mixer for three different LO power levels.

The lower window shows a similar measurement but this time the RF power is held constant, and the LO drive level is changed while the IIP3 is monitored for the three different LO powers as the RF and IF are swept in frequency.

In this plot, it is noteworthy that for this mixer, in the upper window, driving a higher RF drive level produces a higher IIP3, much higher at some frequencies such as the point highlighted by the marker. This implies that the mixer is becoming more linear at higher RF drive levels; such a behavior is consistent with the results of Figure 7.55. Similarly, the change in IIP3 vs frequency for different LO drive levels shows some unexpected results in the lower window. The behavior at marker 2, on the traces for LO = +9 dBm, shows dramatically improved IIP3 or a narrow region even for the low level LO drive. Again, this behavior is anticipated in the results from Figure 7.56 and even from the compression measurements from Figure 7.49.

In practice, any of the RF, LO or IF may be fixed or swept. When making IMD measurements on mixers it is a good practice to set up and calibrate over a broad band so that the distortion performance of the mixer can be characterized over any of the frequency range inputs or outputs of any of the ports, as well as a wide range of RF and LO powers. From the measurement results shown in this and the previous sections, it is clear that the distortion characteristics of mixers do not always follow simple rules, especially for mixers with starved LOs. Thus the designer is required to make a wide range of mixer characterizations to fully understand the mixer behavior in a system.

7.8 Noise Figure in Mixers and Converters

Mixers and frequency converters are often the first component after the antenna in a receiver system, and as such the noise figure of the first converter predominantly sets the noise figure of the system. Thus noise figure measurements on frequency converters are an essential aspect of their full characterization.

Most frequency converters have filters that keep unwanted images out of the response, and have amplifiers either before or after the embedded mixer as well, and so have overall gain and excess noise at the output above the kT_0B noise floor of a passive device. Thus characterizing the noise figure of a frequency converter is quite similar to that of an amplifier (see Section 6.7), where one measures the excess noise, then determines the gain of the MUT, and essentially computes the noise figure from these two values. Y-factor or cold source techniques both work reasonably well on frequency converters. For the next sections, the aspects where mixer characterization differs from amplifier characterization will be highlighted.

7.8.1 Y-Factor Measurements on Mixers

Characterizing the noise figure of stand-alone mixers, in contrast to that of frequency converters, is a substantial challenge due to the fact that many passive mixers have no excess noise and so their effective noise figure is simply the inverse of their conversion gain. If the mixer has no frequency selectivity at the input, the traditional Y-factor technique gives the wrong answer as the gain estimate has a substantial error; for many mixers the Y-factor measurement error for gain, and thus the noise figure error, is around 3 dB! Consider the comparison of measurement of a mixer or frequency converter illustrated in Figure 7.58.

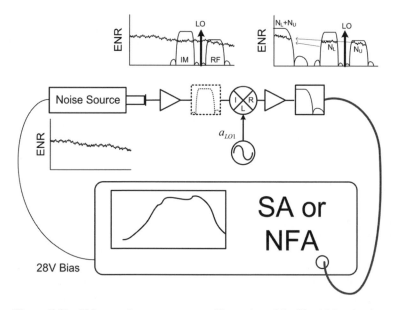

Figure 7.58   Y-factor mixer measurement illustration of double sideband noise.

For the mixer (or a converter without image protection, represented by the dotted filter), the Y-factor technique applies the noise source at the input with the noise source turned on, and measures the output noise; then the noise source is turned off and again the output noise is measured. In this case, a down-converter, the noise source provides excess noise at both the desired RF input (one IF frequency above the LO) and the image input (one IF frequency below the low); both noise powers are converted to the IF, where they add together,and are measured by the NF receiver. To illustrate the point, consider a noise-free mixer ($T_E = 0$) with conversion loss of 6 dB, equally, at the RF and IM frequencies; this would be the case of a broadband mixer. If the ENR noise source has a hot noise temperature of T_h that is constant with frequency, and a cold noise temperature that is T_0, then the noise temperature measured at the output consists of the kT_0B noise plus the hot noise of the noise source at the RF times the conversion gain at the RF, and the hot noise at the IM plus the conversion at the IM to yield

$$T_{H\_Rcvr} = T_{h\_RF} \cdot \left| SC_{2IF,1RF} \right|^2 + T_{h\_IM} \cdot \left| SC_{2IF,1IM} \right|^2 + T_0 = T_h \cdot (0.5) + T_0 \qquad (7.31)$$

The cold measurement simply measures the cold temperature, as the mixer has no excess noise so

$$T_{C\_Rcvr} = T_0 \qquad (7.32)$$

and the noise figure computed from the Y-factor becomes

$$Y = \frac{T_{H\_Rcvr}}{T_{C\_Rcvr}}$$

$$NF_{DSB} = 10 \log_{10} \left(\frac{ENR}{Y-1} \right) = 10 \log_{10} \left(\frac{\dfrac{T_h}{T_0} - 1}{\dfrac{T_h (0.5) + T_0}{T_0} - 1} \right) \qquad (7.33)$$

$$NF_{DSB} \approx 10 \log_{10} \left(\frac{\dfrac{T_h}{T_0}}{\dfrac{T_h (0.5)}{T_0}} \right) = 3dB$$

$$NF_{SSB} = NF_{DBS} + 3dB = 6dB$$

In fact, from first principles, the noise figure should be identical to the conversion loss, or 6 dB, but the implied conversion loss from the measurement is 3 dB too high, due to double sideband conversion of the hot noise power. In most Y-factor measurement instruments, there is a mode to add 3 dB to the measured noise figure for mixers to compensate for double sideband noise effects. If the mixer to be measured is not a bare mixer, but instead has a preamplifier, as in the case shown in the figure above, the amplifier at the input of the mixer may have some significant gain slope, so the noise power from the unwanted image (which is at a lower frequency) may be quite a bit higher than from the desired RF band above the LO. In this case, the error can be substantially more than 3 dB. The only way to know for certain how to correct for the gain in the undesired sideband is to measure the conversion loss of both sides, and compute the noise contribution from each to know how to correct the overall Y-factor.

In short, the cold noise power measurement in a Y-factor method is the correct cold noise, but the hot noise depends upon the sum of the upper and lower sideband conversion. If the mixer is image protected, so that hot noise power from the noise source at the image frequency is blocked from the mixer, then the proper implied conversion gain is measured and the Y-factor will give a reasonable result. In fact, the error could be greater than 3 dB if the sideband of interest has *lower* conversion gain than the undesired sideband. This might happen while evaluating the noise figure of a mixer at the edge of its specified range.

Another measurement technique, the cold source method described below, doesn't have an issue with images in the mixer as the gain is computed independently; the image noise *is* a true source of noise at the output, during the cold power measurement so it must be included. The image noise in a hot measurement is a *source of error* in the Y-factor method, but does not occur in cold source measurements. A comparison is shown in the next section.

7.8.2   Cold Source Measurements on Mixers

For noise figure measurements using a modern VNA, the cold source method can be applied which may give better results than Y-factor for mixers without image protection.

The cold source method follows the same in mixers as in amplifiers, with two measurements required to compute the noise figure. The first measurement is the gain of the mixer, SC21, and is measured as described in Eq. (7.17) for standard mixers and (7.18) for image mixers. The noise power is measured on a noise receiver, which has been calibrated in accordance to the method in Eq. (6.25). It can be converted to incident noise power, and the converter noise figure computed as

$$NF = \frac{T_{Inc}}{T_0} \cdot \frac{|1 - \Gamma_{S\_RF}S_{11\_RF}|^2}{\left(1 - |\Gamma_{S\_RF}|^2\right)|SC_{21}|^2} = \frac{|1 - \Gamma_{S\_RF}S_{11\_RF}|^2}{\left(1 - |\Gamma_{S\_RF}|^2\right)} \cdot \frac{DURRNPI_{IF}}{|SC_{21}|^2} \tag{7.34}$$

For a well-matched system, the noise figure is simply

$$NF = \frac{DUTRNPI_{IF}}{|SC_{21}|^2} \tag{7.35}$$

That is, the mixer noise figure is the measured excess noise power in the IF divided by the square of the conversion gain. In dB terms, it is simply the excess noise power in dB minus the gain in dB.

Figure 7.59 shows a comparison of a Y-factor measurement and a cold source measurement, for a frequency converter with image protection (single sideband conversion), displaying the gain, relative noise power (DUTRNPI), and noise figure. In this case, the Y-factor and cold source gain and DUTRNPI match almost exactly. The noise figure of the Y-factor is about 0.2 dB higher than for the cold source. This is most likely attributable to errors in the source match of the noise source. Since the DUTRNPI and gain are nearly identical, the noise figure should match with one computed by Eq. (7.35). However, the cold source also does a mismatch correction for the effective source match of the system, as in Eq. (7.34), which is why it displays

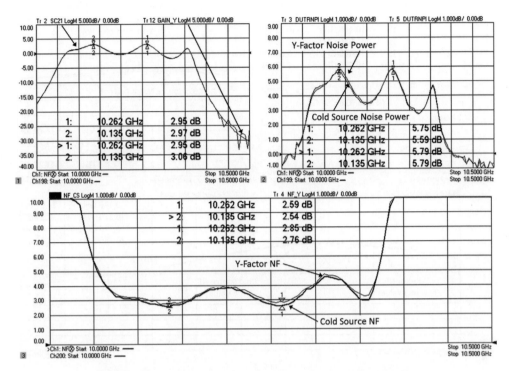

Figure 7.59 Comparison of Y-factor NF, cold source NF, and SC21.

a slightly lower noise figure. In the upper left plot, the gain computed as SC21, and the gain computed as Y-factor are compared. In the passband of the mixer, the two compare nearly identically. In the stopband, however, the Y-factor gain shows significant error.

For this mixer, which has an image filter, both the Y-factor and the cold source give reasonable results. However, if the RF image reject filter is removed from the converter's input, an entirely different result occurs.

The design of this converter has first an LNA, then image filter (this configuration ensures that the loss of the image filter does not hurt the noise figure, and removes the excess LNA noise in the IM band from the mixer), then the mixer followed by another IF low-pass filter to remove any LO leakage. If the IF LPF is not used, the LO feed through from the mixer can overdrive the NF receiver's LNA.

To understand the importance of the image filter, a second set of measurements on the same LNA/mixer/IF filter set was performed, but in this case the RF image filter was removed. One would expect substantially higher noise figure as the LNA noise at the IM frequency will convert to the IF. In this case, the LNA has a large gain slope, so the excess noise from the LNA is even higher in the IM range than the RF range. The results of making both cold source and Y-factor measurements are shown in Figure 7.60.

The measurement of the mixer in Figure 7.60 actually shows more gain variation in the SC21 trace. This is likely because the the image filter at the input enhanced the conversion gain over some frequencies, so one sees about ±3 dB change in gain without the filter. However,

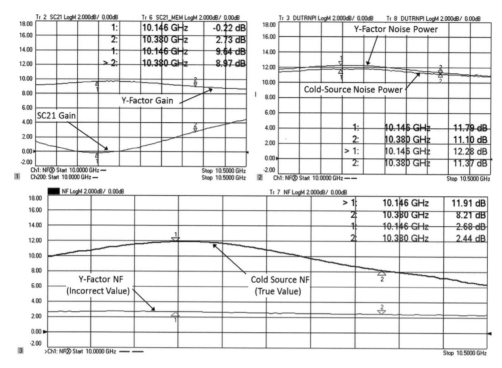

Figure 7.60   Comparison of Y-factor and cold source on a mixer with no image filter.

the Y-factor gain goes up by more than 6 dB! And it is no longer equal to the SC21 gain. Since the SC21 gain is simply computed by input and output power, there is very little uncertainty in this result, indicating that the Y-factor gain has very large errors. Any time the Y-factor gain does not match the SC21 gain, the Y-factor noise figure must be suspect.

It's clear from this comparison that the Y-factor NF is in error by even more than the expected 3 dB associated with the double sideband conversion. In this case, the image frequency is much lower than the RF frequency, and so the gain of the LNA is much higher at the image frequency. Thus the noise power from the unintended sideband, for both hot and cold measurements, is much too high. The DUTRNPI measurement should normally show very similar powers for both cases, but this LNA is sensitive to match, and since the mismatch of the noise source is not identical to the mismatch of the VNA source, there can be some small noise power differences in the upper right plot.

While the noise figure shown for Y-factor of less than 3 dB is appealing, and many engineers would happily accept this result, it is most certainly also incorrect. The SC21 gain (performed using standard mixer measurements as described earlier, Section 7.4.2) shows the true gain of the MUT. The fact that the Y-factor gain is so much higher is the smoking gun for an invalid measurement. Unfortunately, ignoring the effects of the image gain is a common mistake that inexperienced users make when using a Y-factor mixer measurement on an SA or NFA. In this case, only the cold source technique can be used to show the proper noise figure for this mixer which does not have image protection, and it shows it to have degraded by up to 8 dB due to the image noise.

7.8.2.1   Low Gain Mixer NF Measurements

For bare-mixers (or any device with loss) the measurement of noise figure becomes simply a measurement of the gain of the device, or very nearly so. If there is little or no excess noise in a device, the noise power density measured in the NF receiver is very near kT_0B and is usually swamped by the internal noise of the NF receiver. As part of the error-correction, the noise of the NF receiver is subtracted from the overall noise measurement to try to obtain just the DUT noise. However, if the DUT is poorly matched, the NF receiver may be changed by the match so the receiver noise figure measured during calibration is not the same as during the DUT measurement. Some advanced VNA systems do perform noise-parameter extraction on the noise receiver, and so this effect is reduced. Also, because of the jitter in a noise figure measurement, the characterized value for the receiver noise figure is only an estimate; substantial noise averaging (on the order of 100 or more) must be used if one wants to make measurements at kT_0B with low error. In most cases, the error due to jitter swamps other errors, and the noise figure of the bare-mixer is in error by the jitter in the receiver noise characterization, and the jitter in the MUT cold noise measurement. In these cases, a measurement of loss is probably a better estimate of the true noise figure than an actual noise figure measurement. This is quite related to the effort to verify noise figure measurement by measuring passive devices (see Section 6.7.9).

7.8.2.2   Excess LO Noise in Mixers

For measurements of mixers or frequency converters with an LO input, it is assumed that a perfect LO is used to drive the LO port. In some cases, the LO may have significant amplification in the signal path before the DUT, and this noise can add substantial error to the measurement of noise figure of a mixer. Any excess noise from the local oscillator which is present at the LO port can be converted to the IF, as determined by the $SC_{2IF,3RF}$ and $SC_{2IF,3IM}$ terms (see Figure 7.13). Figure 7.61 illustrates the source of this excess noise. Consider this converter which is image protected by a filter after the LNA. The filter removes any excess noise from the amplifier out of band, so only the amplifier's noise in the RF band is converted to the IF. However, if there is excess noise on the LO port, it can be converted to the IF output from either above or below the LO signal, in either the RF or the image band. Thus, at the output one can see approximately twice the LO broadband noise in addition to the noise signal from the converter input. If the amplifier gain is sufficient, the RF noise will overcome the LO noise rendering it negligible. Also, if the conversion of LO noise to IF output is small (the $SC_{2IF,3RF}$ and $SC_{2IF,3IM}$ terms) then the LO noise contribution will be similarly reduced.

To fully understand the effect, consider a numerical example where the external LO has an amplification with 30 dB of excess noise (5 dB noise figure and 25 dB gain) which is called ENP_{LO_RF} and ENP_{LO_IM} as it appears at both the RF and the image frequency, and the conversion terms of RF signal on the LO port to IF output is $SC_{2IF,3RF} = SC_{2IF,3IM} = -16dB$, so that the total excess noise power at the output of the mixer, ENP_{Tot_Hot} (presuming 3 dB conversion gain, and 15 dB excess noise of the noise source) is

$$ENP_{Tot\_Hot} = \left|SC_{2IF,1RF}\right|^2 (ENR_{RF} + NF_{Mix}) + \left|SC_{2IF,1M}\right|^2 (ENR_{IM} + 1)$$
$$+ ENP_{LO\_RF} \left|SC_{2IF,3RF}\right|^2 + ENP_{LO\_IM} \left|SC_{2IF,3IM}\right|^2$$

(7.36)

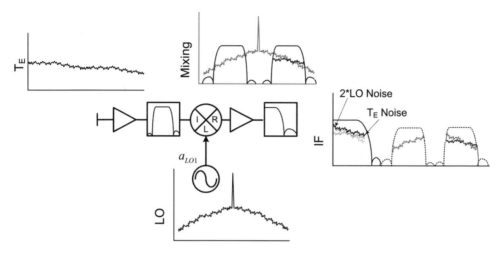

Figure 7.61   Errors due to excess converted noise of the LO.

Which for this example, makes the hot excess noise

$$SC_{2IF,1RF\_dB} = 3\,dB, \ \ N_{F\_Mix} = 3\,dB,$$

$$\left|SC_{2IF,1RF}\right|^2 = 2, \ \ \left|SC_{2IF,1IM}\right|^2 = 0 \ (due\ to\ image\ protection);$$

$$NF_{Mix} = 2, \ ENR = 10^{15/10} = 31.6\,; \ ENP_{LO} = 10^{30/10} = 1000;$$

$$\left|SC_{2IF,3RF}\right|^2 = \left|SC_{2IF,3IM}\right|^2 = 0.00251$$

$$ENP_{Tot\_Hot} = (2)\cdot 33.6 + 2\cdot (0.0251)\cdot 1000 = 67.2 + 50 = 117.4$$

$$ENP_{Tot\_dB\_Hot} = 10\log_{10}(117.2) = 20.6\,dB$$

(7.37)

Here the excess noise from the LO adds about 2.4 dB to the hot noise. And the cold excess noise is

$$ENP_{Tot\_Cold} = \left|SC_{2IF,1RF}\right|^2 (NF_{Mix}) + ENP_{LO\_RF}\left|SC_{2IF,3RF}\right|^2 + ENP_{LO\_IM}\left|SC_{2IF,3IM}\right|$$

(7.38)

which becomes

$$ENP_{Tot\_Cold} = (2)\cdot (2) + 2\cdot (0.025)\cdot 1000 = 4 + 50 = 54$$

$$ENP_{Tot\_dB\_Hot} = 10\log_{10}(54) = 17.3\,dB$$

(7.39)

Here the excess LO noise adds 11.3 dB to the cold noise power! From this, the noise figure for this example can be computed as

$$N_F = 10\log_{10}\left(\frac{ENR}{Y-1}\right) = 10\log_{10}\left(\frac{31.6}{\left(\dfrac{117.4}{54}-1\right)}\right) = 14.3\,dB \qquad (7.40)$$

Thus, in this example, the excess noise on the LO adds about 11.3 dB to the measured noise figure for this mixer. On the other hand, if the excess noise of the converter were not 3 dB, but the converter had more gain, say 20 dB total with 3 dB noise figure, then the excess noise from the mixer would be 23 dB, and the error would be less than 1 dB in the measured noise figure.

This error, excess noise from the LO, also affects the cold source method, as it will also measure a similar error since the added LO noise is unrelated to the single sideband or double sideband nature of the RF to IF conversion of the mixer, but only depends upon the RF and IM conversion of the noise impinging on the LO port. Figure 7.62 demonstrates this effect with two measurements of a mixer, one where the LO is applied directly to the MUT, and another where a narrowband filter is add to the LO path to remove the RF and IM noise frequencies from the LO signal. The second condition is the same as used for Figure 7.59.

This measurement is for a mixer with about 11 dB gain at the input, for an overall gain of 3 dB, and a noise figure of about 3 dB, just as in the example computation of Eq. (7.40). The degradation due to excess noise on the LO port is lessened if there is more gain before

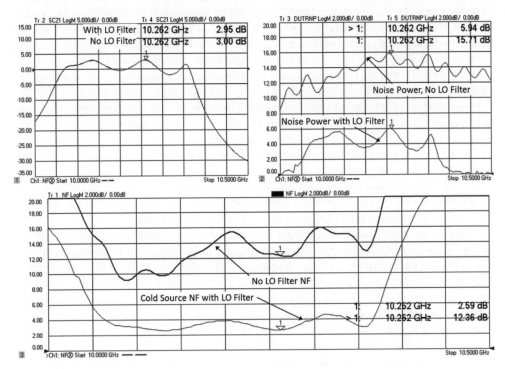

Figure 7.62   Error in noise figure due to excess LO noise.

the mixer or if the MUT is an active mixer (as is common in an IC frequency converter). In such a case, the added noise from the gain and mixer noise figure swamps the contribution of the excess LO noise. Without gain before the mixer, it is *imperative* that the LO be filtered to remove noise that could reconvert to the IF. In most converters, the LO is often embedded, and so any excess noise it adds should be included in the converter noise figure. In some cases, using higher performance signal synthesizers, instead of the built-in VNA sources, may lower the excess noise contribution from the LO, but this depends entirely on the particular synthesizer design.

7.9   Special Cases

There are several special cases in characterizing mixers and frequency converters that fall outside the simple measurements described so far. These include mixers with multipliers in the RF or LO chain, measuring higher-order products, measuring group delay on mixers with a swept LO, measuring frequency converters with an embedded (built-in) LO, and converters with very high gain. Details for improving measurements of these special cases are discussed below.

7.9.1   Mixers with RF or LO Multipliers

Some frequency converters, particularly high microwave or mm-wave converters, are driven with lower frequency LOs, and have an explicit or implicit multiplier in the mixer structure. In many cases, the multiplier is a doubler or a tripler circuit to provide a higher frequency LO, but in some cases, the mixer is operated in second or third, or higher harmonic mode of the LO frequency (odd orders being more common in balanced mixers as even-order products are substantially suppressed). In an ideally switched mixer, the third-order conversion gain is down about 9 dB from the fundamental, and the fifth order is down about 14 dB, if no explicit multiplier is used.

Measurements of mixers with built-in multipliers in the LO are very straightforward, and only require taking into account the multiplying effects of the LO port when setting the RF, IF and LO frequencies. In many VNA mixer applications, an entry is available for providing the LO multiplier value, and the programming of the LO frequency automatically takes into account the effect of the multiplier.

In some cases, particularly in mixers used in radar applications, there can be multipliers in the RF path as well. Again, many VNA applications provide for this case. However, this does make the computation of phase response and group delay more difficult.

A degenerate case is that of a straight frequency-doubler, which is like a mixer which has the input and LO driven with the same frequency. The output of frequency in such a case moves twice as fast as the input frequency, and makes computation of group delay a little more complicated. In the case where a multiplier is used in the RF path, the phase of the output signal must be divided by the multiplying effect in order for the group delay to be computed properly, relative to the sweeping input frequency. Calibration for phase or delay measurements is particularly difficult as there are no reciprocal devices that can be used as frequency multipliers in one direction, and dividers in the other. One *cannot* use a mixer driven with the same frequency in the RF and LO, as sweeping the LO frequency produces an additional phase shift in the mixer – see Eq. (7.4) – that is not part of the RF to IF path.

Currently, one of the few ways to measure the delay of a frequency multiplier is to use the coherent receiver method of Section 7.4.3.3 calibrated with a phase reference method (Section 7.5.3.3). An example magnitude, phase and group delay measurement of a small, solid state frequency-doubler is shown in Figure 7.63. This system was calibrated as per the phase-reference method, and treated as a mixer with a multiplier of two in the RF, and an LO frequency of 0 Hz. The response shows a small delay that is nearly equivalent to the physical size of the converter.

One can verify that the delay measurement is valid by adding a known delay at either the input or output port. If the phase multiplication is properly accounted for, then the change in delay for the doubler will be correct whether the delay element is added at the input or the output. If one used a normal VNA measurement in frequency offset mode, one may see twice the effective delay in the output as in the input, thus demonstrating an error in the measurement.

7.9.2 Segmented Sweeps

One feature of a VNA that is particularly convenient in making mixer measurements is the ability to define arbitrary segments of frequency sweeps, with the sources and receivers being able to be set independently for each segment. This allows a complex characterization of a mixer to be completed with a single-sweep data acquisition and a single calibration. (The

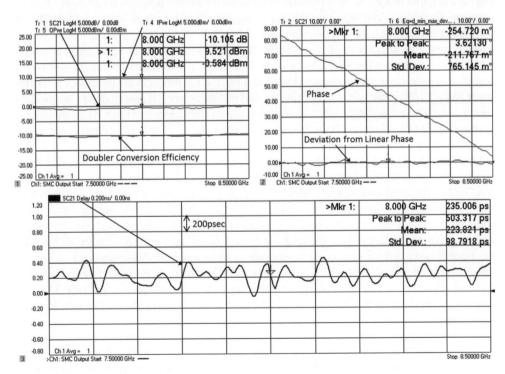

Figure 7.63   Magnitude, phase and delay response of a frequency doubler.

alternative, of course, is setting up many channels, each one representing a single segment condition, but this also requires a separate calibration for each channel.) Specialized display modes in segmented sweeps provide for easy comparison between segment conditions. Some particular examples using segment sweep are discussed in the next sections.

7.9.3 Measuring Higher-Order Products

For mixers and frequency converters used as receivers, it is normal to see some higher-order products cross into the band of interest as the receiver is swept across frequency. Many manufacturers specify the level of higher-order products for particular frequencies, and these are often displayed in an N × M array where the frequency associated with the array position is $M \cdot LO - N \cdot RF$; typically the array elements are in units of dBc below the $M = N = 1$ element, which is the power of the primary IF. Products where $M = N$ are the harmonics of the IF. Using traditional sources and spectrum analyzer techniques, one can directly measure these elements, but the measurements tend to be slow as the SA must be programmed individually for each frequency.

With a VNA, the segmented sweep function can be programed so that the RF and LO frequencies are fixed at a single value, or swept over a single range of frequencies, but each segment of measurement has the receiver swept over the mixer product. Thus, the higher-order products can be directly measured. If one of the segments represents the IF, the other segments can be compared using a second channel to measure the output power and equation editor to compute dBc values. The segment table for a mixer measurement with a 10–10.5 GHz RF, 8 GHz LO and 2–2.5 GHz IF is shown in Figure 7.64.

An example measurement of this type is shown in Figure 7.65, using a segment sweep with the x-axis set to point spacing. Thus, each segment will appear as a direct point order, regardless of measurement frequency. In this display, the RF products are measured at 0 dBm input and −10 dBm input; it is interesting to note that the lower RF power affects the product power only for products that are a multiple of the RF frequency, and not when they are a multiple of the LO frequency. Trace 2 is acquired on a separate sweep, where the output power is measured for each segment (same value as the first segment of trace 1).

One caution when setting up measurements to measure higher-order products is that these products can sometimes go out of range of the measuring instruments (either too high or too low) and so the valid data is not available for all product frequencies. In some cases, it will be up to the user to program segments so that the higher-order products remain in-band. Also, some products may land exactly on other products especially at band edges if round numbers are used. For example, the start of band for the 4:3 spur (last segment) exactly aligns with the main IF tone, and so one sees a large spike in the last segment measurement. A similar effect occurs on the third harmonic of the IF, which exactly lands on the 2:1 spur. Here, the 2:1 spur is higher than the third harmonic. Of course, the LO feed-through products increase with lower RF drive (in dBc terms), as the IF output power is lower but the LO feed-through power remains the same.

From this measurement setup, it is very easy to do other evaluations. For example, it is common wisdom (though often not correct) that increasing the LO drive should make the mixer more linear, and therefore a larger LO drive may lower the spurious products. In

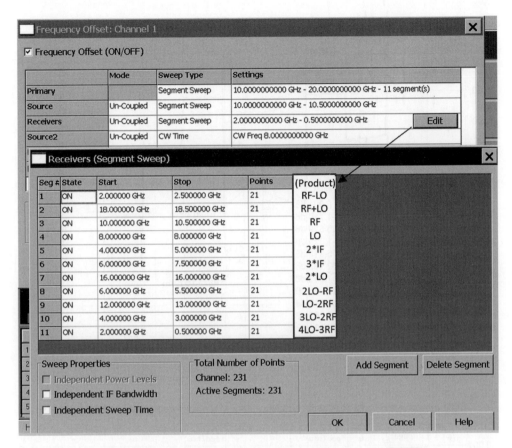

Figure 7.64   Segment table for higher-order products measurement.

Figure 7.66, the LO power is changed while keeping the RF power at a constant level of 0 dBm. It is remarkable that many of the higher-order products actually go up with higher drive. In fact, this makes sense for any products that are a multiple of the LO drive; the harmonics of the LO increase with increasing LO drive, as is clearly demonstrated by the LO feed through elements.

In fact, the only products that are reduced with higher LO drive power are higher-order products of the RF signal, with the IF output third harmonic (3*IF) showing the most (though meager) improvement. The second harmonic of the IF, in fact, gets worse with higher LO drive.

Using similar techniques, a wide range of mixer product measurements can be performed simply and quickly. The entire measurement cycle time for this measurement is less than 250 msec. Because each segment can have an independent IF BW, one can decrease it for higher-order products down to as low as 1 Hz, and obtain more than 100 dB of dynamic range (approximately −120 dBm noise floor) for these measurements, without slowing down the measurements of the other products.

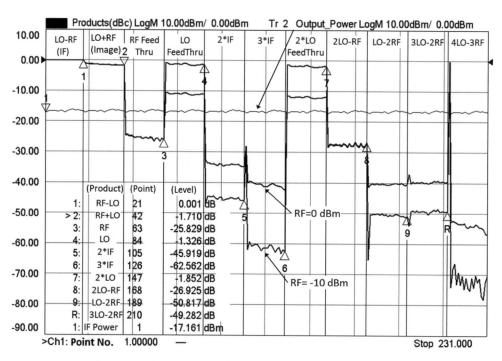

Figure 7.65   Higher-order products measured using overlapped segments.

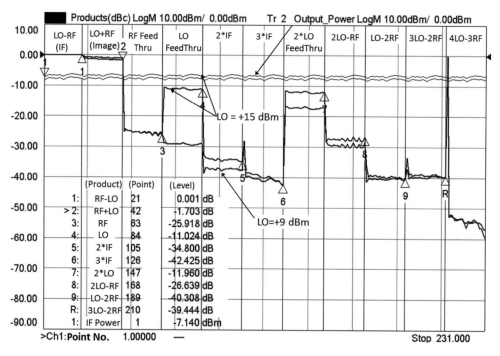

Figure 7.66   Higher-order products as a function of LO drive power.

7.9.3.1 Swept LO Group Delay

Broadband mixers are often tested with sweeping LO and RF frequencies at a fixed IF frequency so that a single measurement can show the behavior of a mixer as a block down-converter. This works fine for amplitude response, but sweeping the LO frequency does not allow valid phase or group delay responses to be measured. Using segment sweeps can solve the problem of swept LO measurements for group delay by creating segments with a fixed LO, swept-RF and swept-IF measurement, over all of the LO frequencies. While tedious to do using the GUI, it is a straightforward setup using remote-programing commands, and will provide the proper measurement of the delay at each LO frequency. An example of such a measurement is shown in Figure 7.67, where a portion of the segmented frequency table is shown along with the group delay result. In this case a 201 point LO sweep was desired, so 201 segments are required.

In this measurement example, the VNA used is an Agilent PNA, and it treats the delay format somewhat differently in a segment sweep mode that is particularly useful in this application. As the frequency spacing, and even LO frequency, can be different in each segment, the group delay computation is performed on a segment-by-segment basis, and so the delta phase/delta frequency computation occurs only in a segment and doesn't cross segment boundaries. In this example, exactly two points are used in each segment, with the frequency spacing of the two points at each segment providing the delay aperture. Each segment computes the delay

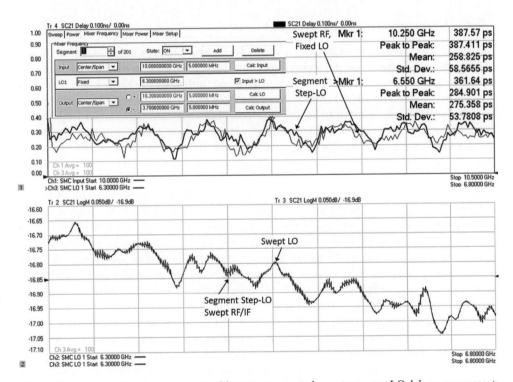

Figure 7.67 (upper) Segment sweep provides a way to properly create a swept LO delay measurement; (lower) the magnitude response for normal sweep, and segmented sweep.

independently, and as in normal delay formatting, either the first or last point is repeated so that the delay format has the same number of points as the underlying measurement. Thus each segment computes a different delay for the different LO, and for a two-point segment, both points are assigned the same value. Displaying the segmented sweep with the LO frequency as the x-axis provides for a direct measurement of group delay versus swept LO frequency, for a single fixed IF measurement. The calibration of this measurement is best performed using the phase reference method, as it does not require any wideband reciprocal mixer, but any of the phase measurement methods can be used as well. Also shown in the upper window is a measurement of the converter delay with a fixed LO, sweeping the RF and IF. The fact that both traces are very similar (within 100 psec) indicates that the RF frequency range contributes most to the ripple and response of the group delay.

The lower window shows an interesting crosscheck of the segmented sweep. One trace is for a swept RF, swept LO, fixed IF measurement and it shows the amplitude response. The other trace is for the segmented sweep. The trace is exactly as expected, as the segment sweep measures the RF to IF response, at each LO frequency, but sweeping the RF from just below to just above the normal RF frequency from the standard measurement. This allows computation of the delay response at each stepped LO frequency, and the fact that the saw-tooth wave exactly crosses the normal swept LO response trace indicates a valid setup, measurement and calibration for this test case.

7.9.4   Mixers with an Embedded LO

All of the mixer measurements presented so far rely on absolute knowledge of the RF, IF and LO frequencies. Some frequency converters (especially ones used in satellite applications) have their own internally supplied local oscillator, and further, this LO is not accessible and may not have a common 10 MHz reference for the use with the test system. This measurement scenario is commonly referred to as an "embedded LO" use case, and further implies that the LO frequency may not be well known. Even in the highest quality instruments, without locking the common references together, their internal RF frequencies may differ by many kHz. Consider that a source with a frequency accuracy of 1 part per million will have an error of 30 kHz at 30 GHz.

When measuring a converter with an embedded LO, the IF frequency can be offset by this difference in frequencies, when the embedded LO is not derived from the same reference as the test system source. For amplitude measurements, this usually doesn't present much of a problem because the IF BW or resolution BW of the measurement receiver can be widened to always capture the IF signal (although it will lead to higher trace noise). In phase measurements, however, even 1 Hz of offset will give 360° of phase change per second of measurement time, and thus virtually no offset can be tolerated between the defined frequencies in the test equipment and the frequency of the embedded LO.

One technique that can overcome the problem of LO offsets in some modern VNAs is the use of a software signal tracking and phase locking mechanism. In this technique, one point of the sweep, typically the center point, is designated as the software locking point. Before each measurement sweep for mixer response occurs, some background sweeps are performed where the LO frequency is determined with great accuracy, and in a very fast manner. This is accomplished by first taking a course sweep of the receiver while holding the input and LO frequency fixed. This gives an approximate IF frequency (to the resolution of the receiver's IF BW filter) from which a first estimate of the LO frequency is derived, as illustrated in Figure 7.68. Next, the receiver is fixed tuned to the estimated IF frequency, and a phase vs time sweep is performed, in a so-called precise mode. The slope of this response is the frequency error between the estimated LO frequency and the actual frequency. The resolution of this sweep is the trace noise of the phase sweep divided by the sweep time.

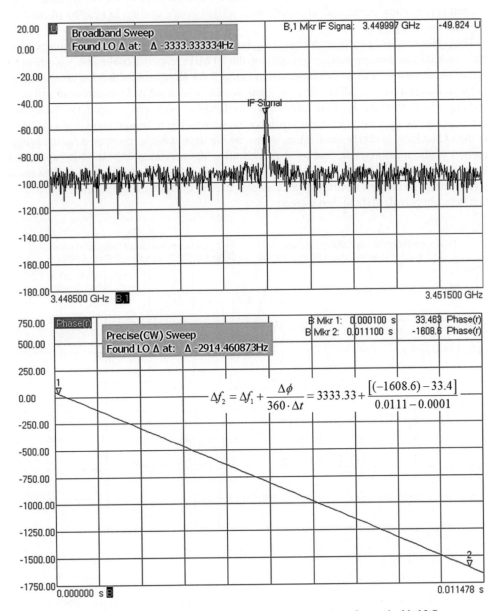

Figure 7.68 Background acquisitions for software locking of an embedded LO.

The sweep time is set by the inverse of the receiver bandwidth and the number of measurement points. If the phase change is large, the offset frequency is recomputed and another phase vs time sweep is acquired; this continues until the desired offset error is achieved. In practice, and offset frequency of less than 0.3 Hz is required to have less than 100 psec of error in absolute delay for a typical measurement. In this implementation, the IF BW is chosen so that 0.3 Hz offset produces less than a 1° change in phase. The phase vs time sweeps are also shown in Figure 7.68. For this implementation, the acquisition of the LO frequency can happen very quickly, perhaps less than 50 msec, so that even a drifting LO can be tracked on a sweep-by-sweep basis.

The performance of a software locking loop is quite good, and at these rates it emulates quite well the hardware locking mechanism of tying the 10 MHz reference of the test equipment to that of the converter's LO, as demonstrated in Figure 7.69. Here a mixer that has an embedded LO which does have a 10 MHz reference input, is measured in the two conditions of software locking, and locking the 10 MHz references together. The measured group delay values are identical and the trace noise of the group delay traces, which is a measurement of the stability of the LO locking, is nearly identical, as shown by the trace statistics computed for the center portion of the measurement. Marker 2 is measuring the maximum deviation between two sweeps, both with a 10 MHz lock. Marker 3 is measuring the maximum deviation between a software lock sweep and a hardware lock sweep; in fact, the maximum deviation is less than that between two 10 MHz locked sweeps.

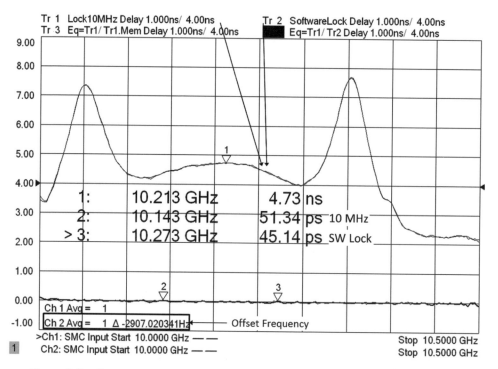

Figure 7.69   Comparison of software vs hardware locking for embedded LO measurements.

Also shown in the Channel 2 display is the offset frequency when the 10 MHz signals are not locked together. This drifts 2–5 Hz over a few seconds, and 5–10 Hz over several minutes. This software locking technique provides an alternative to the modulation measurement technique described in Section 7.4.4, and can be performed with a standard VNA, without requiring a modulated or two-tone source.

7.9.5 High-Gain and High-Power Converters

Another use case that requires special consideration is the measurement of converters with either high gain or high power, or measurements at the end of lossy or long test cables. In these cases, special care must be taken to avoid overdriving the converter, and as such, some of the match correction techniques described in Section 7.5.1 can cause more harm than good. One implication of a high gain converter is that the input power is typically quite low; some satellite converters have input powers below −100 dBm. In this case, the S11 or input match measurement can be very noisy indeed. For high power converters, the output signal may need to be attenuated so that the output does not overdrive the measurement receivers. This attenuation can cause the measurement of S22 to become noisy and thus to embed that noise on the SC21 trace through the output match correction. In these cases, turning off the input and output match corrections can improve the measurement results, and some VNA measurement applications provide this as an optional choice in correction method.

Generating an amplitude calibration with a high-gain converter can be problematic, as the typical method for calibrating the input power is to use a power meter to measure the source power at the RF frequency. But there aren't any commercial power meters that can measure a signal at −100 dBm, so a modification to the normal techniques needs to be used.

First, for very low source power levels, it is best to reverse the port 1 test-port coupler, to lower the drive signal while improving the noise performance of the S11 measurement. Next, the VNA source attenuator needs to be set to a level such that the source ALC will provide the desired output power at its minimum level, then the power meter acquisition can be performed at its maximum level. For most commercial power sensors, even with high sensitivity heads, the minimum power is −70 dBm and one should stay 10–20 dB above the minimum power to avoid noise in the power reading from degrading the power calibration. This means the minimum power for calibration is on the order of −60 to −50 dBm. Most modern VNAs provide about 40 dB of ALC range, and so the minimum power for measurements can achieve the −100 dBm level. The details of these setups for low-power source drive are discussed in Section 6.3.

An alternative method for calibrating the input power at low powers when high sensitivity power sensor is not available, is to calibrate at a higher power using a higher attenuator setting, then change the attenuator setting after calibration. The error due to the change in attenuator setting can be de-embedded from the final measurement by characterizing the difference in attenuator settings using a standard VNA channel, over the input frequency range. This is accomplished by performing a two-port calibration in the low attenuator setting (not 0 dB), and then without changing any other setting, change just the source attenuator and save the S2P file. This now contains the error in the attenuator setting (there is slight error in the source match which can usually be ignored if the low attenuator setting is not 0 dB). The mixer calibration is performed at the low attenuator setting, and then the attenuator can be changed and the S2P file de-embedded from the mixer calibration to create a high quality calibrated measurement at a very low source power setting. This is very similar to the method described in Section 3.14.3.

When measuring high power mixers, it may be necessary to add an attenuator to the output of the mixer to reduce the power applied to the receiver. Calibration in this case can take one of two forms: first the mixer calibration can be performed with the attenuator in place. The difficulty with this method is that if the mixer is also a high-gain mixer, the source drive during calibration will be quite low and the added loss of the attenuator can cause substantial trace noise in the calibration response. In such a case, the only option is to increase averaging or reduce IF BW during calibration.

An alternative method is to first perform a two-port calibration at high power as described above, and then calibrate with the attenuator in place, changing the source attenuator after calibration.

7.10   Conclusions on Mixer Measurements

Mixer and frequency converters can have very complex responses, and are particularly demanding to characterize. Bare-mixers have a wide variety of conversion products that emanate from all the ports, and remixing of these products can cause ripples and other errors in mixer characterizations. The performance of mixers depends in complicated ways on the RF and LO drives, and one must often characterize each desired condition because the behavior is not necessarily well-controlled, especially across a wide frequency band.

Besides gain, the phase and delay responses of mixers require great care in calibration and measurement. Several different test and calibration methods have been discussed, and the interested reader is recommended to pursue further study through the references provided. The test methods for these parameters are being constantly improved, and great progress has been made in understanding and simplifying converter delay tests.

Mixer distortion has likewise seen major improvements in the measurement methods, many of which have been described here. In particular, the ability to quantify distortion as a function of RF or LO drive has been greatly simplified, while the accuracy of the measurements has improved.

Mixer noise figure remains a difficult measurement, and great care is still required in setting up and making these measurements, particularly on low-gain converters. The specific construction of the mixer, image rejection and LO noise considerations must all be taken into account to understand the quality of the NF measurement.

Finally, many special cases of frequency converter testing including high-gain, high-power and embedded LO have been reviewed. Many of the specialized test methods for these conditions are derived from similar requirements on amplifier test, so one may want to review the material from Chapter 6 to gain further insight into these situations.

While demanding, the theoretical basis for mixer behavior is now relatively well understood, and the test engineer facing some new challenge is advised to return to these first principles. The basics of test remain the same: understand the device through pretest, optimize the measurement system, calibrate with appropriate standards and analyze the measurements for consistent, reasonable results.

References

1. Maas, S.A. (1986) *Microwave Mixers*, Artech House, Dedham, MA. Print.
2. Dunsmore, J., Hubert, S., and Williams, D. (2004) Vector mixer characterization for image mixers. Microwave Symposium Digest, 2004 IEEE MTT-S International, vol.3, no., pp. 1743–1746 vol. 3, 6–11 June 2004.
3. Williams, D.F., Ndagijimana, F., Remley, K.A., *et al.* (2005) Scattering-parameter models and representations for microwave mixers. *IEEE Transactions on Microwave Theory and Techniques*, 53(1), 314–321.
4. Dunsmore, J. Understanding uncertainties in mixer measurements. WHWE11 (EuMC 2009 Workshop) Practical Approaches to Achieving Confidence in Microwave Measurements; workshop notes.

8

VNA Balanced Measurements

8.1 Four-Port Differential and Balanced S-Parameters

Traditionally, common RF structures involved single input and single output devices, with a common ground reference. But with the advent of advanced monolithic microwave integrated circuits (MMIC) and the move to using CMOS technology to higher frequencies, more and more RF circuits are being designed using differential devices. Even computer backplanes and clock rates are reaching such speeds that RF and microwave considerations must be taken into account. Because of this, differential or balanced S-parameters have become an important area of RF and microwave research and application. Fortunately, the theoretical basis for differential S-parameter theory has been well founded, and a consistent definition has been accepted [1].

Consider the network described in Figure 8.1. It is a four-port network, with an associated 16-term S-parameter matrix. The input and output waves are defined by the matrix formulation

$$
\begin{bmatrix} b_1 \\ b_2 \\ b_3 \\ b_4 \end{bmatrix} = \begin{bmatrix} S_{11} & S_{12} & S_{13} & S_{14} \\ S_{21} & S_{22} & S_{23} & S_{24} \\ S_{31} & S_{32} & S_{33} & S_{34} \\ S_{41} & S_{42} & S_{43} & S_{44} \end{bmatrix} \cdot \begin{bmatrix} a_1 \\ a_2 \\ a_3 \\ a_4 \end{bmatrix}
\tag{8.1}
$$

But for a differential amplifier, the input differential port is considered to comprise both ports 1 and 3, and the output differential port is composed of both ports 2 and 4. Note that port numbering here is arbitrary, and another choice for definition of input and output differential ports is ports 1 and 2 at the input and ports 3 and 4 at the output. The original reference paper used this definition, but conventional test equipment often puts the input ports as ports 1 and 3, and the output ports as 2 and 4, and this has become commonly accepted. The literature is about equally split on the definition; so, following common industry practice, the odd ports (1 and 3) will be defined here as the inputs, and the even ports (2 and 4) will be defined as the outputs. Also, it must be recognized that whenever a port-pair is described as a differential port, and there exists a common ground node, then there can also exist a common mode signal on the port pair. Thus, all four-port devices that also have a common ground node, must be

Handbook of Microwave Component Measurements: With Advanced VNA Techniques, First Edition. Joel P. Dunsmore.
© 2012 John Wiley & Sons, Ltd. Published 2012 by John Wiley & Sons, Ltd.

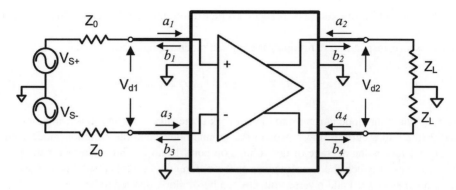

Figure 8.1 A four-port network used as a balanced amplifier.

properly described as having both a differential mode and a common mode on each port pair, which is commonly called mixed mode.

From this description, it must be recognized that a new definition for differential input waves and differential output waves must be created. In this case, one can define the differential incident (or forward) waves and scattered (or reflected) waves from those shown in Figure 8.1 based on the differential input voltage and differential input current

$$V_d^F = \frac{1}{2}(V_d + I_d Z_d), \quad V_d^R = \frac{1}{2}(V_d - I_d Z_d) \tag{8.2}$$

where V_d and I_d are defined as the difference in input voltage and the difference of input currents between ports 1 and 3, as shown in the figure. If the node voltages are V_1 and V_3, and the node currents are I_1 and I_3, then the differential input voltage and current can be defined as

$$V_d = V_1 - V_3, \quad I_d = \frac{1}{2}(I_1 - I_3) \tag{8.3}$$

The definition of V_d is intuitive, but the factor of $^1/_2$ in the definition of I_d is a little less intuitive and needs further explanation. In a single ended ground-referenced circuit, the input voltage is the difference between the voltage on the node and ground, and the current is the current flowing into just the input node; the ground current isn't considered when computing the input current. Similarly, the differential input voltage is the difference between the two inputs, with port 3 playing the role of ground; if the circuit were purely differential, the current into port 1 would equal the current out of port 3. However, the internal connections of the network can allow different current to enter port 1 and exit port 3; for example, port 1 could be connected to ground through a resistor and port 3 could be open circuit. In such a case, there would be no "differential ground" current flowing out port 3 so how does one compute a differential current? This is done by presuming an averaging of the current flowing into port 1 and out of port 3 as the differential current. The "left-over" is due to common mode current flowing into the network and out of the actual ground node. Thus one half of the current flowing into port 1 is differential, and one half is common, and one half the current flowing out of port 3 is differential and one half is common. Therefore the average differential current is one half of port 1 current plus one half of port 3 current. Finally, recognize that differential current out of port 3 is the same as $-I_3$ so the average differential current is should be described as in (8.3).

There is a corresponding definition for common mode voltage and current

$$I_c = I_1 + I_3, \quad V_c = \frac{1}{2}(V_1 + V_3) \tag{8.4}$$

where the common mode current into the network is the sum of the currents going into ports 1 and 3 (the return current flows through the ground node), and the common mode voltage into the network is the average of the voltage on ports 1 and 3. Thus, take one half of each of the voltages, in much the same manner as the differential current was defined above. The common mode forward and reverse voltages can be similarly defined to be

$$V_c^F = \frac{1}{2}(V_c + I_c Z_c), \quad V_c^R = \frac{1}{2}(V_c - I_c Z_c) \tag{8.5}$$

The final aspect of the definition of differential and common mode circuits is in the definition of the differential mode reference impedance as it appears in (8.2), and the common mode reference impedance as it appears in (8.5). The impedance of a single ended termination is simply the node voltage divided by the current into the node. A differential termination can be defined to say that it is the differential voltage divided by the differential current into the node, whereas the common mode impedance is the common mode voltage divided by the common mode current, and thus one finds

$$Z_d = 2Z_0, \quad Z_c = \frac{Z_0}{2} \tag{8.6}$$

assuming that the source impedance for both sources at the port pair is Z_0, and the sources are independent, that is, there is no coupling elements between the sources.

The final thing needed now is a definition of differential and common mode S-parameters, and they are derived from our earlier definitions above, and a new definition of mixed mode a and b waves

$$a_d = \frac{V_d^F}{\sqrt{Z_d}}, \quad b_d = \frac{V_d^R}{\sqrt{Z_d}}, \quad a_c = \frac{V_c^F}{\sqrt{Z_c}}, \quad b_c = \frac{V_c^R}{\sqrt{Z_c}} \tag{8.7}$$

This is presuming Z_d and Z_c are real. From this the S-parameters can be defined as

$$\begin{bmatrix} b_{d1} \\ b_{d2} \end{bmatrix} = \begin{bmatrix} S_{dd11} & S_{dd12} \\ S_{dd21} & S_{dd22} \end{bmatrix} \cdot \begin{bmatrix} a_{d1} \\ a_{d2} \end{bmatrix}, \quad \begin{bmatrix} b_{c1} \\ b_{c2} \end{bmatrix} = \begin{bmatrix} S_{cc11} & S_{cc12} \\ S_{cc21} & S_{cc22} \end{bmatrix} \cdot \begin{bmatrix} a_{c1} \\ a_{c2} \end{bmatrix} \tag{8.8}$$

Here there is also a port numbering associated with the mixed mode waves, where port $d1$ and $c1$ are the differential and common input ports, comprising ports 1 and 3, and $d2$ and $c2$ the differential and common output ports, respectively, comprising ports 2 and 4. In addition to the differential mode S-parameters and the common mode S-parameters of Eq. (8.8), it is entirely reasonable to consider a network that is driven in one mode, but outputs a signal in another mode. The cross mode parameters come in two versions: drive a differential incident wave and measure a common mode scattered wave; or drive a common mode incident wave and measure a differential mode scattered wave. They are defined as

$$
\begin{bmatrix} b_{c1} \\ b_{c2} \end{bmatrix} = \begin{bmatrix} S_{cd11} & S_{cd12} \\ S_{cd21} & S_{cd22} \end{bmatrix} \cdot \begin{bmatrix} a_{d1} \\ a_{d2} \end{bmatrix}, \quad \begin{bmatrix} b_{d1} \\ b_{d2} \end{bmatrix} = \begin{bmatrix} S_{dc11} & S_{dc12} \\ S_{dc21} & S_{dc22} \end{bmatrix} \cdot \begin{bmatrix} a_{c1} \\ a_{c2} \end{bmatrix}
$$
(8.9)

The 16 mixed mode S-parameters are all needed to fully describe a four-port network, and are often shown in matrix form as

$$
\begin{bmatrix} b_{d1} \\ b_{d2} \\ b_{c1} \\ b_{c2} \end{bmatrix} = \begin{bmatrix} S_{dd11} & S_{dd12} & S_{dc11} & S_{dc12} \\ S_{dd21} & S_{dd22} & S_{dc21} & S_{dc22} \\ S_{cd11} & S_{cd12} & S_{cc11} & S_{cc12} \\ S_{cd21} & S_{cd22} & S_{cc21} & S_{cc22} \end{bmatrix} \cdot \begin{bmatrix} a_{d1} \\ a_{c2} \\ a_{c1} \\ a_{c2} \end{bmatrix}
$$
(8.10)

With this, the definition of mixed mode S-parameters is nearly complete, and in fact many references stop here. However, it would be convenient to be able to describe the mixed mode S-parameters in terms of the single ended S-parameters. One can use the previously defined waves to make the following observations; from Eqs. (1.8) and (1.9), and (8.2) and (8.5) one can derive

$$
a_{d1} = \frac{(a_1 - a_3)}{\sqrt{2}}, \quad a_{c1} = \frac{(a_1 + a_3)}{\sqrt{2}}, \quad b_{d1} = \frac{(b_1 - b_3)}{\sqrt{2}}, \quad b_{c1} = \frac{(b_1 + b_3)}{\sqrt{2}}
$$
(8.11)

and

$$
S_{dd11} = \frac{b_{d1}}{a_{d1}} \bigg|_{a_{c1} = a_{d2} = a_{c2} = 0}
$$
(8.12)

And recognizing that $a_{c1} = 0$ implies that $a_3 = -a_1$ so that $a_{d1} = 2a_1$ then

$$
S_{dd11} = \left(\frac{b_1 - b_3}{a_1 - a_3} \right) = \left(\frac{b_1 - b_3}{a_1 - (-a_1)} \right) = \left(\frac{b_1 - b_3}{2a_1} \right)
$$
(8.13)

This is combined with a version of Eq. (1.17) considered for ports 1 and 3 to obtain

$$
S_{dd11} = \left(\frac{b_1 - b_3}{2a_1} \right) = \left(\frac{[S_{11}a_1 + S_{13}a_3] - [S_{31}a_1 + S_{33}a_3]}{2a_1} \right)
$$

$$
= \left(\frac{[S_{11}a_1 - S_{13}a_1] - [S_{31}a_1 - S_{33}a_1]}{2a_1} \right)
$$
(8.14)

And factoring out the common a_1 yields

$$
S_{dd11} = \frac{1}{2}(S_{11} - S_{13} - S_{31} + S_{33})
$$
(8.15)

A similar computation can be performed for each of the mixed mode parameters to find their equivalence in the single ended S-parameters. Sdd21 is the most important attribute of a balanced amplifier and one can compute its value as a function of single ended parameters by recognizing that

$$
S_{dd21} = \left(\frac{b_2 - b_4}{a_1 - a_3} \right) = \left(\frac{b_2 - b_4}{a_1 - (-a_1)} \right) = \left(\frac{b_2 - b_4}{2a_1} \right)
$$
(8.16)

And following the same logic as (8.14)

$$S_{dd21} = \left(\frac{b_2 - b_4}{2a_1}\right) = \left(\frac{[S_{21}a_1 + S_{23}a_3] - [S_{41}a_1 + S_{43}a_3]}{2a_1}\right)$$

$$= \left(\frac{[S_{21}a_1 - S_{23}a_1] - [S_{41}a_1 - S_{43}a_1]}{2a_1}\right)$$

(8.17)

which, when factored, yields

$$S_{dd21} = \frac{1}{2}(S_{21} - S_{23} - S_{41} + S_{43})$$

(8.18)

The mixed mode conversions are sometimes expressed in the form of a matrix transformation, where

$$[S_{MM}] = [M] \cdot [S] \cdot [M]^{-1},$$

$$[M] = \frac{1}{\sqrt{2}}\begin{bmatrix} 1 & 0 & -1 & 0 \\ 0 & 1 & 0 & -1 \\ 1 & 0 & 1 & 0 \\ 0 & 1 & 0 & 1 \end{bmatrix}, \quad [M]^{-1} = \frac{1}{\sqrt{2}}\begin{bmatrix} 1 & 0 & 1 & 0 \\ 0 & 1 & 0 & 1 \\ -1 & 0 & 1 & 0 \\ 0 & -1 & 0 & 1 \end{bmatrix},$$

(8.19)

A complete list is shown in Table 8.1.

For the most common of the mixed mode S-parameters, the interpretation is straightforward: Sdd21 is the differential gain, Scc21 is the common mode gain, but for the other, cross mode

Table 8.1 Mixed mode S-parameters expressed as single ended S-parameters

Differential mode parameter		Cross mode parameter: Common to differential	
Sdd11	(S11 − S13 − S31 + S33)/2	Sdc11	(S11 + S13 − S31 − S33)/2
Sdd12	(S12 − S14 − S32 + S34)/2	Sdc12	(S12 − S32 + S14 − S34)/2
Sdd21	(S21 − S41 − S23 + S43)/2	Sdc21	(S21 − S41 + S23 − S43)/2
Sdd22	(S22 − S42 − S24 + S44)/2	Sdc22	(S22 − S42 + S24 − S44)/2
Cross mode parameter: Differential to common		Common mode parameter	
Scd11	(S11 + S31 − S13 − S33)/2	Scc11	(S11 + S31 + S13 + S33)/2
Scd12	(S12 + S32 − S14 − S34)/2	Scc12	(S12 + S32 + S14 + S34)/2
Scd21	(S21 + S41 − S23 − S43)/2	Scc21	(S21 + S41 + S23 + S43)/2
Scd22	(S22 + S42 − S24 − S44)/2	Scc22	(S22 + S42 + S24 + S44)/2

parameters, the practical meaning is not so obvious. When a device is driven differentially, it is essentially self-shielded, but if it produces common mode signals at its output terminals, it implies that significant current is flowing through the common ground. This condition may lead to radiated emissions from the device. Thus the Scd parameters are sometimes associated with a measure of potential radiated emissions of the device. Similarly, if a device that is intended to have differential outputs produces such an output signal when there is a common mode signal applied, then it implies that it is susceptible to ground currents, and the Sdc parameters are associated with a measure of its potential external-signal immunity.

8.2   Three-Port Balanced Devices

The mixed mode parameters may also be defined for a three-port network, comprising one single ended port and one mixed mode port which is usually defined as the differential or balanced port. But again it must be recognized that for a device referenced to ground, a balanced port always has the possibility to support a common mode signal as well. A schematic of such a device is shown in Figure 8.2.

The mixed mode parameters for a three-port network are defined by

$$
\begin{bmatrix} b_s \\ b_d \\ b_c \end{bmatrix} = \begin{bmatrix} S_{ss} & S_{sd} & S_{sc} \\ S_{ds} & S_{dd} & S_{dc} \\ S_{cs} & S_{cc} & S_{cc} \end{bmatrix} \cdot \begin{bmatrix} a_s \\ a_d \\ a_c \end{bmatrix} \tag{8.20}
$$

Since for the three-port case the ports are unambiguous, there is no need to use port numbers, and one can refer only to the three port modes, but it is also commonly found in the literature to include the port numbers, especially if defining the three-port, single ended (SE) to balanced properties of a four-port network, where it may have more than one SE input. The values of the three-port mixed mode parameters can also be computed in terms of the single-ended parameters.

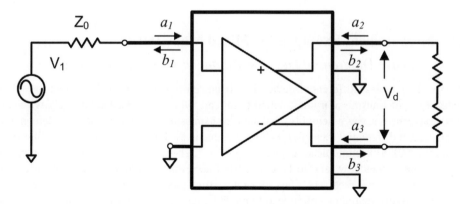

Figure 8.2   A three-port device used as a single ended to balanced device.

The same definitions hold as in Eq. (8.7) for differential and common waves, so that the three-port mixed mode parameters can be computed in terms similarly to Eq. (8.19) but in the three-port case

$$[M] = \frac{1}{\sqrt{2}} \begin{bmatrix} \sqrt{2} & 0 & 0 \\ 0 & 1 & -1 \\ 0 & 1 & 1 \end{bmatrix}, \quad [M]^{-1} = \frac{1}{\sqrt{2}} \begin{bmatrix} \sqrt{2} & 0 & 0 \\ 0 & 1 & 1 \\ 0 & -1 & 1 \end{bmatrix} \tag{8.21}$$

which yields the following conversions for mixed mode parameters

$$\begin{bmatrix} S_{ss} & S_{sd} & S_{sc} \\ S_{ds} & S_{dd} & S_{dc} \\ S_{cs} & S_{cc} & S_{cc} \end{bmatrix} = \begin{bmatrix} S_{11} & \frac{1}{\sqrt{2}}(S_{12} - S_{13}) & \frac{1}{\sqrt{2}}(S_{12} + S_{13}) \\ \frac{1}{\sqrt{2}}(S_{21} - S_{31}) & \frac{1}{2}(S_{22} - S_{23} - S_{32} + S_{33}) & \frac{1}{2}(S_{22} + S_{23} - S_{32} - S_{33}) \\ \frac{1}{\sqrt{2}}(S_{21} + S_{31}) & \frac{1}{2}(S_{22} - S_{23} + S_{32} - S_{33}) & \frac{1}{2}(S_{22} + S_{23} + S_{32} + S_{33}) \end{bmatrix}$$

$$\tag{8.22}$$

The most common use of a single ended to differential device is as a balun (BALanced-to-UNbalanced transformer), used to drive a differential device from a single ended measurement instrument. In the past, when VNAs were limited to only two ports, baluns were used extensively, and sometimes incorrectly, in making differential measurements. With the advent of four-port VNAs, the need for baluns in characterization of linear, passive measurements was essentially eliminated. However, even today they remain a key component for testing other more complicated characteristics such as compression, distortion and noise figure.

The concept of mixed mode parameters is not limited to three- or four-port devices. In fact, they can be extended to an arbitrary number of ports. The convenience of mixed mode parameters as developed is that the familiar formulations for computing with single ended S-parameter applies just as well to mixed mode S-parameters including concepts for maximum power transfer, stability, and the effects of cascading networks and de-embedding networks.

8.3   Measurement Examples for Mixed Mode Devices

8.3.1   Passive Differential Devices: Balanced Transmission Lines

Perhaps the simplest differential device is a balanced transmission line. In fact one of the oldest styles of transmission lines, called twin-lead, which was used extensively with early television receivers, was essentially a balanced transmission line. The modern implementation of balanced or differential transmission lines is found in the high-speed low voltage differential signaling (LVDS) communication paths found in high-speed digital and communications circuits. These lines now tend to be dual PCB traces that interconnect pairs of differential drivers to differential receivers. One very interesting aspect of these signaling lines is that making them differential can often lead to much better signal-to-noise performance, reduced interference and improved frequency response compared to single ended transmission lines. Of this, the self-shielding aspect may be the most important.

Figure 8.3 A test PCB for characterizing differential lines.

Figure 8.3 shows an example of a test board used for evaluating PCB transmission lines characteristics. With current speeds of digital devices approaching 3 GHz clock rates, and data transmission at 10 Gbits/sec, the integrity of these signaling lines as they pass between board layers, around PCB vias, and transition through connectors is becoming critical to understand. This is often called the "physical-layer" of a communications stack which includes modulation, formatting, paging and framing, all of which are higher orders of processing the raw signals to improve reliability of signaling. This area of study has become known as "signal integrity" and is a separate, but closely related field to that of the RF and microwave communications engineer.

This test board has a design artifact where the traces are narrowed in one region, perhaps to represent an area where a smaller footprint on the PCB is required. A full four-port S-parameter characterization was performed, with the 16 S-parameters shown in Figure 8.4.

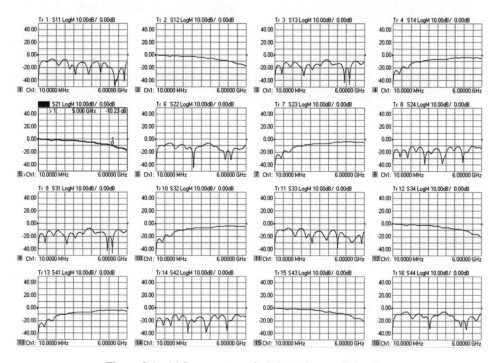

Figure 8.4 16 S-parameters of a balanced transmission line.

In the test boards, the input and output lines are narrowed as the lines form a differential pair, so that the differential mode impedance is maintained at a constant value.

While this display is very busy, it is useful in noting that the one may quickly see if there are any unusual aspects of the measurement. One apparent issue is the high loss in the transmission paths, S21 and S43, on trace 5 (highlighted) and 15. Further, there is very strong coupling between the lines as evinced by S41 and S32 (traces 13 and 10). From a cursory look at the single-ended transmission lines, one might conclude that these traces can only be used up to about 3 GHz, where S21 goes below −3 dB.

However, when the mixed mode S-parameters are examined, a different story unfolds. Figure 8.5 shows a comparison of Sdd21 with the two of the transmission parameters associated with port 2, S21 and S23. From this result, one can see that the differential transmission is very good, less than −3 dB, beyond 5 GHz. In fact, because of the particular phasing of the signals, the coupling of S23 exactly compensates for the loss of S21 so that the differential mode of transmission sees much less loss than the single ended mode.

There is still some ripple and loss in the differential Sdd21 (likely due to the stepped impedance test feature), and evaluation of other differential parameters can shed light on the attributes of the differential line that contribute to the loss in Sdd21.

Figure 8.6 shows a display of all the 16 mixed mode parameters, with each mode in its own window. The cross mode terms show very little signal, indicating that the lines are quite well balanced, which mean that they are well shielded, even if they do have significant mismatch

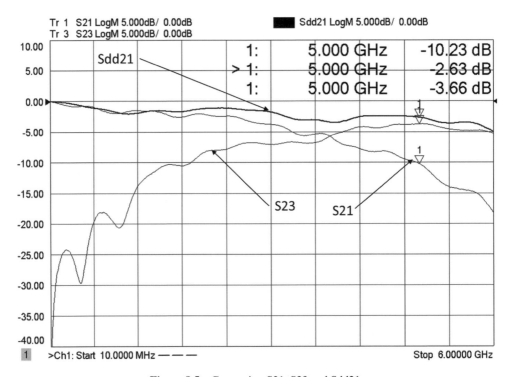

Figure 8.5   Comparing S21, S23 and Sdd21.

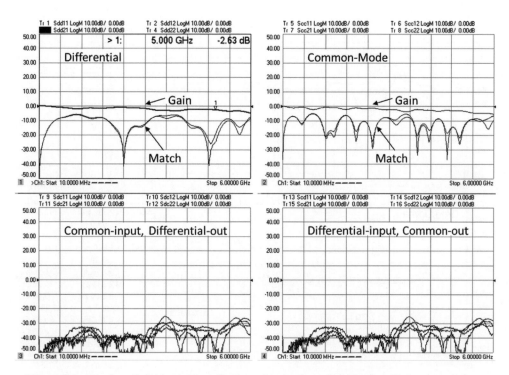

Figure 8.6   Display of all 16 mixed mode parameters; each mode is displayed in its own window.

in the differential mode as exhibited by the Sdd11, and in the common mode, as shown by the Scc11 traces.

Mixed-mode parameters allow analysis in exactly the same manner that single ended S-parameters allow, so one may investigate attributes of differential lines using all the tools of S-parameter analysis, including time domain transforms.

In Figure 8.7 the upper window shows Sdd11 and Scc11, the lower window shows the TDR of each. From the TDR, it is very clear that the ripple in the transmission, and the mismatch in the reflection is due to the discrete change in line impedance that occurs at about 250 mm down the line (as indicated by marker 2). This plot illustrates several interesting points. First, note that the marker readout displays have been changed from default to read the equivalent impedance (essentially a Smith chart marker), and is a convenient marker feature found on some modern VNA. This provides for directly reading the impedance of the lines. One can see that at 0 time (the first graticule line), the differential impedance is nearly matched, but the common mode impedance is *not* matched. This also implies that the single ended result will not be matched either. Because the lines are tightly coupled (the S23 trace of Figure 8.5 demonstrates this), the SE line impedance must be modified to maintain a constant differential impedance. From Figure 8.3, one can tell that the line widths have been narrowed where the differential portion starts (the PCB was measured with differential port 1 on the right).

The step in impedance due to the thin sections of line are clearly shown in both TDR plots, but it is interesting to see that the step is slightly delayed in the common mode transmission, when compared to the differential mode transmission. That is because the field distribution

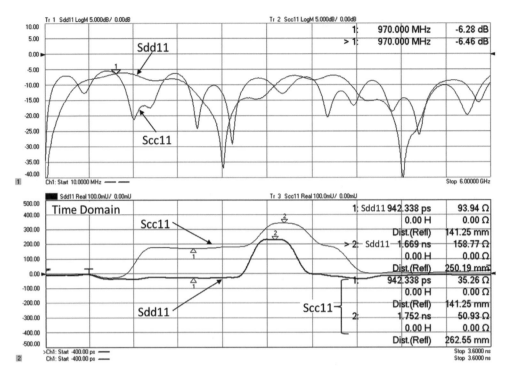

Figure 8.7 (upper) Frequency domain; (lower) time domain of Sdd11 and Scc11.

in the differential line has substantial fields in the air above the lines. The common mode transmission, from line to ground, has the fields substantially more in the PCB, with its higher dielectric constant. Thus, the velocity factor in the common mode is slower than in the differential mode, and this is reflected in the TDR.

For linear, passive devices, making balanced or differential measurements is very straightforward using the newest four-port VNAs. All of the fixturing, de-embedding and impedance transformation that is possible with standard S-parameters is also available for mixed mode parameters. Since devices are linear, making single ended fully corrected S-parameter measurements, and converting to mixer-mode parameters gives exactly correct results. Whether this also holds true for active devices has been the topic of a great deal on speculation, but recent studies have set very good guidelines for applying mixed mode analysis to active devices, as discussed in the next section.

8.3.2 Differential Amplifier Measurements

Differential amplifiers are common in low-frequency electronics; known as operational amplifiers or op-amps, they have extremely high differential gain to a single ended output, and utilize feedback from the output to the input to perform many useful functions. However, their performance characteristics and definitions are almost entirely *unrelated* to the RF and microwave usage of the term "differential amplifier". For most RF and microwave work, a differential amplifier refers to one of moderate gain (~20 dB) with balanced input and output ports, just as shown in Figure 8.1.

As with the amplifier tests described in Chapter 6, all manner of measurements desired for a single ended amplifier are also desired for a differential amplifier. First and foremost are its linear gain attributes.

Figure 8.8 shows the four-port single ended measurement of one kind of differential amplifier. The input and output matches show that it is tuned to about 1 GHz. The forward gain parameters are interesting in that one of them, S21 (called out in the upper-right window) has substantially less gain than the others at only about 1 dB. S41 also has somewhat lower gain, at just 5 dB, where S23 and S43 have gain at about 8 dB. All the reverse isolation paths have reasonable isolation. This amplifier is a prototype low-noise limiting amplifier, and its behavior is typical of what one might find in the early stages of a design project. Because it is not completely symmetrical, many assumptions on differential behavior may not hold, and full differential analysis (also known as true mode measurements) may be required. There is substantial coupling between ports 1 and 3, is indicated by the high level of S31 and S13. In contrast, the output ports have more isolation at low frequencies, but have some coupling at high frequencies. Coupling between the ports is an indication that the common mode match may not be very good. For this example, the amplifier does not follow the expected behavior of a true differential amplifier as each of the forward gains are not identical. Later in this chapter, an example of a normal differential amplifier will also be shown, which is considerably less interesting.

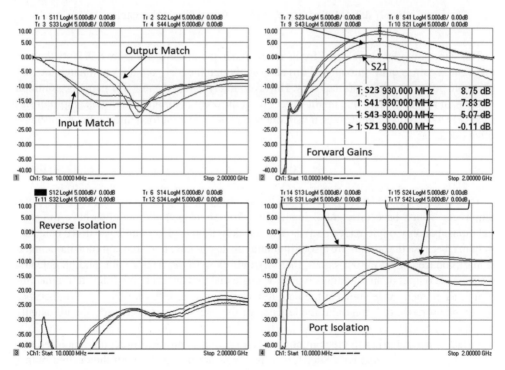

Figure 8.8 Four-port single ended S-parameters of a differential device.

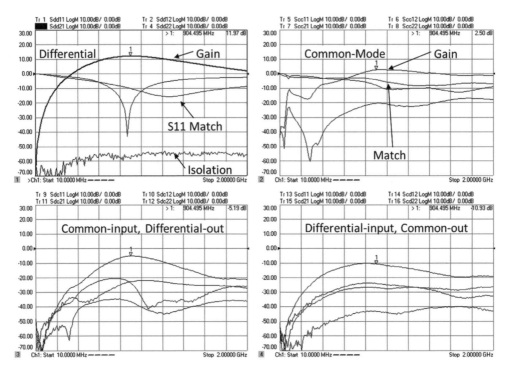

Figure 8.9   Mixed mode parameters for a differential amplifier.

The same amplifier is evaluated for its mixed mode parameters, in Figure 8.9. Note that the differential gain is quite smooth and broad, even though the S21 gain does not match the S23 gain, and peaks at a different frequency. However, likely due to the low S21 gain, the common mode gain is greater than one might expect for a differential amplifier and this amplifier has relatively low common mode rejection. Common mode rejection for a differential amplifier is often quoted as

$$CMRR = \frac{S_{dd21}}{S_{cc21}} \tag{8.23}$$

This definition came from the definition for low-frequency op-amps. Unlike RF differential amplifiers, which have differential inputs and differential outputs, op-amps have differential inputs but only single-ended outputs. But this definition is of very little use in RF balanced systems, because the outputs are also differential. In an op-amp, if a large common mode input signal produced some output signal, then it would flow to the next section and be detected because the output of an op-amp is typically single ended. However, in a full differential amplifier, a common mode signal at the output does *not* propagate through the system, because the common mode to differential mode gains of the following sections (Sdc21) are usually very low.

In fact, when referring to the common mode rejection of an RF differential amplifier, one would be most concerned with Sdc21, as it is a measure of the effect on the desired output signal (the differential output voltage) due to a large common mode input signal. Thus, for full differential amplifiers, the common mode rejection ratio (CMRR) is of very little concern, and Sdc21 should be the proper measure of common mode isolation. In the example of Figure 8.9, the CMRR is about 9.5 dB, but if there were equal common mode and differential mode signals at the input, the effect on the differential output due to the common mode input would be down by 15 dB, the value of Sdc21.

One other interesting attribute is that the differential reverse isolation is much higher than any of the single ended isolation terms, indicating very good balance in the reverse isolation. For these measurements the amplifier is driven with quite low power, and is operating in the linear region, so one expects that taking data with a four-port VNA using single ended S-parameters, and mathematically computing the mixed mode parameters following Eq. (8.19), applies quite well.

8.3.3   *Differential Amplifiers and Non-Linear Operation*

One common assertion is that while the mixed mode parameters of an amplifier may be computed from single ended four-port measurements when the amplifier is operating in a low-power linear mode, the results will not be valid when the amplifier is driven with a large signal, and operated in a non-linear mode. Rather, for non-linear characterization, such as 1 dB compression point, it is presumed that the amplifier must be driven with a true differential signal, usually called a true mode drive.

As it turns out, the truth of this assertion depends upon the configuration of the amplifier, and in most cases, single ended mixed mode drive *does* provide a valid measurement of gain-compression of a differential amplifier, provided that it is has normal differential behavior [2]. Here, normal behavior for a differential amplifier has two main aspects:

1. A normal amplifier, differential or not, usually compresses because of limitations at the output of the amplifier. This is obviously because the RF voltages and currents are much greater at the output than at the input. Some amplifiers may have limiting elements at the input (as does the example amplifier from Figure 8.9), that clip the input signal before the output signal is limited by internal mechanisms in the amplifier, but this would be an unusual case.
2. A differential amplifier should have more differential gain than common mode gain. In some discussions of balanced applications, two single ended amplifiers are combined to create a so-called "balanced amplifier" which will amplify a differential signal; but it will also amplify a common mode signal with the same gain. This is not a differential amplifier in the normally understood definition.

To understand the effects of the non-linear response of differential amplifiers, consider what happens when a differential amplifier is driven with a single ended signal, as illustrated in Figure 8.10. Here, the single ended signal can be decomposed into a differential portion, and a common mode portion. For an amplifier with substantially higher differential mode gain than common mode gain (as one would expect for a normal differential amplifier) the differential portion of the input signal is amplified, while the common mode is suppressed.

In this case, the single ended signal has an input of 1 volt on the positive input, and 0 volts on the negative input, for a differential voltage of 1 volt. The common mode voltage in this case is $^1/_2$ volt. The output is 2 volts differential, and only 0.1 volt common mode. The common mode has been suppressed by the behavior of the amplifier.

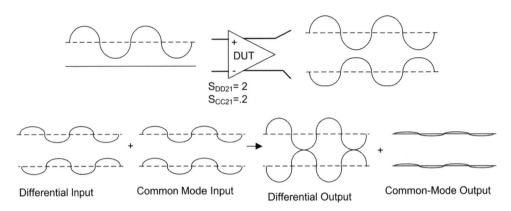

Differential Input   Common Mode Input   Differential Output   Common-Mode Output

Figure 8.10   A differential amplifier driven with a single ended signal.

Many differential amplifiers are multistage, so that there are one or more stages of differential gain at the input, often followed by single ended output buffer stages at the output to obtain higher power performance. An example schematic, in Figure 8.11, shows a design with two stages of differential gain (Input Diff-Pair and 2nd Diff-Pair) and one stage of emitter follower output stage. It is quite typical to have at least the input stage be a differential amplifier using emitter-coupled or source-coupled pairs.

To understand the effect of non-linear operation of a differential amplifier, consider the two cases of the amplifier being driven with a differential mode signal or a single ended signal. The results of driving the amplifier of Figure 8.11 with these two drive inputs are shown in Figure 8.12 (left). Figure 8.12 (right) shows the output from the first stage.

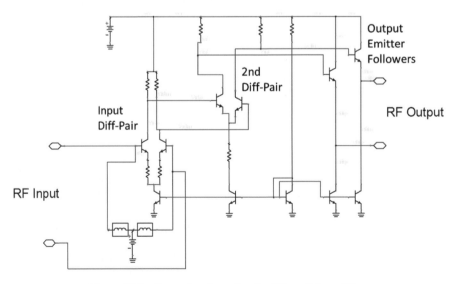

Figure 8.11   Example schematic of a differential amplifier.

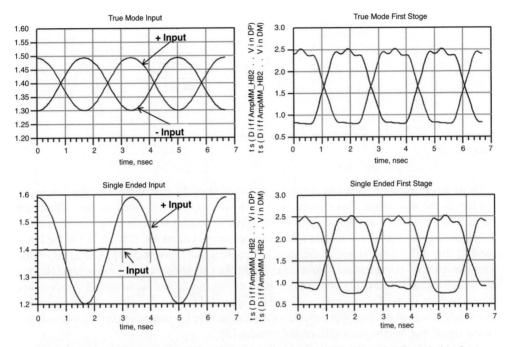

Figure 8.12   Driving an amplifier with true mode (upper) and single ended (lower) signals.

In this case, the differential input voltage is identical at 0.3 volts. The non-linear response of the output of the first stage is nearly identical regardless of the input drive mode, as long as the differential input voltage is the same. This signal then drives the output stage, and since the drive to the next stage is the nearly identical, the output waveform should be identical as well.

Consider next a different design that does not have a differential input stage, or rather, that has a non-linear stage before the differential input stage, as illustrated in Figure 8.13. Here, the input stage is modeled as a balanced-amplifier with equal common mode and differential mode gains, followed by a differential amplifier, and the input amplifier compresses at about 1.5 volts on the positive half cycle.

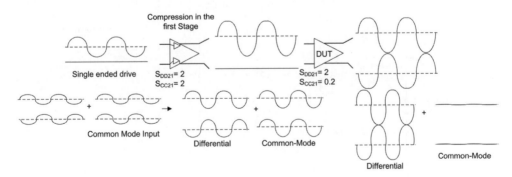

Figure 8.13   An example case where the input is non-linear and non-differential.

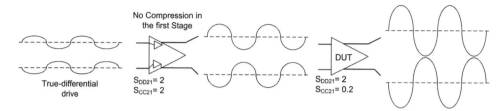

Figure 8.14 Non-linear non-differential input driven with a true differential signal shows less compression.

From this illustration, one can see that compression in the first stage will clip the upper portion of the waveform. This clipped waveform will then be amplified by the differential stage. The common mode portion will be removed by common mode suppression, leaving only a differential signal, but it is now distorted (compressed) on one half cycle. On the other hand, if true differential signal of the same differential amplitude were applied to this same amplifier pair, no compression would occur in the first stage (presuming the compression is due to the output signal rising above 1.5 volt, in this example) and so no output compression would be seen in the final signal, as shown in Figure 8.14 This is an example case where a true mode differential signal will provide a different answer for compression than will a single ended signal that has the same differential content [3].

Finally, while this reasoning clearly applies to sinusoidal inputs used for normal gain and compression testing of amplifiers, a common question is whether the same arguments apply when the input signal has some modulation content. In fact, they do. Consider the amplifier from Figure 8.11 driven with a high power two-tone signal. To further test the conditions on non-linearity, an RF clipping circuit (in the form of a diode string) is added across the input of the amplifier, as shown in Figure 8.15. This now has the condition of being non-linear on the input, but the non-linear mechanism is still differential.

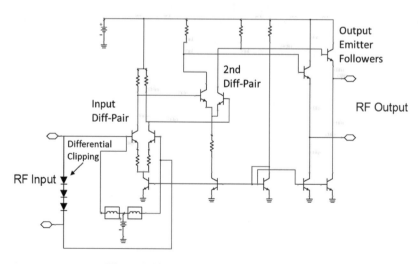

Figure 8.15 Amplifier with input clipping.

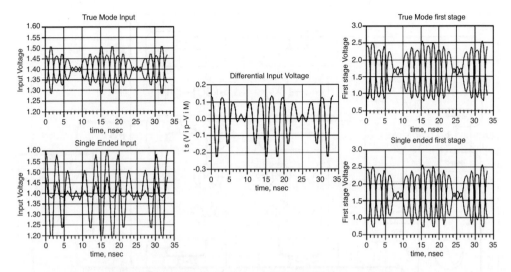

Figure 8.16 Two-tone response of a non-linear differential amplifier.

The result of driving this amplifier with a two-tone signal is shown in Figure 8.16. The graphs on the left show the input signals, with the upper graph representing the true differential drive, and the lower graph representing the single ended drive. Interestingly, the voltage on the negative input of the SE drive is non-zero because the clipping action of the input diodes drives some current into the negative input impedance, and causes some small voltage to appear. However, the differential voltage across the amplifier input is identical, as shown in the middle graph where the differential portion of the input for each of the drive cases are superimposed to see clearly that they are identical. Finally, the output from the first stages shows significant two-tone distortion, but it is identical for both drive cases, so the amplifier's second and final stage response to both signals will be the same.

This demonstrates that when an amplifier has a differential input stage before clipping, or if the clipping itself is differential, then the measurement of an amplifier with a single ended drive or a differential drive will yield the same result. The only time that true differential mode drive is required is when the input stage is not differential, and there is an input non-linearity that is also non-differential.

8.4 True Mode VNA for Non-Linear Testing

While it is true that many differential devices respond similarly to single ended and differential signals, not all do, and one may not know if the device to be tested will behave as a normal differential device. In the past, testing a device for non-linear behavior, particularly compression, required that a balun or hybrid be placed before and after the DUT, so that normal, two-port single ended test equipment could be used to drive a DUT with a true differential drive.

In 2007, the first fully capable true mode VNAs were developed based on four-port network analyzers with dual sources [4]. Previous systems relied on baluns or hybrids to create the true

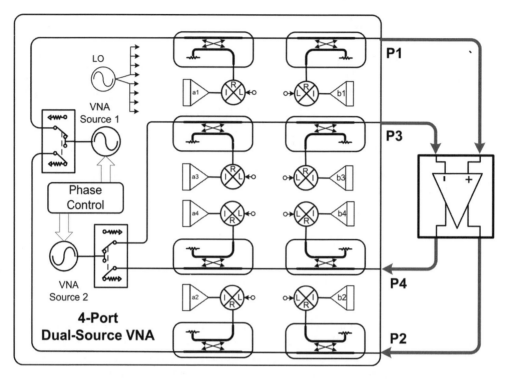

Figure 8.17 Block diagram of a four-port dual source VNA.

mode signals, and relied upon matching of cables and connectors within the system to create a balanced signal. Some modern VNAs offer a means to provide independent amplitude and phase control of the two sources. In Chapter 6, these independent sources were utilized to create an active load to present to the output of a DUT. Here, the independent sources allow the system to create either a pure differential signal or a pure common mode signal, at either the input or output of the DUT. A representative block diagram is shown in Figure 8.17. Key to this system is two independent sources, each driving two of the ports, to allow independent signals at ports 1 and 3, and ports 2 and 4. As the magnitude and phase of the sources can be precisely controlled electronically, any deviation in the cables and connectors of the test system can be completely accounted for. Interestingly, the mismatch of the DUT may provide the largest source of imbalance in the drive signals.

In single ended two- or four-port VNA usage, the phase of the source is of no concern because all the parameters are relative to the single-source phase. In a true mode VNA, the relative phase of the source at port 1 and port 3 is very important, as it sets the absolute differential amplitude. Chapter 3 described a means for characterizing and correcting the VNA receivers for absolute power. Now, these receivers can be used to characterize the power and relative phase of the sources in a VNA.

If one presumes that a full four-port calibration can be performed on the VNA, then the corrected ratio of a_1 to a_3 should give exactly the relationship of the signals for the reference plane at the input of the DUT. For a true mode drive, two input states are required. In the first state, differential drive, the ratio must be

$$\frac{a_1}{a_3} = 1 \cdot e^{j\pi} \tag{8.24}$$

that is, equal amplitude and 180° out of phase.

For the second state, common mode drive or zero phase, the ratio must be

$$\frac{a_1}{a_3} = 1 \cdot e^{j0} \tag{8.25}$$

The ratio of the a_1 and a_3 waves at the DUT interface is not the same as the raw-measured waves at the a_1 and a_3 receivers. However, the corrected waves can be computed directly as

$$\frac{a_1}{a_3} = \frac{(a_{1M}ERF_1 + b_{1M}ESF_1 - a_{1M}ESF_1 \cdot EDF_1)\,ETF_{31}}{(a_{3M}ERF_3 + b_{3M}ESF_3 - a_{3M}ESF_3 \cdot EDF_3)\,ERF_1} \tag{8.26}$$

where the subscript indicates the port for the reflection tracking (ERF), source-match (ESF), directivity (EDF) and transmission tracking (ETF). The subscript M identifies that as a measured wave. For the reverse direction, a similar formulation can be used to compute the ratio of a_2/a_4 as

$$\frac{a_2}{a_4} = \frac{(a_{2M}ERF_2 + b_{2M}ESF_2 - a_{2M}ESF_2 \cdot EDF_2)\,ETF_{24}}{(a_{4M}ERF_4 + b_{4M}ESF_4 - a_{4M}ESF_4 \cdot EDF_4)\,ERF_2} \tag{8.27}$$

This computation of the ratio of the incident waves at the DUT interface provides the proper correction including considerations of the mismatch of the DUT. Surprisingly, no direct computation of the input match is required, but one should recognize that the measured b waves contain information about the reflections from the DUT ports.

In practice, one of the sources is turned on, and the other source is adjusted until the resulting ratio of a_1/a_3 meets the requirements of Eqs. (8.24) or (8.25) for the differential or common mode drive, respectively. Typically, the sources must be iterated to achieve the desired drive as the changing drive on the DUT can cause the match of the DUT to change, thus affecting the ratio. The mismatch correction of the a_1 and a_3 waves is important if the true mode drive is to be maintained. Figure 8.18 shows the error in the phase of the drive signal if the mismatch of a DUT is ignored. In this case, the DUT had in input match of about -15 dB.

The full measurement requires four total source stimulus settings: differential drive on port 1, common mode drive on port 1, differential drive on port 2 and common mode drive on port 2. For each drive, the two reference and all four test receivers must be measured. From these measurements, the values either single ended S-parameters or mixed mode S-parameters may be computed from Eqs. (8.1) or (8.10), respectively. Of course, each of the raw measurements must be corrected appropriately. Equation (8.1) provides four equations that describes the relationship between a and b waves. Taking four stimulus conditions provides 16 simultaneous equations that can be solved to find the 16 raw S-parameters, since

$$[\mathbf{b}] = [S] \cdot [\mathbf{a}] \tag{8.28}$$

where $[\mathbf{b}]$ and $[\mathbf{a}]$ are four by four matrixes representing the various stimulus conditions, then

$$[S] = [\mathbf{b}] \cdot [\mathbf{a}]^{-1} \tag{8.29}$$

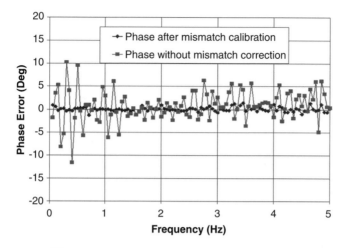

Figure 8.18   Error in phase due to DUT mismatch.

Once the raw S-parameters are known, normal error correction methods can be applied to compute the corrected S-parameters, and from these the mixed mode parameters are easily computed.

However, following this path, there are a few modifications to the normal S-parameter correction that must be made. First, since port 3 is active when port 1 is active, the port 3 load-match term ELF_{31} is not used; rather the source match ESF_3 is used. A similar substitution must occur for port 1, with ELF_{13} being replaced with ESF_1. And of course the same substitution is needed for ports 2 and 4, when they are driven. Finally, since the load match of the paired port is not the same as when the single ended error correction terms are derived, the transmission tracking between paired ports must be modified, according to

$$ETF_{ji\_TrueMode} = ETF_{ji} \cdot \left(\frac{ERF_j}{ERF_j + EDF_j \cdot ELF_{ji} - EDF_j \cdot ESF_j} \right) \qquad (8.30)$$

One should recognize that similar substitutions must occur for the reverse tracking terms.

8.4.1   True Mode Measurements

8.4.1.1   Measuring a Limiting Amplifier

With this understanding, the true mode response of the amplifier from Figure 8.9 is compared with the single ended measured mixed mode parameters in Figure 8.19, for the differential parameters for the case of -25 dBm input and -5 dBm input. At -25 dBm input, there is no difference in the SE or true mode derived S-parameters. At -5 dBm input, the differential gain is compressed 1.7 dB for the true mode stimulus, but 2.7 dB for the SE stimulus.

The same two conditions are measured in Figure 8.20 for common mode parameters.

From the figures above, it is clear that this amplifier does respond differently to true mode stimulus; in fact this is a limiting amplifier designed to clip the input signal on the each input port. However, only the forward parameters change, and this makes sense as there is no gain in

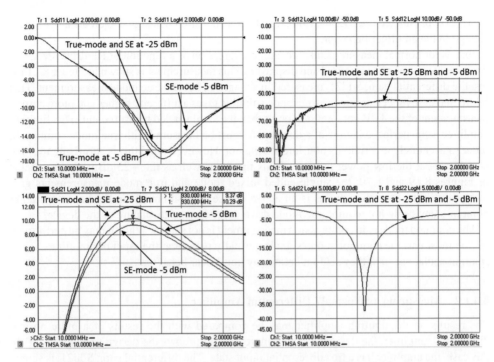

Figure 8.19 True mode vs single ended measurements for differential S-parameter.

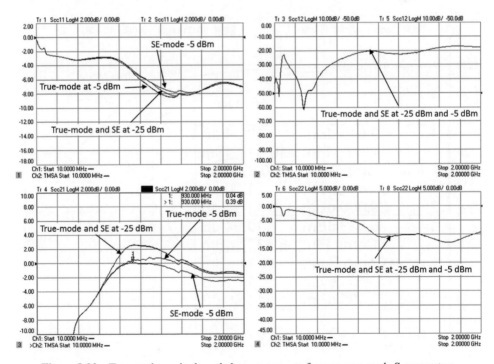

Figure 8.20 True mode vs single ended measurements for common mode S-parameters.

the reverse parameters, Sdd12 and Sdd22, so there is not sufficient signal to create a non-linear change in the amplifier behavior.

Just as for normal measurements, the gain compression as a function of RF drive level, at a fixed frequency, can be measured and displayed. This measurement is shown in Figure 8.21 for just the differential gain, Sdd21, measured in a single ended and true mode measurement. This figure makes it very clear that the true mode stimulus gives a different, higher and more correct measurement of the 1 dB compression point than the single ended measurement. This limiting amplifier is the perfect example of a device that must be measured with true mode stimulus in order to accurately ascertain its non-linear behavior. However – and this is surprising to some engineers – at the low power portion of this figure, as well as the −25 dBm measurements of Figures 8.19 and 8.20, the single ended mixed mode and true mode measurements give identical results, even though this is an active device. Measurements in the linear operating point of a device do not require true mode drive, regardless of the nature of the device.

The common mode gain compression measurement is not shown. While it may be performed, for the most part the common mode signals normally present at the input of amplifiers are so small in actual use, that they seldom create any non-linear behavior.

8.4.1.2 Measuring a "Normal" Differential Amplifier

The results on the limiting amplifier may be compared with the results of a more normal differential amplifier, the design of which more closely follows the design of Figure 8.11. In this case, the amplifier has a true differential input state. The differential gain, Sdd21, is shown in Figure 8.22, measured in both single ended and true mode, at both low power (light trace)

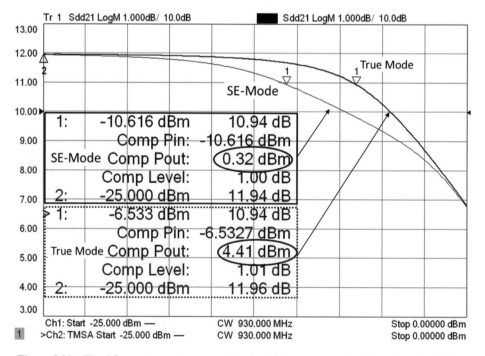

Figure 8.21 Fixed frequency swept power measurement showing differential gain compression.

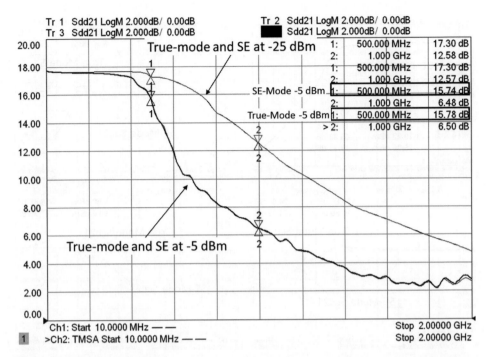

Figure 8.22 Differential gain for a normal differential amplifier, measured in single ended and true mode.

and high power (dark trace). It is clear from these measurements that the non-linear differential response of this amplifier is the same regardless of whether true mode is used or not, with only a slight difference at very high frequency, where the gain is low. Establishing the nature of the differential amplifier's response to single ended drive is critical if one wished to make other, more advanced non-linear measurements such as two-tone IMD measurements.

For completeness, the four modes of forward gain, Sdd21, Scc21, Sdc21 and Scd21 are measured and compared in a power-sweep measurement, in Figure 8.23. As expected, the differential gain remains the same, when the drive level of the SE drive is set to achieve the same differential input drive voltage. The common mode gain does compress differently between true mode drive and single ended drive, as the common mode signal is likely to be clipping at the input, since the output signal of the common mode drive is very low, due to high suppression of common mode gain. Similarly, the cross mode terms see similar effects, where cross mode terms with differential drive (Scd21) shows the same compression as single ended drives, and cross mode drives with common mode inputs show a change in compression between single ended drive and true mode drive. In fact, since the amplifier has very little common mode gain, there is almost no compression in the common mode drive parameters, since there is not sufficient output power to generate non-linear behavior. In the SE drive, when evaluating common mode non-linear performance, the input signal has significant differential content, and so some non-linear behavior is evident.

The non-linear behavior of a differential amplifier may affect the phase response as well as the amplitude response. For this case of a normal differential amplifier, the phase and amplitude of Sdd21 was measured in both SE mode and true mode, and is displayed in Figure 8.24.

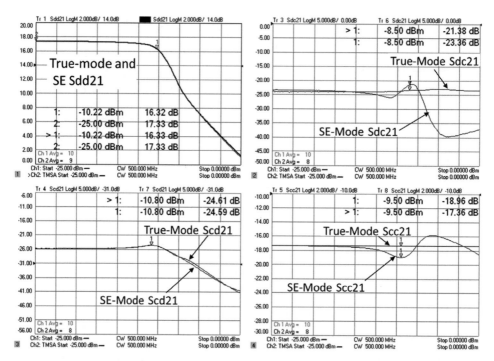

Figure 8.23   Swept power measurements of the mixed mode transmission parameters.

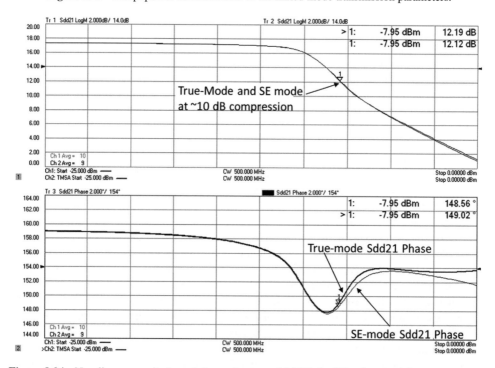

Figure 8.24   Non-linear magnitude and phase response of Sdd21 for SE and true mode power sweeps.

Here one can see that the phase response is also identical, regardless of drive mode, up to the point where the compression is more than 10 dB.

In all the measurements of the single ended drive, the drive power must be increased by 3 dB over the differential mode drive to ensure that the same differential voltage is applied in the single ended case as in the true mode case. This is because when power is applied in true mode, half the voltage is applied to the positive input, and opposite sign voltage is applied to the negative input of the amplifier. For single ended to have the save total voltage across the input port pair, it must have 3 dB higher power. While it might seem that 3 dB higher power, which is twice the power, generates four times the voltage, in the single ended case the load resistance is 50 ohms, versus 100 ohms in the differential case, so that 3 dB more single ended power gives two times the voltage across the equivalent 100 ohm resistance. In the previous graphs, the VNA had a source offset applied to account for this. If one does not use the built-in offset feature of a VNA, the SE drive can simply be set with 3 dB higher start and stop powers.

8.4.2   Determining the Phase-Skew of a Differential Device

Differential devices are designed to have gain for differential inputs, and loss for common mode inputs, but layout and other errors in the design can cause some phase skew in the design such that the gain is not maximized for a 180° differential input signal.

Some VNAs with true mode drives facilitate investigating the phase-skew effects by allowing the phase of the drive sources to sweep over user-defined range of phases, while monitoring the differential gain of the DUT. In fact, in the normal sense, the gain of the DUT does not change with the drive signals, that is, without special consideration, any offsets in the drive-phase will be removed through the normal measurement and correction process, and the differential gain of the DUT, presuming that it is in the linear mode of operation, will return a constant value. However, for the purpose of measuring phase skew, the offset in phase of the source is also de-embedded from the correction array to give the effect of adding a phase shifter in front of one port of the DUT. This virtual phase shifter allows the investigation of phase skew simply and conveniently.

Figure 8.25 shows one example of phase control for a true mode VNA. In this example, the phase sweep settings are highlighted. Special note is made of the selection "Offset as fixture". It is this selection that allows the phase shift to be applied as a virtual phase shifter, which allows one to directly measure phase skew.

If the one suspects the phase skew may be due to external influences, such as uncompensated phase offsets in the test fixture, it is possible to take the value found in the phase-skew test and apply it as a fixed offset under the settings for balanced port offset. In this case, the offsets should also be treated like a fixture, as they are compensating for externally induced effects. For completeness, offsets are available for both amplitude and phase in this example user interface.

A phase skew test was performed on the amplifier used in Figure 8.22. Here the phase was swept from 0 to 360°. One might expect that the maximum differential gain should occur at 180°, but the offset here is offset from the expected phase, thus 0° offset for a differential drive means that the signals are exactly 180° out of phase. When the offset for differential is 180°, the drive signals are both in phase, that is, they are common mode. A marker search is used to find the maximum Sdd21 gain; the marker's stimulus value represents the phase skew of the DUT, as shown in Figure 8.26.

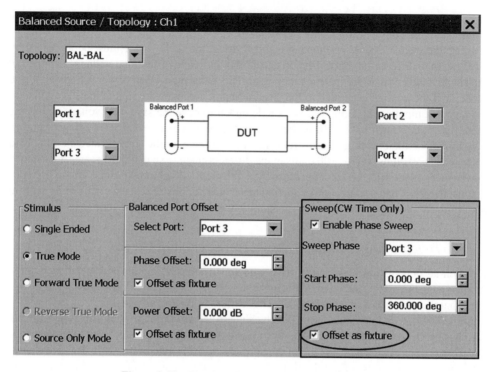

Figure 8.25   User interface for setting phase sweep.

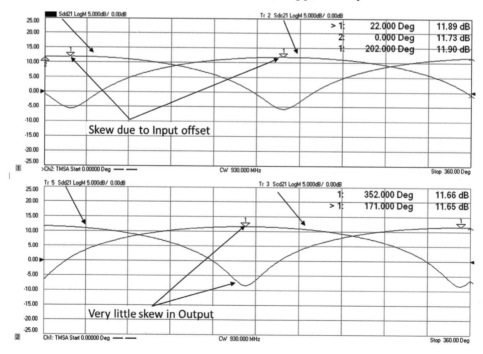

Figure 8.26   Phase skew test on a differential amplifier.

In the upper window, two traces are shown while port3 is swept from 0 to 360°. Sdd21 and Sdc21 are sensitive to the skew in the input signal. The peak of Sdd21 shows an offset of 22°, and the peak of Sdc21 shows an offset of $180 + 22°$. One would expect to see the Sdc21 take on a large value when the common mode drive becomes offset by 180°, thereby stimulating the differential mode fully. The offset in phase here is a confirming indication that phase skew is at the input. For this test example, a small delay adapter was added to one of the inputs to generate some skew; its delay is about 60 psec, so at 930 MHz one would expect about 20° of skew.

In the lower window, port 4 is phase swept from 0 to 360° for the same configuration. Since the output has no additional delay, no skew would be expected. In fact, there appears to be a few degrees of skew in the output portion of this device – about 8 or 9°.

Skew in the input reduces the effective input differential signal, and thus the resulting output power is lower and so is the gain. Skew in the output reduces the differential portion of the output power, and so also reduces the gain. Sweeping both the input and the output independently will show the skew values of each.

8.5 Differential Testing Using Baluns, Hybrids and Transformers

In some cases, such as when a true mode four-port VNA is not available, one must resort to the use of baluns to characterize differential devices. While they have been used for many years, it is only just recently that advances in understanding and advanced de-embedding capability have simplified and improved the measurements made with baluns. Any device which converts a single ended signal into a balanced signal is considered a balun, and they come in two main forms: hybrids and balun transformers. Most commonly, even though hybrids provide a balanced signal, they are not usually referred to as baluns.

8.5.1 Transformers vs Hybrids

Transformers and baluns are terms that are often used interchangeably, and usually refer to three-port devices that change a single ended signal into a balanced signal. In many cases these are actual small transformers, but occasionally transmission line structures are used. Baluns have at least one single ended input and a differential output pair. Several examples are shown in Figure 8.27. The common mode impedance of a balanced port of a balun is often not defined, but is generally either very high (open) for an ungrounded transformer (circuit a), or very low (short) for a center-tapped transformer (circuit c). Some baluns are created with series structures, and the common mode impedance is set by the parasitic inductance of the ground leg of the balun (circuit b).

Some baluns are designed as four-port devices, and actually separate the mixed mode on the right side to two discrete ports on the left side: a delta port which measures the differential mode and a sum port which measures the common mode. This structure (circuit d) is commonly used in simulation to create two separate ports for analysis of the two modes independently. Each transformer is 1:1 and the upper one is center-tapped to provide the common mode connection. Normal mixed mode S-parameter definitions have the differential port impedance as 100 ohms, and the common mode port impedance as 25 ohms. This definition ensures that the single ended view of the balanced port is 50 ohms each to ground. Depending upon

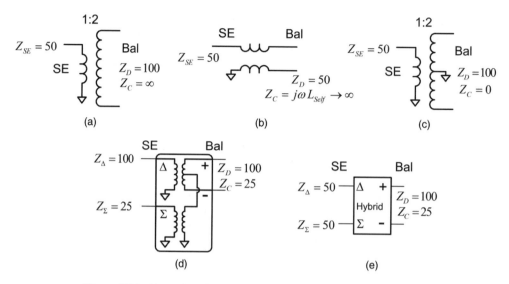

Figure 8.27   Examples of several RF transformers, baluns and a hybrid.

the nature of the balun, it can either supply impedance transformation or not. If the balun does not supply impedance transformation, one must decide the appropriate definition for the differential port impedance. For example, the balanced port of transformer (b) above presents a balanced 50 ohm impedance to a differential device. This structure is commonly fashioned by a coaxial cable with an RF choke around the outer conductor. In such a case, the common mode impedance is set by the self-inductance of the core. Many VNAs supply fixturing software which allows one to redefine the port impedance mathematically, transforming the measured S-parameters into any desired impedance.

Hybrids are four-port devices which have an in-phase port and difference port input (often called the sum or Σ port, and the delta or Δ port) and a pair of balanced ports as outputs. When the sum port is connected to an RF source, the output from the balanced pair is common mode. When the difference port is connected to the source, the output from the balanced pair is differential mode. Most hybrids require that the unused port be terminated in 50 ohms. Hybrids have the nice attribute that the balanced ports are also isolated, and hybrids provide a 50 ohm match on each one of the balanced pair ports, when the sum and difference port is terminated in 50 ohms. This makes the balanced-port differential impedance 100 ohms, and the common mode impedance 25 ohms. This implies that the hybrid is not a 1:1 transformer as in circuit (d) above, since the impedance on the sum and delta ports are not Z_d and Z_c, but 50 ohms. Using hybrids as baluns give results similar to the true mode VNA, without requiring any additional impedance transformations.

The drive signal from the source is divided by the balun, so that half the power (0.707 times the voltage) goes to each port. So when driving from a single ended port, the differential power will be the same as the single ended input power, but the power is into both loads of the balanced device. Since each load sees half the power, the total power is the same. In terms of voltage the plus port sees 0.707 times voltage, and the negative port will see -0.707 times the voltage, so the differential voltage will be 1.4 times the single ended voltage. But the differential impedance is two times greater, so the differential power is the same as

$$P_{Diff} = \frac{V_{Diff}^2}{Z_{Diff}} = \frac{\left(\sqrt{2} \cdot V_S\right)^2}{100} = \frac{2V_S^2}{100} = \frac{V_S^2}{50} = P_S \tag{8.31}$$

Of course, all real hybrids have some non-zero additional loss as well as phase-skew, so that their outputs are not perfectly balanced. Proper calibration can account for some of these effects if the hybrid is measured as a three-port, its single ended to differential S-parameters computed, and de-embedded from the overall measurement. Some modern VNAs allow saving the mixed mode S-parameters directly as an S2P file, which greatly simplifies the task of creating the de-embedding file. Simply create a three- or four-port calibration, measure the single ended to balanced parameters, and save the desired 2 × 2 matrix (SE-differential or SE-common) to an S2P file. An example user interface for this save function is shown in Figure 8.28. In this case, a hybrid was measured with two different SE inputs, one that gives differential signals at the output and one that gives common mode signals at the output. Here the matrix for single ended to differential de-embedding was chosen.

Examples of the measurements of an RF hybrid are shown in Figures 8.29 and 8.30. The first figure shows the measurement results from the delta input to the balanced port. In the upper left window is the loss from the delta input to the balanced port differential mode, and the isolation of the common mode. In the upper right is the match of the delta input, and balanced

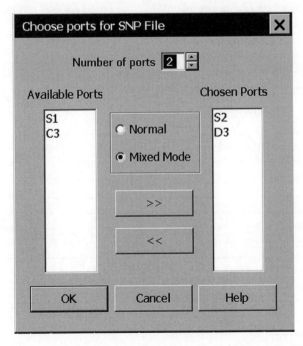

Figure 8.28   User interface to save mixed mode parameters to an S2P file.

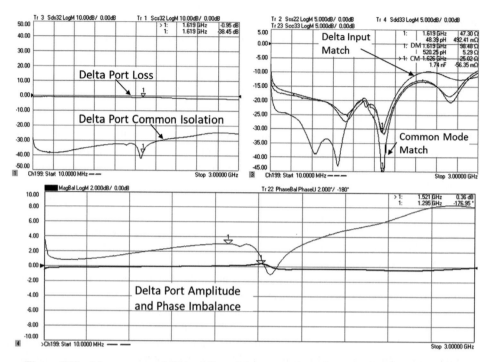

Figure 8.29   Measurement of SE to differential response from the difference port of a hybrid.

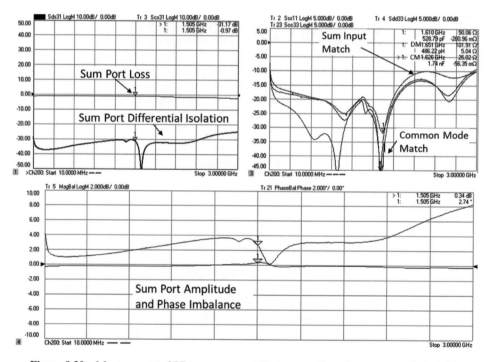

Figure 8.30   Measurement of SE to common mode response from the sum port of a hybrid.

differential mode and common mode. Notice that the differential impedance is near 100 ohms, and the common mode is near 25 ohms. In the second window is a direct measurement from the input port to the plus and minus outputs, displayed as amplitude or magnitude balance and phase balance. Since this is driven from the delta port, the phase balance is relative to 180°.

Similar measurements were made from the sum or common input port, and the loss and isolation are nearly the same, Fig. 8.30. Any imbalance in the hybrid usually shows up similarly in both the sum and delta ports. Note that in the sum-port input case, the amplitude balance is nearly the same, but of course the phase balance is relative to a 0° phase offset.

8.5.2 Using Hybrids and Baluns with a Two-Port VNA

If one has measured a hybrid with a three-port or four-port analyzer, and the single ended to balanced parameters can be saved as a two-port equivalent S2P file, then in many modern VNAs it is a simple task to use the built-in fixturing functions to de-embed the S2P file of a pair of hybrids from the input and output ports of a two-port VNA. This has the effect of changing the two-port VNA into a two-port differential (or common if the VNA is connected to the sum port) network analyzer. Only the differential mode (to the limit of the common mode isolation, about −30 dB in the case of the hybrid from Figure 8.29) drives the DUT and only differential mode output signals are measured from the DUT. An example setup is shown in Figure 8.31.

This allows a very good measurement to be made on differential devices. A hybrid is ideally suited for this case as the terminating impedance is 50 ohms for both inputs. Properly de-embedding these hybrids gives good results, even in the case where the device is in compression, and the device is one that requires true mode drive. As an example, the device from Figure 8.19, which was shown to require true mode drive, was measured with both the

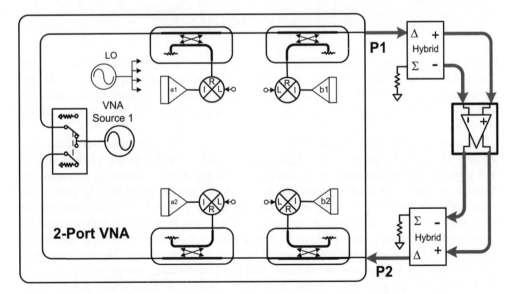

Figure 8.31 Test setup for using hybrids to test differential parameters.

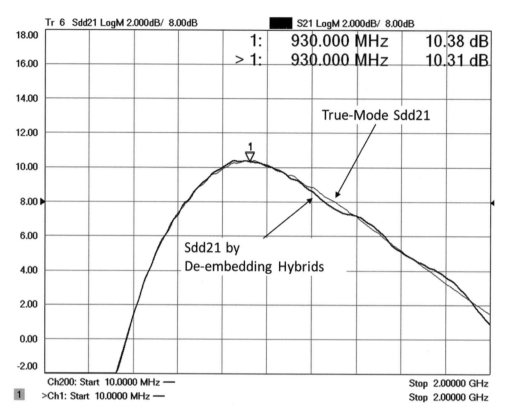

Figure 8.32 Frequency response for differential parameters using de-embedding for −5 dBm drive (1 dB compression).

full four-port true mode system and the system of Figure 8.31, for the case where the DUT was driven to the 1 dB compression point at 930 MHz. As can be seen in Figure 8.32, the measurement with the de-embedded hybrids is nearly identical to the true mode measurement, with only about 0.1–0.2 dB variation. This variation is consistent with the −30 dB isolation of the common mode signal.

The same two conditions, true mode drive and de-embedded hybrids, were used to measure a power sweep, in Figure 8.33. This figure shows almost perfect response when compared with a true mode measurement. At the same time, a single ended mixed mode measurement shows a substantial error when measuring this limiting amplifier. Note that the gain in the linear region of the power sweep agrees within 0.04 dB comparing the de-embedding and the SE method to the true mode result. In the compression region, near 1 dB compression, the SE mode gives much lower compressed power, as noted earlier, but in the case of the de-embedded hybrids, the 1 dB compression level is within 0.01 dB of the true mode measurement. This is remarkably good agreement and indicates that the de-embedding computations do a very good job of accounting for all the effects of the hybrid. Details on de-embedding are discussed in Chapter 9.

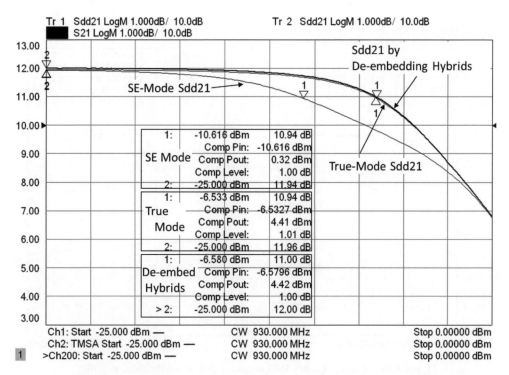

Tr 1 Sdd21 LogM 1.000dB/ 10.0dB   Tr 2 Sdd21 LogM 1.000dB/ 10.0dB
      S21 LogM 1.000dB/ 10.0dB

	1:	-10.616 dBm	10.94 dB
		Comp Pin:	-10.616 dBm
SE Mode	Comp Pout:		0.32 dBm
	Comp Level:		1.00 dB
	2:	-25.000 dBm	11.94 dB
	1:	-6.533 dBm	10.94 dB
True		Comp Pin:	-6.5327 dBm
Mode	Comp Pout:		4.41 dBm
	Comp Level:		1.01 dB
	2:	-25.000 dBm	11.96 dB
	1:	-6.580 dBm	11.00 dB
De-embed	Comp Pin:		-6.5796 dBm
Hybrids	Comp Pout:		4.42 dBm
	Comp Level:		1.00 dB
	> 2:	-25.000 dBm	12.00 dB

SE-Mode Sdd21

Sdd21 by De-embedding Hybrids

True-Mode Sdd21

Ch1: Start -25.000 dBm —   CW 930.000 MHz   Stop 0.00000 dBm
Ch2: TMSA Start -25.000 dBm —   CW 930.000 MHz   Stop 0.00000 dBm
>Ch200: Start -25.000 dBm —   CW 930.000 MHz   Stop 0.00000 dBm

Figure 8.33 Non-linear response to a power sweep using hybrids to test Sdd21.

With this confirmation that de-embedding hybrids creates a measurement system that provides good differential measurement capability, one can use the same setup to make distortion and noise measurements on differential devices, as discussed in the next two sections.

8.6   Distortion Measurements of Differential Devices

Distortion measurements on differential devices can be quite difficult to perform, especially if true mode drive is required, as is the case for the limiting amplifier. But newer VNAs, with built-in applications and calibration for IMD measurements, allow the same de-embedding computations as the S-parameter gain and power. Thus, one may calibrate a two-port VNA for IMD measurements, apply the hybrids as shown in Figure 8.31, and proceed to read the IMD results simply and directly from the VNA. In Figure 8.34, IMD measurements of the limiting amplifier are made across frequency, with an input power near the 1 dB compression point. A typical level for 1 dB compression of about −26 dBc IM3 holds true for this differential amplifier as well. The accuracy is ensured here as the input and output hybrids are de-embedded from the measurement result, including the appropriate source power offset to account for the loss in the port 1 hybrid.

The power spectrum of the IMD output can be shown for this power level, at a single output frequency of 930 MHz, with the result matching the measurements shown for the marker position of this sweep, in Figure 8.35.

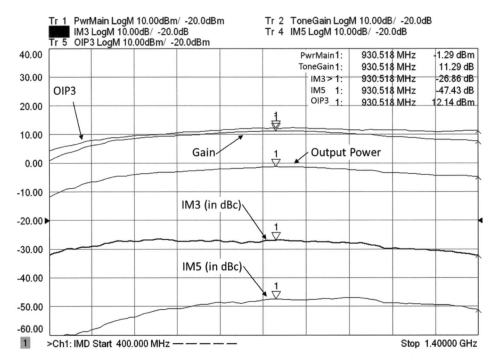

Figure 8.34   Swept frequency IMD measurements of a differential amplifier.

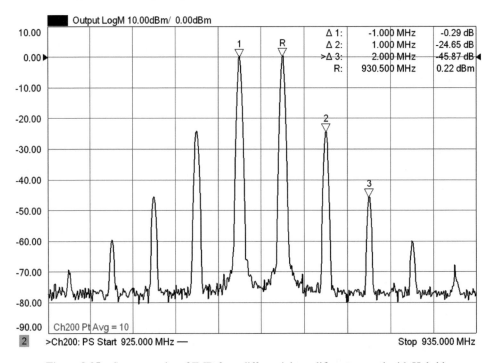

Figure 8.35   Spectrum plot of IMD for a differential amplifier measured with Hybrids.

Here the markers show a slight skew in the output tones, likely due to a difference in gain or match in the amplifier. Marker 1 shows the low tone to be about 0.4 dB lower than the high tone, for an average (in dB) power of −1.1 dBm and matches closely with the power from the swept frequency plot. Also, the IM3 is within about 0.5 dB and the IM5 is within about 0.5 dB as well, which is within the amplitude accuracy of the SA mode in this VNA.

Finally, the IMD behavior of the amplifier with respect to drive power is also easily shown using a swept power IMD measurement, illustrated in Figure 8.36.

From this plot, it is clear that the OIP3 level starts to degrade for powers above about −21 dBm input, and near the 1 dB compression point, the IM3 has increased substantially over the value one would project using the normal 2:1 degradation for IM3 dBc level versus RF drive. The value at −11 dBm is only −22.8 dBc, but the normal expectation would be −28.8 dBc, based on the level at −21 dBm. The fact that the IM5 tone goes down at high power might indicate that some additional IM3 is created by conversion of the IM5 power.

8.6.1   Comparing Single Ended IMD Measurement to True Mode Measurements

For a normal differential amplifier, one would expect that driving the single ended input and measuring a single ended output should give a very similar result for IMD as measurements

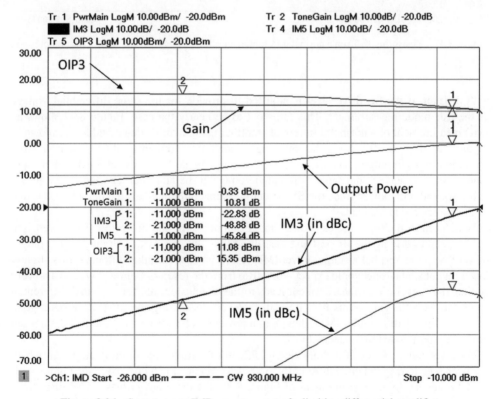

Figure 8.36 Swept power IMD measurement of a limiting differential amplifier.

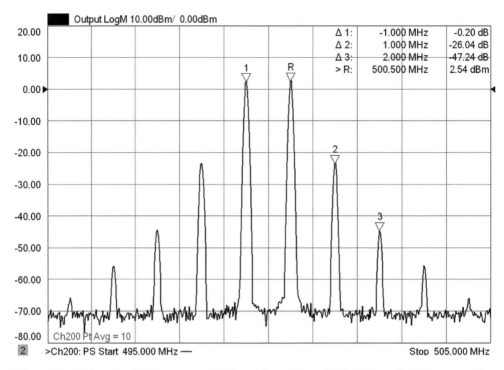

Figure 8.37   Measuring IMD on a normal differential amplifier, using hybrids, at the 1 dB compression power.

with a balun, if the SE drive level is adjusted to provide for the same differential voltage as the true mode measurement. This turns out to be nearly the case. Figure 8.37 shows an IMD measurement of a normal differential amplifier using hybrids de-embedded as shown in Figure 8.31. Figure 8.38 shows the four single ended measurements without a hybrid, from each input to each output. The SE measurement must be performed with 3 dB higher input tone powers, to achieve the same differential voltage at the amplifier, and the output power will also read 3 dB lower since the SE output measures only one-half the output power, that is, the power in one SE load.

Here, each of the input drives shows similar output power and distortion. The lower-right window shows driving the -In port and measuring the -Out port, with a slightly higher output power (the R marker) but a similar lower IMD result. Each of the other outputs shows slightly lower output signal (one would expect −0.5 dBm from the result of Figure 8.37) and also very similar IMD levels. This measurement leads to the conclusion that for a normal differential amplifier, with SE gain terms that are well matched, measuring the SE IMD gives a good prediction of the differential IMD, though at a power level 3 dB lower. Note that the unused ports were terminated in a matched load.

However, the same result is dramatically different in case of the limiting amplifier, which has been seen to respond quite differently to SE drive versus true mode drive (see Figure 8.33) for CW signals. In the measurement of Figure 8.35, the drive power was −11.7 dBm for 1 dB compression in the DUT, which gives the resulting −27 dB IM3. Figure 8.39 shows the

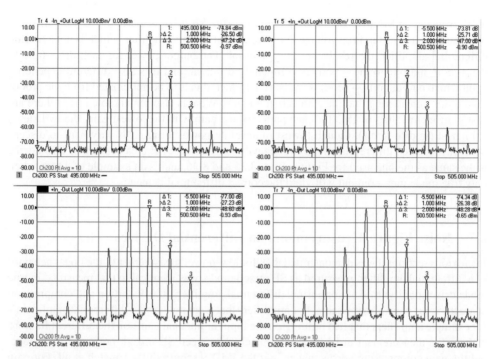

Figure 8.38 Measurement of IMD for a normal differential amplifier with SE drive and SE measurement.

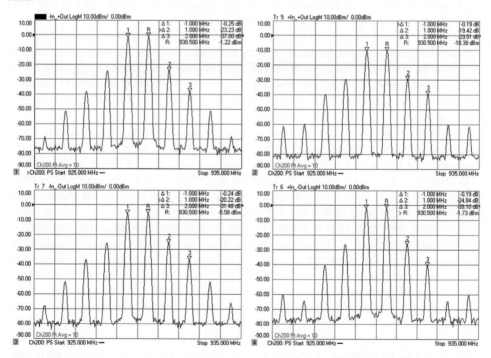

Figure 8.39 The SE measurements of a limiting amplifier show poor prediction of differential behavior.

result of using an SE drive on the same amplifier; the SE drive is increased to −8.7 dBm, to produce the same differential voltage at the input. But for this amplifier, which is not a normal differential amplifier, the results are very unusual. The output power for each drive (one to the + port and one to the − port) is completely different, and it is also different depending upon which output port (+ or −) is measured. In fact, this is to be anticipated from the plot of Figure 8.8, considering that the gains are not equal. What is also remarkable is that there are vastly different values of IM3 for each path. For the path with the lowest gain, the IMD level is highest (in dBc terms), though lowest in absolute power terms. It is very hard to discern what the expected differential mode IM behavior is from these SE measurements. This device is the same as the one used for Figures 8.34, 8.35 and 8.36, and is an example of a case when only true mode measurement will yield the correct differential behavior.

8.7 Noise Figure Measurements on Differential Devices

In previous sections it was demonstrated that, for low-power measurements, the linear behavior of a differential amplifier, measured as SE mixed mode, was identical to that measured in true mode drive conditions. And for normal differential amplifiers, the distortion and compression characteristics were also the same. For amplifiers that were not normal differential amplifiers – for example amplifiers that had limiting on the input or unequal gains on the different forward paths – the measurement at lower power matched between SE mixed mode and true mode, but the gain compression versus power, and the IMD response did not match at all. For gain compression, the four-port true mode method could easily be use, and it was demonstrated that it matched exactly with the non-linear response of properly de-embedding a hybrid. This hybrid was then used to characterize the IMD response.

Noise figure is generally considered a small signal measurement, and the base measurements, gain and noise power measurements, are always performed when the amplifier is operating in a linear mode. However, an analysis of the definition of noise figure makes it very clear that at the least, a balun or hybrid must be used at the output of a differential amplifier or the noise figure results may not be valid. This is true for any type of differential amplifier, whether it is normal or limiting or anything in between.

Consider the schematic diagram of an amplifier with internal noise sources shown in Figure 8.40. This amplifier has both common mode sources of noise (such as noise in the common-leg of the current source) and differential sources of noise (noise at the input of the differential pair). In this system, the common mode noise may not be amplified as much as the differential noise, but common mode noise can be injected in any of the stages, which makes estimating the noise figure very difficult.

Also adding to the difficulties is that just as there are four gains in a differential amplifier, so too might one expect four different noise figures, depending upon the usage of the amplifier. The noise power at the output may be partially correlated (differential) or not, depending upon the internal characteristics of the DUT. In general, one cannot rely on single ended measurements of the noise figure of an amplifier because there is no way to distinguish the differential noise (which will affect other stages) from the common mode noise (which will not). Thus some method is needed to provide true mode noise figure measurements.

There have been other methods proposed to make differential noise-figure measurements, based on the Y-factor technique, of multiport devices. In particular, a paper by Randa [5] defines

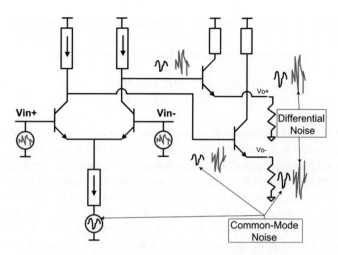

Figure 8.40 Amplifier with internal noise sources.

differential noise figure but does so only for the case where the detection or measurement port is purely differential, which does not apply to real-world measurements. That is, in the theoretical configuration, the differential port is considered in isolation from the single ended ports which comprise it. This is convenient for mathematical purposes, but is not realizable in practice, unless one uses a hybrid or balun to separate the differential signal from the common mode or SE signal. There have been proposals for an N-port representation of noise correlation matrix applied to a distributed amplifier, but not a differential amplifier [6]. Other authors have proposed a de-embedding technique, based also on the Y-factor technique, which does not address the mixed mode nature of the amplifier, and in fact, the amplifier that was used for proof of concept was not a differential amplifier in the commonly understood sense, in that the common mode gain and differential mode gain were identical, for a CMRR of 0 dB [7]. This also does not represent an interesting case for differential amplifiers. Further, Y-factor provides only a noise figure measurement, and does not give any indication of the noise parameters.

8.7.1 Mixed Mode Noise Figure

It is desirable to have a more complete measurement method for differential amplifiers that follows the ideas of a mixed mode approach to noise figure of differential devices. Further, it is desirable that these noise characterizations be expressed in terms of noise parameters, thus illuminating the aspects of minimum noise figure, and optimum match for noise performance, in terms of mixed mode parameters.

The previous discussion of differential signal waves, a_{d1}, a_{c1}, b_{d1}, b_{c1} naturally leads to the concept of differential noise waves, and further to a differential noise parameter matrix. From this groundwork, mixed mode noise parameters can now be defined as

$$NFmn_{xy} = \frac{Smn_y/Nmn_y}{Smn_x/Nmn_x} \qquad (8.32)$$

Where *NF* is the noise figure, *m* and *n* are the modes at the output and input of the device, respectively, and *x* and *y* are the output and input of the device, respectively. This can be reconfigured, for example, to represent the differential noise figure as

$$NF_{dd21} = \frac{DUTRNPI_{d2}}{S_{dd21}}$$ (8.33)

where $DUTRNPI_{d2}$ is the differential relative noise power incident to a differential load connected to balanced port 2 (as in other discussions of noise power, RNP is relative noise power, relative to kT_0B noise power at the input), and S_{dd21} is the mixed mode differential gain. This configuration implies that to measure a mixed mode noise figure, all that is necessary is a measurement of noise at the output, and another measurement of the device gain, each in the proper mode. This formulation represents the so-called "cold-source" method of noise figure measurement, as discussed in Chapter 6, extended to differential amplifiers. This methodology removes the need for any noise sources to be used during the measurement process, which is used to advantage in developing the mixed mode noise figure measurement.

The method described below provides for full vector error correction of source and load effects, as well as vector noise receiver correction, because it employs the equivalent of a noise tuner (in the form of an Ecal module) during the noise calibration and measurement process. In fact, this system measures the noise-parameters of the DUT, and from this computes the noise figure in the system impedance.

One key attribute of the cold-source method is the ability to move the noise-calibration planes (which utilize coaxial noise sources or power meters during calibration) to any reference plane for which an S-parameter calibration can be provided. Just as with compression and IMD, the two-port noise calibration, combined with a hybrid de-embedding, provides a system which can characterize the mixed mode noise-parameters and noise figure of the amplifier under test.

8.7.2   Measurement Setup

The setup from Figure 8.31 can be used for the differential noise measurement system, using scalar cold-source techniques, as described in Section 6.7. A tuner such as an Ecal module may be added behind the port-1 test-port coupler for full vector noise measurements. The baluns are hybrid circuits which have both sum and difference inputs. Note that the noise figure and noise-parameter calibration plane will be before the hybrid, and thus uses only standard calibration procedures to provide a calibrated noise figure at this two-port coaxial plane.

In this system, the noise parameters of the VNA noise receiver are fully corrected by using the tuner during the calibration stage to "pull" the noise receiver and establish noise parameters or noise correlation matrix for it. Likewise, an S-parameter calibration is performed at the same reference plane, which provides a way to fully characterize the tuner for its impedances.

Then for differential measurements, one can use the two-port equivalent of the hybrid input and output networks, based on the three-port mixed mode parameters of the hybrid, as shown in Figure 8.29, and de-embed them from the noise calibration. When a noise measurement is performed, only the modes (common or differential) of the input and output hybrid are exhibited in the results. Example S-parameter and noise-figure measurements are shown in Figure 8.41.

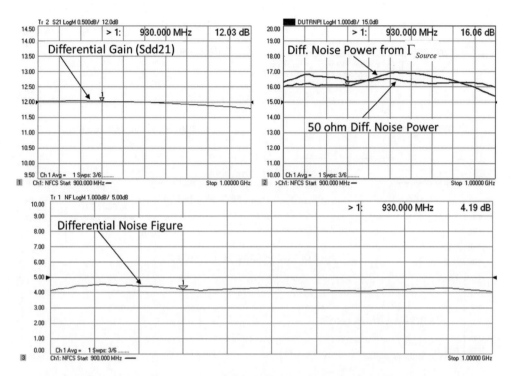

Figure 8.41 Noise figure and gain, and DUTRNPI measurement for a differential amplifier.

Also shown is the DUT relative noise power incident (DUTRNPI). This is the noise power delivered to a non-radiating, non-reflecting load, relative to the kT_0B noise floor. Highlighted in the upper right window, two versions of DUTRNPI are shown. The first is the normal measurements, made with the tuner set to its nominal match condition; unfortunately the source match is rather high in this setup and so the DUTRNPI has some high variation at higher frequencies. Also shown is a measurement of the DUTRNPI with the differential input of the hybrid loaded with a well matched load. Here the response is more in line with the expected response, given the noise figure and gain. In practice, DUTRNPI should be equal to the S21 gain plus the input noise figure of the device (in dB), but DUTRNPI is not vector corrected for input mismatch. This parameter is useful in comparing noise powers out of a DUT in various conditions, and can be thought of as the excess noise (above kT_0B) at the output of the DUT.

The noise figure is approximately 4.2 dB at 930 MHz and qualifies this as a relatively low-noise amplifier. This represents the case where the differential mode input/output of the amplifier is measured.

In Figure 8.42, these same parameters are measured for the common mode case of the amplifier. This is accomplished by using the same method, but in this case using the common mode (or summing) input of the hybrids. What is interesting here is that even though the common mode noise figure is considerably higher (8.41 dB), the common mode noise power, DUTRNPI (the excess noise), is in fact about 5.5 dB lower than the noise power measured the differential mode.

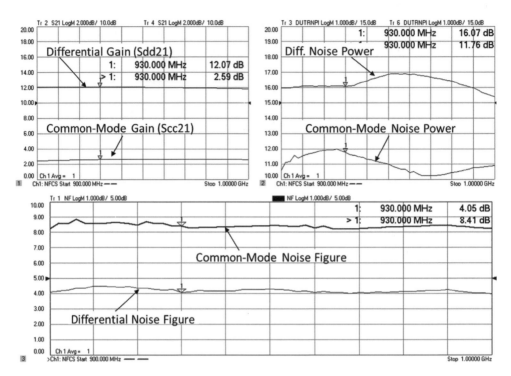

Figure 8.42　Common mode gain and noise figure.

Still, the common mode noise is much higher than one might expect, as the CMRR (differential gain/common mode gain) is about 10 dB. If the noise was generated only at the input, and was equal in each mode, we would expect the DUTRPNI to be 10 dB lower in the common mode case: only 6 dB above the kT_0B noise floor. This result implies significant uncorrelated (in the differential sense) noise in the output stage, or rather, a significant amount of noise correlated in the common mode.

Figure 8.43 shows the noise figure of the amplifier for a single ended case, where the input is at port 3, and the output is from port 2 of the amplifier, which is the highest gain path from Figure 8.8, with the other ports terminated in 50 ohms.

The gain in this noise figure measurement is 3 dB lower than the differential gain, but surprisingly, the noise figure is higher by about 1.7 dB. A common misconception is that measuring the single ended noise figure will be a good estimate of the differential noise figure of an amplifier, but this experiment shows that such is certainly not the case. Even though the gain is 3 dB lower, the noise power is only 1.5 dB lower. One can estimate the contribution of the common mode noise to the SE noise by taking half the differential noise, and adding it to half the common mode noise (adding in the linear sense), since the noise power is the single ended case is divided between two load resistors, then converting it back to dB, as in

$$DUTRNPI_{SE\_estimate} = 10\log_{10}\left(\frac{10^{\frac{16.07}{10}}}{2} + \frac{10^{\frac{11.8}{10}}}{2}\right) = 14.44 \text{ dB} \qquad (8.34)$$

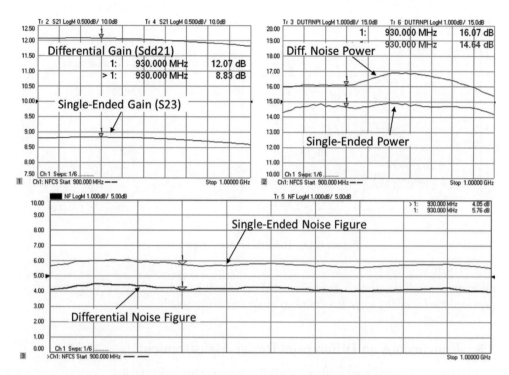

Figure 8.43   Noise figure measured on single ended inputs.

This estimate gives a slightly lower than measured estimate of the SE excess noise, perhaps because the gain for this amplifier is quite uneven between the port pairs. Another reasonable source of difference may be because the DUTRNPI has some ripple error that changes as a function of source match, as illustrated by the upper-right plot in Figure 8.41.

Finally the noise parameters of a differential amplifier are shown in Figure 8.44. These can be extracted from the noise-correlation matrix of the vector noise figure correction. From this measurement one can see that the amplifier is very nearly matched for minimum noise figure over the gain region, but could be improved by 0.3–0.4 dB. Because the tuner is applied to the differential input of the hybrid, only the differential impedance is varied during the measurement, yielding truly differential noise parameters, without depending on correlated noise sources. In this case, the noise calibration method uses a power meter for characterizing the receiver, so no noise source is needed at all.

From these noise parameters, it appears that some small series capacitance could optimize the noise figure for the matched $Z_d = 100$ ohm case, on the order of

$$C_{Match} = \frac{1}{2\pi f \cdot Z_{Match}} = \frac{1}{2\pi \cdot 930 \cdot 10^6 \cdot 54} = 3.17 \text{ pF} \qquad (8.35)$$

Similar experiments on normal differential amplifiers also yield unexpected excess common mode noise in the output. From these experiments, it is clear that one must detect only the differentially correlated noise at the output to have a good estimate of the noise figure for a

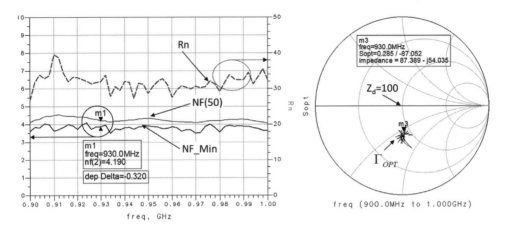

Figure 8.44   Differential noise parameters.

differential amplifier, as only the differential noise will be detected by the following stage, provided the following stage has common mode rejection as well.

8.8   Conclusions on Differential Device Measurement

In this chapter, measurement methods were discussed for the differential device characterization, from which a few conclusions can be drawn:

For normal differential devices, ones that have common mode rejection, and do not have non-linear behavior in the input sections, the results from measuring single ended mixed mode measurements and from fully differential true-mode measurements are nearly the same. But this is true only for measurement methods (like the VNA) which can provide a mathematical difference (in the vector sense) for the output measurements.

For measurements that do not provide a magnitude and phase result, such as IMD or noise power and noise figure, one must sense the differential output using a hybrid or balun. Using modern techniques, the non-ideal hybrid effects can be almost completely de-embedded from the measurement. If the DUT has significant common mode or cross mode terms, and does not have a good common mode termination, then there may be some resulting error due to common mode reflection and mode conversion.

For devices that are *not* normal differential amplifiers – for example an amplifier with limiting on the input stage or before a differential stage – they must be measured with true mode methods, which include true-mode-stimulus VNAs, or incorporating the use of hybrids or baluns to generate and detect the true mode signals. With these techniques, complete characterization of differential devices is possible.

References

1. Bockelman, E. and Eisenstadt, W.R. (1995) Combined differential and common mode scattering parameters: Theory and simulation. *IEEE Trans. on MTT*, 43(7), 1530–1539.
2. Dunsmore, J. (2003) New methods and non-linear measurements for active differential devices. microwave sym. Digest, 2003 IEEE MTT-S, Volume: 3, 8–13 June 2003, pp. 1655–1658.

3. Dunsmore, J. (2004) New measurement results and models for non-linear differential amplifier characterization. Conference Proc. 3fourth European Microwave Conference, 2004, pt. 2, vol. 2, pp. 689–692.
4. Dunsmore, J., Anderson, K., and Blackham, D. (2008) Complete True mode Measurement System. Symp Digest, 2008 IEEE MTT-S June 2008.
5. Randa, J. (2001) Noise characterization of multiport amplifiers. *IEEE Trans. Microwave Theory Tech.*, 49(10), 1757–1763.
6. Moura, L., Monteiro, P.P., and Darwazeh, I. (2005) Generalized noise analysis technique for four-port linear networks. *IEEE Transactions on Circuits and Systems I: Regular Papers*, 52(3), 631–640.
7. Abidi, A.A. and Leete, J.C. (1999) De-embedding the noise figure of differential amplifiers. *IEEE Journal of Solid-State Circuits*, 34(6), 882–885.

9

Advanced Measurement Techniques

For all the measurement techniques discussed, stable and reliable calibrations are the key to accuracy. There are many advanced situations where the basic coaxial calibration techniques are inadequate, usually because of the addition of external components or fixtures. A compendium of advanced techniques is discussed in this chapter.

9.1 Creating Your Own Cal Kits

RF and microwave test engineers are often called upon to test a device that does not have connectors for which standard calibration kits exist. The most common case is for parts that are mounted on PCBs, followed by parts with specialized connectors such as blind-mate or push-on connectors. While there are many ways to deal with this situation, the most straightforward is to create a calibration kit for the particular interface of the DUT.

The two main types of calibration methods, SOLT and TRL, each require different calibration standards. TRL is very straightforward requiring only an unknown reflect and two transmission lines of known impedance, and it is often applied to PCB test situations during R&D development where designing test boards with different length lines is possible. For low-frequency measurements, TRL is replaced with TRM, and an ideal load element is needed as well, since very long line lengths become impractical. In manufacturing situations, where the test fixture may be embedded into a part-handler, there may not be a way to create sufficiently long lines, so TRL cannot be applied.

SOLT or SOLR (unknown thru) require known standards for reflection, typically an open, short and load, although sets of offset shorts can be used as well. For in-fixture calibration, where the fixture is short, the method of TRL is not practical, and so fixed standards are then required. If a thru with good impedance can be created, TRM is preferred; if not, then SOLR using unknown thru is preferred. In this section, a complete PCB calibration kit is defined and various methods are used to determine the quality of the calibration that results.

Handbook of Microwave Component Measurements: With Advanced VNA Techniques, First Edition. Joel P. Dunsmore.
© 2012 John Wiley & Sons, Ltd. Published 2012 by John Wiley & Sons, Ltd.

9.1.1   PCB Example

An example PCB used for SMT part characterization is shown in Figure 9.1. This board has a short, two loads and an open. It is not wise to have an open line near the short standard, as coupling between equal-length lines can cause resonances to appear in the standards. For this example, two different loads were created. The upper load was created with two SMT 100 ohm resistors (1206 package) to ground, making a 50 ohm impedance. The lower load was created using one 50 ohm SMT resistor to ground. The SMA connectors for each device are soldered to top and bottom grounds. For this example, the single-resistor load will be used.

The open standard typically sets the reference plane. The thru standard is designed to be exactly twice the length of the open, but sometimes the modeling of the open adds delay to the open so that the values for fringing capacitance are reasonable. In this case the thru might be longer than the open. To ensure a gap between the load and the open, the board dimension is increase for the portion of the board containing open and load, relative to the thru. For this board, the load was set to a shorter length on the assumption that the load is zero reflection and so its delay doesn't matter. But the evaluation will show that the load has significant reflection; by matching this to a model for the load, even a poor load can be used as a precision calibration standard. In such a case, the modeling method to incorporate the load inductance requires that the load position sets the reference plane. Also shown (far right in the figure) is a shunt device connected to a line that matches the thru line. The short standard is made by placing PCB vias at the same relative reference as the open.

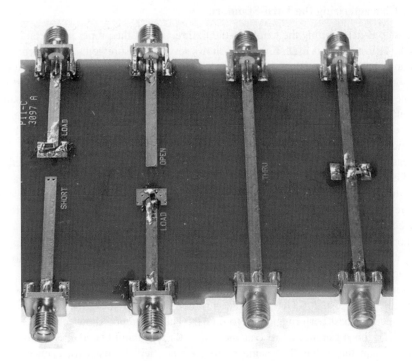

Figure 9.1   PCB designed for SMT part characterization.

9.1.2   Evaluating PCB Fixtures

The investigation of PCB fixtures and standards really must start at the PCB fixture itself, more specifically, the PCB connector. A very common method of creating PCB fixtures is to create a board as in Figure 9.1 with sets of standards, lines, sometimes to evaluate PCB RF features like an embedded filter, and then measure each and try to determine their characteristics. But a critical and often ignored element in this work is the quality and especially repeatability of the SMA-to-PCB connector. Here, quality refers primarily to its return loss, and repeatability refers to the how one SMA connector compares to each of its neighbors. In all the work on PCB fixtures, it is always assumed that once they are calibrated out, the connector can be ignored; but this is only the case if they are identical. So, the first step in evaluating PCB fixtures is to look carefully at the connector repeatability. In the example above, there are six total connectors for the standards. One can readily compare them by using a time domain approach.

The general approach is to use time domain gating to isolate the return loss of each connector. Using one connector as a reference, the gated response of each other connector will be compared to the reference connector to determine its vector difference. This difference displayed in dB will represent the connector repeatability. This fixture is intended to be used up to 6 GHz, but for proper time domain analysis, a much broader frequency sweep is used, in this case 26.5 GHz. Using a frequency sweep of four times the intended operating frequency gives reasonable resolution of important features.

9.1.2.1   Characterizing the Thru-Standard

The first step is determining the length of the fixture, so that the proper gates can be set, and for this the thru standard is used. Figure 9.2 shows several measurements of the thru: the upper left window shows the time domain response of transmission (S21) with a marker at the peak (this is sometimes called T21). This is the total length of the fixture. The lower left window shows the time domain response of S11 and S22. Note that marker 1 is set to the same time value (half the physical distance) as the S21 trace; this time represents exactly halfway through the fixture. Remember that in reflection, signals travel the twice the distance to S11 and S22 as they must go forth and back from the reflection.

For reasons described in Chapter 4, it is best that the time gate be as wide as practical, and centered on the reflection to be gated around. A band-pass gate is used, with the gate center time set to 100 psec and the gate span set to 1040 psec, so the gate stop is near the center of the fixture. This gate will show the response of just one connector, the input connector for the S11 trace and the output connector for the S22 trace.

The normal S11 and S22 frequency response data for the thru is shown in the upper right window, and the gated response for S11 and S22 is shown in the lower right window, along with a trace that is the difference between the gated responses.

This shows us that each connector is similar, with about 28 dB return loss at 3 GHz, and about 20 dB return loss at 6 GHz. For the purposes of this exercise, the calibration kit will be determined for best performance up to 6 GHz. The difference shows about 33 dB return loss at 6 GHz; this difference represents the connector repeatability and one cannot get better calibrations than this value. The same reference trace will be used for each of the other standards to determine their repeatability to the thru standard.

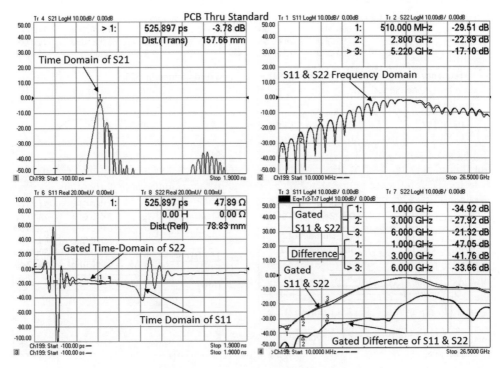

Figure 9.2   Analyzing the thru standard.

9.1.2.2   Characterizing the One-Port Standards

The investigation of the one-port PCB standards starts with the load, which sets the quality of the calibration for both the SOLR and the TRM calibrations. Figure 9.3 shows a similar measurement as above for the connector on one of the load standards. The upper left is S11 of the connector and load element. The lower left is the time domain response, which shows the load response occurs just slightly before the halfway point of the thru (70.9 vs 78.8 mm). The time domain response peak gives a good first estimate of the delay to the load, but it is not necessarily the best way to establish the reference plane, as will be shown below.

The upper right shows the gated S11 response for the connector on the load standard. Here, the response is slightly worse than the gated response for the thru connector, shown in the lower right window, along with the vector difference. In this case, the difference appears to be worse than between the two ends of the thru connector.

Further investigation is needed to understand the difference. From Figure 9.2 lower left, the time domain response shows a peak and dip near the connector. The response from Figure 9.3 shows only a dip. The peak-and-dip response of the thru connector may be improving the return loss in the low frequency region, at the expense of much worse return loss near 14 GHz, where it approaches 0 dB. A common mistake is to leave a gap in the backside ground between the connector and the ground plane of the PCB; this gap is typically required for normal PC mounting, but will often cause inconsistency between standards and this shows up as larger residual directivity and residual match. Inspection of the PCB, Figure 9.4, reveals that the

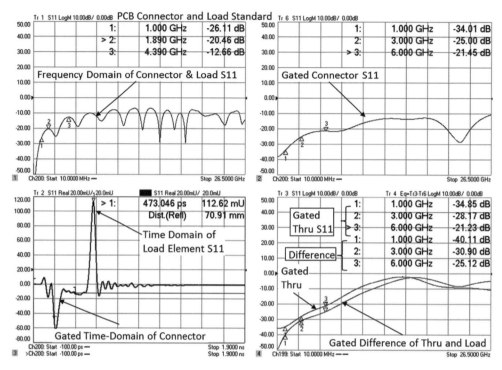

Figure 9.3 Comparing the load connector to the thru connector.

ground plane on the back of the connector for the thru is not soldered (as is normal) but the load is. The load-connector has better high frequency performance and a lower deviation in the time domain.

The difference between the load connector and the thru connector is quite large, even though the S11 traces are not that much different, because of the phasing of the reflections. Here, the best-case residual connector repeatability is only −25 dB, rather than −40 dB above.

Figure 9.4 Comparing load connector soldering to thru connector.

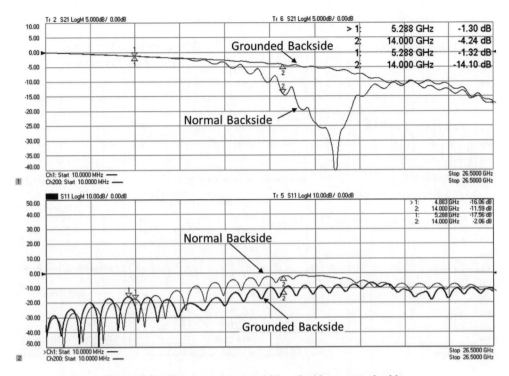

Figure 9.5 Thru measurements with and without ground solder.

The thru element should be modified to match the load element. In general, great care must be used to investigate and repair any differences between PCB fixture connectors in order to obtain good calibrations with them. The thru standard was measured before modifying the ground, and again after the ground of the SMA was soldered; Figure 9.5 shows the comparison.

The normal backside grounding actually shows better low frequency response, up to 6 GHz, but has much worse response at higher frequencies, where it has a very large resonant loss in S21. It is likely that the large inductive peak in the lower left window of Figure 9.2 compensates for excess capacitive response, canceling somewhat the reflection at lower frequencies. With the thru connector re-soldered, it can now be compared again with the load connector, as shown in Figure 9.6. Here one can see that at 3 GHz, the difference is much improved to a worst case of −38 dB repeatability; however, the 6 GHz performance is nearly the same with one connector achieving −30 dB but the other still at around −25 dB. It appears that this level may be the limit for these connectors using normal care in assembly.

9.1.2.3   Investigating the Load-Standard

From Figure 9.3, further investigation into the load is needed, as its time domain response is very high relative to the connector. Figure 9.7 shows the result of applying a notch gate to

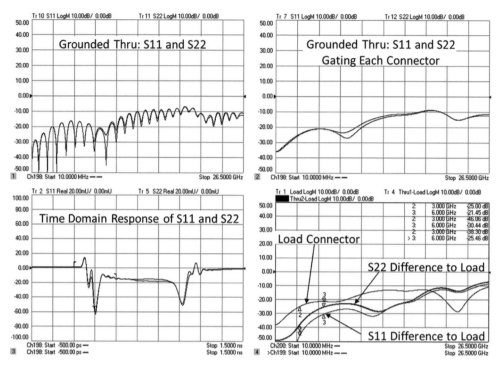

Figure 9.6   Difference between each end of the thru and the load connector.

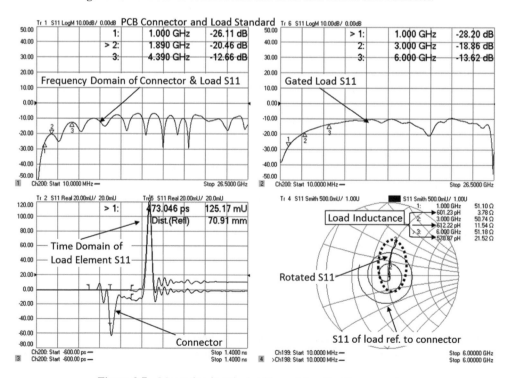

Figure 9.7   Measuring just the load by notch gating the connector.

the connector, to remove its effect on the load. The upper right shows the resulting S11 trace with rather poor load response at 6 GHz. One might normally think that this load would be unacceptable as calibration load, but in fact it may be satisfactory if it can be characterized. This is the single SMT resistor, and previous publications indicate that such a configuration is typically too inductive.

The lower right window of the figure shows a Smith chart plot of the gated S11 of the load, referenced to the input connector. For this plot, data is displayed only up to 6 GHz. Electrical delay of 462 psec was added to rotate the S11 plot until the values of the real part at several frequencies align as closely as possible (to nearly 50 ohms in this case) so the trace follows a trajectory of constant resistances in series with an inductor. The equivalent value of the inductance is highlighted, and it is reasonably constant, at about 590 pH $\pm$ 3%. This is a very good estimate of the series inductance of the load. With this information, one can make an equivalent model of the load that works in almost all VNAs.

The model of a load was given in Chapter 3 as an offset impedance transmission line in series with a resistor. The impedance of the transmission line and its length can be modified to account for the series inductance of the load found in the figure above. The impedance and velocity factor of a transmission line are given as

$$Z_{Line} = \sqrt{\frac{l}{c}}, \quad v = \frac{1}{\sqrt{l \cdot c}} \tag{9.1}$$

where l and c are the inductance and capacitance per unit length. The delay of a transmission line is given as

$$\tau = \frac{d}{v} \tag{9.2}$$

where d is the distance or length of the transmission line. From these, one can compute an equivalent inductance to a line by

$$L_{Equivalent} = l \cdot d$$

$$= \sqrt{l}\sqrt{l}\frac{\sqrt{c}}{\sqrt{c}}d = \sqrt{\frac{l}{c}} \cdot \left(\frac{\sqrt{l \cdot c}}{1}\right) d = Z\frac{d}{v} \tag{9.3}$$

$$L_{Equivalent} = Z \cdot \tau$$

Thus, one can simply define a line length and impedance to compute an equivalent inductance. The approximation is most accurate for high impedances; most VNAs allow up to 500 ohms for the offset impedance of a load, so the delay for a given inductance is computed as

$$\tau_{Load\_L} = \frac{L_{Equivalent}}{Z_{Offset\_Line}} = \frac{L_{Equivalent}}{500} \tag{9.4}$$

Similarly, for a condition where the load has some shunt capacitance, the offset delay can be set to a very low impedance as

$$\frac{1}{Zv} = \sqrt{\frac{c}{l}} \cdot \frac{\sqrt{l \cdot c}}{1} = c \quad F\big/_m$$

$$C_{Equivalent} = c \cdot d = \frac{1}{Z} \frac{d}{v} = \frac{\tau}{Z}$$

(9.5)

For this example load, the equivalent inductance is 590 pH, giving an offset length of 1.18 psec for an offset impedance of 500 ohms. The key to the quality of calibration is match between the actual load and the model of the load. A simulation of such an offset load with a 500 ohm line terminated in 51 ohms was created and the data loaded into a trace on the VNA. Figure 9.8 shows the measured response of the time-gated load and the model of the calibration load on a Smith chart in the left window, and the return loss (in dB) of each, as well as the return loss of the vector difference, in the right window.

The characterization of this load is better than 40 dB up to 6 GHz, so it is a fine load for use in calibration, especially considering that the connector repeatability is no better than −25 dB, from Figure 9.6.

9.1.2.4 Open/Short Characterization and Modeling

The final elements to be analyzed are the open and the short. The connector repeatability for each of these is compared to the load, with the results shown in Figure 9.9, with the worst case being on the order of −28 dBc at 6 GHz, and around −33 dBc at 3 GHz. These values determine the overall quality of the PCB cal. This uses a time gate around the input connector. Using time gating when the line is terminated in a very large reflection, such as an open or short, is somewhat unreliable because the assumption that the response returns to zero as time tends to infinity is not really valid. The reason for this is the periodic nature of the VNA transform. Since the response must be periodic, signals from outside the time range of $1/\Delta f$ can fold back in an aliased way and distort the response of interest.

The final task is to determine the offset delay for the short and the open, as well as any excess fringing for the open. The same data is used as for Figure 9.9, but in this case a notch

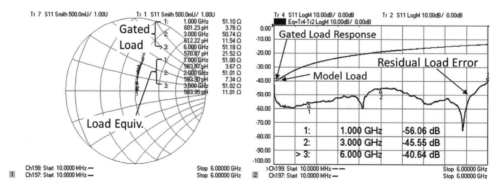

Figure 9.8 Residual load error from equivalent load.

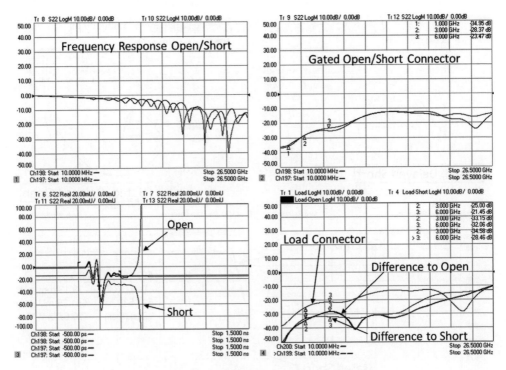

Figure 9.9 Open and short connector repeatability.

gate is applied to remove the effects of the connector, and the responses are shown in the Smith chart, Figure 9.10. Each response is rotated until the markers read a consistent value. For the short, we expect nearly zero inductance. For the open, the capacitance should be small, typically less than 100 fF. To find a model for the open and the short, first the approximate delay to the each must be determined. A simple way to do this is to use a marker search to find the point in the time domain trace where the magnitude is 0.5 (−0.5 for the short). This is in the middle of the rise time of the standard, as illustrated in the lower window of Figure 9.10. A time domain gate is placed around this region, to remove the effects of the connector, and any re-reflection further in time. In this case, the gate center is at 525 psec, with 450 psec of gate span.

The short is shown in the upper left window, and the open in the upper right. The gated responses have been rotated by 525 psec of electrical delay. This value was chosen so that the marker values seem reasonable; this is slightly less than the time found in the time domain trace marker, because there is some extra fringing capacitance for the open, and some excess inductance for the short. The exact setting is up to the engineer's judgment, but it is good practice to try to achieve a small but constant reactance. In this case, it was not possible, and the open and short were set to similar values that kept the marker phase close to zero or just slightly positive for the short, and just slightly negative for the open, without causing loops or clockwise rotations of the Smith chart. More delay would cause the traces to reverse; less delay results in excessive capacitance value. The loss of the open and short don't play a major role in the determination of the fringing elements, but using some form of loss compensation

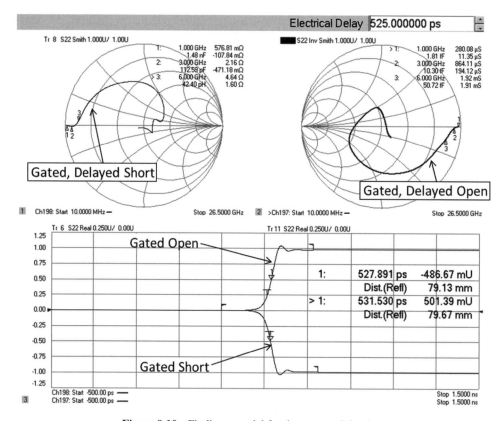

Figure 9.10 Finding a model for the open and the short.

can help to generate a more consistent fringing element. Loss compensation is discussed in Section 9.4.

Now the models for the Cal Kit can be determined.

9.1.2.5 Creating Cal Kit Models

Since it is not ideal, the load element will set the reference plane. The open and short can be offset by an arbitrary amount using a Z_0 line, but the load must use the impedance of the offset line to accommodate its series inductance, and only a single offset is offered in most Cal Kits, so the reference of the load will be the reference of the kit. The load element values from above are $R = 51\,\Omega$, and $Z_{Offset\_delay} = 500\,\Omega$ and the offset delay is $\tau = 1.18$ psec. The offset also provides for a loss, but at these levels using the default loss is fine. These values, along with the open, short and thru are tabulated in Table 9.1.

In rotating the load to find the proper reactive element, 462 psec of electrical delay was added. For the open and short, 525 psec of delay is added to achieve the plot in Figure 9.10. Values for the open and short were approximated, and matched to the equivalent models as described in Section 3.3.2. In this case, only three of the four coefficients were used, and in an

Table 9.1 Cal Kit values for example PCB

Standard	Element	Offset Z_0	Offset Delay (psec)	Equivalent
Load	R = 51	500	1.18	585 pH
Open	C0 = 0, C1 = 200, C2 = 1000, C3 = 0	50	31.5	
Short	L1 = 0, L2 = 0, L3 = 1000, L4 = 0	50	31.5	
Thru		50	63	

example such as this, using even that many elements may be pushing the validity of the data. Since the measurements of the standards are lossy, the true values are also somewhat different from those shown in the figure.

The most straightforward method to find the model elements is to take three or four values, and use them to solve directly for the coefficients, but this can sometimes lead to ill-conditioned values especially in the case of the inductor above, where at the first two frequencies the reactance is actually slightly negative. In this example, some matching to a three element model yielded the curves found in Figure 9.11 for the values given in Table 9.1; these values were adjusted from the directly computed values to avoid the situation of having two very large coefficients with opposites signs, and so the fit is not exact, but the curve's response is well controlled. Values computed directly from measurements using matrix techniques will create a curve fit that exactly matches the measured values, but can have very large coefficients – some positive and some negative – which almost completely cancel to give an appropriate curve between data points, but any extrapolation is very poor. Thus, unless one uses measured points at the extremes of the frequency range, one must be careful that the polynomial model does not "blow up" outside the exact measured ranges.

The offset delay for the open and short is half the difference between the electrical delay of the load and of the open or short. One must use half the delay, as electrical delay is set as the total round trip delay, but the standards use only the one-way transmission delay. If port extensions are used to rotate the phase (port extensions operate on all traces; electrical delay operates independently on each trace) then the two-way response is implied in the extension delay and the value read from the port extension is used directly. The offset delay values are given in Table 9.1 as well.

Finally, since the physical thru-standard is twice as long as the open or short, and longer than twice the load reference, a defined delay must be added, and in this case it is simply twice the open or short delay.

Using this calibration kit for the PCB calibration will give the reference plane at the load port, and measuring the thru, one will see 63 psec of delay. If it is desired to have the reference plane at the center of the thru line, one may simply add 31.5 psec of port extension to each port.

9.1.2.6 Measurement Results with a PCB Cal Kit

The PCB and Cal Kit definitions can now be used to calibrate the VNA to the PCB connectors, and measurement on components at that reference can be performed. It is highly recommended to calibrate over a very broad range at first, to see if there are any issues with the Cal Kit or standards. A common problem is a Cal Kit defined over a narrow range, for measuring a

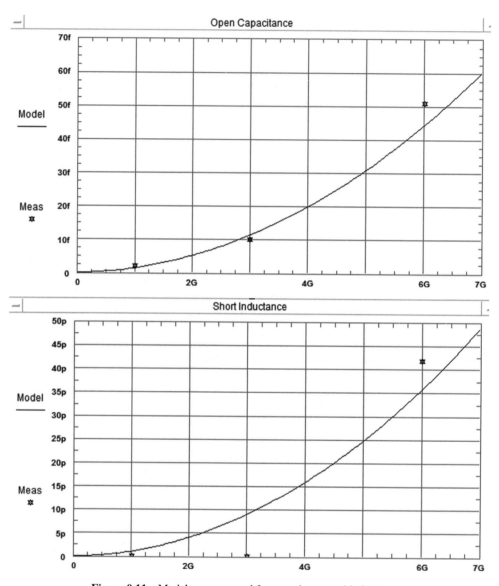

Figure 9.11   Models vs measured for capacitance and inductance.

particular component in a particular application. Later, the same Cal Kit is used over a different range without the user realizing that the values had been optimized over only a narrow range. In particular, the fringing polynomials may extrapolate poorly if they were defined over only a narrow range. The results of using this calibration kit on a 26.5 GHz sweep are shown in Figure 9.12, after performing an unknown thru (SOLR) calibration. In this case, there are large spikes in the results above 20 GHz, likely due to the open and short standards not maintaining sufficient separation at high frequencies. In such a case, the reflection from each might overlap,

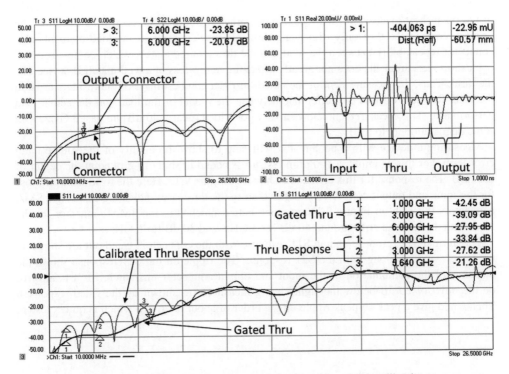

Figure 9.12   Measurement of the thru after an unknown thru calibration.

and the calibration becomes undefined at those points. This wideband sweep shows no issues well past twice the expected frequency of operation, 6 GHz.

Since the open, short and load are all defined in the Cal Kit, the unknown thru calibration can be used. After calibration, re-measuring the response of the thru can show a lot about the quality of the calibration, since it is essentially an independent standard. For these types of evaluations, it is still important to use wide frequency sweeps even if the Cal Kit is only intended for use at a lower frequency range; for example, this one is optimized for 6 GHz but 26.5 GHz sweeps allow very good resolution in the time domain for further analysis. Figure 9.12 shows several views of the thru, with a key one – the time domain response of S11 – shown in the upper right window. In the lower plot, the calibrated thru response is shown for S11 of the thru. The definition of the thru should be 0 dB insertion loss and no return loss, but here one sees a return loss of 21 dB near 6 GHz. Also shown in the plot is the time gated response of the thru. Here now one sees a much better response, when the effects of the input and output connector are removed.

These connectors should have been removed by the calibration process, but because of connector-to-connector repeatability, residual connector match remains. The upper left plot shows the results of time gating the input connector from the S11 trace and the output connector from the S22 trace. These are similar, though slightly worse, to the some of the individual estimates of connector repeatability in the previous figures. These represent not only the connector repeatability between connectors, but also any errors in the estimation of the open/short/load, which will cause the characterization of the port to be in error. The primary error in the thru only (time gate thru response) is error in the load element, plus connector repeatability between the load and the thru (source match, which comes from the open/short measurements does not greatly affect the return-loss of a well matched device like the thru). Thus, the measurement of the gated-thru is a good estimate of the residual error in the load. Residual error is the difference between the model of the load and the actual load performance. This measurement shows a very good load (nearly 40 dB) up past 3 GHz, with some degradation up to 6 GHz.

Measuring a device with a poor match will help to show the quality of the source math error term. For this example, a 100 ohm resistor is added in shunt to ground at the center of the PCB. The response for this element is shown in Figure 9.13. Here the residual return loss for the input and output connector is better than in the thru case, indicating this set of SMA connectors (far right in Figure 9.1) are much better matched to the SMA connector for the load, with the residual error around 40 dB for 3 GHz, and 30 dB for 6 GHz.

The normal S11 response and an S11 response with the connectors gated out are shown in the lower window, along with the S21 response. The ripple on the S11 response shows the evidence of the combination of the input and output connector mismatch, for a total error on the order of 20 dB below the signal being measured (as shown by the gated response). This would give a peak-to-peak ripple on the order of about ±1.6 dB, which is quite close to

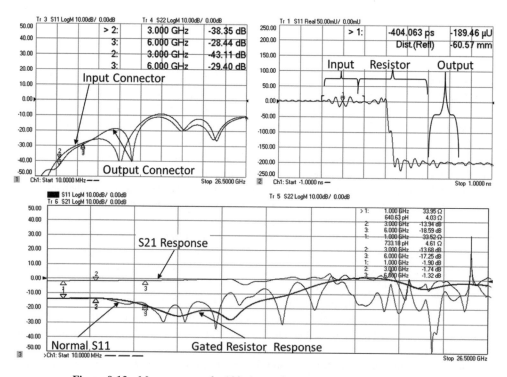

Figure 9.13 Measurement of a 100 ohm resistor connected shunt to ground.

what is measured between 3 and 6 GHz. Thus, the ripple in this measurement may be entirely attributed to the non-repeatable SMA connectors.

As a final evaluation, the same component was measured to just 6 GHz, using the PCB calibration kit, and a simulation model was created composed of a shunt resistor with series parasitic inductance, and with a short length of transmission line to represent the offset from the center of the thru to the reference plane, as shown in Figure 9.14. This simple model gives an indication of the expected S21 curve and S11 response.

The model's inductance was matched to the DUT measurement using the S21 trace. The transmission line delay was matched using the S11 trace on the Smith chart. The upper plot shows the S11, S22 and model S11 on the same graph. Here again it is clear that source match is causing the ripples that produce the deviation from the model. This ripple can be evaluated as shown in Chapter 3, Eq. (3.85) to estimate the sum of the directivity and source match error. In this case, there is 0.45 dB ripple at 3 GHz, and some offset from the model. The ripple is very likely measurement error because the model shows S11 = S22 for all frequencies. Of course, the simple model may not be an exactly correct estimate of the match of the shunt resistor, so the additional offset from the model may be due to the very simple nature of the model.

From the ripple of 0.45 dB, one can estimate the residual errors as about −40 dB. Some of this is from the connectors, but some might be because the definition of the open and short is not correct in the Cal Kit, causing an error in the source match. Since the directivity error

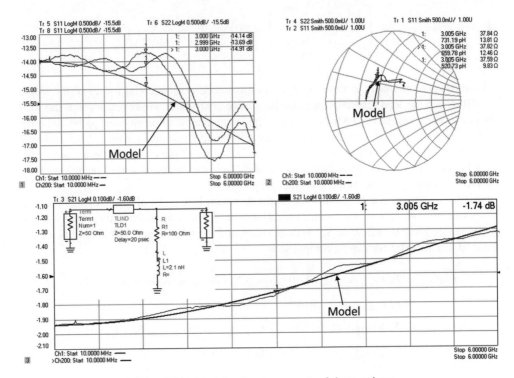

Figure 9.14 Model and measurements of shunt resistor.

is estimated from the gated time domain response in Figure 9.12 to be about -39 dB, this might entirely account for the offset. The connector residual match is about -38 dB at the input and -43 at the output, at 3 GHz. These can be added to the S11 value (estimated as the center of the ripple, about -14 dB at 3 GHz) to see the total error effect. If the sum of the ripples is evaluated from the residual directivity and each connector is smaller than the total ripple observed, the remainder must be source match; if there is no remainder, then the source match error (due to improperly defined opens and shorts) must be small. These errors as a contribution to the total ripple are computed as

$$EDF_{Linear} = 10^{-39/20} = 0.011, \quad \Delta_{Conn1} = 10^{-38/20} = 0.013, \quad \Delta_{Conn1} = 10^{-43/20} = 0.007$$

$$S11_{Linear} = 0.2, \quad S21_{Linear} = 0.82$$

$$Error_{EDF} = \frac{20\log_{10}}{2}\left(\frac{1 + \dfrac{\Delta EDF}{S11}}{1 - \dfrac{\Delta EDF}{S11}}\right) = 20\log_{10}\left(\frac{1 + 0.055}{1 - 0.055}\right) = 0.48 \text{ dB}$$

$$Error_{Conn1} = \frac{20\log_{10}}{2}\left(\frac{1 + \dfrac{\Delta_{Conn1}}{S11}}{1 - \dfrac{\Delta_{Conn1}}{S11}}\right) = 20\log_{10}\left(\frac{1 + 0.013}{1 - 0.013}\right) = 0.56 \text{ dB}$$

$$Error_{Conn2} = \frac{20\log_{10}}{2}\left(\frac{1 + \dfrac{S21^2\Delta_{Conn2}}{S11}}{1 - \dfrac{S21^2\Delta_{Conn2}}{S11}}\right) = 20\log_{10}\left(\frac{1 + 0.005}{1 - 0.0005}\right) = 0.2 \text{ dB}$$

$$(9.6)$$

From this computation, it shows that the first connector can account for all of the ripple at 3 GHz, and the load response may also be a large contributor. However, because the load standard and the test DUT (100 ohm resistor) are located in the same physical plane, the error in the directivity term due to the load model error would show up as a very slowly moving error versus frequency, and so probably explains the offset of the trace. The fine grain ripple most likely represents the connector errors as they have the electrical length separation from the DUT to cause plus and minus phasing of the error.

This measurement points out that for moderate return loss (-14 dB), even small residual errors can give substantial errors in the overall measurement. Since each of these match terms is reasonably good, the effect on S21 is quite small, and shows as a small ripple deviation from the model. The S21 error can be computed as

$$ESF_{Raw} \approx ELF_{Raw} \approx -12.66 \text{ dB}, \quad \text{or } 0.23 \text{ linear, (worst case 3–6 GHz)}$$

$$ESF_{Residual} \approx ELF_{Residual} \approx -40 \text{ dB}, \quad \text{or } 0.01 \text{ linear}$$

$$\Delta S_{21\_p-p} \approx 2 \cdot 20\log_{10}\left(1 + ESF_{Raw}\,ELF_{Residual} + ESF_{Residual}\,ELF_{Raw}\right)$$

$$(9.7)$$

$$= 2 \cdot 20\log_{10}\left[1 + (0.23) \cdot (0.01) + (0.01) \cdot (0.23)\right] = 0.08 \text{ dB}$$

which falls very well in line with the ripple seen in Figure 9.13.

9.1.2.7 Conclusions on PC Board Fixtures

Creating fixture calibration standards and kits, whether for PCBs or other fixtures, is a reasonable approach to improving component measurements. With some care, custom Cal Kits can be created and characterized to provide a custom in-fixture calibration. This section presented a variety of approaches to characterize and verify the performance of an example calibration kit, including measurements of components and comparison to estimated model responses. Using SOLT with the unknown thru method allows one to use the thru as a verification standard, and with time domain gating, one can verify the connector repeatability and the load model response. In this example, additional verification was attempted using a model of a well-controlled shunt element (100 ohm resistor) to demonstrate the quality of the calibration.

In many cases, in-fixture calibration is not practical or possible, so other fixture compensation methods must be used, as discussed below.

9.2 Fixturing and De-embedding

Another approach to dealing with component fixtures is to calibrate in a common coaxial connector, such as SMA, and then attach the fixture and measure the combination of the fixture and the DUT. If some information about the fixture can be obtained, through independent measurements or modeling, then the fixture effects can be removed and the DUT response can be de-embedded from the fixtured measurements. If in-fixture standards are available it is relatively straightforward to characterize the fixture and determine its S-parameters (see Section 9.3.1).

But, it is often difficult to create and model in-fixture standards and so techniques have been developed to measure the combination of the fixture and some limited cal standards, and determine at least some of the fixture effects. This then provides a means so that the S-parameters of the fixture are known, or at least estimated. Once the S-parameters are determined, de-embedding of the fixture can be accomplished by modifying the calibration coefficients such that the error-corrected results remove the fixture effects as well as the effects of match, directivity and tracking of the VNA. Techniques for determining the fixture characteristics will be discussed in later sections.

Another use for de-embedding comes from the need for multiport testing. In many cases, the DUT has a $2 \times N$ path structure, such that the measurements are desired on any two-ports of an N-port DUT. While normal calibration processes would require $(N-1)(N/2)$ calibration paths (e.g., a 10-port DUT requires 45 two-port calibrations), using de-embedding techniques, the calibration can be reduced to one two-port calibration and $(N-2)$ one-port calibrations.

De-embedding techniques can also be used to modify the effective port impedance of the VNA, allowing measurements made on a DUT in the system Z_0 to be translated to any other impedance. This is particularly useful in balanced devices where the input impedances are often not matched to the VNA system Z_0.

Related to impedance transformation, port-matching is a de-embedding technique that allows the measured S-parameters to reflect the condition of additional virtual components added to the input or output of the DUT; this is called port matching or embedding. It is common in cases where devices such as ICs are not well matched with the intention that off-chip components will perform the matching function; it desired to know the response of the IC after the matching element is added, as was done in Section 6.1.5.1. In the case of a DUT being used in a cascade of other devices, the S-parameters of which are already known, the response of the overall system can be modeled by measuring only the DUT and using port matching to embed the characteristics of the other elements of the system. Often attributes of the DUT, such as bias, are adjusted to obtain the desired overall system performance. An example of this is using port matching to add the effect of a CATV cable loss to the measurement of a CATV amplifier. The specifications on the amplifier state a particular frequency flatness when driven into a long length of cable (up to a km or more). In legacy measurement systems, actual rolls of CATV cables were used, but each cable had slightly different loss, and results were not reliable from one test station to another. Emulating the loss of a cable, using embedding, removes one source of non-repeatable error from the results.

9.2.1   De-embedding Mathematics

The most common method for de-embedding fixture effects from a DUT is to modify the calibration coefficients to account for the effects of the fixture. The signal flow graph for the forward error terms is shown in Figure 9.15.

Computing the fixture effects on the systematic error terms can be accomplished through a variety of methods applied to the entire flow graph, but a convenient method that is not often discussed is through the use of intermediate variables in the flow graph and through the use of a select case for the S-parameters of the DUT. Consider the first case where the DUT is well matched and unilateral, that is $S11 = S12 = S22 = 0$ and $S21 = 1$, which results in the flow graph of Figure 9.16. The choice of DUT has no effect on the error terms, so a convenient choice can be made to simplify the computation of the error terms.

For the fixturing computations, the error terms are converted to the equivalent simplified flow graph of Figure 3.23, the fixtured terms replacing the nominal terms. Note that here the source loss and the reference tracking are included in the signal flow diagram. During the de-embedding process, all of the error terms are modified such that the new error model includes the effects of the fixture. While this technique has been used for many years [1], the application to source and receiver power calibrations were first presented in commercial VNAs in early 2011.

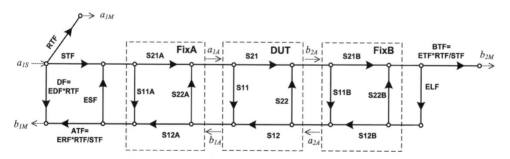

Figure 9.15   Signal flow diagram for a DUT in a fixture.

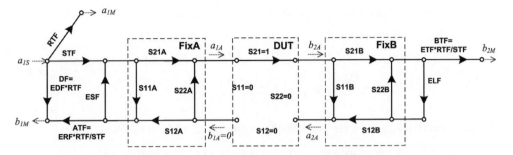

Figure 9.16 Fixture flow graph for a unilateral DUT.

In the case of Figure 9.16, $b_{1A} = 0$. With this diagram, the only loop terms are from ESF and S11A on the input side, and ELF and S22A on the output side. By inspection one can see that

$$DF_{Fix} = \left. \frac{b_{1M}}{a_{1S}} \right|_{S11=S12=0} = DF + \frac{STF \cdot S11A \cdot ATF}{1 - ESF \cdot S11A} \tag{9.8}$$

$$STF_{Fix} = \frac{STF \cdot S21A}{1 - ESF \cdot S11A} \tag{9.9}$$

$$ATF_{Fix} = \frac{ATF \cdot S12A}{1 - ESF \cdot S11A} \tag{9.10}$$

$$ESF_{Fix} = S22A + \frac{S12A \cdot ESF \cdot S21A}{1 - ESF \cdot S11A} \tag{9.11}$$

Because the fixture has no effect on the signal to the reference channel, the value for RTF is unchanged

$$RTF_{Fix} = RTF \tag{9.12}$$

From the output side, the error terms modified by the output fixture are computed as

$$BTF_{Fix} = \frac{S21B \cdot BTF}{1 - S22B \cdot ELF} \tag{9.13}$$

$$ELF_{Fix} = S11B + \frac{S21B \cdot ELF \cdot S12B}{1 - S22B \cdot ELF} \tag{9.14}$$

The traditional 12-term error model values are easily derived from Eqs. (9.8) (9.9), (9.10) and (9.13). The error term DF differs from the traditional EDF error term by the loss to the reference coupler

$$EDF = \frac{DF}{RTF} \tag{9.15}$$

The fixture version of EDF can be related to the 12-term model as

$$EDF_{Fix} = \frac{DF_{Fix}}{RTF} = \frac{DF}{RTF} + \frac{STF \cdot S11A \cdot ATF}{RTF\,(1 - ESF \cdot S11A)}$$

$$= EDF + \frac{ERF \cdot S11A}{(1 - ESF \cdot S11A)}$$

(9.16)

The reflection tracking term can get defined in terms of the 12-term model as

$$ERF_{Fix} = \frac{ATF_{Fix} \cdot STF_{Fix}}{RTF_{Fix}} = \frac{1}{RTF} \cdot \frac{ATF \cdot S12A}{1 - ESF \cdot S11A} \cdot \frac{STF \cdot S21A}{1 - ESF \cdot S11A}$$

$$= \frac{ATF \cdot STF}{RTF} \cdot \frac{S21A \cdot S12A}{(1 - ESF \cdot S11A)^2}$$

(9.17)

$$= \frac{ERF \cdot S21A \cdot S12A}{(1 - ESF \cdot S11A)^2}$$

And ETF of the fixtured calibration as

$$ETF_{Fix} = \frac{BTF_{Fix} \cdot STF_{Fix}}{RTF_{Fix}} = \frac{1}{RTF} \cdot \frac{BTF \cdot S21B}{1 - S22B \cdot ELF} \cdot \frac{STF \cdot S21A}{1 - ESF \cdot S11A}$$

$$= \frac{BTF \cdot STF}{RTF} \cdot \frac{S21A \cdot S21B}{(1 - ESF \cdot S11A)\,(1 - S22B \cdot ELF)}$$

(9.18)

$$= \frac{ETF \cdot S21A \cdot S21B}{(1 - ESF \cdot S11A)\,(1 - S22B \cdot ELF)}$$

And for completeness, the final term is the crosstalk term, which like the RTF term, is not affected by the fixture

$$EXF_{Fix} = EXF$$

(9.19)

The error terms for the reverse direction are computed in a similar manner.

9.3   Determining S-Parameters for Fixtures

The best possible way to de-embed the fixture is to determine the exact S-parameters of the fixture. But determining the exact S-parameters for fixtures can be quite difficult if a calibration kit for each fixture port is not available. One approach is to model the fixture in a linear simulator or 3-D EM structure simulator, and this approach is often used to remove effects of connection pads to IC devices from on-wafer probe tests. Another approach is to characterize the loss and delay of the fixture using port extension techniques. Still other techniques use time domain gating applied to a pair of fixtures connected by a thru, with special mathematical functions use to split the fixture effects apart. These methods are discussed in detail in the next section.

9.3.1 Fixture Characterization Using One-Port Calibrations

The simplest technique for characterizing a fixture or adapter is to directly measure its characteristics using a two-port calibration. Often, however, this is not possible as the full set of calibrations standards may not be available, or the physical limitations may make it impossible to do a full two-port calibrated measurement on the adapter or fixture. However, the exact S-parameters of the fixture can be determined if a one-port calibration can be applied before and after the fixture. Formulations from Eqs. (9.11), (9.16) and (9.17) provide the basis for computing the fixture's S-parameters.

These error term equations, while derived from the premise of adding a fixture to an already calibrated test port, could equally well be used to relate the S-parameters of an unknown fixture to a one-port calibration before and after the fixture. However, while there are four unknown S-parameters, a one-port calibration presents only the three error term equations. The order of the problem is reduced by presuming that the fixture or adapter is passive and bilateral, such that $S21 = S12$. With this choice there are only three equations and three unknowns. If the first tier calibration (at the input of the fixture or adapter) has error terms EDF, ERF and ESF, and the second tier calibration (at the output of the fixture or adapter) has error terms EDF_{Fix}, ERF_{Fix} and ESF_{Fix} then the S-parameters of the fixture or adapter are computed as

$$S_{11\_Fix} = \frac{(EDF_{Fix} - EDF)}{[ERF + ESF \cdot (EDF_{Fix} - EDF)]}$$

$$S_{21\_Fix} = S_{12\_Fix} = \frac{\sqrt{ERF \cdot ERF_{Fix}}}{[ERF + ESF \cdot (EDF_{Fix} - EDF)]} \qquad (9.20)$$

$$S_{22\_Fix} = ESF_{Fix} + \frac{ESF \cdot ERF_{Fix}}{[ERF + ESF (EDF_{Fix} - EDF)]}$$

These are identical formulations to the adapter removal calibration as described in Chapter 3.

9.3.1.1 Computing the Square Root of a Complex Frequency Response

There is a subtle point to defining the phase of S21: the square root function can have a positive or negative result; in complex form, this manifests itself into a 180° uncertainty about the phase of the S21 term. For an isolated point, there is no way to determine the proper root, but if only a little information is known, particularly its nominal delay, then the proper root is chosen so that the phase response lies nearest to the phase predicted by the nominal delay. An alternative method for defining the phase response is to plot the unwrapped phase response of $S21^2$ as a function of frequency, and divide it by two. This gives the correct phase and delay provided that the data used has less than 180° phase change per frequency point. The only remaining uncertainty is the phase of the first point. This is found by determining the slope of the phase and projecting it back to DC, or to put it another way, using the delay of the adapter to predict the approximate phase offset at the first point of the data trace. Mathematically, computing the phase versus frequency with the proper offset from the S21 phase data is derived as

$$\phi_{(f)Offset} = \frac{\phi_{(f)\_Unwrap}}{2} + \left(Int \left[\frac{(\phi_N - \phi_0)}{(f_N - f_0)} \cdot \frac{f_0}{360} \right] \right) \cdot \frac{360}{2} \qquad (9.21)$$

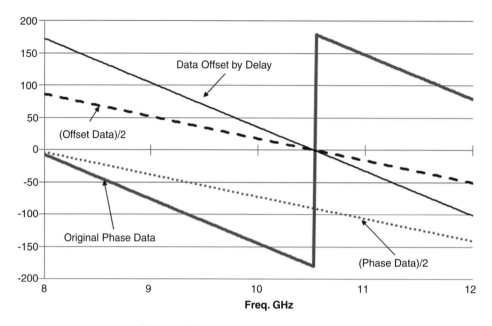

Figure 9.17　Determining the phase of S21.

where N is some point chosen in the trace to compute the group delay. Using the entire trace produces the lowest noise in the group delay estimation, but may not be practical if the fixture has non-constant delay at higher frequencies. This works well for normal fixtures or adapters that have a nearly constant group delay, but can have some difficulties when the adapter is band limited, as in the case of waveguide adapters. In particular, the delay of the waveguide portion of the adapter is not constant with frequency but the delay of the coax portion is constant with frequency.

An example of the phase response computation is shown in Figure 9.17, where an adapter is characterized over a narrow frequency span, starting at 8 GHz. The trace shows the normal (wrapped) phase response of the adapter, "Original Phase Data", and the square root phase based on dividing the unwrapped original phase data by 2, "(Phase Data)/2". The projection of the 8 GHz point back to DC indicates that a 180° offset is needed if the original phase is to cross the DC value at 0° phase, and the original unwrapped phase, offset by 180°, is shown as "Data Offset by Delay". Finally, the correct square root phase is computed by dividing the phase response by two to obtain the final result labeled "(Offset Data)/2". This result can be validated by measuring the S21 response of the adapter, if possible, or by doing the characterization over a broader frequency range extending down to DC, so that no phase wraps are neglected. In fact, when using relatively broadband devices, it is always good practice to measure them over an extended frequency range. Going lower in frequency avoids phase wrapping issues as discussed here, and going higher in frequency can illuminate other effects, such as intermittent connections, more easily.

One final note for characterizing adapters for the purpose of de-embedding is that while the characterization results in the four S-parameters, the numbering of the ports is sometimes a point of confusion. In common practice, port 1 of the de-embedding network faces the VNA test port. This makes perfect sense for de-embedding from port 1, as that is the natural choice based on how the measurement of the adapter is performed. However, when the same device is measured in the same way, with the intention of de-embedding it from port 2, then the natural choice is port 2 of the adapter facing the test port. But this logic fails in the case of more than two ports for the VNA. Thus, the convention of de-embedding networks always having port 1 facing the test port ensures a common practice for 1, 2 or N port VNAs. However, this requires the data for an adapter to sometimes be reversed, exchanging port1 for port 2, so S11 becomes S22, with no effect on passive adapters as S21 = S12. In the most modern VNAs, provisions are made in the de-embedding settings to reverse the de-embedding networks ports, if desired, an example of which is shown in Figure 9.18.

Another important consideration when de-embedding is to ensure that the de-embedding network matches the frequency range of the measurement to which it is applied. In this case, the S2P file range only covers a portion of the measurement range. Newer VNAs provide the ability to extrapolate S2P data to cover the entire frequency range. For some devices, such as attenuators, this is not too unreasonable, but one should carefully consider whether extrapolation is valid for the particular adapter or fixture used.

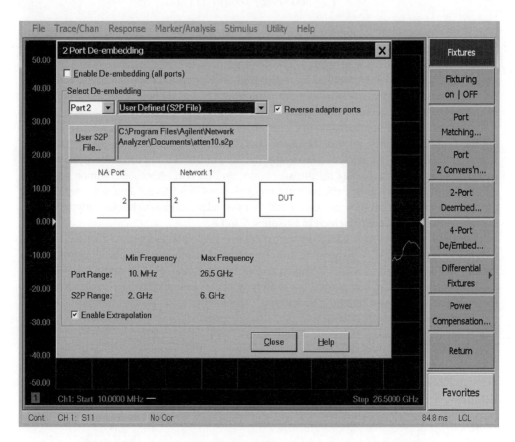

Figure 9.18 De-embedding setup dialog allowing arbitrary port selection.

9.3.1.2   Port Extensions

When measuring microwave components, it is common to calibrate in a coaxial reference plane first, then add some small adapter or fixture to connect to the DUT. Often, these adapters are well matched, but measurements of the DUT's S-parameters are in error by the phase shift or electrical delay of the adapter. The first VNAs used a mechanical line stretcher to compensate for delay in test fixtures or adapters, sometimes called trombone lines as the structure's center-conductor looked like a trombone when disassembled. These line stretchers were used in one path of the VNA ratio-measurement to add delay to the reference channel to compensate for electrical delay in the test port due to some adapter or fixture.

Starting with the HP8510A, the function of the line stretcher was replaced with a mathematical function that added or subtracted delay to the phase response of the DUT; this function was called port extension. Port extensions worked by adding a phase shift, equivalent to an amount of electrical delay specified in seconds, to the measurement's phase data. The phase shift for a given delay is computed as

$$\phi = \begin{cases} 360 \cdot Freq \cdot Delay : S21, S12 \\ 2 \cdot 360 \cdot Freq \cdot Delay : S11, S22 \end{cases} \tag{9.22}$$

That is, the phase shift is applied twice per port for reflection measurements at that port, accounting properly for a signal's two-way transit through the adapter for reflection measurements. The phase shift is added once per port for transmission measurements, so that the phase response of S21 or S12 is modified by the sum of port extensions applied at each port. For multiport measurements, any transmission parameter's phase change is similarly dependent on the port extension for each of the ports associated with that parameter.

One common, but improper, use for port extensions is to remove the effective delay of the DUT, such as a filter, so that the phase deviation from linear is more easily perceived. The "electrical delay" function is a more appropriate choice as it applies only to the selected parameter rather than affecting all measurements on that port. Another common case is to remove small delay mismatch effects between drive ports in a balanced measurement, where the effective drive must represent exactly 180° for the balanced gain to be correctly measured. Details on these cases are presented in Chapters 5 and 8.

More recently, VNAs have added the concept of loss to the port extension function, allowing the small loss associated with an adapter to be removed, along with its phase response. The loss factors are entered as a known loss at one or two frequencies. If a single known loss is used, the formulation for computing loss at all frequencies follows the classic square root loss function as described in Chapter 1, which is

$$Loss(f) = A \cdot \sqrt{f} \tag{9.23}$$

Once a single loss factor is known, the loss at any frequency can be computed. However, this loss curve does not represent the loss of non-ideal coaxial, microstrip or many other types of transmission lines. In the case of transmission lines other than airline, a more flexible function is used

$$Loss(f) = A \cdot f^b \tag{9.24}$$

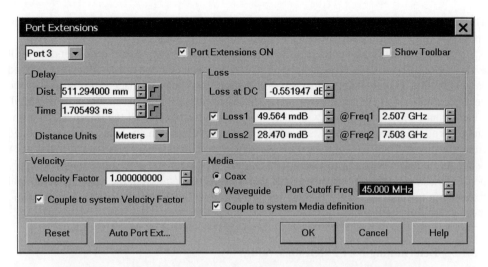

Figure 9.19   Port extension dialog including loss and waveguide compensation.

where A and b are two factors computed by solving for each from the known loss at two select frequencies. Both of the above equations presume no loss at DC, but occasionally a fixture will have some small resistive loss, which is accommodated by a DC loss term in the port extension setup; an example is shown for the port extension dialog of the Agilent PNA series of VNAs, Figure 9.19.

For convenience, newer analyzers provide either a time delay or a distance delay as the port extension entry, where the relationship between distance and time delay is

$$Dist_{Delay} = VF \cdot c \cdot Time_{Delay} \tag{9.25}$$

where c is the free-space velocity of light, and VF is the velocity factor of the media, 1 being the VF for a vacuum (and very nearly air).

Because waveguide transmission lines have strong dispersion, it is necessary to account properly for different phase shifts at different frequencies for a line that has a single port extension delay time. If the media choice is waveguide, the delay is based on the physical length of the guide, with the phase computed according to Eq. (3.36) as

$$\phi_{(f)} = \left(360f \sqrt{1 - \left(\frac{f_c}{f} \right)^2} \right) \cdot Time_{delay} \tag{9.26}$$

The losses and delay of the port extension are sometimes implemented as a de-embedding network that is applied to the calibration factor. While older analyzers applied only the phase offset term, and only as a scaling function in the display as the last step in data processing, modern analyzers have other fixturing attributes and the proper use of port extensions sometimes requires that the port extension math comes before other de-embedding functions such as port matching or impedance transformation. If the other fixturing and de-embedding functions are applied to the calibration set, the port extension must also be applied to the calibration set.

This leads to a functional difference in how measurements are made and port extensions are applied between older, legacy VNAs and more modern versions.

As a scaling function, the port extensions simply adds phase shift to the final parameter result as described by Eq. (9.22), and can be applied to raw or uncalibrated results. But when applied as a calibration de-embedding, a full two-port measurement must be made to accommodate the necessary terms to apply port extensions. In this case, even if a raw measurement is displayed, a background unity cal set (one with idealized error terms of 1 for all the transmission terms and 0 for all the reflection terms) is created, and the port extension attributes are de-embedded from that calset. Thus, an S21 trace with port extension will engender two sweeps in the VNA using the fixturing function to apply port extensions. In general it is more robust to apply port extensions in this way, and it ensures that proper computation in the case that additional fixturing is applied. Normally, turning on any fixturing function induces the de-embedding method for port extension.

9.3.1.3   Determining Port Extensions Values

For most circumstances, it is best to perform a full correction (one-, two- or N-port) at the test port, before applying the fixture or adapter and using port extensions. The one-port calibration on each port removes all the mismatch and directivity effects up to the fixture. If the fixture is well designed, it will have small mismatches. A well designed fixture should have less than 26 dB mismatch for RF frequencies and less than 20 dB mismatch for microwave frequencies.

One of the difficulties in using port extensions to correct for adapters or fixtures is determining the correct value for the port extension delay. In older analyzers, the loss could not be compensated, and must simply be accepted. The delay value was often determined by looking at the phase trace of a fixture with nothing connected to the DUT plane. This open reflection should have approximately 0° of phase across the band. In actuality, this yielded a delay that was slightly too long, as the expected phase shift for an open circuit is a few degrees at low frequency due to fringing capacitance, and can be much larger at microwave frequencies. One can account for the open circuit fringing by estimating the fixture's fringing capacitance based on its similarity to other connectors (in the case of a coax adapter) or by using some modeling to estimate the fringing capacitance of the open fixture (as in the case of a PCB microstrip fixture).

Many analyzers have an "electrical delay" scaling factor, which adds or subtracts phase according the entered delay value. This delay offset is done as part of the formatting, and does not include a two-way response as port extensions do for reflection measurements. But many VNAs include a "marker to delay" function that computes the group delay within a span around a marker on the phase trace (typically ±10%) and then uses that delay value as the electrical delay. If using this in a reflection measurement, half of this value should be used as the port extension. If the S11 of the open fixture is displayed on the Smith chart, the electrical delay or port extension value can be adjusted until the marker reads 0°, or better still, reads the same value as the expected fringing capacitance. This process must be repeated for each of the ports for which port extension compensation is desired.

9.4   Automatic Port Extensions

The process for determining the proper value for port extensions on fixtures can be tedious, and the values obtained can vary due to noise or small perturbations of the S11 trace from mismatch effects. Recently, the process for determining port extension values has been substantially automated, so that it has become essentially as simple as a second tier response calibration.

When the fixture is connected to the test port, a trace with ripple in magnitude and a rapidly changing phase response is the common result. The ripple is almost always due to the mismatch at the coax to PCB transition, re-reflecting with the open of the fixture. In cases where the DUT has ground connections in the same plane as the signal connections, it is a simple thing to create a "shorting block" that is essentially a piece of conductor (metal) the same size as the DUT package. Sometimes a short provides a more reliable reference for port extensions than an open, due to radiation of an unshielded open.

When using automatic port extensions (APE), the VNA measures either an open or a short, or both, and uses a least squares method to fit the delay values to the phase response. An example setup dialog for APE is shown in Figure 9.20. If an open is used, the target phase response is $0°$; if a short is used, the target phase response is $180°$. If both an open and short are used, the measurements are averaged (accounting for the $180°$ phase shift of the short) and the least squares fit is applied to the averaged phase response.

The dialog provides for doing the least squares fitting over the current span of the measurement, or over a user specified span within the current measurements. Sometimes, the fixture used for the DUT has some wildly varying phase response at the upper band edges, but the interest for the measurement is only at a particular portion of the band, usually the center or lower frequency portion. This is the case for fixtures used to measure small ceramic or SAW filters used in cellular phone handsets. For these cases, it is convenient to use a narrow band

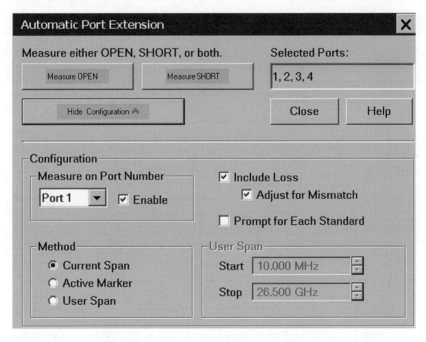

Figure 9.20 Automatic port extension setup.

around the center of the passband to set the port extension value. A third choice uses just the active marker, and computes the delay value at just one point in the trace.

A similar process is used for the magnitude response of the fixture, to determine the loss. For magnitude, the least squares fitting attempts to find the best fit to Eq. (9.24). Once the best fit is found, the values for the curve are determined at three frequency points: one-quarter and three-quarters of the span above the start frequency and a DC point.

Figure 9.21, upper window, shows the measurement of an open and a short in a fixture. The ripple in the magnitude trace is apparent, and the open ripple is nearly the inverse of the short ripple. For wideband measurements, using the average of the open and the short or using the least square fit through either the open or the short yields nearly the same result. For narrow band measurements, using both an open and a short can avoid an offset error due to the ripple. In this way, the effect of the mismatch at the input of the fixture on the port extension loss is minimized. A similar process is used for the phase response. The lower window shows the result of the automatic port extension, with the resulting delay and loss computations shown in the port-extension tool bar above the figures.

Unfortunately, this straightforward approach has the drawback that if the loss is compensated for directly, the ripple in the open or short response will cause the S11 to go above 0 dB, that is, the magnitude of S11 will be greater than 1. While minimizing the maximum error, the result can cause a great deal of difficulty if the data for the DUT also contains high reflections and is used later in circuit simulators. For example, amplifiers often have prematching networks

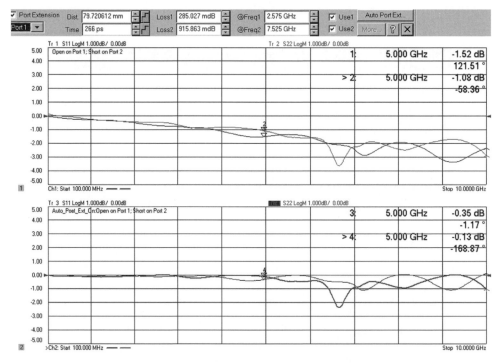

Figure 9.21 (upper) Measurement of an open and shorted fixture; (lower) after automatic port extensions.

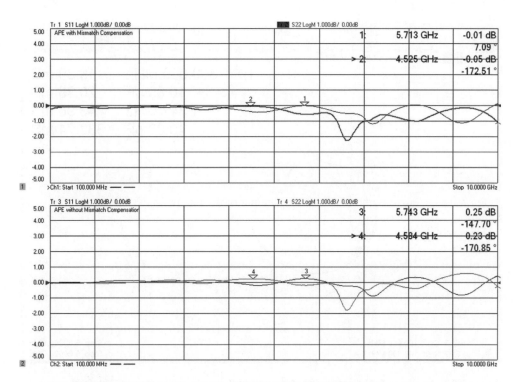

Figure 9.22   S11 response after APE with and without mismatch compensation.

that can have high reflections out of the band of interest. If the S11 goes above 1, many computations "blow up", such as stability factor, and optimizations can become difficult. That is because all passives and most active devices have S11 less than 1.

A special feature of APE allows the user to choose between ignoring the mismatch of the port extension or offsetting the loss computed by the mismatch ripple to ensure that loss compensation does not result in S11 greater than 1. Figure 9.22 shows the result of simple APE compensation, using direct least squares fit (lower window), and a result where APE compensates for mismatch (upper window). The algorithm for mismatch compensation essentially modifies the values in the loss table so that the peaks of the measured response (open or short) are below the S11 = 1 (or 0 dB) reference line. The theoretical max error is greater with mismatch compensation, with the positive error tending toward zero and the negative error being twice the ripple; however, the S11 trace is always well behaved and this usually provides a better case for simulations.

APE is especially convenient for fixtured parts in a multiport or balanced fixture. It is even possible to use APE in a PCB fixture that has a DUT part soldered in place. Figure 9.23 shows the S11 measurement of a shunt 10 pF bypass capacitor, mounted from the center of the fixture thru trace to ground. The light gray trace shows the uncompensated S11, from which it is impossible to infer any attributes of the DUT. The dark trace shows the result with APE applied. From the low frequency response, near 300 MHz where the trace crosses the j50 ohm susceptance line (that is, where the susceptance is 0.02 siemens), marker 1 shows the

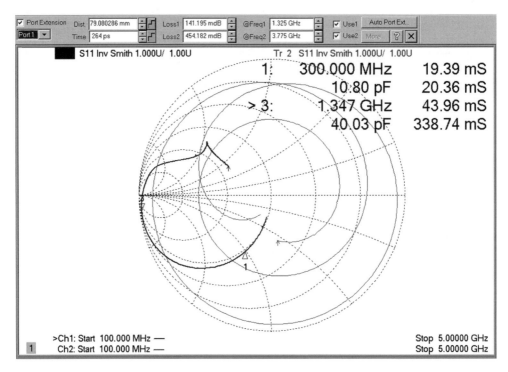

Figure 9.23 Measurement of a shunt 10 pF capacitor; (light gray) as measured in the fixture; (dark) with APE applied.

correct value for the shunt capacitance. Marker 2 is positioned by searching for the maximum reflection, and shows the self-resonant frequency of approximately 1.35 GHz. From this, we can compute the effective series inductance of this capacitance as

$$L_{SRF} = \frac{1}{(2\pi f_{SRF})^2 \cdot C} = \frac{1}{\left(2\pi \left(1.35 \cdot 10^9\right)\right)^2 \cdot 10 \cdot 10^{-12}} = 1.4\,\text{nH} \tag{9.27}$$

This is quite in line with other estimates made for the series inductance of an SMT capacitor. Using APE allowed a simple approach to removing the fixture effects for this device.

The complete S-parameters in LogMag format are shown Figure 9.24. As a side note for designers, from the S21 plot (dark trace, lower window), it is clear that this 10 pF capacitor does not provide a wide range of low impedance bypass, because at even low RF frequencies, the series inductance dominates the response. And when used as a filter element, the effects of the series inductance must be included in the design or the filter response is likely to be shifted much lower in frequency.

Most RF parts have high reflections when unpowered, so the APE process can be performed even on a fixture even if the DUT part is attached, provided the power is removed. The phase compensation will be valid if the part has a reasonably high reflection. In this case, the loss compensation should be disabled unless it is known that the part's impedance in the unpowered state provides a full reflection.

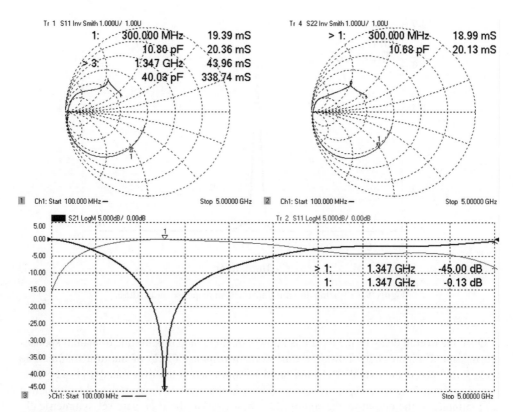

Figure 9.24 Complete measurement (S21 is dark trace) of a shunt 10 pF capacitance with APE.

9.5 AFR: Fixture Removal Using Time Domain

Automatic port extension is a simple way to compensate for a fixture's loss and delay, and works on one-port fixtures. Another common method for compensating for loss in PCB or other fixtures is to create an identical test fixture as the DUT fixture, but provide a thru connection. The simplest compensation using a thru fixture is to calibrate in a coax standard such as SMA, then measure the thru and normalize the trace to the thru using data into memory and data over memory. While this provides a degree of normalization, mismatch in the input and the output of the test fixture can cause significant errors, up to ±1 dB on a transmission measurement.

In recent years, advanced automatic fixture removal (AFR) techniques have been developed which make use of time domain measurements (for more information on time domain, see Chapter 4) on PCB fixtures to compensate for input and output mismatch, as well as loss, even if the input and output mismatch are not the same [2].

The time domain fixture removal starts by measuring the fixture thru response in the time domain, as shown in Figure 9.25. To achieve the best resolution, the widest possible span should be used, even if the fixture will be used only over a narrow span. The peak of the response represents the overall delay of the fixture. Alternatively, the average group delay response could be used as well. In many cases, the input and output of the fixture are designed to be the same length, with the DUT reference plane directly in the middle of the fixture.

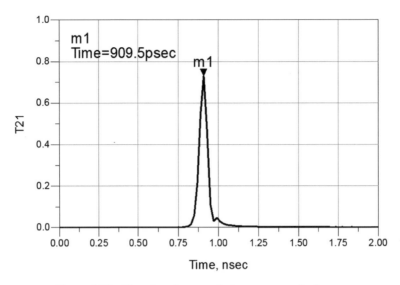

Figure 9.25   Time domain transmission response of a fixture.

Once the overall delay of the fixture is known, the input and output time domain responses are measured. Figure 9.26 shows the time domain response (T11) of the thru fixture. The wide gray trace is the overall T11, and the dark thin trace is the time gated T11. The time domain response shows a capacitive discontinuity at the input of the fixture and an inductive discontinuity at the output of the fixture. It is best to have the time gate set symmetrically about the first reflection: the time difference from the first reflection (here about 46 psec) to the center (at 909 psec) is computed and is subtracted from the first reflection, to set the start time gate at −817 psec. The gated S11 response is shown in the figure as a narrow, dark trace. It is clear that after the stop gate, the trace has a constant value. The small offset in the baseline is a result of the DC loss of the transmission line in the fixture; a review of the fixture shows about 1.5 ohms of DC loss, which represents a reflection coefficient of 0.015, almost exactly matching the baseline offset of Figure 9.26.

This gated response represents the time domain measurement of S11 of the left fixture, or fixture A from Figure 9.15.

In Figure 9.27, the overall response of the thru fixture (a fixture with a thru in place of the DUT) is shown in a light, narrow trace, and shows substantial ripple. Also shown is the gated S11 response of the thru fixture, S11A (dark trace) and the independently obtained actual measurement of S11A of fixture A (wide, light-gray trace). It is clear that the gated response very closely matches the actual value of the fixture.

The value of S22B of fixture B, the output portion of the thru fixture, is found by gating the S22 response of the thru fixture in a similar manner. In this way, six known values are obtained: S11A of fixture A, S22B of fixture B, and the four S-parameters of the thru fixture, which are designated as S_{11T}, S_{21T}, S_{12T}, S_{22T} of the thru fixture. This leaves three unknown S-parameters for each fixture.

The rest of the values of S-parameter for fixture A and fixture B can be obtained by presuming that the S21A = S12A, S21B = S12B, which leaves only four total unknowns:

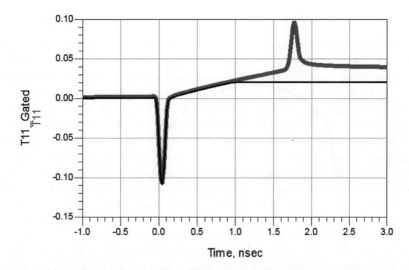

Figure 9.26 Time domain response (gray, T11) and gated response (black, T11_Gated).

S21A, S21B, S22A, S11B. The four S-parameter measurements of the original thru give sufficient independent equations now to solve for these values.

Figure 9.28 shows the computed value for S22A of the sample fixture as a dark trace, and the independently measured actual value as a wide gray trace. The results overlap almost completely with only slight difference at the edges of the bands.

Figure 9.29 shows the comparison of the value for S21A (S21A_AFR, narrow, dark trace) computed using the AFR technique, with that of the independently measured value for S21A of fixture A (S21_FixA, wide light-gray trace), with almost perfect results.

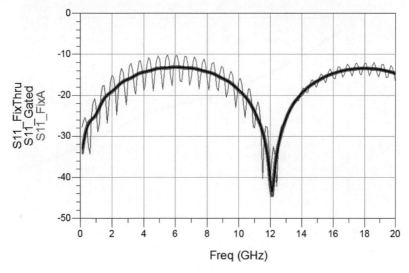

Figure 9.27 Frequency response of the thru (S11_FixThru, narrow light trace), the gated S11 (S11_Gated, wide black) and the actual fixture S11 (S11_FixA, wide, gray trace).

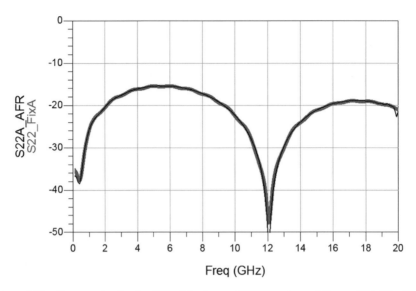

Figure 9.28 Computed value for S22 of the fixture (S22A) and actual fixture S22 (Fix_S22).

Thus, even for a fixture with non-symmetric mismatch, it is possible to determine the individual values for the input fixture (fixture A), and output fixture (fixture B) using only a thru measurement. An example is shown in Figure 9.30 showing a measurement of filter in a fixture (Filter_Fix11, wide gray trace), and the same measurement but with AFR applied (Filter_AFR, dark narrow trace), as well as an independent measurement of the actual filter (Filter_Actual, narrow, light-gray trace) for S11 on the left, and S21 on the right. The filter response is greatly

Figure 9.29 Computed value of S21 (S21A_AFR) and actual fixture S21 (S21_FixA).

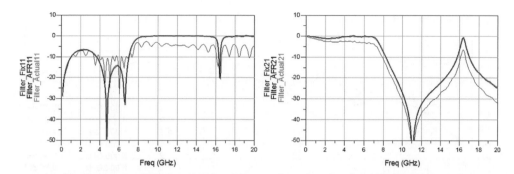

Figure 9.30 Filter measurement comparing the actual filter, filter in a fixture and with AFR applied.

improved over the fixtured response, using nothing more than the measurements of the thru fixture and AFR techniques for compensation.

In some cases, the DUT will not be centered in the fixture, so the loss and delay of fixture A and fixture B will not be identical. In these cases, the offset loss and delay can be easily accommodated by first performing AFR on a thru standard in the fixture, then adding an open measurement for the fixture, where the open fixture occurs the same physical point as where the DUT will be inserted. Using the open response, an APE is applied as described in Section 9.4 to obtain a loss and delay offset from the fixtured results obtained using a thru that was presumed to be centered. This will result in a small positive port extension for one port and an equal but opposite port extension for the other port.

These fixture removal techniques represent the state-of-the-art in dealing with PCB and similar fixtures. Further, the same techniques can be applied to balanced measurements, where the balanced parameters are substituted for the single ended parameters.

9.5.1.1 AFR Measurement Example

A simple and effective example for evaluating the AFR technique in a real application is to use the same example PCB and DUT as in Section 9.1.1. In this case, a calibration is first performed over a broad frequency and the thru standard is measured. The AFR technique of above is applied and the individual PCB input and output fixture is computed. The resulting S-parameters of these fixtures are shown in Figure 9.31.

The values for the input and output return loss for the fixture match quite closely with the gated estimates of the thru standard input and output match from Figure 9.2.

The final step for AFR is to use the same calibration as when used to characterize the thru standard, and de-embed the input and output fixtures from this calset. After the de-embedding step, the sample 100 ohm shunt resistor from the first example is re-measured, the comparison of the results being shown in Figure 9.32.

In some ways this is a remarkable result in that only one standard is used, the thru standard, but the S11 and S22 measurements, as well as the S21 measurements follow very closely to the measurements performed with the PCB Cal Kit.

The residual difference between the AFR measurements and the PCB Cal Kit measurements is less than −40 dB below 3 GHz, and less than −30 dB up to 6 GHz. These residuals are

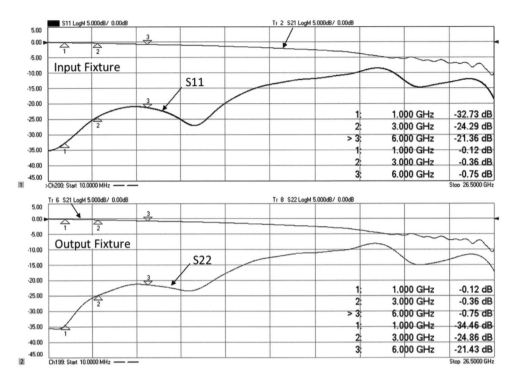

Figure 9.31 Using AFR to compute PCB input and output fixtures.

quite small in the normal sense, and on the same order as the absolute errors found for the PCB Cal Kit. This measurement validates the AFR technique on a real-world device.

9.6 Embedding Port-Matching Elements

While de-embedding involves measuring a DUT in the presence of other elements such as a test fixture, port matching and embedding involves measuring a DUT, directly at the VNA ports, but adding the effects of other virtual elements to the measured response of the DUT, as used in the stability matching of Section 6.1.5.1.

A common example of this application is an amplifier which may require external matching networks to bring the optimum power match condition from 50 ohms down to some lower impedance. The amplifier designer may stipulate that the gain of the amplifier is to be measured when some exact value of external components, such as an inductor and/or capacitor, are placed in the input or output port path of the amplifier. In practice, providing a consistent value of external component can present a problem when multiple test systems are involved in manufacturing. An even greater difficulty occurs if the port matching should be done with an on-wafer device. Some probe manufacturers will create custom probes with port matching elements but these can be quite expensive, have long lead times, and it is difficult to verify their correct values.

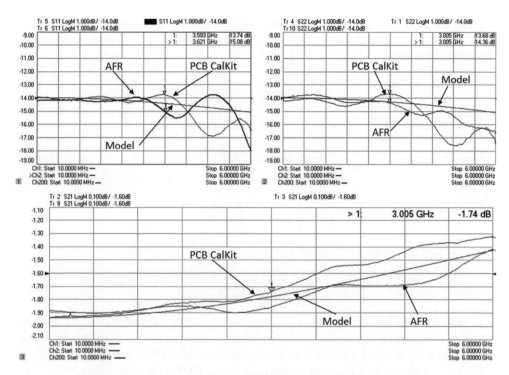

Figure 9.32 Comparing AFR measurement with in-fixture cal standards.

Instead, a virtual port matching can be performed mathematically using techniques quite similar to de-embedding. In practice, de-embedding techniques can be directly used by recognizing that embedding the effects of an S-parameter network is the same as de-embedding the parameters of the equivalent anti-network. The anti-network, S^A, is defined as that network, which when cascaded with the desired network S^N, forms a unity S-parameter matrix S^U, which has the characteristics of S11 = S22 = 0, and S21 = S22 = 1; the flow graph for this is shown in Figure 9.33. Note that as an alternative definition, an anti-network could be defined such that the anti-network follows the desired network, rather than preceding it as below [3], and will result in different values for the anti-network.

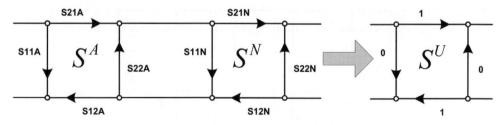

Figure 9.33 Signal flow for an anti-network.

Using normal flow graph analysis to correspond the left flow graph with the unity flow graph on the right, the values of the anti-network are computed. The output term of the anti-network is computed first as

$$S_{22}^A = \frac{S_{22}^N}{\left(S_{11}^N \cdot S_{22}^N - S_{21}^N \cdot S_{12}^N\right)} \tag{9.28}$$

and used to derive all the transmission terms for anti-network values, as

$$S_{12}^A = \frac{\left(1 - S_{11}^N \cdot S_{22}^A\right)}{S_{12}^N} \tag{9.29}$$

$$S_{21}^A = \frac{\left(1 - S_{11}^N \cdot S_{22}^A\right)}{S_{21}^N} \tag{9.30}$$

which are then in turn used to compute the input term of the anti-network as

$$S_{11}^A = \frac{S_{21}^A \cdot S_{12}^A \cdot S_{11}^N}{\left(S_{11}^N \cdot S_{22}^A - 1\right)} \tag{9.31}$$

In other publications, it is common to show the anti-network following the desired network, but such a configuration is not commensurate with the common understanding of de-embedding requiring that the network which is to be de-embedded has port 1 facing the port of the VNA.

Figure 9.34 shows the equivalent representation of a DUT with an anti-network/network pair preceding and following the DUT. Note that the S-parameters of the final cascade before de-embedding are identical to those of the original DUT. From this diagram, it is clear that if the anti-network of the port 1 matching circuit is de-embedded from port 1, and that of port 2

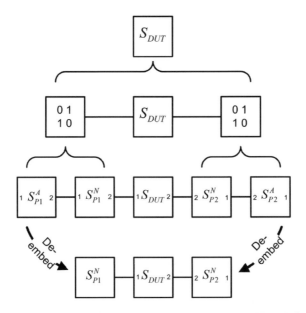

Figure 9.34 Representation of a DUT with input and output port matching, before de-embedding.

is de-embedded from port 2, the S-parameter calculation will be exactly the DUT with the port 1 and port 2 networks' effects embedded in the result.

Using this technique allows port matching of networks to be accomplished using de-embedding functions. Modern VNAs typically include both de-embedding and embedding functions (often called "port matching"), so the step of computing the anti-networks is not required. One last note on computing anti-networks: because the computations include the difference between two values in the denominator (particularly the computation of S22A), the values of the anti-networks can become undefined or infinite at some particular frequencies. Alternative computations, for example using T-parameters or ABCD networks, while having their own numerical issues, may be computable where anti-networks are not.

9.7 Impedance Transformations

Most VNAs are available only with 50 ohm nominal impedance, although some RF models also provide a 75 ohm version. However, in many cases, it is desired to know the S-parameters of a DUT referenced to other than 50 ohms. One method is to transform the S-parameters into some other parameter that does not have a port impedance dependency, such as Z-parameters (see Chapter 2). However, fixturing and de-embedding techniques can provide a simple alternative, by recognizing that the impedance transformation can be implemented by creating an S-parameter model of an ideal, lossless transformer and then embedding that into a measurement. To transform from one impedance to another, the S-parameters of an ideal transformer [3], represented in Figure 9.35 are computed based on the ratio of impedances.

For a transformer, the impedance transformation effect goes as n^2 so that to transform from the reference impedance Z_0 to some load impedance Z_L, the turns ratio is computed as

$$n = \sqrt{\frac{Z_L}{Z_0}} \tag{9.32}$$

and the S-parameters are computed as

$$S_{11}^T = \frac{1-n^2}{n^2+1}, \quad S_{12}^T = \frac{2n}{n^2+1}$$

$$S_{21}^T = \frac{2n}{n^2+1}, \quad S_{22}^T = \frac{n^2-1}{n^2+1} \tag{9.33}$$

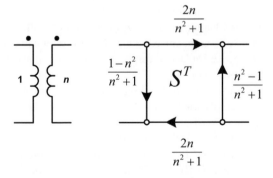

Figure 9.35   S-parameters of an ideal transformer.

To change the effective value of a VNA test port impedance, one may calibrate in one impedance, compute the S-parameters for an ideal transformer to another impedance, and de-embed those S-parameters to allow the VNA readings to match those that would have been obtained if the VNA had been of the desired effective impedance. The S-parameters are correct, but to get the Smith chart to read the correct values of impedance, one must also change the system Z_0, which sets the value for the center of the Smith chart.

But, of course, the actual impedance of the VNA port is unchanged. It is not likely to be exactly equal to either impedance, but rather its impedance will be the raw impedance of the test port. For RF VNAs, at 50 ohms, the actual test port impedance typically ranges from 40 to 60 ohms. For high frequency VNAs, operating up to 50 or 70 GHz, mismatches in the VNA have a larger effect, and it is not uncommon for the VNA impedance to range from 25 to 100 ohms. One should recognize that the raw S-parameters, while measured in whatever impedance the test port provides, are transformed to a 50 ohm reference through the error-correction process. If this transformation is valid, it is also valid to transform to some other impedance.

9.8   De-embedding High-Loss Devices

It is common in the measurement of high power devices to add a large attenuator at the output of the DUT to reduce the power level to the VNA to avoid damage. Because of the loss, the match measurements at the output port can be quite poor, and normal two-port calibration techniques fail to give good results. For example, Ecal modules cannot orient themselves because there is too little difference between impedance states apparent at the VNA test port. A good rule of thumb is that de-embedding devices or fixtures with up to 10 dB insertion loss (one way) can be easily performed. Up to 15 dB of loss, and low reflections (an attenuator would qualify, but not a filter in its stopband) can be de-embedded with care, using averaging and low IF BWs. If the device to be de-embedded has more than 20 dB insertion loss, de-embedding will likely yield poor results.

Consider the case of de-embedding a 20 dB attenuator placed at port 2, at the output of an amplifier. The S21 trace would have a large signal, due to the amplifier gain, even with the 20 dB of attenuation. But the S22 signal would be quite small, and noisy. Likely, the reflection from the attenuator match itself would be larger than the reflection signal from the amplifier. After applying the de-embedding math, the value for ERF would be very small, and the error correction would have the effect of subtracting two very large numbers, one of them noisy, to generate the S21 trace; that is, the noise and error in S22 would be imposed on the S21 trace (this is discussed in detail in Chapter 6). The error correction for the port 2 match would be adding more error than it is removing, since the match of the attenuator presented to the DUT amplifier is typically quite good. Enhanced response calibration, discussed in Chapter 3, can be used in such a case to good effect.

As an example of the limits of de-embedding, consider Figure 9.36. The upper window shows the S11 and S21 of a 20 dB attenuator. The return loss is quite good at −34 dB, and the insertion loss is pretty flat at −20 dB.

This adapter is added to port 2 of a VNA, and is de-embedded; a thru connection is measured and displayed in the middle window. In this case, de-embedding appears to work well, and the S21 measurement is very close to 0 dB. The S11 measurement is essentially the measurement

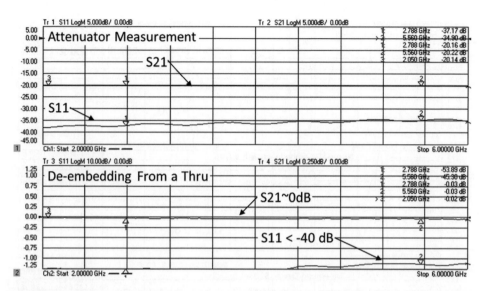

Figure 9.36 (upper) Attenuator S11 and S21; (middle) de-embedding the attenuator; (lower) de-embedding while setting s22 = 0.

of the S22 of the attenuator on port 2 (remembering that the common practice is to label as port 1 the port of the de-embedded device facing VNA port 2).

However, the measurement from port 2 of S22 of the DUT (a thru in this case) is very unstable. Thus the correction for mismatch at port 2 between the attenuator and the DUT is also unstable. In many cases the instability is worse than the errors being removed, and in these cases, changing the value of S22 of the de-embedded device to zero will actually improve the results.

The mismatch error effect would be even larger in the case of de-embedding a device with both large loss and high reflections; a filter in its stopband is a case in point. In such a case the raw S22 is quite large, due to the reflection back to port 2 of the filter to be de-embedded, and the filter's S22 (remember that port 2 of the de-embedded device faces the DUT) presents a very poor load match to the DUT. Any inaccuracy in measuring the error-terms would yield large errors to the S21 measurements, because very large raw errors are being compensated.

One method to allow reasonable de-embedding of a transmission measurement but avoid difficulties in the match correction with very lossy networks is to set the reflection term of the de-embedding network that faces the DUT (normally S22) to zero. When this is set to zero, and the loss is high, the mathematical effect of the load match for the fixture is nearly zero as well, and very little S22 match correction effect is applied. De-embedding of a filter can clearly demonstrate this effect.

Figure 9.37 shows the S11 and S21 response of a filter to be de-embedding. Its measurement is saved as an S2P file. The filter is then de-embedded from port 1. The resulting middle S21 trace should be a flat line over all frequency, but noise, drift and errors in measurement cause the S21 trace to deviate from a flat line when the loss of the filter is greater than about −10 dB, as shown by the markers at the lower and upper passband edges. The fact that the filter cannot be de-embedded successfully for even moderate loss is often a surprise to many users. The

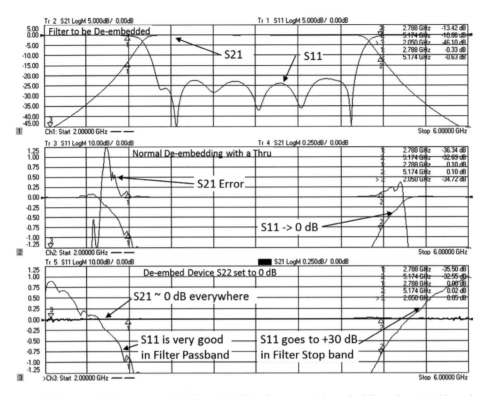

Figure 9.37 (upper) S-parameter of a filter; (middle) after normal de-embedding; (lower) with setting S22 to 0.

actual return loss of the filter is nearly 0 dB (total reflection) at these band edges, so this case is very sensitive to match correction.

Remarkably, a simple change to the filter S-parameters can greatly improve the measurement of S21. The lowest trace is the same network de-embedding, but this time with S22 of the de-embedded network (the filter) set to zero. Here it is clear that in the lossy region of the filter, the quality of the S21 trace with this de-embedding is greatly improved. While normal de-embedding gives an error greater than 0.1 dB when the loss of the network is near 10 dB and greater than 1 dB for losses of 20 dB, the lower trace has less than 0.05 dB error even though the loss is greater than 45 dB. Of course, the de-embedded value for match is in substantial error outside the passband of the filter, but here the transmission measurement quality is the key figure of merit. Further, the normal de-embedding gives terrible results for the S11 as well; in this case the DUT is a thru line with −50 dB return loss, and both de-embedding functions give S11 as 0 dB at the 20 dB of filter loss; it is merely academic that the new method of de-embedding is worse (providing positive S11) at even higher losses.

Using the method of setting S22 of the de-embedded network to zero can remove much instability in the resulting measurement. Looking at the result from a mathematical aspect, it is quite similar to turning the two-port cal with de-embedding into an enhanced response cal, since the load match correction is essentially removed with S22 of the de-embedding network set to zero. As discussed in Chapter 6, enhanced response calibration is a very good method for dealing with high-loss at port 2.

9.9 Understanding System Stability

System stability describes how well the system maintains its calibrated measurement accuracy over time, temperature and measurement connections. While the VNA system itself is often questioned as to its long-term stability, without question the largest source of instability is the test port cables. The next section describes methods to evaluate the quality and stability of test port cables.

9.9.1 Determining Cable Transmission Stability

The stability of a cable can be tested in three ways to determine its effect on the VNA test system. The first test is a simple test of transmission stability. A cable is connected between ports of a VNA, and two traces of S21 are displayed, one for magnitude and one for phase. The cable response is normalized (for example, using data into memory and data over memory). The cable is then flexed. If desired, it can be disconnected and reconnected. The worst case magnitude and phase deviation are recorded. Some VNAs provide functions that allow tracking the minimum or maximum of a response automatically; for example, in the Agilent PNA, the equation editor function Maxhold(mag(S21)) will give the worst case deviation at each frequency. An example is shown in Figure 9.38 for magnitude stability of both

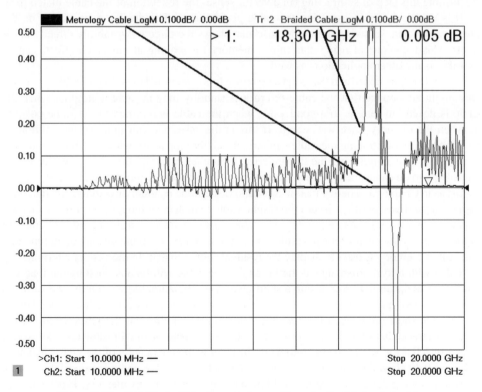

Figure 9.38 Stability in S21 of a metrology cable and a flexible braded cable.

a metrology grade cable and a lower cost flexible braided cable. Each cable was normalized, and each flexed 10 times then reconnected. The metrology cable has almost no change in transmission (less than 0.01 dB). The braided cable has approximately 0.1 dB deviation over much of the frequency response, with a large deviation (likely due to a cable mode) around 16 GHz. Even without the mode, the metrology cable is more than 20 times more stable than the braided cable.

9.9.2   Determining Cable Mismatch Stability

While transmission stability is typically specified for a cable, it is usually the match stability that causes most of the problems when doing fully corrected measurements, especially reflection measurements. The match stability of cables is often not specified, and measuring the match stability is not something that is well documented or widely agreed upon. A very good way of measuring match stability is to terminate the cable with various loads, and then flex the cable while looking at the response.

The cable's input match gives a reflection that must be calibrated out. This reflection becomes part of the raw directivity, and the stability of this match adds directly to the residual directivity. To test for match stability, the cable is terminated in a load, and then the response of the cable is put into memory, and the display math is set to data minus memory, or Data-Mem. Performing this step of subtracting (in a vector sense) the reflection of the cable match from itself is a critical step in assessing the cable performance. The display math occurs on the linear vector data, so taking the LogMag format shows the directivity stability directly in dB terms. This technique of using data minus memory for looking at mismatch effects can be useful in many other applications as well.

The cable is then flexed and the worst case return loss is recorded over several cable positions. One might also notice that the cable return loss stability may become quite poor when the cable is flexed but returns to a good value when the cable is returned to the same position as when the memory trace was stored. If this is the case, one can improve the calibration performance by positioning the cable similarly for calibration as for when it will be connected to the DUT, thus minimizing the flexure error. Examples of cable directivity stability with a metrology grade cable and a braided flexible cable are shown the plot of Figure 9.39.

Just as in Chapter 3, where a load was used to discern the residual directivity, and measurement of a short was used to discern the residual source match, so too here a short can be used to determine the source match stability of a cable. Just like the load, a short is added to the end of the cable and the measurement result is stored in memory. Then the display math function, data minus memory, is used to display the residual source match. In the case of a terminating the cable with a load, mismatch at the far end of the cable will be absorbed by the load and not appear as a stability error. Using a short gives a combination of the directivity error from above, plus any mismatch error as well as any error in phase or delay of the cable. This is because the short produces a large vector reflection. If nothing but the phase changes, the vector difference between the memory and the flexed measurement will show up as a residual source match error, where the magnitude of the error is the arc tangent of the phase change.

Figure 9.40 shows the result of flexing both a metrology cable and a braided cable with a short terminating the cable, and the display math set to data minus memory. It is clear here that the result is far worse than the directivity stability. This exactly illustrates that the source

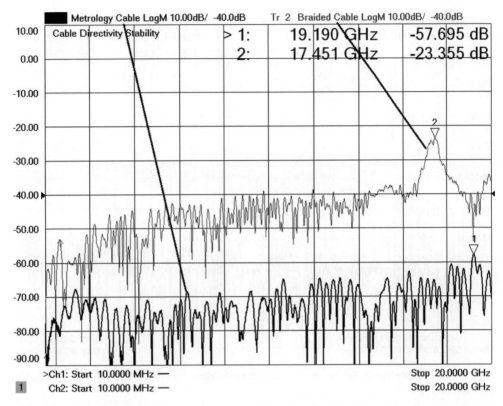

Figure 9.39 Directivity stability of a metrology cable with a load (dark); stability of a braided cable with a load (light).

match stability is the sum of residual directivity plus errors in the open (or short) measurement. Source match stability is critical to measurement of high-reflection devices, as is clear from the error analysis of Eq. (3.68) that shows the effect of residual source match, of which source match stability is a key component. From that equation, the error from the source match goes as the square of the reflection coefficient. For the measurement of a load or other device with a very good match, the effect of source match stability is negligible.

9.9.3 Reflection Tracking Stability

Finally, with the same short attached, the display math can be changed to show data divided by memory, with an S11 trace of LogMag and one of phase displayed (both using data/mem). In this measurement, any change in the loss or the phase of the cable is directly shown by the two-way path measurement after terminating in a short. The overall trend should compare well with the transmission stability measurements and reflections measurements are sometimes easier to do as only one end of the cable needs to be connected to the VNA, so the other end is free to be flexed. However, if the cable has any mismatch, additional ripple will appear as the

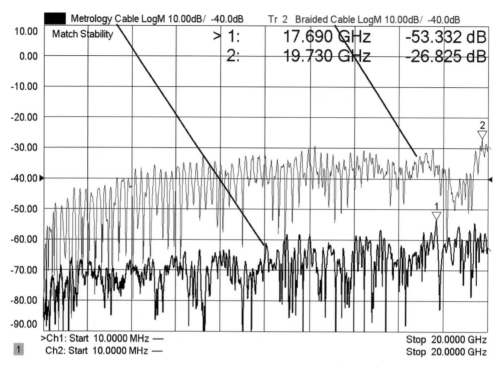

Figure 9.40 Source match stability of a metrology cable (dark); stability of a braided flexible cable (light).

phasing of the mismatch to the short reflection will change. Figure 9.41 shows the reflection tracking phase stability of the metrology cable and the braided flexible cable.

Note that the ripple portion of the phase stability is in line with the implied phase stability of the mismatch measurement. The phase stability should predict a mismatch-stability on the order of the tangent of the phase stability, when measuring a short. In the figure, 3° of phase stability ripple translates to −26 dB of source match stability. Usually, the stability of the cable is the final limitation in the quality of the calibration.

9.10 Some Final Comments on Advanced Techniques and Measurements

In this final chapter, many advanced techniques are discussed to improve the microwave component measurement results for real-world situations where fixtures, cables and adapters would otherwise distort the measurement results.

Users of older, legacy VNAs can use the mathematical functions described in this chapter to do offline processing of measured data, but for most users of modern VNAs, these functions are enabled directly in the user-interface of the VNA.

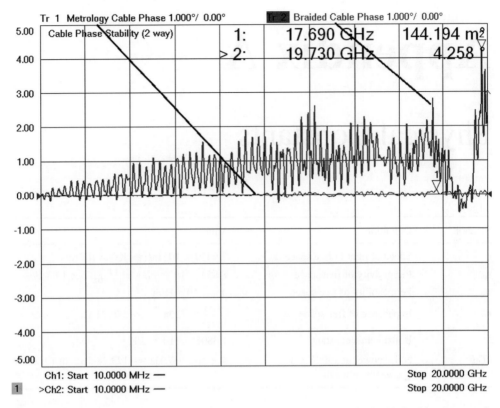

Figure 9.41 Reflection tracking phase stability of a metrology cable and a braided cable when terminated in a short.

With only a little care, and the knowledge provided in this and previous chapters, the R&D or test engineer can create and optimize measurement results for almost any conceivable test condition.

Modern VNAs are rapidly replacing entire racks of test equipment, and using these advanced measurement methods, the test accuracy and test speeds are increasing at a fantastic rate, while the overall size and cost of the test systems are being reduced. Companies and test engineers that fail to utilize these new methods put themselves at a competitive disadvantage. The author hopes that the material presented in this book helps level the playing field.

References

1. Agilent AN 1287-9 In-Fixture Measurements Using Vector Network Analyzers Application Note; available at http://www.icmicrowave.com/pdf/AN%201287-9.pdf.
2. Dunsmore, J., Cheng, N., and Zhang, Y.-X. (2011) Characterizations of asymmetric fixtures with a two-gate approach. Microwave Measurement Conference (ARFTG), 2011 77th ARFTG, pp. 1–6, June 2011.
3. Hong, J.-S. and Lancaster, M.J. (2001) *Microstrip Filters for RF/microwave Applications*, Wiley, New York. Print.

Appendix A

Physical Constants

Symbol	Definition	Value
c	Speed of light in free space	2.9979×10^{8} m/s $\approx 300 \times 10^{6}$ m/s
ε_0	Permittivity of free space	8.854×10^{-12} F/m $\approx \left(\frac{1}{36\pi} \right) \times 10^{-9}$ F/m
μ_0	Permeability of free space	$4\pi \times 10^{-7}$ H/m
η_0	Impedance of free space	$\sqrt{\dfrac{\mu_0}{\varepsilon_0}} \approx 120\pi$ $\Omega \approx 376.73$ Ω
k	Boltzmann's constant	1.38065×10^{-23} J/K
kT_0B	Noise power at 290 K in a given bandwidth	4×10^{-21} W/Hz ≈ -173.98 dBm in 1 Hz BW

Handbook of Microwave Component Measurements: With Advanced VNA Techniques, First Edition. Joel P. Dunsmore.
© 2012 John Wiley & Sons, Ltd. Published 2012 by John Wiley & Sons, Ltd.

Appendix B

Common RF and Microwave Connectors

Name/Notes	Outer Conductor Diameter (air dielectric only)	Rated Frequency (GHz)	First Mode (air dielectric only)	Maximum Useable Frequency
7/16	16 mm	7.5	8.1 GHz	7.5 GHz
Type N (50 ohm) Precision	7 mm	18	18.6 GHz	26.5 GHz[a]
Type N (50 ohm) Commercial	7 mm	12	12.5 GHz	15 GHz
Type N (75 ohm) Precision	7 mm	18	18.6 GHz	18 GHz
Type N (75 ohm) Commercial	7 mm	12	12.5	15 GHz
7 mm (e.g., APC-7)	7 mm	18	18.6 GHz	18 GHz
BNC	Dielectric Interface	4	N/A	11 GHz
TNC (Threaded BNC)	Dielectric Interface	4	N/A	11 GHz
SMA	Dielectric Interface	18	N/A	22 GHz
QMA (Snap-on SMA)	Dielectric Interface	6	N/A	~18 GHz
SMB (Snap-on)	Dielectric Interface	4	N/A	10
SMC (Threaded SMB)	Dielectric Interface	4	N/A	10
3.5 mm	3.5 mm	26.5	28 GHz	33 GHz
SSMA	Dielectric Interface	36	N/A	36 GHZ
2.92 mm ("K")	2.4 mm	40	44 GHz	44 GHz
2.4 mm	2.4 mm	50	52 GHz	55 GHz
1.85 mm ("V")	1.85 mm	67	68.5 GHz	70 GHz
1 mm	1 mm	110	120 GHz	~125 GHZ

[a]Some instrument manufacturers place this connector on 26.5 GHz instruments because it is very rugged; it has the same first modes as type N and 7 mm.

Handbook of Microwave Component Measurements: With Advanced VNA Techniques, First Edition. Joel P. Dunsmore.
© 2012 John Wiley & Sons, Ltd. Published 2012 by John Wiley & Sons, Ltd.

Appendix C

Common Waveguides

E.I.A Waveguide Designation	U.S. Operating Band	Normal Frequency (GHz)	Lower Cutoff Frequency (GHz)	Next Mode Cutoff Frequency (GHz)
WR-284	S (part)	2.60–3.95	2.08	4.16
WR-187	C (part)	3.95–5.85	3.15	6.31
WR-137	C (part)	5.85–8.20	4.30	8.6
WR-90	X	8.2–12.4	6.56	13.11
WR-62	Ku, P[a]	12.4–18.0	9.49	18.98
WR-42	K	18.0–26.5	14.05	28.10
WR-28	Ka, R[a]	26.5–40.0	21.08	42.15
WR-22	Q	33.0–50.0	26.35	52.69
WR-19	U	40.0–60.0	31.39	62.78
WR-15	V	50.0–75.0	39.88	79.75
WR-12	E	60.0–90.0	48.37	96.75
WR-10	W	75.0–110.0	59.01	118.03
WR-8	F	90.0–140.0	73.77	147.54
WR-6	D	110.0–170.0	90.79	181.58
WR-5	G	140.0–220.0	115.714	231.43
WR-4	Y	170.0–260.0	137.242	274.49
WR-3	J	220.0–325.0	173.571	347.14
WR-2		325–500	295.07	590.14
WR-1.5		500–750	393	786
WR-1		750–1100	590	1180

[a]Alternative band designation.

Handbook of Microwave Component Measurements: With Advanced VNA Techniques, First Edition. Joel P. Dunsmore.
© 2012 John Wiley & Sons, Ltd. Published 2012 by John Wiley & Sons, Ltd.

Appendix D

Some Definitions for Calibration Kit Opens and Shorts

Connector	Standard	Definition
7/16 mm (Agilent 85038A)	Open (Male or Female) Short (Male or Female)	Delay = 66.734 ps, Loss = 0.63 GΩ/s $C0 = 32 \times 10^{-15}$ F $C1 = 100 \times 10^{-27}$ F/Hz $C2 = -50 \times 10^{-36}$ F/Hz^2 $C3 = 100 \times 10^{-45}$ F/Hz^3 Delay = 66.734 ps, Loss = 0.63 GΩ/s $L0 = 0 \times 10^{-12}$ H $L1 = 0 \times 10^{-24}$ H/Hz $L2 = 0 \times 10^{-33}$ H/Hz^2 $L3 = 0 \times 10^{-42}$ H/Hz^3
7 mm (Agilent 85050D)	Open Short	Delay = 0 ps, Loss = 0.7 GΩ/s $C0 = 90.4799 \times 10^{-15}$ F $C1 = 763.303 \times 10^{-27}$ F/Hz $C2 = -63.8176 \times 10^{-36}$ F/Hz^2 $C3 = 6.4337 \times 10^{-45}$ F/Hz^3 Delay = 0 ps, Loss = 0.7 GΩ/s $L0 = 0.3566 \times 10^{-12}$ H $L1 = -33.392 \times 10^{-24}$ H/Hz $L2 = 1.7542 \times 10^{-33}$ H/Hz^2 $L3 = -0.0336 \times 10^{-42}$ H/Hz^3

(Continued)

Handbook of Microwave Component Measurements: With Advanced VNA Techniques, First Edition. Joel P. Dunsmore.
© 2012 John Wiley & Sons, Ltd. Published 2012 by John Wiley & Sons, Ltd.

Connector	Standard	Definition
Type N Precision (Agilent 85054B)	Open (Male)	Delay = 57.993 ps, Loss = 0.93 GΩ/s $C0 = 89.939 \times 10^{-15}$ F $C1 = 2536.8 \times 10^{-27}$ F/Hz $C2 = -264.99 \times 10^{-36}$ F/Hz^2 $C3 = 13.4 \times 10^{-45}$ F/Hz^3
	Open (Female)	Delay = 22.905 ps, Loss = 0.93 GΩ/s $C0 = 104.13 \times 10^{-15}$ F $C1 = -1943.4 \times 10^{-27}$ F/Hz $C2 = 144.62 \times 10^{-36}$ F/Hz^2 $C3 = 2.2258 \times 10^{-45}$ F/Hz^3
	Short (Male)	Delay = 63.078, Loss = 1.1273 GΩ/s $L0 = 0.7563 \times 10^{-12}$ H $L1 = 459.88 \times 10^{-24}$ H/Hz $L2 = -52.429 \times 10^{-33}$ H/Hz^2 $L3 = 1.5846 \times 10^{-42}$ H/Hz^3
	Short (Female)	Delay = 27.99, Loss = 1.3651GΩ/s $L0 = -0.1315 \times 10^{-12}$ H $L1 = 606.21 \times 10^{-24}$ H/Hz $L2 = -68.405 \times 10^{-33}$ H/Hz^2 $L3 = 2.0206 \times 10^{-42}$ H/Hz^3
3.5 mm (Agilent 85052D)	Open (Male or Female)	Delay = 29.243 ps, Loss = 2.2 GΩ/s $C0 = 49.433 \times 10^{-15}$ F $C1 = -310.13 \times 10^{-27}$ F/Hz $C2 = 23.168 \times 10^{-36}$ F/Hz^2 $C3 = -0.15966 \times 10^{-45}$ F/Hz^3
	Short (Male or Female)	Delay = 31.785 ps, Loss = 2.36 GΩ/s $L0 = 2.0765 \times 10^{-12}$ H $L1 = -108.54 \times 10^{-24}$ H/Hz $L2 = 2.1705 \times 10^{-33}$ H/Hz^2 $L3 = -0.01 \times 10^{-42}$ H/Hz^3
2.92 mm (Maury 8770D)	Open (Male)	Delay = 14.982 ps, Loss = 1.8 GΩ/s $C0 = 47.5 \times 10^{-15}$ F $C1 = 0 \times 10^{-27}$ F/Hz $C2 = 3.8 \times 10^{-36}$ F/Hz^2 $C3 = 0.19 \times 10^{-45}$ F/Hz^3
	Open (Female)	Delay = 14.883 ps, Loss = 1.8 GΩ/s $C0 = 45.5 \times 10^{-15}$ F $C1 = 100 \times 10^{-27}$ F/Hz $C2 = 0.3 \times 10^{-36}$ F/Hz^2 $C3 = 0.21 \times 10^{-45}$ F/Hz^3
	Short (Male or Female)	Delay = 16.83 ps, Loss = 1.8 GΩ/s $L0 = 0 \times 10^{-12}$ H $L1 = 0 \times 10^{-24}$ H/Hz $L2 = 0 \times 10^{-33}$ H/Hz^2 $L3 = 0 \times 10^{-42}$ H/Hz^3

Connector	Standard	Definition
2.4 mm	Open (Male or Female)	Delay $= 20.837$ ps, Loss $= 3.23$ GΩ/s $C0 = 29.722 \times 10^{-15}$ F $C1 = 165.78 \times 10^{-27}$ F/Hz $C2 = -3.5386 \times 10^{-36}$ F/Hz^2 $C3 = 0.071 \times 10^{-45}$ F/Hz^3
	Short (Male or Female)	Delay $= 22.548$ ps, Loss $= 3.554$ GΩ/s $L0 = 2.1636 \times 10^{-12}$ H $L1 = -146.35 \times 10^{-24}$ H/Hz $L2 = 4.0443 \times 10^{-33}$ H/Hz^2 $L3 = -0.0363 \times 10^{-42}$ H/Hz^3
1.85 mm	All	Data based standards[a]
1 mm	All	Data based standards[a]
Type-N 75 ohm (Agilent 85036B/E)	Open (Male)	Delay $= 17.544$ ps, Loss $= 1.13$ GΩ/s $C0 = 41 \times 10^{-15}$ F $C1 = 40 \times 10^{-27}$ F/Hz $C2 = 5 \times 10^{-36}$ F/Hz^2 $C3 = 0 \times 10^{-45}$ F/Hz^3
	Open (Female)	Delay $= 0$ ps, Loss $= 1.13$ GΩ/s $C0 = 63.5 \times 10^{-15}$ F $C1 = 84 \times 10^{-27}$ F/Hz $C2 = 56 \times 10^{-36}$ F/Hz^2 $C3 = 0 \times 10^{-45}$ F/Hz^3
	Short (Male)	Delay $= 17.544$, Loss $= 1.13$ GΩ/s $L0 = 0 \times 10^{-12}$ H $L1 = 0 \times 10^{-24}$ H/Hz $L2 = 0 \times 10^{-33}$ H/Hz^2 $L3 = 0 \times 10^{-42}$ H/Hz^3
	Short (Female)	Delay $= 0.93$, Loss $= 1.13$ GΩ/s $L0 = 0 \times 10^{-12}$ H $L1 = 0 \times 10^{-24}$ H/Hz $L2 = 0 \times 10^{-33}$ H/Hz^2 $L3 = 0 \times 10^{-42}$ H/Hz^3

[a]Data based standards do not follow the normal polynomial fit, but use a data file based on an explicit reflection versus frequency for both amplitude and phase.

Index

Handbook of Microwave Component Measurements: With Advanced VNA Techniques, First Edition. Joel P. Dunsmore.
© 2012 John Wiley & Sons, Ltd. Published 2012 by John Wiley & Sons, Ltd.